More QRP POWER

The best recent QRP articles from *QST* and *QEX*

Compiled by: **Mark Wilson, K1RO**

Production: **Shelly Bloom, WB1ENT**

Jodi Morin, KA1JPA

Cover Design: **Sue Fagan**

Published by :

ARRL The national association for AMATEUR RADIO

Newington, CT 06111-1494

Printed in the USA

ISBN: 0-87259-965-5

1st Edition
1st Printing

Contents

A Note About Contact Information and Resources

www.arrl.org/tis/tisfind.html

www.arrl.org/tis/info/qrphome.html

Foreword

The enduring popularity of QRP—operating with 5 W or less—is demonstrated once again by the variety of articles on the topic that regularly appear in *QST* and *QEX*. Like *QRP Power* and its predecessor, *QRP Classics*, this book brings together the best recent articles on QRP equipment, accessories and antennas. These articles have all been published since the original *QRP Power* was compiled in 1996 and complement the articles in the previous two editions.

Many people enjoy building their own transceivers, transmitters or receivers, and we have included a number of interesting projects covering a variety of bands. Some are simple, others sophisticated. All will provide immense satisfaction for the builder.

For those QRPers who prefer to build a radio from one of the many fine kits available today or purchase one of the popular commercial transceivers, we have included a wide variety of station accessories and compact portable antennas. Whether you build or buy, these projects will help you round out a station that you can use at home or take anywhere.

This book would not have been possible without the hard work and dedication of the authors who took the time to write up their projects for publication in *QST* or *QEX*. If you enjoy an article or find a particular project useful, please drop the author a note.

We hope that you enjoy the articles presented here. Give QRP a try. It's challenging and rewarding, and is sure to inject a little magic into your Amateur Radio experience.

David Summer, K1ZZ
Chief Executive Officer
Newington, Connecticut
January 2006

About the ARRL

The national association for Amateur Radio

The seed for Amateur Radio was planted in the 1890s, when Guglielmo Marconi began his experiments in wireless telegraphy. Soon he was joined by dozens, then hundreds, of others who were enthusiastic about sending and receiving messages through the air—some with a commercial interest, but others solely out of a love for this new communications medium. The United States government began licensing Amateur Radio operators in 1912.

By 1914, there were thousands of Amateur Radio operators—hams—in the United States. Hiram Percy Maxim, a leading Hartford, Connecticut inventor and industrialist, saw the need for an organization to band together this fledgling group of radio experimenters. In May 1914 he founded the American Radio Relay League (ARRL) to meet that need.

Today ARRL, with approximately 170,000 members, is the largest organization of radio amateurs in the United States. The ARRL is a not-for-profit organization that:

- promotes interest in Amateur Radio communications and experimentation
- represents US radio amateurs in legislative matters, and
- maintains fraternalism and a high standard of conduct among Amateur Radio operators.

At ARRL headquarters in the Hartford suburb of Newington, the staff helps serve the needs of members. ARRL is also International Secretariat for the International Amateur Radio Union, which is made up of similar societies in 150 countries around the world.

ARRL publishes the monthly journal *QST*, as well as newsletters and many publications covering all aspects of Amateur Radio. Its headquarters station, W1AW, transmits bulletins of interest to radio amateurs and Morse code practice sessions. The ARRL also coordinates an extensive field organization, which includes volunteers who provide technical information and other support services for radio amateurs as well as communications for public-service activities. In addition, ARRL represents US amateurs with the Federal Communications Commission and other government agencies in the US and abroad.

Membership in ARRL means much more than receiving *QST* each month. In addition to the services already described, ARRL offers membership services on a personal level, such as the ARRL Volunteer Examiner Coordinator Program and a QSL bureau.

Full ARRL membership (available only to licensed radio amateurs) gives you a voice in how the affairs of the organization are governed. ARRL policy is set by a Board of Directors (one from each of 15 Divisions). Each year, one-third of the ARRL Board of Directors stands for election by the full members they represent. The day-to-day operation of ARRL HQ is managed by an Executive Vice President and his staff.

No matter what aspect of Amateur Radio attracts you, ARRL membership is relevant and important. There would be no Amateur Radio as we know it today were it not for the ARRL. We would be happy to welcome you as a member! (An Amateur Radio license is not required for Associate Membership.) For more information about ARRL and answers to any questions you may have about Amateur Radio, write or call:

ARRL—The national association for Amateur Radio
225 Main Street
Newington CT 06111-1494
Voice: 860-594-0200
Fax: 860-594-0259
E-mail: **hq@arrl.org**
Internet: **www.arrl.org/**

Prospective new amateurs call (toll-free):
800-32-NEW HAM (800-326-3942)
You can also contact us via e-mail at **newham@arrl.org**
or check out *ARRLWeb* at **http://www.arrl.org/**

Chapter 1

Construction Practices

By Sam Ulbing, N4UAU

From *QST*, April 1999

Surface Mount Technology—You *Can* Work with It!

Part 1—Start building your projects with surface-mount devices! I'll show you how!

As I look through the various electronic manufacturing companies' product datasheets, three things strike me. First, the large number of available ICs that perform functions formerly requiring several ICs. Second, the continuing shift to lower-power requirements, smaller size and usability at higher operating frequencies. Finally, the increasing number of new products are available *only* in surface-mount packages. It all fits together: Products today are smaller and more energy efficient. Look at modern H-Ts, cell phones, GPS equipment, laptop computers, microwave ovens, intelligent electronic ovens, TV remote controls and pocket calculators: One thing they have in common is their use of surface-mount (SM) ICs.

On the other hand, when I look at Amateur Radio projects, I see continued use of many discrete components and bulky DIP ICs that perform limited functions. Recently, I saw a voltage-controller project based on the use of transistors and relays! Frankly, it bothers me that there seems to be a growing divergence between the technology used by industry and that used by hams. The *Maxim Engineering Journal Vol. 29*, for instance, showcases such new ICs as an image-reject RF transceiver, a low-phase-noise RF oscillator that replaces VCO modules, a 3 V, 1 W, 900 MHz RF power transistor, a direct-conversion down-converter IC that replaces an IF mixer, an IF LO and SAW filter, and a low-voltage IF transceiver that includes the FM limiter and RSSI. All these multifunction ICs are available *only* in SM packages! I think hams are being left behind because they feel that SMT (surface-mount technology) is something they can't handle.

Since I built my first SM project two years ago, I have assembled a dozen others. I find that my skill levels have increased tremendously with practice, and I now routinely tackle projects I never thought possible just a year ago. Based on my experience, I know that amateurs *can* work with SMT. Perhaps when we show this ability, there will be more truly state-of-the-art projects in the amateur publications. How about a very small 2 meter rig, or a 900 MHz personal communicator? The ICs already exist and we need to adapt them to ham use. First however, it is necessary to develop a few basic building skills. This article series will help you develop those skills by showing what I have learned and presenting several useful and easy-to-build projects. Once you have built these, you will be able to handle most of the SM ICs I have seen used in the industry.

N4UAU

[1]Notes appear on page 1-6.

Nothing New

The concept of surface mounting parts is not new to Amateur Radio. In a September 1979 *QST* article,[1] Doug DeMaw, W1FB (SK), discusses a quick and easy circuit-board design that was basically SMT; Doug also proposed a universal PC-board layout for this kind of construction. You may think that there will *always* be DIP versions of all the SM ICs so engineers can experiment, but even today, many manufacturers are making evaluation boards available to designers so they can test the part using SM devices! I suspect it's cheaper for them to sell evaluation boards than to set up a production line to make a very limited number of DIP ICs when their real volume is in SM devices.

Some of the advantages of building with SM devices include:

- Smaller projects: I built a time-out switch that fits on a PC board one-sixth the size of a postage stamp! I was able to put the circuit into the battery compartment of a voltmeter I had so it could automatically power itself down.[2]
- Many SM versions of devices outperform the original DIP versions. Lower operating voltages and quiescent currents in the microampere range offer more efficient operation.
- Most RF projects require the use of short signal leads. SM capacitors are often recommended for use as bypass capacitors because they can be placed close to an IC and exhibit very low lead inductance. Nearly all VHF projects benefit from the use of SM devices.
- Once you've had some experience in working with SM devices, you'll feel more con-

fident about repairing your own gear.

- Making a PC board for SM devices is easier than for through-hole parts because no component-mounting holes need to be drilled.
- Many new SM ICs have entire modules built into them making it much easier to build a complex circuit than with older ICs.[3]

Equipment Needed

Many people think you need lots of expensive equipment to work with SM devices.[4] Not so! You don't need an eagle's eyesight, either! My optometrist describes my eyesight as "moderately near-sighted, needing bifocals (2$^1/_2$ diopters)." My wife thinks I am as blind as a bat.

- A fundamental piece of equipment for SM work is an illuminated magnifying glass. I use an inexpensive one with a 5-inch-diameter lens (see the accompanying trio of tools photographs). I use the magnifier for *all* my soldering work, not just for SM use. Such magnifiers are widely available (see the sidebar "Manufacturers and Distributors of SMT Equipment and Parts") and range in price from about $25 to several hundred dollars. Most offer a 3× magnification and have a built-in circular light.
- A low-power soldering iron is necessary; one that is temperature-controlled (such as the Weller WCC100) practically eliminates the possibility of overheating a part. Use a soldering iron with a grounded tip as most SM parts are CMOS devices and are subject to possible ESD (static) failure. I have found the Weller $^1/_{16}$-inch (EJA) screwdriver tip works well. I used to use an ETJ with its finer conical tip, but it does not seem to transfer the heat as well as the screwdriver tip.
- Use of thin (0.020-inch diameter) rosin-core solder is preferred because the parts are so small that regular 0.031-inch diameter solder will flood a solder pad and cause bridging.
- A wet sponge for cleaning the soldering-iron tip.
- A flux pen comes in handy for applying just a little flux at a needed spot. I find that RadioShack's flux is too sticky and it leaves a messy residue. The Circuit Works CW8200 flux pen with a type R flux is much cleaner.
- Good desoldering braid is necessary to remove excess solder if you get too much on a pad. Chem-Wik Lite 0.100-inch wide works well.
- ESD protective devices such as wrist straps may be necessary if you live in a dry area and static is a problem. I live in humid Florida, have never used these and have not had a problem.
- Tweezers help pick up parts and position them. I find that a pair of nonmagnetic, stainless-steel drafting dividers work well as tweezers. They have two very sharp needle-like points that allow me to pick up the smallest parts; and the parts seem less likely to slip from grasp perhaps because I use less force to hold them. The sharp points are useful tools for marking the PC-board copper foil before cutting out traces (more on this later). The nonmagnetic property of stainless steel means the chip doesn't get attracted to the dividers.
- If you want to make your own SM PC boards, I recommend using a Dremel Mototool (or something similar) and some ultra-fine cutting wheels.

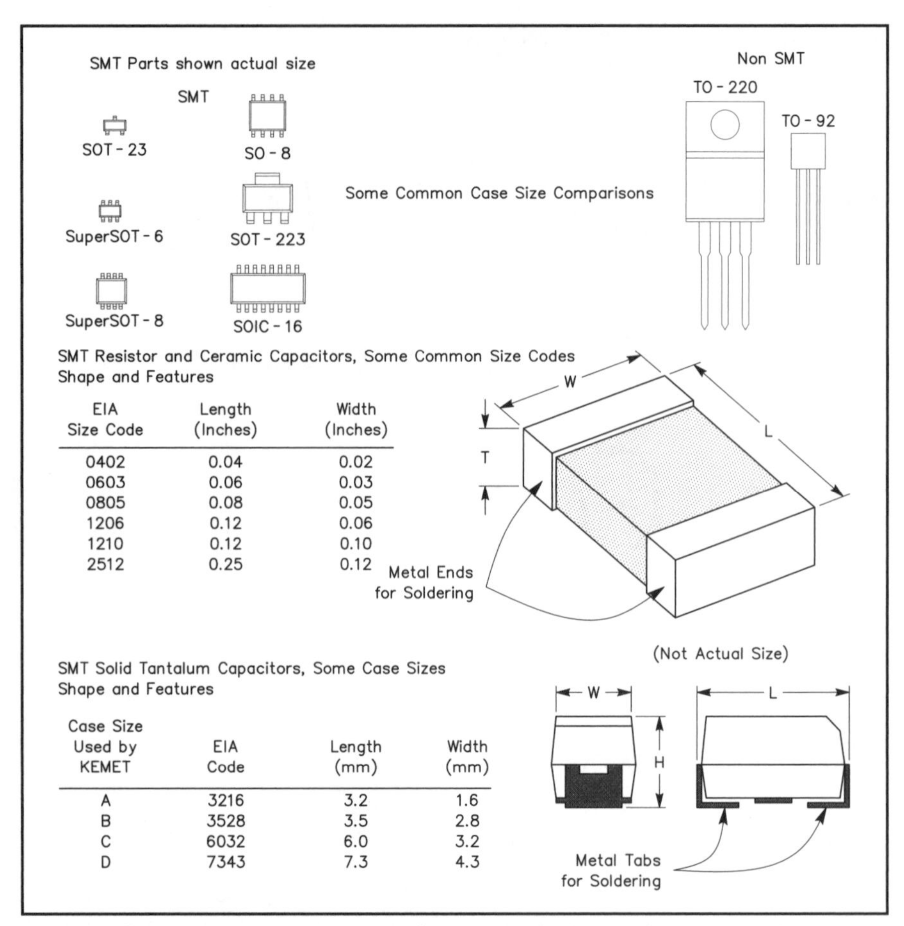

EIA Size Code	Length (Inches)	Width (Inches)
0402	0.04	0.02
0603	0.06	0.03
0805	0.08	0.05
1206	0.12	0.06
1210	0.12	0.10
2512	0.25	0.12

Case Size Used by KEMET	EIA Code	Length (mm)	Width (mm)
A	3216	3.2	1.6
B	3528	3.5	2.8
C	6032	6.0	3.2
D	7343	7.3	4.3

Figure 1—Size comparisons of some surface-mount devices and their dimensions.

Parts

Figure 1 shows some common SM parts. Resistors and ceramic capacitors come in many different sizes, and it is important to know the part size for two reasons: Working with SM devices by hand is easier if you use the larger parts; and it is important that the PC-board pad size is larger than the part. Tantalum capacitors are one of the larger SM parts. Their case code, which is usually a letter, often varies from manufacturer to manufacturer because of different thicknesses. As you can see from Figure 1, the EIA code for ceramic capacitors and resistors is a measurement of the length and width in inches, but for tantalum parts, those measurements are in millimeters times 10! Keep in mind that tantalum capacitors are *polarized*; the case usually has a mark or stripe to indicate the positive end. Nearly any part that is used in through-hole technology is available in a SM package.[5,6]

SMT Soldering Basics

Use a little solder to pre-tin the PC board. The trick is to add just enough solder so that when you reheat it, it flows to the IC, but not so much that you wind up with a solder bridge. Putting a little flux on the board and the IC legs makes for better solder flow, providing a smooth layer. You can tell if you have the proper soldering-iron tip temperature if the solder melts within 1.5 and 3.5 seconds.[7] I use my dividers (or my fingers) to push and prod the chip into position. Because the IC is so small and light, it tends to stick to the soldering iron and pull away from the PC board. To prevent this, use the dividers to

LODER BROOKS, KD4AKW

Figure 2—An LM386 audio amplifier built on a homemade PC board. The board's isolated pads are made by using a hobby tool to grind separating lines through the copper foil.

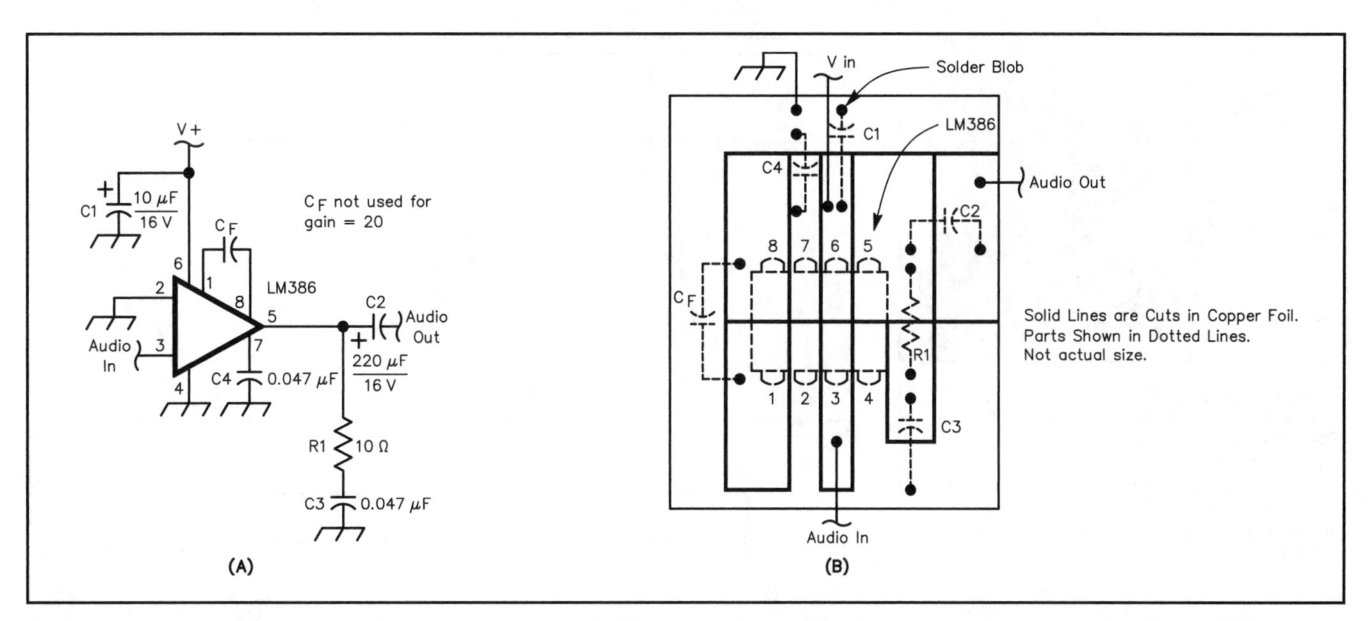

Figure 3—At A, the schematic of the LM386 audio amplifier. The component layout and PC board are shown at B. The solid, heavy lines indicate cuts made in the copper foil. This drawing is not to scale. The board is 1 inch long by 3/4 inch wide. No SM parts are used in this project, but my board-making method is shown. It allows one to get a feel for the process before tackling the smaller SM chips.

C1—10 µF, 16 V
C2—220 µF, 16 V
C3, C4—0.047 µF, 50 V ceramic
Cf—For overall circuit gains greater than 20, use 10 µF, 16 V
U1—LM386N (8-pin DIP)

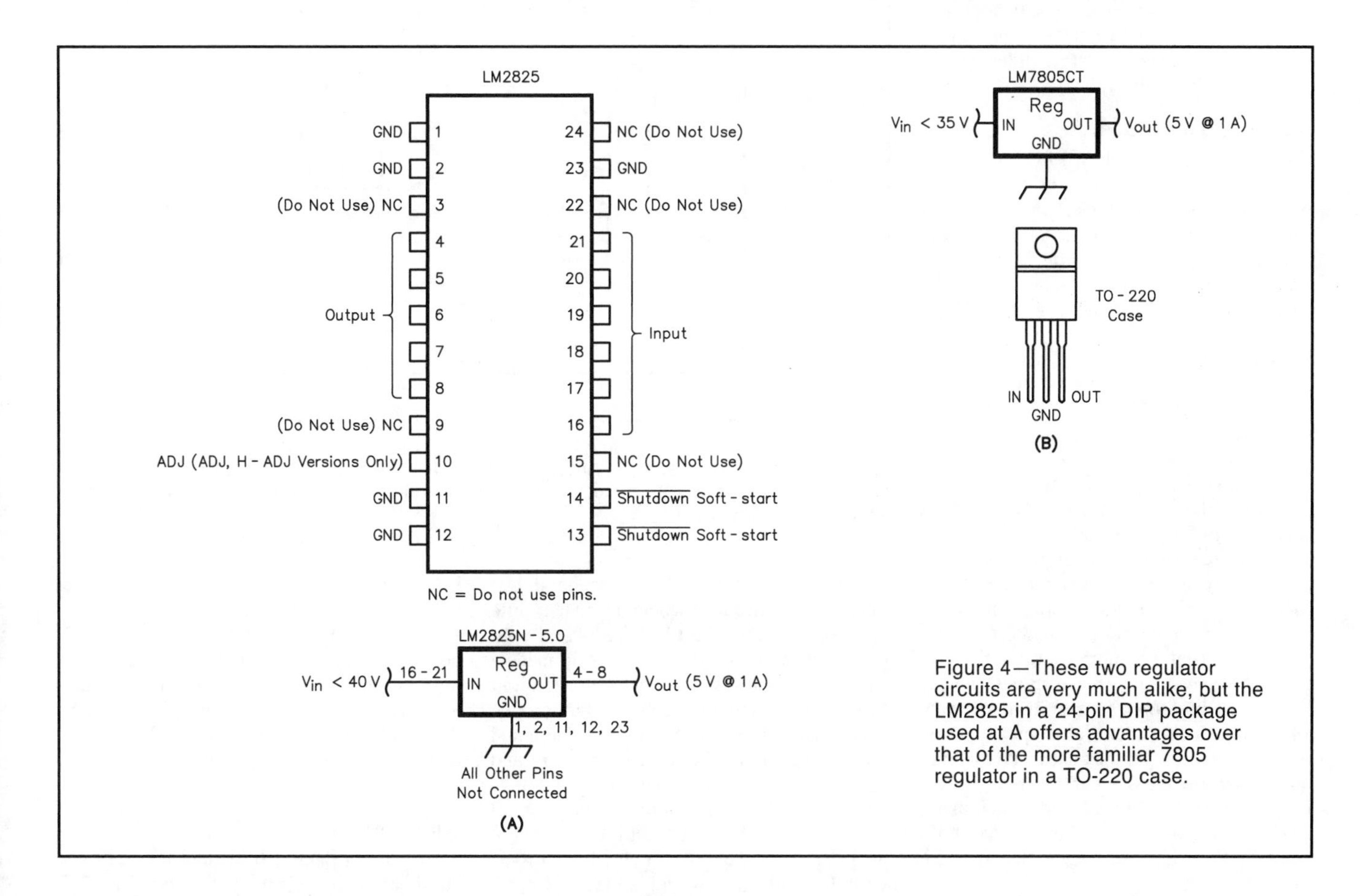

Figure 4—These two regulator circuits are very much alike, but the LM2825 in a 24-pin DIP package used at A offers advantages over that of the more familiar 7805 regulator in a TO-220 case.

hold the chip down while tacking two IC legs at diagonally opposite corners. After each tack, check that the part is still aligned. With a dry and *clean* soldering iron, heat the *PC board* near the leg.[8] If you do it right, you will see the solder flow to the IC.

The legs of the IC must lie flat on the board. The legs bend easily, so don't press down too hard. Check each connection with a continuity checker placing one tip on the board the other on the IC leg. Check all adjacent pins to ensure there's no bridging. It is easier to correct errors early on, so I recommend performing this check often. If you find that you did not have enough solder on the board for it to flow to the part, add a little solder. I find it best to put a drop on the trace near the part, then heat the trace and slide the iron and melted solder toward the part. This reduces the chance of creating a bridge. Soldering resistors and capacitors is similar to soldering an IC's leads, except the resistors and capacitors don't have exposed leads. My reflow method works well for these parts, too.

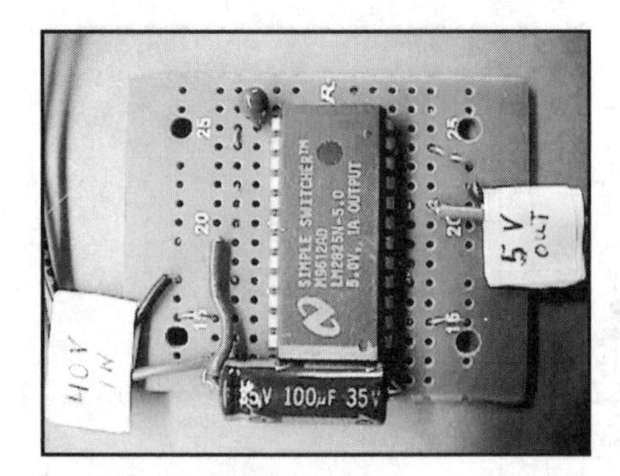

Figure 5—LM2825 circuit is constructed on readily available RadioShack perfboard (RS 276-150 or similar). Photo shows optional input filter capacitor and optional soft-start capacitor. Refer to the datasheet at the National Semiconductor Web page for suggested values for these optional components.

Attaching wires that connect to points off the board can be a bit of a challenge because even #24 stranded wire is large in comparison to the SM parts. First, make sure all the wire strands are close together, then pre-tin the wire. Carefully place the wire on the pre-tinned pad and heat it with the soldering iron until the solder melts.[9]

Making a SM PC Board

It is possible to etch SM PC boards just like a conventional board, but I recall Doug (W1FB) DeMaw's comment on etching: "If you don't mind a few brown stains here and there on your garments, etching is one way to make the board." Evidently he, like I, *did* mind, and he proposed a strong-arm method of using a hacksaw to cut square pads in the board foil. Hacksaws are too large and wide for SM use; I use a Dremel Mototool and a thin cutoff wheel. With these, I can cut a line as narrow as 0.005 inch, which lets me build with most of the available SM ICs.

To make such a "PC" board, start by sketching a layout for it. Don't worry about drawing it to scale, but make the sketch large enough to see what is happening. Normally, we think in terms of *connections between parts* because schematics show lines from point to point, representing the interconnecting wires. I find it is more useful to think in terms of the *spaces between the lines* because I am removing copper material to separate traces, not adding material to make traces. Where wires attach to the board, leave a large surface because the wires are relatively large. When making cuts, it is easiest to do it using a large piece of material that you can hold securely. Cut the board to size after you have cut all the traces.

Once I have the layout drawn, I hold the IC to the copper and used a fine-pointed tool (a 0.5 mm pencil or my dividers) to mark the location of the cuts on the PC board. I then remove the IC and use my Dremel tool to cut the copper along the marks. For critical cuts between an IC's closely spaced leads, I make one cut, then reposition the IC on the board and verify that the remaining marks are still

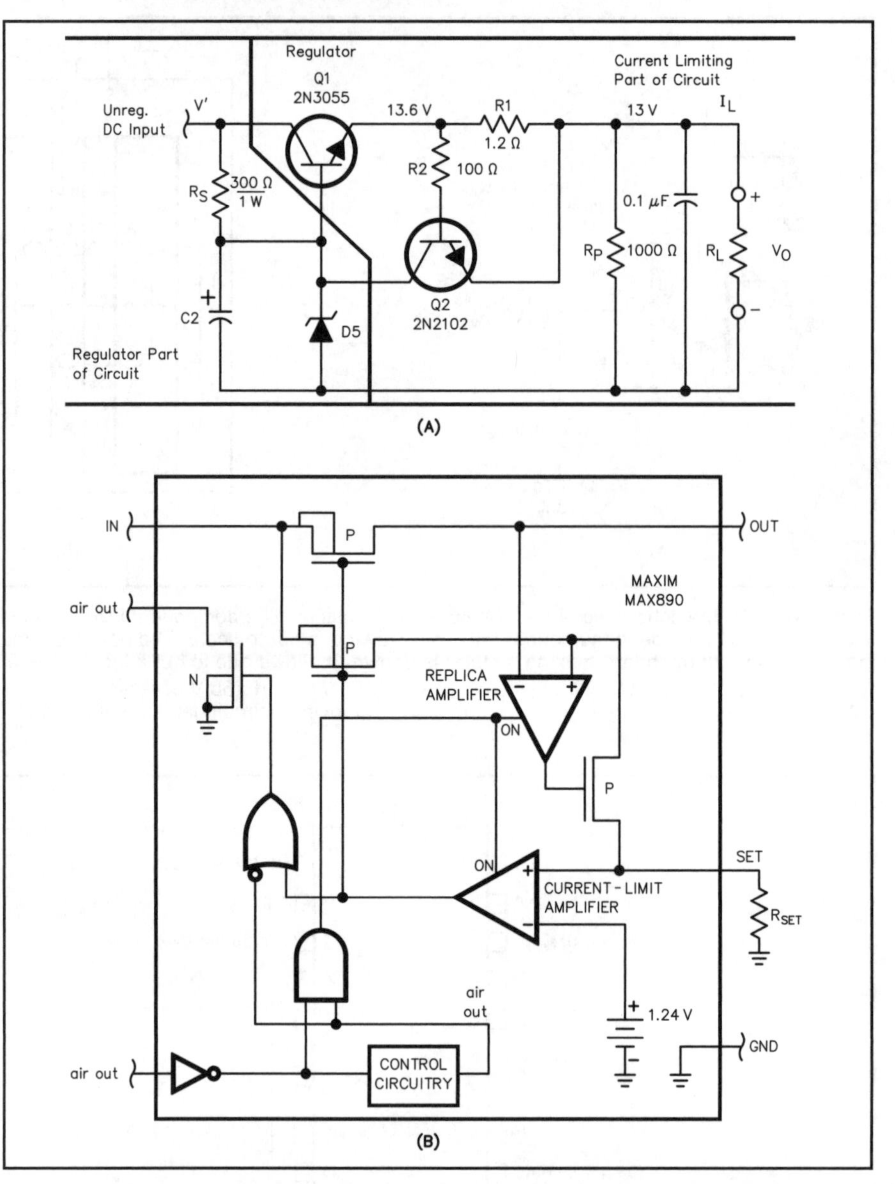

Figure 6—At A, a current-limiting circuit using older technology. Within the confines of the tiny MAX890, (B) newer technology offers a 1-A, P-channel MOSFET switch, a comparator, a voltage reference, a current-measuring circuit, control and fault-indicator circuits! For Project 1 they are:

correctly aligned. I do all this freehand. Using an ultra-thin cutoff wheel, I find it is quite easy to cut in a straight line. At first, I used a fine cutting bit (#108), but that tip made a wider cut and it was difficult to cut a straight line. Dedeco (see the sidebar "Manufacturers and Distributors of SMT Equipment and Parts"), supplier of tools for jewelers and dentists, makes two ultra-thin wheels—0.009 and 0.005 inch. For the very small ICs—those in SOT23-5 and SuperSOT-8 packages—I use a 0.005-inch wheel, otherwise the 0.009-inch wheel is ideal. Be careful when handling these wheels as they break easily. Also, don't cut too deeply into the board material. At the intersection of the cuts, take care not to cut too far. Sometimes I cut close to an intersection, then use a razor blade to finish the job. A quick sanding deburrs the cuts. Run through the cuts with a small screwdriver or pen knife to ensure they are cleanly cut and without burrs. Finally, use your ohmmeter to verify that the islands/pads aren't connected.[10]

I recommend you make your own PC boards: They're easier to produce than through-hole boards, and you'll then be able to experiment with *your own* projects rather than waiting for others' projects to come along. You can use this method with SM or non-SM projects.

The Projects

All the projects I'll present are easy-to-build beginner projects, yet each offers significant advantages over similar projects based on the old (DIP) technology. As you build each project, you'll develop SM skills and wind up with some useful gadgets. I have tried to arrange the projects by degree of skill required. For those who want to make their own PC boards, I describe my layout. Ready-made PC boards are available for all of these projects except the first, Project 0A.[11]

Project 0A—Getting a Feel for SM Techniques

This audio amplifier is a good starter project for those who want to learn to work with SM devices because the technique is the same, but the parts are physically larger because no SM parts are used. I made the layout, cut the board and assembled this project in a little over an hour. Try doing that with etching and through-hole construction! I think you'll agree that the finished product looks as good as if it were assembled on a commercially made PC board (see Figure 2).

This project is shown in *The 1996 ARRL Handbook* (and subsequent editions) on page 25.8 using "dead-bug" construction.[12] All the parts are mounted on a groundplane with no component-mounting holes. It is easy to duplicate this project using SM techniques. Figures 3A and B show the schematic and board layout, respectively. I bent and trimmed the pins on the LM386 so that they look like a large SO-8 package. You can make the cuts with a 0.015-inch wheel. When cutting the ends of the traces to pins 1 and 3, be careful that you don't cut too far and run through the cut from pin 2.

An LM386 is *not* state of the art. If you want to see the difference between it and a state-of-the-art amplifier, build the SMALL.[13] It uses an LM4861, which is available only in an SO-8 package. In addition to its smaller size, the SMALL has more power output, far better fidelity and the ability to work with low-voltage power sources.

Project 0B—The World's Easiest Surface-Mount Project

You may be curious about comparing SMT with conventional technology, but not want to solder those small ICs. If so, this project is for you. It is based on the LM2825, a large DIP 5 V regulator used in the circuit shown in Figure 4A. Next, build a conventional 5 V regulator using an LM7805 in a TO-220 case, Figure 4B. Both can be built on a RadioShack universal PC board; the LM2825 project is shown in the photograph (Figure 5). Although the circuits look nearly identical, if you use a 12-V source to power both of them and put a load of 0.5A or more on each, you'll see that the LM7805 gets *very hot*, while the LM2825 stays cool. That's because the LM2825 is a sophisticated *switching regulator* with all of the tiny SM parts packed in a DIP case.

Out with the Old...

The (*1996 ARRL Handbook*) current-limiting circuit of Figure 6A uses a resistor (R1) and series pass transistor (Q1) in series with the load. R1 detects the current flow and Q1 limits it when necessary. This design has a voltage drop from input to output of 600 to 1200 millivolts depending on the load (before any overload). Its voltage regulation is poor and its efficiency is low.

...In with the New

By contrast, Maxim's MAX890 (Figure 6B) operates with voltage levels from 2.7 to 5.5 V (6 V maximum) with a current drain of only 15 µA. On this tiny chip are a 1-A, P-channel MOSFET switch, a comparator, a voltage reference, a current-measuring circuit and control and fault-indicator circuits! The maximum voltage drop across the switch is only 90 mV unless an overcurrent condition exists. Instead of using a series resistor to monitor current, the MAX890 uses a current replica circuit that controls the MOSFET limiting switch. For a short circuit—or for a large initial surge current—the circuit shuts off the switch in just five microseconds, then slowly turns it on while limiting the current to 1.5 times the maximum current. For prolonged overcurrent situations, there is a large amount of power dissipated in the MOSFET. To combat this, the chip has a thermal shutdown circuit that cycles the switch on and off, if necessary, to keep the temperature within a safe range.

Project 1—The SmartSwitch

This project is based on Maxim's MAX890, available in a common and fairly large SO-8 SM package that is relatively easy to work with. The switch is smart because it limits the current it passes to an amount you *preset*. This device not only protects your expensive electronic projects against a short circuit, but extends their life by limiting inrush current, a major cause of component

Figure 7—Here are two versions of the SmartSwitch compared in size to a TO-92 package transistor. The board on the right is homemade; the one to the left is available from N4UAU; see Note 11.

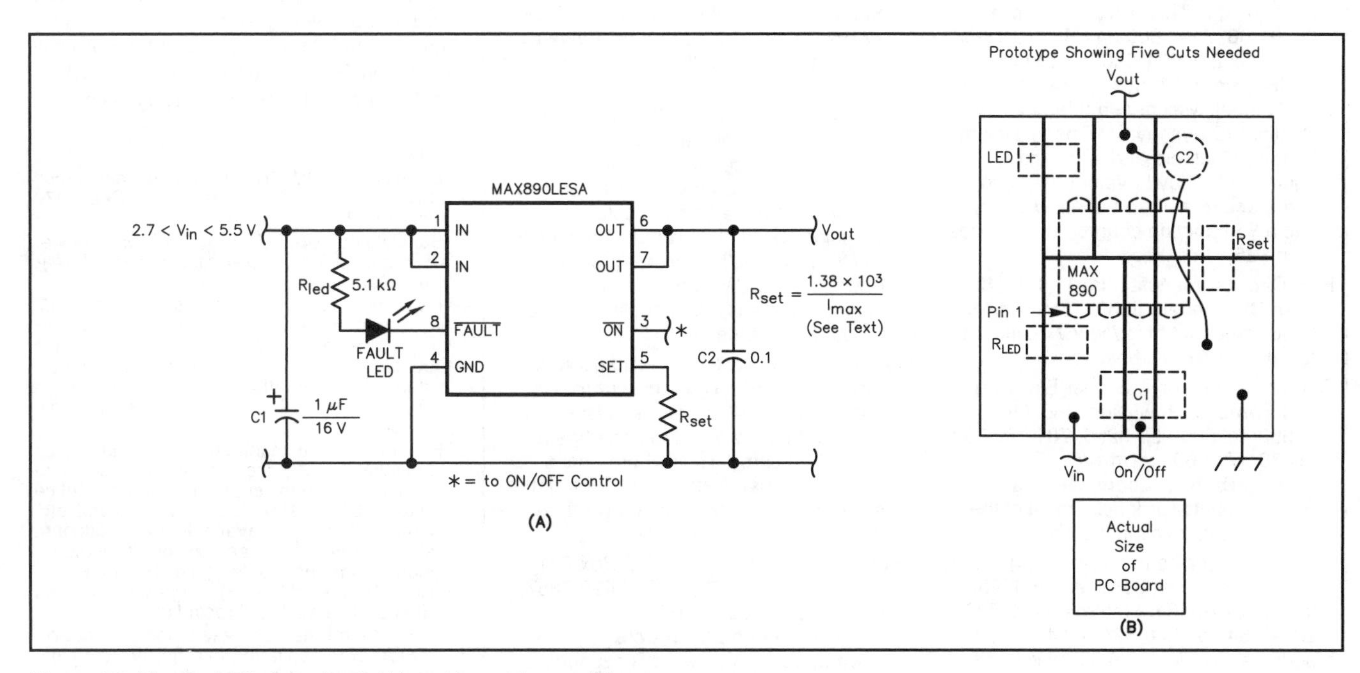

Figure 8— The SmartSwitch circuit (A) and board layout (B).

U1—MAX890LESA
Rset—See Eq 1 in text.
Rled—5.1 kΩ
C1—1 µF tantalum capacitor, 10 V or greater
C2—0.1 µF ceramic

failure. The IC has an output that can be set to trigger a fault indicator, such as an LED or bell. The **ON/OFF** pin exhibits a high impedance and can be controlled by a computer or low-output sensor such as a photoelectric cell. Building the SmartSwitch is straightforward and relatively easy. Figure 7 shows how physically small the switch is.

Figure 8A is the SmartSwitch schematic; the board layout is shown in Figure 8B. Circuit operation is simple: Power connections are made at pins 1 and 2, the high side of the switch and switched power are available at pins 6 and 7, respectively. **Rset** sets the trip current:

$$Ilimit = 1.38 \times 10^3 / \mathbf{Rset} \qquad \text{(Eq 1)}$$

where *Ilimit* is the trip current in amperes, and **Rset** is the controlling resistance value in ohms.

I used a 2.2 kΩ resistor at **Rset** to establish a current limit of 625 mA. (Current-trip levels can be set to values between 200 mA and 1 A.) C1, the input capacitor, prevents input-voltage drop with current surges; in many cases, C1 can probably be eliminated. Output capacitor C2 protects the circuit against inductive spikes. When a current or thermal overload trips the switch, **FAULT** pin 8 goes low. I put a SM LED on the board to indicate when a fault occurs. Pin 8 is not intended to sink a lot of current, so I used a 5.1 kΩ resistor to limit the LED current to about 1 mA. You could use a100 kΩ pull-up resistor instead and an external high-impedance indicator.

Manufacturers and Distributors of SMT Equipment and Parts

AAVID (manufacturer)—143 North Main St, Suite 206, Concord, NH 03301; tel 603-224-9988; fax 603-223-1738; **http://www.aavid.com**; heat sinks, information about them.

AVX (manufacturer)—**http://www.avxcorp.com/products/capacitors/smtc.htm**; low-ESR capacitors

Bourns (manufacturer)—**http://www.bourns.com/**; resistors and potentiometers.

Chemtronics (manufacturer)—8125 Cobb Ctr Dr, Kennesaw, GA 30152-4386; tel 800-645-5244, 770-424-4888; **http://www.chemtronics.com**; soldering paste, solder, solder wick.

Contact East (distributor)—tel 800-225-5370, 888-925-2960, fax 800-743-8141; **http://www.contacteast.com**; flux pens, soldering equipment, illuminated magnifying glasses.

Dedeco International (manufacturer)—Long Eddy, NY 12760; tel 800-964-6616; **http://www.dedeco.com**. Manufactures the cutoff wheels I use. My 0.005-inch wheel is #5190, the 0.009-inch wheel is #5187. I found an assortment of Dedeco wheels at Home Depot, but they did not include the 0.005-inch wheel.

Digi-Key (distributor)—701 Brooks Ave S, PO Box 677, Thief River Falls, MN 56701-0677; tel 800-344-4539, 218-681-6674, fax 218-681-3380; **http://www.digikey.com**. Carries a wide selection of National, Maxim and International Rectifier ICs, many SMT parts, lithium batteries, holders and soldering equipment. They have good links to manufacturers' Web pages. Digi-Key has a $5 handling charge on orders less than $25.

FAR Circuits (manufacturer)—18N640 Field Ct, Dundee, IL 60118; tel 847-836-9148 voice/fax; **http://www.cl.ais.net/farcir**; custom PC boards.

Gerber (distributor)—Gerber Electronics, 128 Carnegie Row, Norwood, MA 02062; tel 800-225-8290, 781-769-6000, fax 781-762-8931; **http://www.gerberelect.com**. National Semiconductor products, most of the new ICs; $25 minimum order.

Hosfelt Electronics Inc (distributor)—2700 Sunset Blvd, Steubenville, OH 43952; tel 800-524-6464, 888-264-6464, 740-264-6464, fax 800-524-5414; (no e-mail address, no Web site); tilt switches and some SMT parts, 3 V lithium batteries and battery holders.

International Rectifier (manufacturer)—233 Kansas St, El Segundo, CA 90245; tel 310-726-8000, fax 310-322-3332; **http://www.irf.com**; IRF7201, IRLML2402, IRFZ46 and other MOSFETs, diodes, etc.

Kemet (manufacturer)—PO Box 5928, Greenville, SC 29606; **http://www.kemet.com**; capacitors; lots of technical information at this site.

Keystone Electronic Corp—(manufacturer), 31-07 20th Rd, Astoria, NY 11105; tel 718-956-8900; **http://www.keyelco.com**. Manufactures a complete line of battery holders, components and hardware.

Maxim (manufacturer)—120 San Gabriel Dr, Sunnyvale, CA 94086; tel 800-998-8800, 408-737-7600; **http://www.maxim-ic.com**; MAX871, MAX890 and other ICs.

Micrel (manufacturer)—1849 Fortune Dr, San Jose, CA 95131; tel 408-944-0800; **http://www.micrel.com**; MIC1555 and other ICs.

N4UAU (distributor)—5200 NW 43rd St, Suite 102-177, Gainesville, FL 32606; supplies parts kits for most of *QST* projects.

National Semiconductor (manufacturer)—2900 Semiconductor Dr, PO Box 58090, Santa Clara, CA 95052-8090; tel 408-721-5000, 800-272-9959; **http://www.national.com/**; LM2662 and many other ICs.

Newark (distributor)—tel 800-463-9275; call this number to get the phone and fax information of the representative in your area; **http://www.newark.com**. Carries products from many manufacturers including National, Maxim, International Rectifier, Micrel, Motorola, Sprague, Bourns. Many SMT parts, batteries, holders. There is a $5 handling charge for orders less than $25.

Motorola (manufacturer)—**http://www.mot-sps.com/sps/General/chips-nav.html** MC14020 and almost every other IC in the world. Motorola has a large Web site. This is where I have found the most useful information. If the site does not have what you want, try the links to other of its sites.

Sprague (manufacturer)—PO Box 231, Sanford, ME 04073; tel 207-490-7257, fax 207-324-7223; **http://www.vishay.com/products/capacitors.html**; low-ESR capacitors.

Star Micronics (manufacturer)—**http://www.starmicronics.com**—information on buzzers.

Construction Comments

To make this project's PC board, I used a Dremel tool and a 0.009-inch disc. For my prototype, I found it easiest to use a monolithic (non-SM) capacitor for C2, mounting it across the top of the IC. (There is no rule that prohibits you from mixing technologies, and this made construction easier.) Notice how large the capacitors are compared to the IC. As is true with most SM projects, circuit layout is important: Short leads offer low inductance to promote fast switching in the event of a current overload. In case of a short circuit, the board's ground plane helps dissipate heat.

Tune In Again...

Next month, we will look at two chips that turn a positive voltage into a negative voltage and are only available in SM cases. One of these is in the large SO-8 case (as in Project 1); the other is in a smaller SOT-23 case. I hope you build Projects 0A and 1 because the skills you develop working with them will be useful in completing the projects to come.

Notes

[1]Doug DeMaw, W1FB, "Quick-and-Easy Circuit Boards for the Beginner," *QST*, Sep 1979 pp 30-32.

[2]Sam Ulbing, N4UAU, "Mega-Mini Micropower Timeout Switch," *73 Amateur Radio Today*, Jul 1998, pp 42-48.

[3]I was intrigued to come across an engineer's comment in an industry magazine: "RF circuits are readily available as easy to use building blocks, so you needn't fully understand their operation to employ them in an application." Perhaps he had Amateur Radio builders in mind!

[4]Flex-mounted illuminated magnifying lenses are available at office-supply stores and electronic-component suppliers such as Office Depot, Office Max, Digi-Key, Newark, etc. Dremel tools are available from discount stores, Home Depot and Lowe's. Thin 0.020-inch diameter solder can be found at RadioShack (#64-013). Digi-Key, Contact East and Newark sell rosin flux pens.

[5]I have found the best way to locate state-of-the-art parts is via the Internet. Virtually every manufacturer has their component datasheets, applications notes and other information posted. It's a design engineer's dream! No longer do you need lots of databooks. Distributors, too, have catalogs

on-line. If you want to know if a company stocks the Maxim 890 for instance, you need only go to the Maxim home page, check out who their distributors are, then go to those sites and see if they have the part. It's true that some distributors have large minimum quantities for orders, but others don't. If you want more information on the parts in this project, see the sidebar "Manufacturers and Distributors of SMT Equipment and Parts."

[6]You might wonder "How small can they go?" National Semiconductor has recently introduced a device (the LMC6035) in a Micro-SMD package that is one-quarter the size of an SOT-23 package! According to National, the package is only slightly larger than the die itself: "This time we may have reached the packaging limits with the smallest possible footprint." Paul McGoldrick, Senior Technology Editor for *EDTN* said he "...expects to see a lot of licenses being sought in the next months for other manufacturers seeking to take advantage of this huge jump in process 'packaging' and in the lower costs associated with it," *EDTN*, Sep 1998. This is available for viewing at **http://www.EDTN.com/analog/prod194.htm/**.

[7]Per Kemet Electronics Corp monograph F-2103A, *Repair Touch Up Hand Solder—Can These Be Controlled*, by Jim Bergenthal. This and other free literature can be obtained from Kemet Electronics at their Web site (**http://www.kemet.com**). In the upper-left-hand corner of the page, select **Literature Request** after clicking on **Tantalum Capacitors**, then fill in the information form. Finally, click on **Request Selected Literature**. Or, use the Kemet mailing address given in the sidebar "Manufacturers and Distributors of SMT Equipment and Parts."

[8]See Note 7. Kemet emphasizes that: "UNDER NO CONDITIONS SHOULD THE IRON TOUCH THE PART. This is a major cause of part damage." I have touched parts often while soldering them and they have not sustained damage. Perhaps I have been lucky!

[9]Another approach to SMT soldering was suggested to me by Fred, W3ITO. He uses solder paste and a hot plate. He believes it is the only reliable method for amateur SMT (but he was dealing with equipment that had to meet military standards). I have not tried this approach as it appears to need fairly accurate temperature control and the solder paste is difficult to locate, expensive and must be specially stored in a cool dry environment. I would be interested to hear from others who may have tried this method.

[10]Universal SM prototype boards are also available from FAR Circuits. See Paul Pagel, N1FB, "Breadboards from FAR Circuits," *QST*, Nov 1998, p 74.

[11]If you are interested in learning to make your own boards as described, I have a limited number of parts kits consisting of a 3¥6-inch double-sided, copper-clad board, eight cut-off wheels (two 0.005 inch, four 0.009 inch and two 0.025 inch) and the special mandrel recommended for use with the ultra-fine cut-off wheels. This kit allows you to make boards for all the projects in this series and many more. Price $13. (Florida residents must add sales tax. For orders outside the US, please add $3 for shipping.)

Project #0B, Gerber Electronics has agreed to sell this chip to readers of this article at a special price of $12.50 ($8 less than the normal unit price) and waive their normal $25 order minimum. Be sure to identify yourself as a *QST* reader to qualify for this price.

Project #1, A limited number of parts kits are available from me for $6, without a PC board. If you want a premade PC board add $1.50. (Florida residents must add sales tax. For orders outside the US, add $1 for shipping.)

Order from Sam Ulbing, N4UAU, 5200 NW 43rd St, Suite 102-177, Gainesville, FL 32606; **n4uau@afn.org**. Credit cards are *not* accepted.

[12]I omitted the feedback capacitor between pin 1 and pin 8 to reduce the gain but my layout allows for it to be added if desired.

[13]Sam Ulbing, N4UAU, "SMALL—The Surface Mount Amplifier that is Little and Loud," *QST*, Jun 1996, pp 41-42 and 68.

Sam Ulbing, N4UAU, studied electronics in the 1960s, but spent his work career in the financial area. Since he retired in 1986, Sam has enjoyed exploring the opportunities offered to the amateur builder by the new ICs. He feels that electronic design for amateurs has become much easier than it used to be. Sam recalls how in the '60s, he spent hours sweating over complex equations to design even simple circuits. Now, although he has forgotten almost all of his math, the circuits he has built with the new electronics do very sophisticated functions and best of all they work! Presently, Sam is playing with three projects, choosing to build all of them using his "surface-mount style" because "It's just more fun to do it that way." You can contact Sam at 5200 NW 43rd St, Suite 702-177, Gainesville, FL 32606; **n4uau@afn.org**.

By Sam Ulbing, N4UAU

From *QST*, May 1999

Surface Mount Technology—You *Can* Work with It!

Part 2—Last month, we built a couple of simple projects with surface-mount devices. This month's inverter projects go a bit farther.

Projects 2A and 2B—Two 5 V Inverters

A low-current, negative 5 V supply is often a handy item to have on the workbench. Many amplifier circuits are simpler to design using positive and negative voltage sources. Perhaps you have an alphanumeric LCD and found it needs a negative voltage on the **CONTRAST** pin to work. A simple way to supply this negative voltage is to use an ICL7660 voltage inverter, which has been around for a long time. (I'll present another voltage-inverter application in Project 4.) Advances in technology have improved on the '7660. Two ICs I know of that offer significant improvements over their precedents, but both are available only in SM cases: The LM2662 by National is in an SO-8 package and Maxim's MAX871 is available only in SOT-23. Certainly it is possible for manufacturers to make these improved IC versions in a DIP, but neither National nor Maxim have chosen to do so. This appears to me as another signal that the industry is moving toward SM-only parts.

The Technology

Figure 9 shows how these voltage-inverter ICs operate internally. Each consists of four CMOS switches (S1 through S4) sequentially operated by an internal oscillator. During the first time interval, S1 and S3 are closed and S2 and S4 are open; the +5 V input charges C1 with its + terminal being positive and the opposite terminal at ground. At time interval two, S1 and S3 are open and S2 and S4 are closed. There is still 5 V across C1 with the pin 2 side being positive, but pin 4 is no longer at ground potential. The 5 V charge across C1 is transferred to C2—and since C2's positive side is connected to ground—the other side must be 5 V lower than ground, or –5 V. The reason the SM switches can

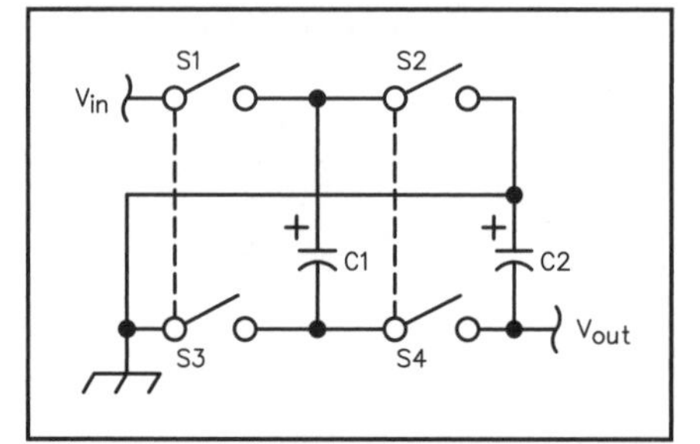

Figure 9—Diagram of the internal workings of the three voltage inverters. See the text for an operational explanation.

handle more current and still be physically smaller is related to their lower resistance and, since both operate at higher frequencies, smaller-value capacitors can be used for a given current output. For best efficiency, low ESR (equivalent series resistance) capacitors should be used. An input bypass capacitor (the value of which depends on the IC and application) improves performance if the power source has a high impedance.

With the trend toward smaller ICs and fewer IC pins, there are often families of nearly identical but specialized ICs. The LM2662 is one of two nearly identical inverters described in the same data sheet. The other, the LM2663, uses pin 1 as a shut-down control instead of a frequency control. This is a common feature with the new technology because of the ever-increasing use of battery power sources, and is especially useful when the inverter is computer controlled. During shut-down, the IC's current drain is reduced to only 10 μA. The MAX871, like the LM2662, has a brother described in the same data sheet. The MAX870 is identical to the MAX871n except that it runs at 125 kHz, and although it needs larger capacitors, it draws only 0.7 mA.

Because large-value capacitors increase a circuit's physical size, it's good to know

LODER BROOKS, KD4AKW

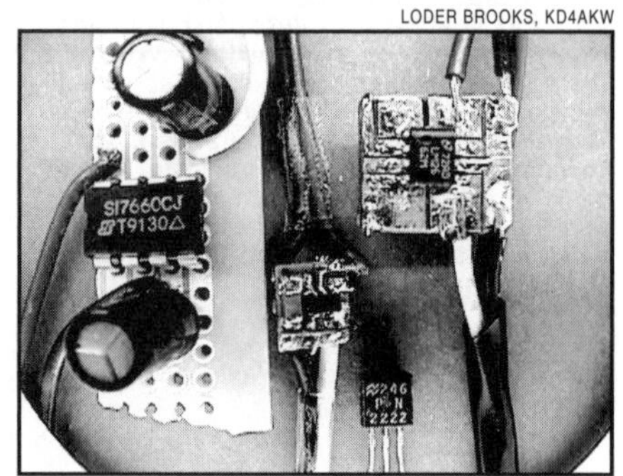

Figure 10—Here, the three voltage-inverter projects are compared in size to a PN2222 transistor.

the minimum capacitance value you can use. This depends on the frequency of operation and the ESR of the capacitor. Nonpolarized capacitor types commonly recommended are Sprague series 593D or 595D, AVX series TPS and the ceramic X7R series. Unfortunately, a capacitor's ESR is often not given in a parts catalog and you may have to consult a data sheet. If you want to try other capacitor values, use the following formulas to calculate output resistance and ripple. Note that C1's resistance is four times as important for reducing resistance as C2, but C1 has no effect on ripple.

$$R_{out} = 2R_{sw} + 1/f \times C1 + 4ESR1 + ESR2 \quad \text{(Eq 2)}$$

$$V_{ripple} = I_{load}/f \times C2 + 2I_{load} \times ESR2 \quad \text{(Eq 3)}$$

where

R_{out} = output resistance of the circuit
R_{sw} = sum of the *on* resistances of the internal switches
f = frequency of the oscillator driving the inverter
ESR1 = equivalent series resistance of C1
ESR2 = equivalent series resistance of C2

[1]Notes appear on page 1-10.

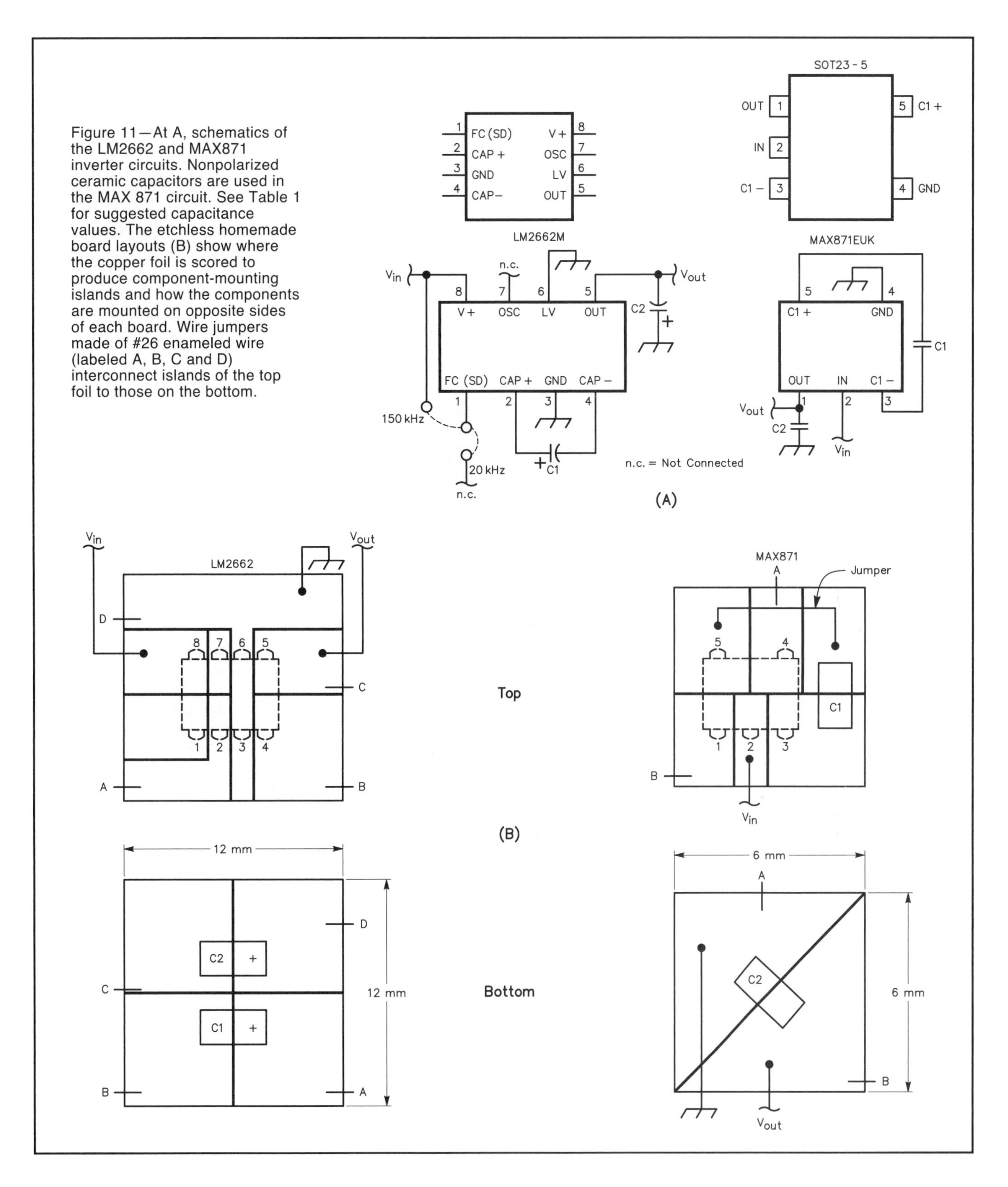

Figure 11—At A, schematics of the LM2662 and MAX871 inverter circuits. Nonpolarized ceramic capacitors are used in the MAX 871 circuit. See Table 1 for suggested capacitance values. The etchless homemade board layouts (B) show where the copper foil is scored to produce component-mounting islands and how the components are mounted on opposite sides of each board. Wire jumpers made of #26 enameled wire (labeled A, B, C and D) interconnect islands of the top foil to those on the bottom.

V_{ripple} = peak-to-peak ripple voltage at the output

I_{load} = load current delivered by the inverter

All three ICs can be used in other modes, such as voltage doublers, connected in cascade to increase output voltage, or connected in parallel to increase output current. For information on circuits to use and more information about design consi-derations, refer to the device data sheets.

The SOT-23 is a popular IC size and it is important to develop the skills to work with it if you want to make full use of the new technology. When you build the MAX871 project, set it aside because you may find it useful in Project 4.

Table 1 summarizes some features of the ICs mentioned, and Figure 10 shows you what the three completed circuits look like. You can see that the LM2662 circuit is somewhat smaller than the '7660, yet it provides *10 times* the current output! The MAX871 circuit is extremely small and outperforms the ICL7660.

Figures 11A and 11B show the schematics and board layouts, respectively. The circuits are simple, each requiring but two capacitors and one IC.[17]

Table 1

	ICL7660	LM2662	MAX871
Package	DIP, SO-8, Can	SO-8	SOT-23
Circuit Resistance (Ohms)	55	3.5	25
Osc Frequency (kHz)	10	20	500
Recommended Cap (μF)	10	100	0.2
V_{out}* @ I = 0	10.0	10.0	10.0
I = 14 mA	9.41	9.97	9.71
R = 100W	6.30	9.66	8.04
I_{supply} (mA)	0.17	0.30	2.7

*These figures are based on actual circuit measurements with the load connected between the positive and negative outputs.

Building the LM2662 Circuit

To save space, I put the IC on one side of a double-sided board, mounting the capacitors on the opposite side. Interconnections between the two board sides are made by short pieces of #26 enameled wire. The wires (labeled A, B, C, D in Figure 11B) bend around the edge of the board. If you have built Project 1, you will have no problem with this one. Be careful to observe capacitor polarity. Even though the LM2662 is smaller than the ICL7660, it offers more features. Pin 1 (which is not used in the ICL7660) controls the LM2662's internal oscillator. The inverter runs at 20 kHz when this pin is left unconnected, and at 150 kHz when connected to V_{CC}. If you want the circuit to operate at 150 kHz, add a jumper between pins 1 and 8 of the IC. This allows you to use smaller capacitors, but at the price of a higher supply current.

Building the MAX871 Circuit

The first time you see this project, you may think "It's too small to build by hand!" But I've built four different circuits this size and made a PC board for each one—so can you! Because the SOT-23 package is smaller than the SO-8, I used a 0.005-inch wheel to make the island-separating cuts on my PC board. Although the IC's pins are small and closely spaced, the SOT-23-5 board requires only two critical cuts: those between pins 1 and 2 and between pins 2 and 3. The spacing between pins 4 and 5 is as large as that of an SO-8 package. Mounting C2 beneath the board makes component layout much easier.

SOT-23 packaged devices are too small for manufacturers to imprint the part number on them—MAX890EUK just will not fit! Instead of MAX890EUK, Maxim uses the marking **ABZO**. If you get two SOTs mixed up, you will have to consult the data sheets to determine which is which.

Next Month

In Part 3, we'll look at a low-voltage battery protection switch that makes use of a few SM ICs: three SO-8s and one SOT-23.

Notes

[16]Part 1 of this four-part series appears in the April 1999 issue of *QST*, pp 33-39.

[17]Obtaining the parts—Project #2A: Gerber Electronics stocks the LM2662 and Newark Electronics stocks low-ESR tantalum SM capacitors. If you cannot find an LM2662, use the LM2660, Maxim MAX660 or the Linear Technology LTC660; all have similar characteristics and identical pin outs. Digi-Key carries some of these ICs, but does not stock the low ESR SM capacitors. Low-ESR SM capacitors are quite expensive, so you may want to use standard tantalum capacitors instead. These are available from most suppliers. I have a PC board for the layout described; price: $1.50. Contact Sam Ulbing, N4UAU, 5200 NW 43rd St, Suite 102-177, Gainesville, FL 32606; **n4uau@afn.org**. Credit cards are *not* accepted.

Project #2B: A limited number of parts kits, with hard-to-find 1 μF ceramic capacitors (to permit maximum current output with minimum ripple) are available from me for $6 *without* a PC board. If you want a pre-made PC board add, $1.25. (Florida residents add sales tax.)

If you are interested in making your own boards as described, I have a limited number of parts kits consisting of a 3×6-inch double-sided, copper-clad board, eight cutoff wheels (two 0.005 inch, four 0.009 inch and two 0.025 inch diameter) and the special mandrel recommended for use with the ultra-fine cutoff wheels. Price: $13. This kit allows you to make the boards for all the projects in this series and more. (Florida residents must add sales tax. For orders outside the US, please add $3 for shipping.)

You can contact Sam Ulbing, N4UAU, at 5200 NW 43rd St, Suite 102-177, Gainesville, FL 32606; **n4uau@afn.org**.

By Sam Ulbing, N4UAU

Surface Mount Technology—You *Can* Work with It!

Part 3—This more-complex SM project employs a total of four ICs and seven other parts on a PC board three-quarters of an inch square! Despite its small size, it can control current levels of up to 10 A—without using mechanical relays!

When Hurricane Georges came through in 1998, my friend, Dave, NØLSK, had left his boat at a marina in the Keys. Although he had the boat tied up well, he forgot that his refrigerator shifts to battery power if the ac-line power is lost. When he returned to the boat after the storm, his boat was okay, but its battery was dead. With the switch about to be described, Dave wouldn't have lost his expensive battery.

Project 3—A Low-Voltage Battery-Protection Switch

This switch, based on a MAX835 (available only in an SOT-23 case), is a *latching* voltage monitor—ideal for controlling a switch. A recent *QST* project used a MAX8211 (a DIP IC) in an undervoltage circuit,[18,19] but because it doesn't latch, that chip wouldn't work well for controlling a switch. Here's why: Every time the voltage dropped low enough to trip the monitor, it would disconnect the load, and the voltage would rise and turn the monitor back on. Such cycling could injure the equipment.

Figure 12 is a schematic of the switch. When V_{CC} drops below 12 V, and the voltage at U1 pin 4 goes below 1.2 V. That causes the output voltage on pin 5 to drop from about 5 V to 0 V. Pin 2 of U2, an MIC5014, accepts a logic-level signal and uses it to control an on-board charge pump. Q1 and Q2 are N-channel MOSFETs used as a 10 A high-side switch. (See the sidebar "Selecting a MOSFET for power control.") To turn the switch on, the gate must be at least 10 V above the source voltage (5 V for logic-level MOSFETs). That means you need 22 V to turn on the switch. This voltage is supplied by the charge pump in U2. U2 also acts as a buffer for U1, which cannot operate at 12 V. R1 through R4 provide a nominal 5 V power source for U1 and provide the voltage-level input signal to pin 4 of U1. Pull-down resistor R5 prevents unplanned resetting. Although the data sheet doesn't show that R5 is required, the very high impedance of this pin (the current drain is 1 nA) makes a pull-down resistor a wise investment, especially in an RF environment.

Depending on your circuit needs, you will use a variety of values for R1 to R4. Here's how I selected my values: U1 can operate with voltages of 2.5 to 11 V, so the R1-R4 divider must keep the voltage at pin 3 in this range as V_{CC} changes. The maximum V_{CC} I ever expected to encounter was 15 V, and the least, 11 V. U1 draws 2 μA,

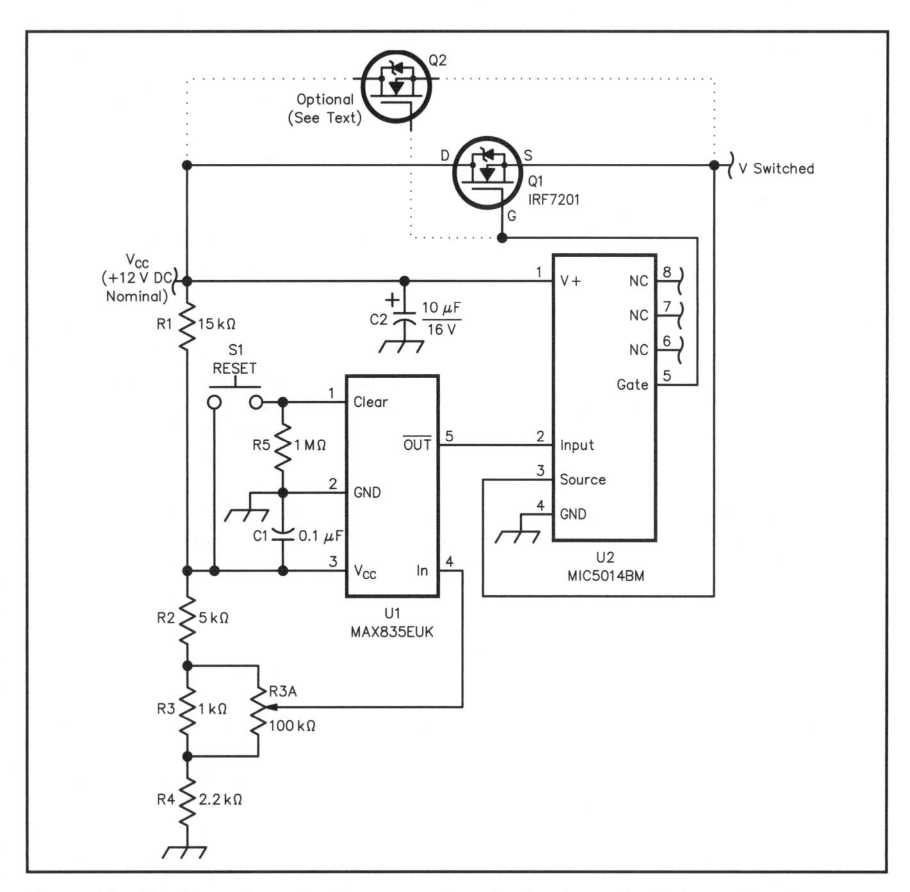

Figure 12—The low-voltage battery-protection circuit schematic. Unless otherwise specified, resistors are 1/8 W chip resistors in a 1206 package. Equivalent parts can be substituted. See Note 21 and notes and sidebar in Part 1 for parts availability.

C2—10 μF, 16 V tantalum
R3A—100 kΩ pot
S1—SPST pushbutton
Q1, Q2—IRF7201
U1—MAX835EUK
U2—MIC5014BM

[1]Notes appear on page 1-13.

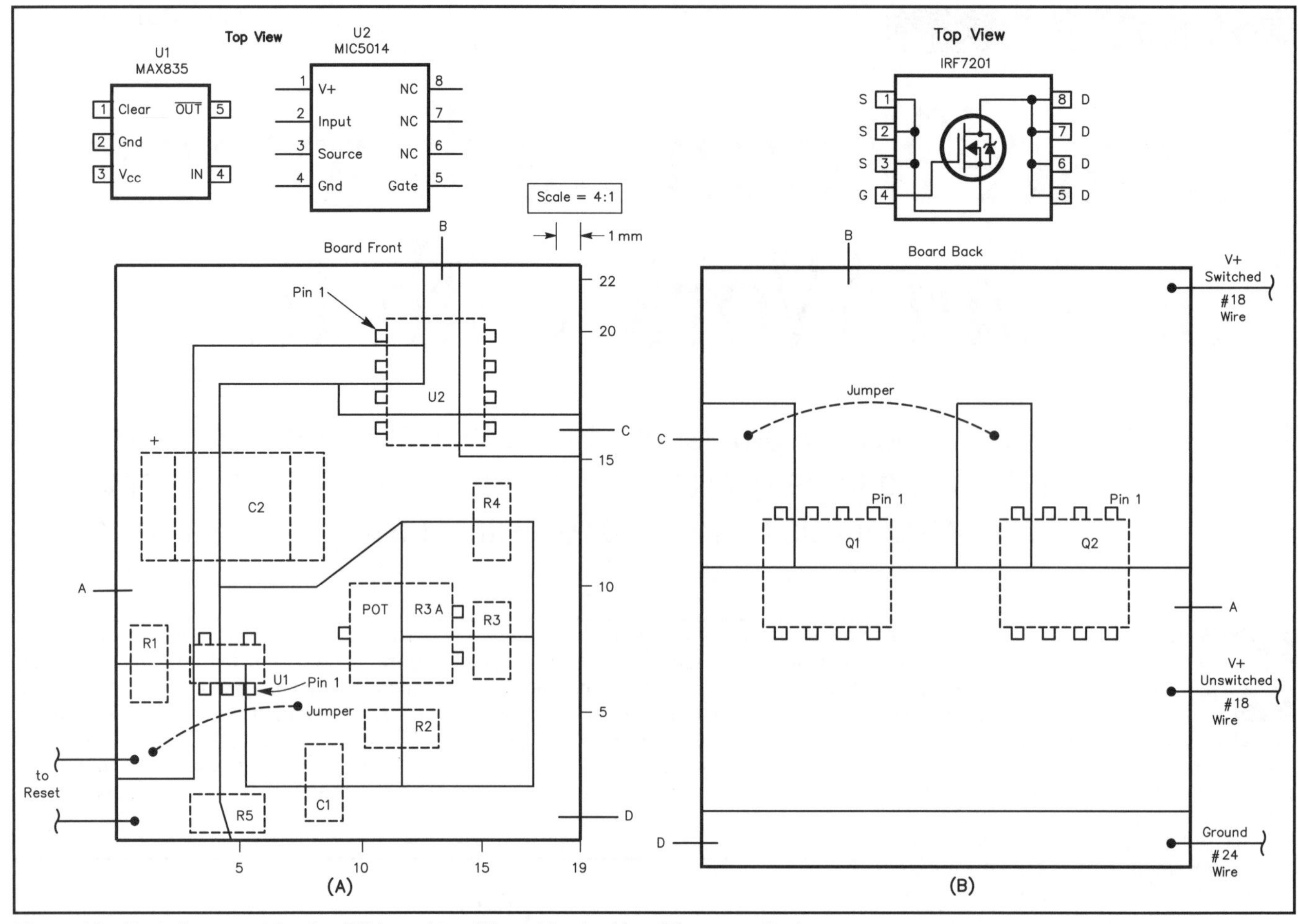

Figure 13—Part placement for the PC-board top (A) and back sides (B), respectively.

so to keep a stiff supply, I wanted the current through the voltage divider to be at least 200 µA. This calls for a total resistance of not more than 50 kΩ. As current drain is not important, I decided to use a total resistance of about 20 kΩ. Using an Excel spreadsheet, I calculated the values shown. The voltage on pin 3 is 5.3 V for V_{CC} of 15 V, and 3.8 V for V_{CC} of 11 V. I used four fixed-value resistors and a SM potentiometer (R3B) in parallel with R3 to allow better control when setting the trip level at 12.0 V. You can run U1 at a lower voltage, but keep in mind that U2 needs at least 2 V to trigger it.

This basic circuit can be optimized for other uses. To handle more current, you need only replace the MOSFET with a more-robust one. I use an IRFZ46 and a heat sink with my ICOM IC-735. If you want to control a low-voltage NiCd-powered circuit, you could use U1 alone, connecting it directly to a logic level *low-side* N-channel MOSFET. In that case, increase the resistances of R1 to R4 for minimum current drain.

The Technology

There is a lot of new technology in this simple circuit. The entire project—including the 10 A MOSFET switch—is on a PC board smaller than the MAX8211 project. Because the quiescent current of U1 is only 2 µA—and it has a wide operating-voltage range (2.7 to 11 V)—it's possible to power it from a resistive divider rather than a 5 V regulator. U2 is a single-chip charge pump that requires no external parts. Like U1, it can operate over a wide voltage range (2.75 to 30 V) and draws only a few microamperes. U2 is designed to let low-level signals control high-voltage and high-current circuits through low resistance *N-channel* MOSFETs used on the *high side*. This arrangement is important for at least two reasons: High-side switches are usually needed with Amateur Radio applications because there is almost always more than one path to ground: the antenna, keyer, computer etc. A low-side switch would force the current to go through one of those connections rather than shutting off the rig. Second, N-channel MOSFETs have much lower resistance than equivalent P-channel MOSFETs (typically, 2.5 times less[20]). This permits the use of smaller MOSFETs for a given current.

MOSFET technology has advanced dramatically. The circuit shown uses two small SO-8 MOSFETs in parallel to control the power to my Kenwood TM-241, which draws a maximum of 11 A on high power. The two MOSFETs in parallel have an *on* resistance of only 15 mΩ (milliohms). The voltage drop across the plug connections and fuses is greater than the drop across the MOSFET! Unlike power transistors, no equalization resistors are needed when paralleling MOSFETs. That's because MOSFET *on* resistance increases with temperature, so they tend to be self-equalizing. MOSFETs make better circuit breakers than fuses or relays because they have no moving parts, are resettable, do not arc or bounce, emit less EMI and are much faster than relays or fuses. The latter can be important in an overcurrent situation. Typical MOSFET shut-down time is less than a microsecond (excluding circuit delays). The blow time on a fast-acting fuse is usually longer than 1000 µs.

U1 and U2 have families. U1 has a push-pull output; the output is internally held at either 0 or V_{CC}. The other version (MAX834) has an open-drain output that requires a pull-up resistor to provide the logic-high output. This is a common family in the SM world. One advantage of the open-drain output is that you can control a circuit with a voltage level different than the V_{CC} of the IC itself. U2's brother is the MIC5015, which operates exactly like the

Figure 14—Close-up views of the low-voltage battery-protection switch using a Maxim MAX835 SM IC. At A (left) is a trial board made using the MAX835 and non-SM parts. The MAX835 can be seen above the large pot and to the right of the 1/4-W resistor. At B (right) is the top side of a PC board made using the hobby tool and all SM parts. The MAX835 is above the SM pot and to the left of the tantalum capacitor. The board size can be compared to the TO-92 case transistor above it. The MIC5014 is at the bottom right of the board. The SM MOSFETs are on the bottom side of the board and not shown.

'5014, but 0 V at the input turns it on, and a high level turns it off.

Making the PC Board

Figures 13A and B show the part placement for the top and back sides of the PC board for this project.[21] Before I made the all-SM version (shown in Figure 14B), I built a quick-and-dirty prototype using two PC boards. Except for the ICs, I used all standard-size leaded parts (one side of one of these boards is shown in Figure 14A), so the only critical cutting area was around the IC; all the rest was old-fashioned pad construction such as used in Project 0A. If you want to use an SM-only IC but don't need small size, this is an easy way to do it. You can also add solid wires as leads from the board and plug the entire circuit into a protoboard to use the subcircuit in a larger through-hole circuit. If you realize that SM projects needn't require *only* SM devices, experimentation becomes easier.

Once more, four jumpers (A, B, C and D) make connections between the top and bottom sides of the board where necessary. Because the board has parts on both sides, soldering is a little trickier than dealing with a board with parts mounted only on one side. Once you solder parts to one side of the board and turn the board over to solder the backside parts, it won't lie flat. Here's where a small vise can help by holding the board steady. I place parts on the more-congested side first. It isn't difficult to use the two SO-8 chips because they are so large; soldering SOT-23 packages would be more of a challenge.

Selecting A MOSFET for Power Control

The MIC5014 can drive just about any N-channel MOSFET. Which MOSFET you use depends on your current load. My circuit uses two small SO-8 MOSFETs in parallel. Although the specs show the maximum current for each as 7 A, a check of the I^2R (power) loss and thermal resistance shows that 4 or 5 A is a more reasonable amount when the chip is mounted on a small PC board. Using two MOSFETs in parallel, the circuit has no problem passing 10 A continuously. When selecting a different MOSFET, *calculate its heat loss* and don't be fooled by the maximum-current figure which—for nonsurface-mounted MOSFETs—is achievable only with a perfect heat sink.

The data sheets give the thermal resistance as temperature rise per watt of heat dissipated in the MOSFET (°C/W). For surface-mount MOSFETs, the data sheet gives a single number: junction-to-ambient thermal resistance. For the IRF7201 used in the project, that is 50°C /W. At 10 A total, each MOSFET carries 5 A, and the I^2R loss is $25 \times 0.030 = 0.75$ W, giving a temperature rise of 37.5°C (100°F) above room temperature.

For nonSMT MOSFETs, the junction-to-ambient figure applies *only* if you are *not* using a heat sink. To gain the most from the MOSFET, you need to use a heat sink. In this case, you can determine the thermal resistance by adding the thermal resistances of the junction to case, the case to sink and the sink to ambient, which depends on the heat sink used. (This is just like electrical circuits: Resistances in series are added.) When in doubt, try a heat sink and see if things get hot. If so, use a larger heat sink, or add another MOSFET in parallel to reduce the current through each one.*—*Sam Ulbing, N4UAU*

*If you want to use a junk-box MOSFET, be sure to check its *on* resistance as it will probably be much higher than those of the MOSFETs I am using. The IRF510, a common MOSFET in a TO-220 package, has an *on* resistance of 0.54 Ω. Even with a rather large heat sink, the *maximum* current it can pass is about 4 A.

Tune In Again

Before we wrap up this series next month, we'll look at a project with a large number of small parts mounted on both sides of the board. This project is one you can use as an appeasement gift to your loved ones for spending so much time at the workbench!

Notes

[18]Parts 1 and 2 of this series appear in the April and May 1999 issues of *QST*, pages 33-39 and 48-50, respectively.

[19]Donald G. Varner, WB3ECH, "A Battery-Voltage Indicator," *QST*, October 1998, pp 50-51.

[20]Micrel Applications Note 5 (Micrel, 1849 Fortune Dr, San Jose, CA 95131; tel 408-944-0800; **http:www.micrel.com**).

[21]If you are interested in learning to make your own boards as described in this series, I have a limited number of parts kits available. Each consists of a 3 × 6-inch double-sided, copper-clad board, eight cut-off wheels (two 0.005 inch, four 0.009 inch and two 0.025 inch) and the special mandrel recommended for use with the ultra-fine cut-off wheels. Order from Sam Ulbing, N4UAU, 5200 NW 43rd St, Suite 102-177, Gainesville, FL 32606; **n4uau@afn.org**. Price $13. (Florida residents must add sales tax). For orders outside the US, please add $3 for shipping.

A limited number of parts kits for Project 3 are available from me for $12 without a PC board. If you want a premade PC board, add $1.50. (Florida residents add sales tax). The kit includes only one IRF7201 MOSFET. If you want to parallel more MOSFETs or try an IRFZ46, Digi-Key, Newark and other suppliers carry those parts.

You can contact Sam Ulbing, N4UAU, at 5200 NW 43rd St, Suite 102-177, Gainesville, FL 32606; **n4uau@afn.org**. QST

By Sam Ulbing, N4UAU

From *QST*, July 1999

Surface Mount Technology —You *Can* Work with It!

Part 4—This month, we wrap up the series. Before we do, though, here's that project I mentioned last month...

The first three parts of this article[22] have described rather easy-to-build projects. This one is a bit more complex. If you like to experiment, you have the opportunity to tailor this project to your specific needs and optimize its operation. Build it for a loved one and impress them with your skills! If you spend as much time working on electronic projects as I do, that loved one might appreciate a little project like this made just for them!

Project 4—The Hourglass 10-Minute Timer

This month's project is a modernized version of "A Simple 10-Minute ID Timer," that appears in *The ARRL Handbook*.[23] You can use the Hourglass as an egg timer, or to remind you to move the sprinkler, or put the laundry in the dryer, or as a two-hour timer to remind your teenager it's time to get off the telephone! You start the timer by *turning it upside down*, just like a sand hourglass! As you'll see, the operations of the old and new circuits are similar, but not exactly the same.

[1]Notes appear on page 1-17.

The Old-Technology Circuit

The *Handbook* circuit (Figure 15) is specified for use with a 12 V supply, which could limit its portability and application. LM555 timer U1 is set up for a short duty cycle: 1 second *on* and 59 seconds *off*. Pin 3 of the 4017 counter, U2, triggers after 10 cycles, increasing the time delay to 600 seconds. The alarm sounds, the circuit resets and starts counting again. Ten minutes is about the maximum practical time delay of this circuit.

The New-Technology Circuit

Surface-mount technology allows us to build this month's project (including its power supply) on a board that fits inside a 35-mm film canister (see Figure 16), so it's completely portable.[24] The low voltage and current demands of the ICs allow powering the circuit with a 3-V lithium battery.

Refer to Figure 17. An RC controlled timer, U1, is routed to a counter, U2, to extend the time base to 10 minutes. When

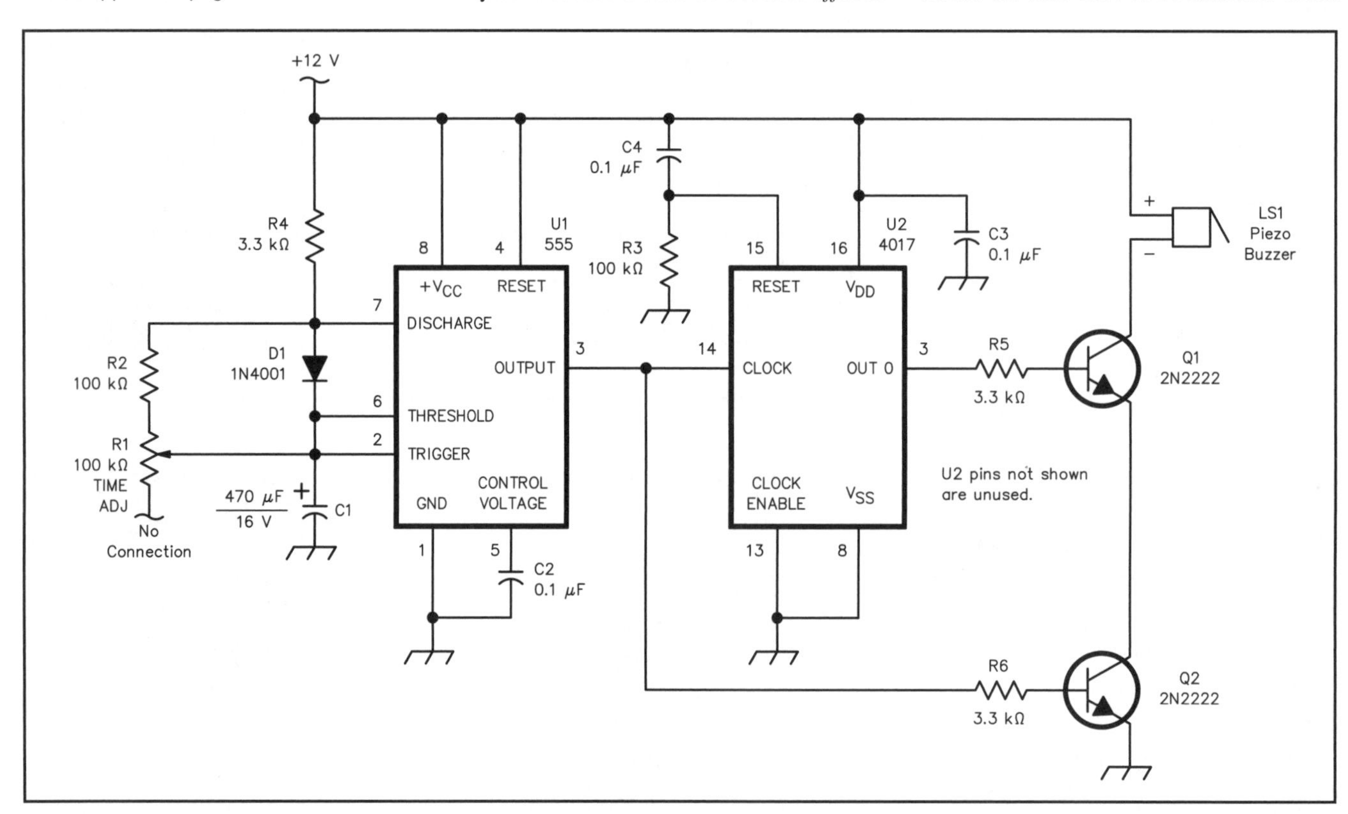

Figure 15—Schematic of the older 10-minute timer. Unless otherwise specified, resistors are 1/4 W, 5% tolerance carbon-composition or film units. Equivalent parts can be substituted.

PHOTO BY LODER BROOKS, KD4AKW

Figure 16—A top view of the SM version of the timer described in the text. The 3-V battery that powers the circuit is mounted on the bottom side of the board.

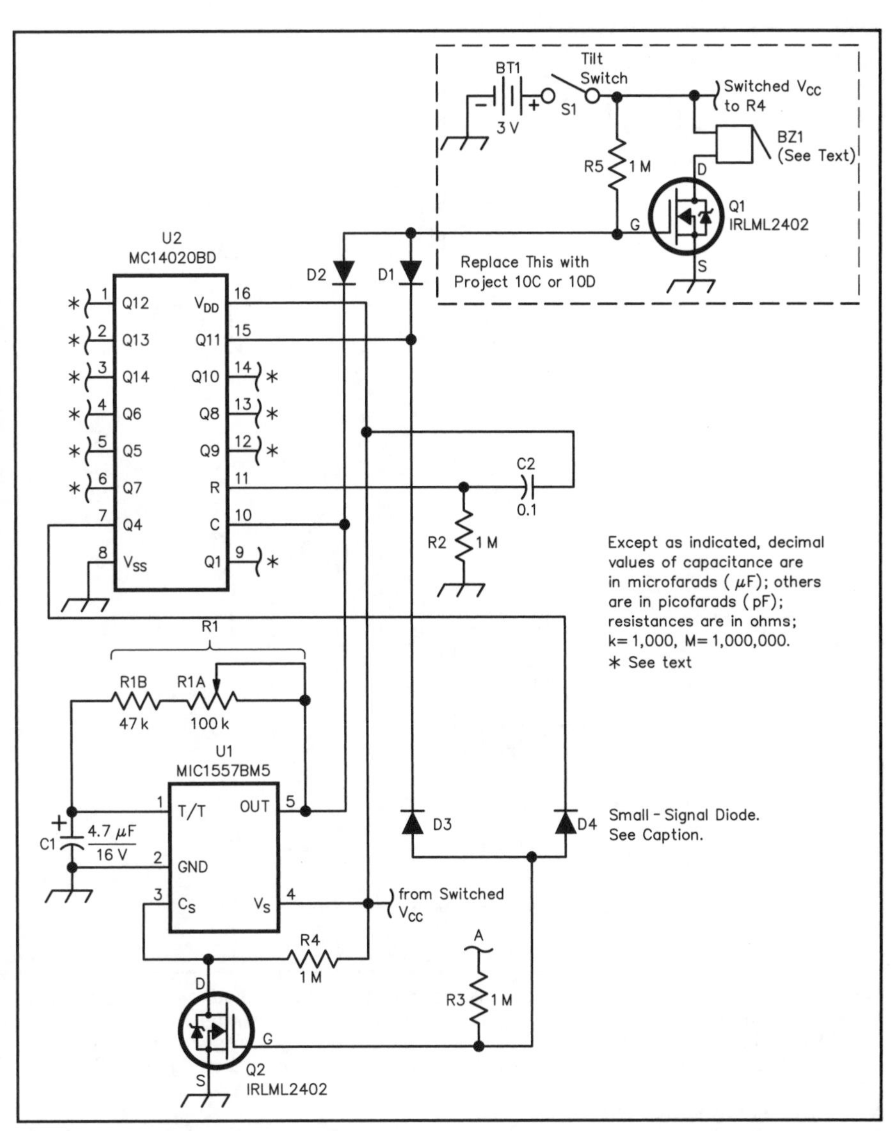

Figure 17—Schematic of the SM "hourglass" 10-minute timer. A 3-V lithium battery powers the circuit. The section of the circuit enclosed in dashed lines can be replaced by either of the circuits shown in Figures 18A and B. Unless otherwise specified, resistors are 5% tolerance SM units. The resistors I used are SM devices in 1206 cases. Equivalent parts can be substituted. See the sidebar "Manufacturers and Distributors of SMT Equipment and Parts," in Part 1, *QST*, May 1999, for a list of suppliers.

BT1—3-V, lithium CR2032, etc.
BZ1—Piezo buzzer (see text)
C1—4.7 µF, 16 V tantalum.
C2—0.1 µF ceramic (I used a SM device in a 0805 case).
D1/D2, D3/D4—BAW56LT1 (common-anode diode pairs in an SOT23 case); pairs of 1N914 or 1N4148 diodes can be substituted.
Q1, Q2—IRLM2402 MOSFET
R1A—100-kΩ pot (Bourns 3364W)
SW1—Encapsulated tilt switch (available from author)
U1—MIC1557BM5, Micrel IttyBitty RC timer/oscillator
U2—MC14020BD, 14-bit binary counter
Misc: Battery holder, Keystone #3002.

the 10 minute limit is reached, the appropriate pin on U2 goes high, turning on a switch, Q1, which sets off an alarm.

U1 of Figure 17 is an MIC1557. Dubbed the "IttyBitty RC Timer" by the manufacturer,[25] it's an SOT-23 version of the 555. R1 and C1 set the cycle time. (R1 is composed of a pot, R1A, and a fixed-value resistor, R1B.) I use a 50%-duty-cycle timer because it requires fewer parts than an asymmetrical-duty-cycle timer. I selected a cycle time of about one second because the data sheets for the LM555 and the MIC1557 indicate that capacitor leakage affects the accuracy of periods longer than 10 seconds. With just a one-second cycle time, it's necessary to use a longer delay in U2, so I added an MC14020, a 14-bit binary counter that can count up to 16,384. By using a count of 1024—and adjusting the values of R1 and C1—I achieved an accurate 10-minute delay.

This flexible circuit can be modified for longer or shorter delays, from as little as a few seconds to as long as 24 hours! (See **Experimenting with the Timer** later.) I had a difficult time finding counters in SM packages, and as you can see in the photo, the chip is "huge." (Perhaps this indicates there's a better way to handle delay circuits with SMT.)

The output at pin 15 of U2 triggers Q1 through D1. Q1 is not just any MOSFET—the IRLML2402 is a state-of-the-art device. Its gate turn-on voltage is only 1.5 V, and at 3 V, the MOSFET is fully on. (Not too many years ago, MOSFETs required 10 or 12 V to turn on. Most logic-level MOSFETs today still require 5 V, which makes them useless in a 3-V supply project.) Although the IRLML2402 is packaged in a Micro 3 package (which is smaller than an SOT-23 package), its *on* resistance is only 0.25 Ω and it can switch current levels up to 1 A.

You might ask, "Why not use a bipolar transistor instead of a MOSFET?" There are several reasons. Transistors require bias current, MOSFETs do not. A small transistor with a 30 mA load develops a 300 mV drop. The MOSFET has only a 4 mV drop, an important consideration when the supply voltage is only 3 V.[26] Also, a MOSFET can be used as a comparator. At levels less than 1 V (for this device), the MOSFET is *off*, and for levels above 1.5 V, it is *on*.

I wanted to use an **AND** condition to sound the buzzer, BZ1. D2 connects the gate of Q1 to the output of U1. This arrangement turns on the buzzer only when U2 pin 15 is positive *and* pin 5 of U1 is positive. Because the level at U1 pin 5 changes at about one cycle per second, the result is a pulsating buzzer that is more noticeable and uses less power than a continuously sounding buzzer. Another reason I could not use a transistor at Q1 is because the voltage drops of D1 and D2 result in a low voltage level of 0.6 V at Q1; that is too high to turn off a transistor.

With a battery-powered device, I didn't want the timer to cycle continually; that would deplete the battery if I forgot to shut

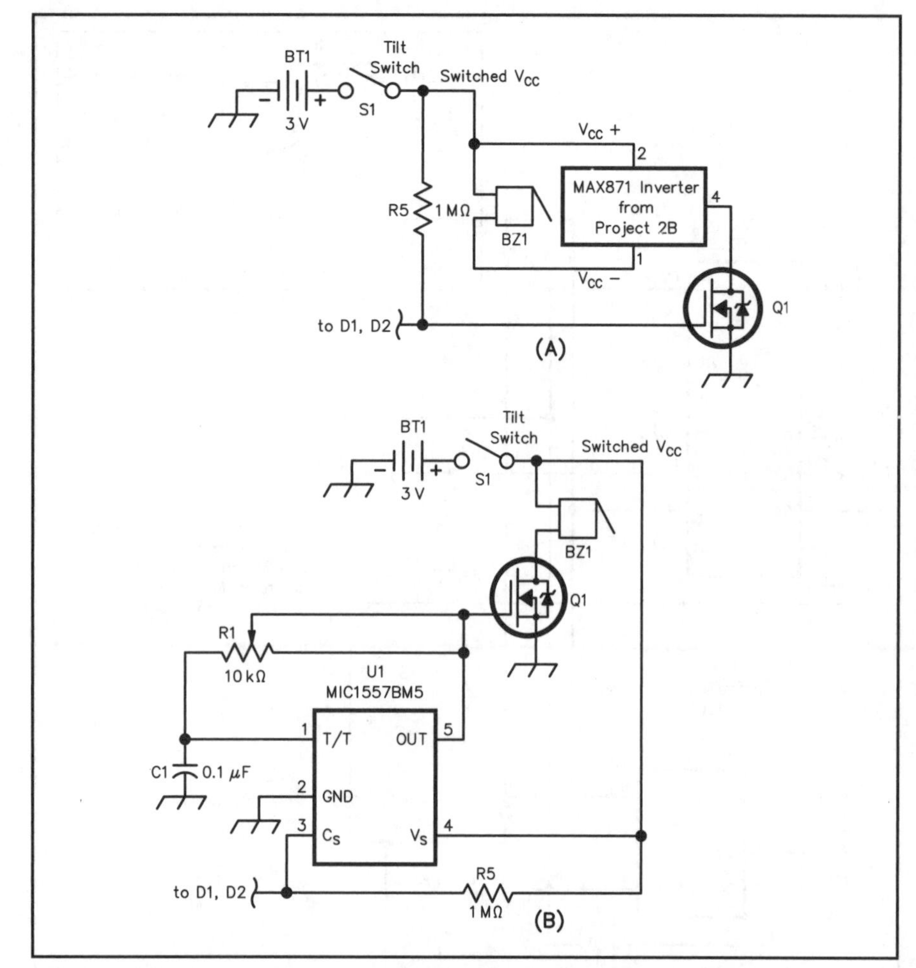

Figure 18—Two modifications you can make to the basic hourglass timer of Figure 17. At A, use of a piezo buzzer requiring a higher supply voltage (6 V) can take advantage of the MAX871 inverter circuit described in Project 2B. R5 is a 1 MΩ resistor in a 1206 SM case. An externally driven buzzer can employ the circuit shown at B using an MIC1557. Component identifications are those given in Figure 17. Note the value change of R1.
R1—10-kΩ pot (Bourns 3364W)

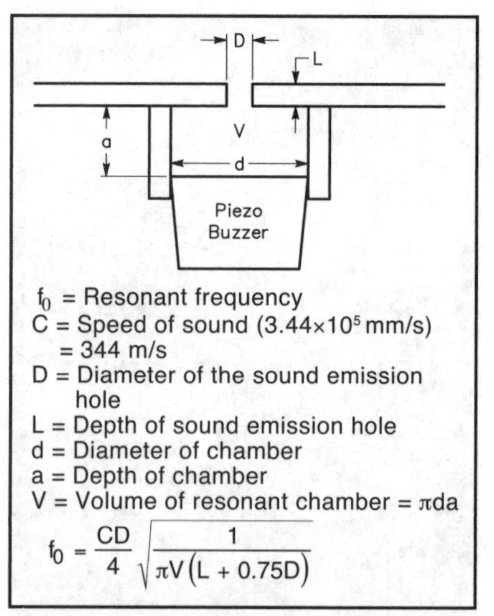

Figure 19—Here's the fundamental approach to constructing a resonator to improve the loudness and purity of the buzzer's tone. The resonant frequency (f_0) should equal twice the frequency of the buzzer to increase sound pressure. Do not make dimension D too small, or the acoustic resistance will increase. The equation is a starting point; experimentation will optimize your results (see text). Dimensions are in millimeters unless otherwise noted.

off the timer. This circuit shuts itself off. The buzzer sounds for about three seconds, and if it is not restarted, the circuit goes to sleep. It works like this: D3 and D4 form an **AND** gate controlling Q2. After pin 15 of U1 goes high and the buzzer sounds, the timer continues to count until U2 pin 7 also goes high. Then, Q2 turns on and pin 3 of U1 goes low. Pin 3 is U1's **CHIP SELECT** pin; when it goes low, U1 stops running and its current drain is reduced to 1 μA. With U1 sleeping, its output goes low. That shuts off the buzzer via D2. Total current drain while sleeping is about 5 μA. Under these conditions, a lithium 2032 battery should last several years.

To restart the timer (from sleep mode or when it is buzzing), just turn it upside down and then right side up. The tilt switch turns the power off, then on. C2 and R2 form a power-up reset that restarts U2 at 0 with a positive pulse to pin 11 through C2.

Experimenting with the Timer

Using the right audio transducer makes a major difference in audibility. Most transducers require more than 3 V to operate. I tried a RadioShack 273-074 transducer and it worked, but its output level was quite low. One way to raise the sound level is to raise the buzzer voltage. I did that with the circuit of Project 2B, as shown in Figure 18A. Some parts catalogs list piezo transducers that are externally driven and operate at 1.5 or 3 V. (The RadioShack buzzer mentioned earlier is internally driven: It has a square-wave generator built into it). I used a piezo transducer driven by an MIC1557, as shown in Figure 18B. It has a loud signal, but I found that setting the exact frequency needed for maximum sound was tricky. The best signal I could obtain came from a TMB-05[27] buzzer that I placed in a resonator and drove with the MAX871 circuit.

A Resonator

A neat way to improve the loudness and purity of the buzzer's tone (some piezo resonators have a harsh note) is to place the buzzer in a Helmholtz resonator.[28] This is a cylinder or tube designed to resonate at a certain frequency. Every resonator I used made the sound available from the transducer louder and clearer. The information in Figure 19 can help you design a resonator. If the math bothers you, try using a simple resonator made from a half-inch water pipe PVC end cap and drill a 3-mm diameter hole in the end; it worked well for me. I ground down the material surrounding the hole to make it thinner (smaller L) and adjusted the distance (A) for maximum sound. Best results are obtained when the tube's resonant frequency is about twice that of the piezo transducer's frequency.

Other Time Delays

In Figure 17, instead of connecting D1 to pin 15 of U2, you can attach it to another pin to obtain a different time delay. Table 2 shows the delays you can achieve when using a one-second cycle time at U1. By adjusting the values of R1 and C1, you can obtain nearly any time delay you want. For the arrangement to work correctly, the U2 pin you use to trigger Q1 must have a greater number of counts than the pin you use to shut down the circuit, which is why the data in Table 2 starts at 16 counts.

If you make your own PC board, you can customize it as needed. The premade PC board (see Note 24) is designed so you can add the circuits of Figure 18A or B on a separate board to drive the buzzer.

Construction Comments

I used a 0.005-inch wheel for the critical cuts at U1, Q1 and Q2 (see Figure 20). For the other cuts, I switched to a 0.009-inch wheel. The 0.005-inch cut is so narrow that

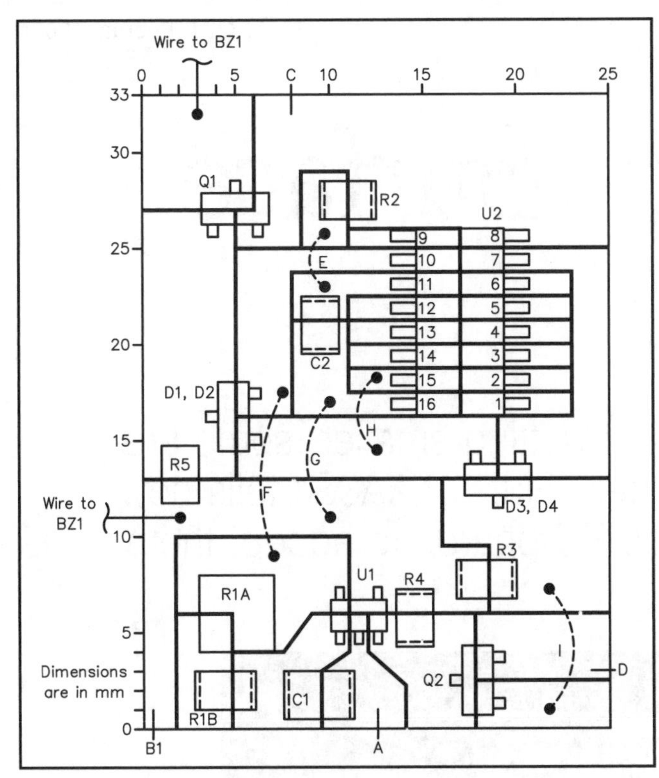

Figure 20—PC board layout of the 10-minute timer. Heavy lines designate cuts made in the foil to create component-mounting islands as described in Part 1 of this series.

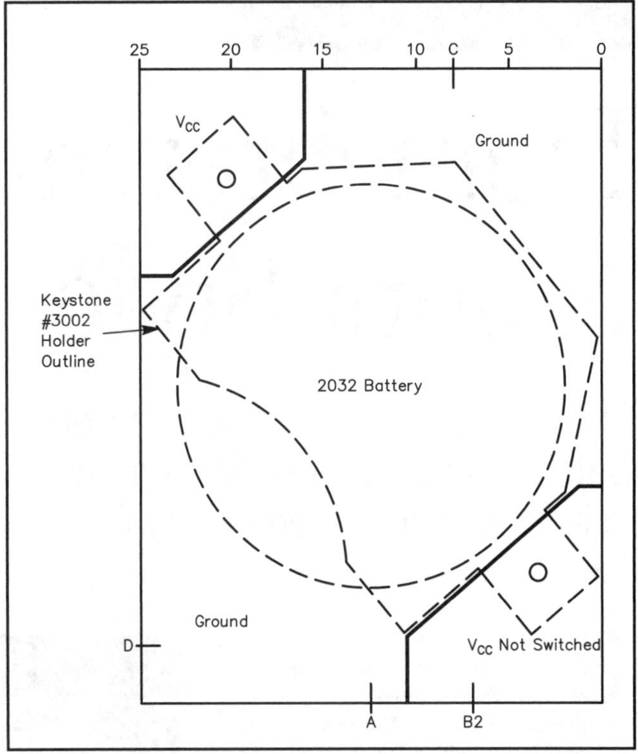

Figure 21—The 3-V battery occupies the bottom side of the 10-minute timer. Again, the heavy lines indicate where cuts are made in the PC-board foil.

solder tends to bridge the gaps. With a 0.009-inch cut, bridging is much less likely to occur. The circuit is on the board's top side; the battery and holder are on the bottom as shown in Figure 20. The tilt switch is connected between B1 of Figure 20 and B2 of Figure 21.

Summary

After completing these projects described over the past months, you should feel comfortable working with SMT devices. And, as I do, you'll probably be turning the pages of *QST* looking for a neat SMT radio project. A couple of the projects I would like to see include: a small, inexpensive VHF transceiver and a pocket-size HF receiver. In addition to the Maxim parts I mentioned earlier, Phillips Semiconductor sells a single-chip SMT AM receiver, MicroChip has an SMT microprocessor and Texas Instruments (TI) has a highly efficient SMT Class-D stereo amplifier. The parts are there. I hope we amateurs start to make use of them. [Let's see some of those projects! *QST* depends on readers and authors such as Sam and you for projects. Send your manuscripts to Steve Ford, WB8IMY, 225 Main St, Newington, CT 06111; **sford@arrl.org**.—*Ed.*]

Table 2
Timer Delay for a U1 Cycle Time of One Second

Pin	*Counts*	*Time*
5	16	16 s
4	32	32 s
6	64	1 m 4 s
13	128	2 m 8 s
12	265	4 m 16 s
14	512	8 m 32 s
15	1024	17 m 4 s
1	2048	34 m 8 s
2	4096	1 h 8 m 16 s
3	8192	2 h 16 m 32 s

Time shown in hours, minutes and seconds.

Notes

[22]Parts 1 through 3 of this series appear in the April, May and June 1999 issues of *QST*, pages 33-39, 48-50 and 34-36, respectively.

[23]R. Dean Straw, N6BV, Ed., *The 1999 ARRL Handbook* (Newington: ARRL, 1998), p 22.58

[24]A limited number of parts kits are available from me for $11, which includes all the parts (including the hard-to-find tilt switch) except for the buzzer and PC board. If you want a premade PC board, add $2 (Florida residents must add sales tax). Piezo buzzers are widely available at places like RadioShack or many of the parts sources listed in the article.

If you are interested in learning to make your own boards as I described in Part 1, I have a limited number of parts kits available. These consist of a 3 × 6-inch double-sided copper-clad board, eight cutoff wheels (two 0.005 inch, four 0.009 inch and two 0.025 inch) and the special mandrel recommended for use with the ultra-fine cutoff wheels; price: $13. (Florida residents must add sales tax.)

[25]The MIC1557 has a brother, the MIC1555, optimized for monostable operation. It is described in the same data sheet.

[26]These are the results of measurements I made.

[27]The TMB-05 internally driven buzzer is made by Star Micronics, 70-D Ethel Rd West, Piscataway, NJ 08854; tel 800-782-7636 (X512), fax 732-572-5095; **sales@starus.com; http://www.starmicronics.com/product/audio/index.cfm**. See the Star Micronics Buzzers and Transducers catalog, page 5.

[28]A resonator based on this principle is described by Wally Millard, K4JVT, "A Resonant Speaker for CW," Hints and Kinks, *QST*, Dec 1987, p 43—*Ed.*

You can contact Sam Ulbing, N4UAU, at 5200 NW 43rd St, Suite 102-177, Gainesville, FL 32606; **n4uau@afn.org**.

By Ed Kessler, AA3SJ

From *QST*, February 2004

Homebrewing—Surface Mount Style

There's no denying that components are getting smaller; some are obtainable only in surface mount devices (SMDs). AA3SJ tells how to use a common Dremel tool to make the PC boards to accept them.

Building one's own rigs and accessories is a rewarding part of ham radio, but current surface mount technology (SMT) can be daunting and prohibitive, even for the enthusiast. Yet, with a bit of patience and experience, a builder can utilize surface mount devices confidently. Surface mounted devices are designed for use with printed circuit boards. Generally, they are useful only when a builder is willing to fabricate a board to accept them. In this regard, several available kits using surface mount devices provide a helpful arena for practice.[1] These are of little help, however, for the experimenter wanting to try new circuits with a "one of a kind" board.

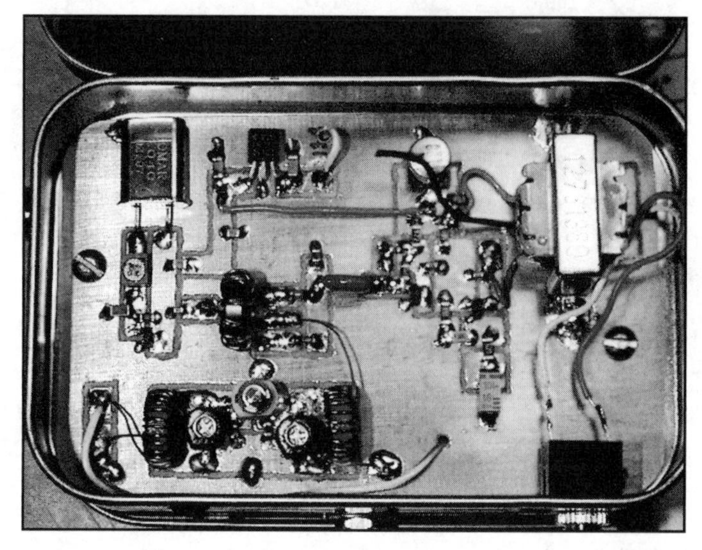

Figure 1— A direct conversion receiver built using the techniques described by the author.

While the components are smaller, good construction practices should apply to any technique used. Whether the builder tackles a project using "ugly construction," gluing pads onto a board (commonly known as "Manhattan" construction) or etching one's own circuit board, time is well spent studying the circuit and working out an effective layout.

Using A Dremel Tool to Make SMD Boards

One technique for building a project comprised of SMDs is to use a Dremel brand rotary tool to cut pads into printed circuit board material. This technique evolved naturally from the methods mentioned above. The process involves considerably more time devoted to board layout and preparation than the actual installation of the components themselves. What follows are some suggestions for the builder who would like to experiment with surface mounted components using this technique. The aim is to shorten the learning curve.[2]

Dremel bits come in a variety of shapes and sizes. My first attempts at grinding pads involved 3/32 and 1/8 inch round carbide burrs. They were used successfully in making a few pads; however, the carbide burrs have several drawbacks. They do not cut well using the tip of the bit where the cutting edges are fine and more closely spaced. If one lays the bit on its side, however, they cut more effectively. There is also the tendency to apply more pressure while cutting, and when the "teeth" of the bit finally engage the copper material, the result can be more of a gouge than a finely engraved line.

Dremel also markets diamond impregnated burrs[3] which cut more consistently across the entire bit and require less pressure on the cutting tool. A 1.8 mm round diamond burr works well for cutting long broad lines and the 1 mm version produces a fine line that can be bridged by even a well-placed 0603 sized component (1.6 mm × 0.8 mm).[4] For example, a capacitor or resistor can be connected from a pad to the ground plane this way. Even so, the diamond bits still cut better when laid on their sides.

Regardless of which bit one uses, there is a helpful maxim—let the Dremel do the work. A light touch with more time and more patience does a better job of cutting the lines on the substrate. I discovered that, if I try to force the cutting process, the tip will bind in the copper, producing an uneven line in both thickness and depth. It is even possible to cut completely through the board, if one pushes too hard, although with care that seldom happens.

The copper used on PC boards also varies in thickness and thus in ease of re-

[1]Notes appear on page 1-20.

moval. While the thinner and/or softer copper substrate lends itself to easier cutting with any of the Dremel bits mentioned above, the diamond bits are far less frustrating when working with harder material.

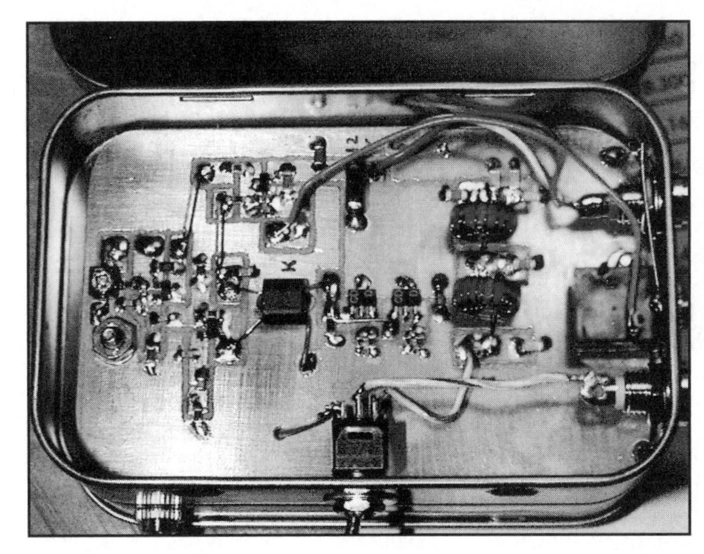

Figure 2—A matching low power transmitter built with SMT parts.

Practical Suggestions For Using the Dremel Technique

Here is a description of the process. Begin with the schematic, a ruler and an extra fine tip Sharpie pen and draw the pads on the copper clad board. An ink eraser removes unwanted lines. Keep a selection of components nearby to use as templates for pad layout, especially an SOT-23 transistor or diode, an SOT-89 transistor and the common integrated circuits. (Putting together a small bag of components labeled "Layout" that are used only for constructing a board is helpful.) Another construction practice is to use double-sided board material and keep the bottom side for the primary or additional ground plane connections. A small hole can be drilled from a pad through the board and a wire jumper soldered on both sides. This is not always necessary if the layout lends itself to a sufficient "top surface" ground plane. With smaller projects the builder can lay out the circuit directly onto the copper board, but for more extensive circuits it may be helpful to use a slightly different technique.[5]

Larger projects require a different technique. Use a piece of graph paper that has a grid that accommodates the necessary pad size (0.25 inch grids work well) and draw the pads with a pencil and the necessary eraser. When you're satisfied that the pad placement will work, transfer the grid onto a piece of PC board using a square and a pencil. The next step is to transfer the pad placement onto the board itself, using a fine-tipped Sharpie. I have built receivers and transmitters this way. Small boards, with one or two stages on each and interconnected after construction, work well for larger systems. An 80 meter CW transceiver with a superhet receiver and a 3 pole crystal filter was built using this method.

When using the Dremel "freehand," lay out the circuit one section at a time and be very careful to check the corners of pads. These can be likely culprits for shorts to the ground plane. After each section has been cut, use a hard ink eraser or Scotch Brite to buff the board. Be sure to blow or brush off the copper grit. On several projects it was noticed that fresh solder flux tended to attract the copper grit. Additional pads can be inserted later, if you desire to modify the circuit, provided there is sufficient space remaining on the board itself. While no shorts have been experienced because of this, extra care may be advised when adding pads after some components have been installed. Always check each isolated pad for continuity to ground and to other adjacent pads with an ohmmeter. Use magnifier headsets to inspect the pads and to inspect the installed components.

Place the copper board on a surface higher than the workbench itself. As mentioned earlier, the Dremel burrs cut most efficiently on their sides, not on the tip. A 3/4 inch piece of pine board can be used as a secondary work surface, while you rest your Dremel hand on the lower surface of the workbench.

Fine solder, such as 0.020 Kester NoClean does a nice job. However, since there is little flux in 0.020 solder, it sometimes is helpful to tin the pad first, using desoldering braid (wick) to remove the excess solder. Components can be held on the board with fine-tipped tweezers while soldering. Sometimes the tweezer tips tend to pick up solder flux, making them sticky. Trying to manipulate an SMT part with sticky tweezers is interesting, to say the least! It's helpful to frequently clean the tweezers with isopropyl alcohol to remove the flux. If one is using metal tweezers, degaussing is periodically necessary to eliminate magnetic effects. Controlling a tiny SMT part with a magnetized tool can be equally frustrating. Alternately, one could use a ceramic or plastic pair of tweezers.

A neat job can be accomplished by placing connecting wires and jumpers on the bottom of the board. This helps keep wires out of the way and makes layout simpler. Drill a very small hole through the appropriate pads from the top of the board through to the bottom. Be sure to use the Dremel to clear the copper around the hole on the bottom of the board to prevent shorting the line to ground. In a 40 meter superhet receiver, the crystals for the filter (5 pole) were placed on the bottom of a single-sided board, mounting only the SMT capacitors on the top. Although this worked, I did run out of room and had to put one crystal on top.

Keeping common components sorted with small Ziploc bags speeds up the construction process. At first, all the surface mount resistors were kept in their respective tapes in one large box. After spending 15 minutes looking for one needed resistor value, I decided that it would save time if the common values were sorted, and each placed in its own separate bag. Alternately, parts can be stored in small "coin" envelopes that come in various sizes.

Working with small SMD components and using the Dremel to make fine cuts on a board does seem to tax one's energy level more severely. When a builder has trouble concentrating or feels tired or bleary-eyed, it's time to take a hike, get on the air or go to bed.

Don't be afraid to use leaded components when they will fit better into your layout or when you simply do not have a needed value in SMD form. Unless you're a purist, it doesn't detract from the project. My first projects used about 50% SMDs and the last used all SMDs, with the exception of a 9 V voltage regulator that I didn't have in SMT form, a couple of toroidal inductors and a few 7-125 pF trimmer caps.

How Small Can You Go?

SOIC (small outline integrated circuit) components, with their 0.05 inch lead spacing, present a challenge to the Dremel process, but with care they can be used successfully. It's easier, however, to bend the leads of an IC in a DIP (dual in-line) package to imitate a surface mounted part because of its 0.1 inch pin spacing. The DIP package is not much larger or taller and the corresponding pads are sized nicely for SMT resistors and capacitors. Another technique is to use an SOIC part but place it on a small board of its own and use wires to connect it to the main board.

To be certain, the smallest SMDs are impractical for the home builder. There are limits to how small a device we can handle and install using this Dremel technique. MSOP (miniature small outline package) parts with 0.025 inch lead spacing are best left to etched boards. But don't be afraid to think small! When my first VFO using SMDs was attempted, I was amazed at how much board space had been "wasted." A board half as small could have been used with room to spare.

Even with that said, don't attempt anything too small at first. Your boards will begin to shrink in size as you gain confidence and experience. It is often beneficial to begin the layout process at the center of the board and then cut it to final form after the pads have been cut. When placing a 0603-sized capacitor, don't sneeze!

Adding an X-Y Table

One colleague (W7ZOI) has modified this procedure for his SMT breadboards. His shop includes a Dremel drill-press stand. The Dremel tool is vertical for normal use with this stand and pushing on the level causes the tool to move downward to drill a hole. He's built a fixture from wood scraps that holds the Dremel tool, in the stand, in a horizontal position. A cut can then be made in a copper surface when the Dremel is loaded with a rotary cutting tool.

This setup is enhanced with a simple X-Y table made from wood and aluminum scraps. A small piece of circuit board material is clamped to the table. The rotating cutting wheel is then lowered until the copper is being cut. It is held in this position while the table is moved, allowing a long cut to be made. The process is then repeated after the "target" has been adjusted, allowing another cut to be made close to and parallel with the first. These methods are described on the Web.[6]

The Benefits

Several benefits emerge justifying the use of surface mount components and the techniques described above. The industry standard is now focused on leadless components. They are thus readily available and generally cost less than their leaded counterparts. A more compact design is realized because of the smaller size and lighter weight of the finished circuit. A 40 meter superhet transceiver I built weighs about 6 ounces and is about the size of a small paperback book (including the SMT keyer!). The MicroR1[7] and its companion transmitter, shown in Figures 1 and 2, weigh just 5 ounces and an 80 meter transceiver, although larger in size, weighs less than a pound. Since I like to backpack with my rigs, this is ideal.

All of these projects were built at home using simple shop tools and nonhazardous etchants. All the circuits were built over a healthy ground plane that provided optimum performance. Modifying an existing rig or adding a keyer or additional filtering is easier using this technique because the boards are much smaller. They also can be more easily placed inside an already full enclosure. This Dremel method adapts well to most of the common sizes of SMT components. I find that I rarely think about whether the resistor or capacitor I use will be 0603, 0805 or 1206 in size. Generally, any of the larger sizes will fit just fine and that is quite helpful when shopping for SMT components in the surplus market. And finally, it's just good old ham radio fun.

Notes

[1]Embedded Research (**www.embres.com**) currently markets a surface mount keyer kit.

[2]See **www.dremel.com** for a description of their product line.

[3]I bought the diamond impregnated bits at a local hardware store, but I have also used "generic" versions that I bought from a vendor at a hamfest.

[4]The 4 digit code of an SMD component refers to its approximate size in hundreds of an inch. Length is first, followed by width. A 0603 component is thus about 0.06 inches long by 0.03 inches wide.—*Ed.*

[5]Figures 1 and 2 show projects that were drawn freehand, directly on the board material.

[6]Hayward's Web site information can be accessed at: **users.easystreet.com/w7zoi/smtbb.html.**

[7]Hayward, Campbell, Larkin, "Experimental Methods of RF Design," page 8.4. Available from the ARRL Bookstore. Order no. 8799. Telephone toll-free in the US 888-277-5289 or 860-594-0355, fax 860-594-0303; **www.arrl.org/shop/**; **pubsales@arrl.org**.

All photos by the author.

Ed Kessler, AA3SJ, was first licensed as a Novice in the 1970s at age 14, but was deterred by other pursuits until the fall of 1998 when he re-entered the Amateur ranks and earned his General class ticket. An Extra class license followed a year later. Having always enjoyed homebrew construction and design, he became an ardent low power (QRP) advocate under the guidance of several mentors, including W3TS, AA3PX and W7ZOI. Ed is a clergyman by profession and is part-time adjunct professor of biblical studies at Messiah College in Grantham, Pennsylvania. He can be contacted at 950 Woodside Station Rd, Millersburg, PA 17061; **edkess@pa.net**. *Ed maintains a Web page at* **www.qsl.net/aa3sj**.

By Rick Littlefield, K1BQT

From *QST*, July 2000

Build a Simple SMD Workstation

Tired of chasing surface-mount parts with a toothpick? This "helping hand" is a better solution!

Working with surface-mount devices (SMD) isn't as difficult as you might imagine, especially with the right tools. The handy little workstation described here will help you conquer the most difficult task of all—holding flea-sized parts in place while soldering. It works like a tiny spring-loaded finger that moves on three axes over the circuit board. Want to mount a part? Simply position it with tweezers, place the stylus on top (to hold it) and solder away!

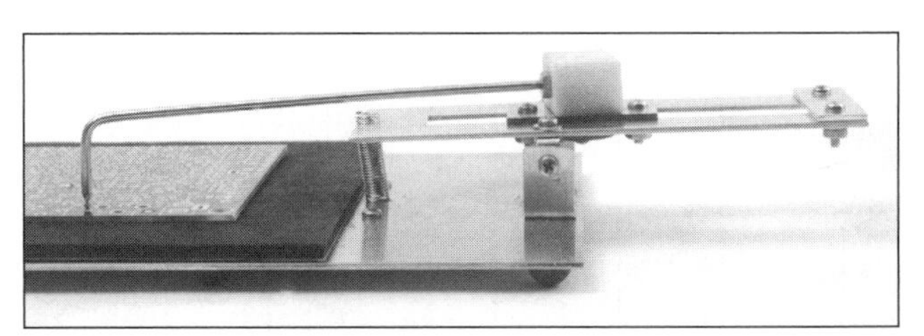

Side view of the completed stylus arm assembly.

Description

This second-generation workstation features several improvements over my first attempt.[1] By studying Figure 1 you'll see immediately how it works. The stylus arm swings side to side and slides back and forth to cover a 12 1/2 square-inch work area. It also tilts vertically, with a tension spring supplying the downward force needed to hold SMD parts in place. The arm's slide rule-style mechanism was chosen because it provides smooth motion while under spring tension—and that's important for precise stylus control. Small in size, the mini-workstation stores easily in a drawer or toolbox and can be used for SMD repair or prototype construction.

A close-up view of the pivot-block portion of the stylus arm.

Construction

This was a true junk-box project—I used whatever shop scraps I could find to put it together. Dimensions aren't critical, and your version may be scaled to preference. Most parts were cut from leftover pieces of 1/16-inch aluminum or G-10 circuit board. The base plate is a 5- × 7-inch panel-stock remnant. Its non-skid work surface was trimmed from an old mouse pad. A discarded length of 0.1-inch stainless antenna rod became the stylus. The 90° 4-40 threaded angle brackets that hold the arm onto the tilt bracket were recycled from an old hobby box (also easily made from scrap angle stock).

Assembly is straightforward and much easier to visualize than to explain (see Figure 2). The slider mechanism is the workstation's most critical part. This item, made from three pieces of PC board sandwiched together, rides along the arm mechanism's slot much like a slide rule.

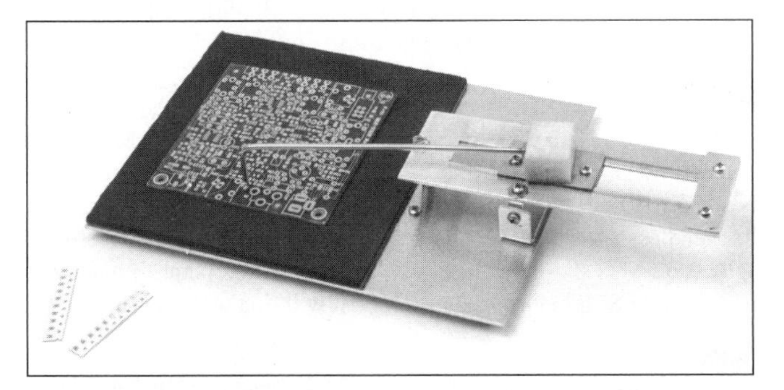

Figure 1— The SMD workstation. The stylus arm swings side to side and slides back and forth to cover a 12 1/2 square-inch work area. It also tilts vertically, with a tension spring supplying the downward force needed to hold SMD parts in place.

Before assembling, place the slider's spacer plate inside the slot and confirm that it moves from end to end without binding (dress surfaces as needed). Next, install the top and bottom slider plates onto the spacer with the fiberglass surfaces facing each other.

When assembled, install the slider into the arm track, making sure it moves smoothly. Shim the center spacer with a sheet of paper if added vertical clearance is needed to reduce friction.

When installing the pivot block on the slider, do not over tighten its 6-32 mounting screw—this should allow free side-to-side arm movement. By the same token, install the screws attaching the arm to the tilt bracket loosely to permit unrestricted up-and-down arm motion. If needed, secure the threads with Locktite or contact cement to prevent the screws from backing out.

When selecting a compression spring, look for one made with small-gauge wire that has a large number of compacted coils—this will ensure a gentle even pull over the arm's full range of motion. Adjusting stylus pressure may require some trial and error. Too much tension will bind the slider and may cause the stylus to eject parts off the PC board. Too little will allow the SMD part to slide out of position as you feed solder onto the pad. When you find the right compromise, trim off the unused coils.

The foam work pad—a discarded mouse pad given new life—is held in place with contact cement (the non-skid surface should

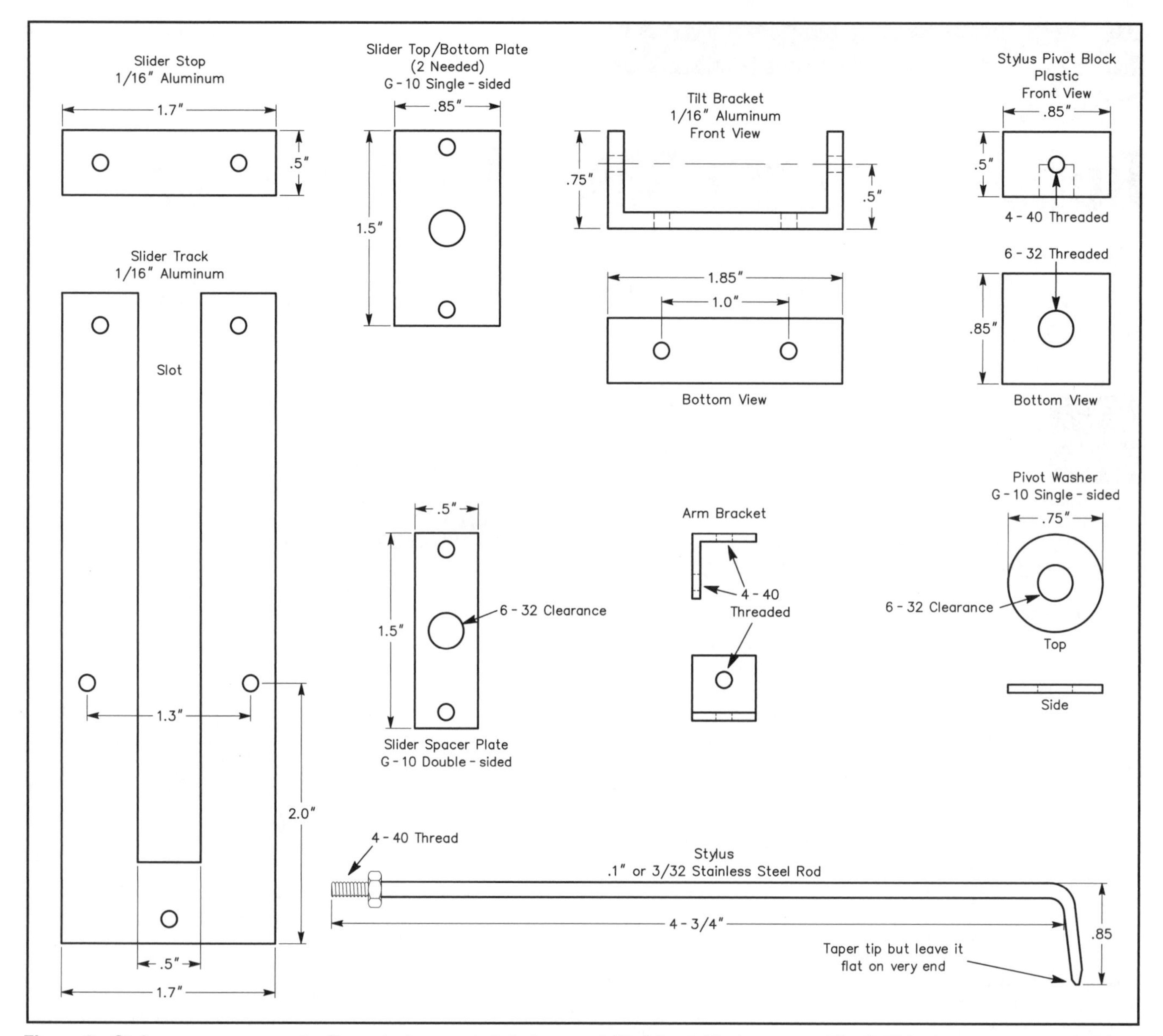

Figure 2—Stylus arm components. Except where noted, holes are drilled for 4-40 clearance. The 6-32 holes are exaggerated in size for clarity—not to scale.

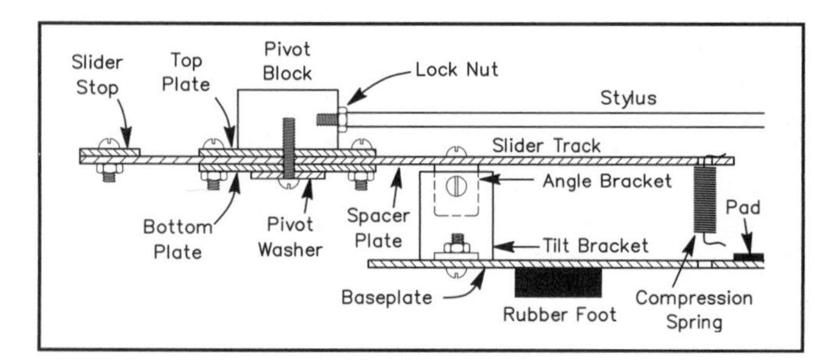

Figure 3—The stylus arm assembly diagram.

face up). Rubber feet keep the base plate stable and protect the bench's work surface from protruding hardware.

Operation

Using the workstation is easy. First, position an SMD part on the PC board using tweezers, then carefully place the stylus on the center of the part to hold it in place. Apply heat to the pad you wish to solder. Once the pad is heated, gently apply solder—allowing the solder to flow onto the pad and onto the component's contact surface. Be careful not to tap the component with the tip or the iron or push the part out of position with still-solid solder. After securing one end of the part (or two leads of an IC) with solder, remove the stylus and complete the remaining connections.

Conclusion

Most seasoned experimenters will tell you that SMD hand construction isn't any slower or more difficult than through-hole methods. To do it successfully, however, you must be able to fully immobilize leadless parts on the PC board while keeping both hands free for soldering. This simple workstation will help you do that. It made my transition to SMD much easier. I hope it will do the same for you!

Notes

[1] R. Littlefield, "A Low-cost Mini-workstation for SMD Construction," *Communications Quarterly*, Spring 1996, pp 56-58.

109A McDaniel Shore Dr
Barrington, NH 03825
k1bqt@aol.com

By Bill Sepulveda, K5LN

From *QST*, December 2002

Panel Layout with Microsoft *PowerPoint*

Labeling headache? Cure it with a common software program.

Have you ever struggled to figure out how to label the front panel of a great project so it can be made to look even better?

Struggle no longer! With some double-sided rug tape, clear contact shelf paper, a computer and Microsoft *PowerPoint*, you can complete a professional looking front panel that you can be proud of. Here's what you'll need—

Tape: Some type of double-sided adhesive tape to secure the label is required. Double-sided rug tape can be purchased from any hardware store but other, similar type tape will also work.

Clear Cover: Matte finish, clear contact shelf paper with adhesive backing can be purchased at most craft stores. A second type of clear material (thicker) can be purchased from many drafting supply stores. It is a glossy clear film with contact adhesive backing. Both of these coatings work well. Simply spraying paper with a clear plastic coat is not recommended. It may cause the ink to bleed and the paper to wrinkle.

Paper: Either white or colored paper may be used to enhance the labeling effects. [Avery makes a variety of ink jet and laser printer labels that might be suitable if you don't want to deal with double-sided tape.—*Ed.*]

Program Setup

After opening *PowerPoint*, select a new Presentation. When the "New Slide" window opens and asks you to "Choose an AutoLayout," select a blank page and click "OK." This will open a blank presentation slide. Depending on the size of the project, you may wish to change the page layout from Landscape to Portrait. To do this, click on "File," then click "Page Setup" and change the page orientation.

The ruler within *PowerPoint* should be used to help create the proper panel sizes, component locations and labeling information. If the ruler isn't displayed, click on "View" and when the drop-down window appears, click on "Ruler."

The final step in the setup is to display the Drawing toolbar so it can be used to create all the information needed to complete the project. Move the mouse to the light gray area in the upper right side of the window (to the right of the Standard and Formatting toolbars). Click the right mouse button and a drop-down window will appear. Click on the word "Drawing." This will display a Drawing toolbar on the screen.

Planning

The project layout can now be started. As an example, a CW keyer project, which was installed in an aluminum box, will be described.

The box is 1.25 inches high by 3.00 inches deep by 5.50 inches wide. In this example, I wanted to make the height of the box the front and rear of the label and use the top of the box (the depth) for keyer instruction labeling. The label size will thus be the sum of the front box dimension, the top box dimension and the rear box dimension. The total label size is thus 5.50 inches wide by 5.50 inches high. For details of this example, see Figure 1.

Making the Label

1. Make the label outline by using the "Box" or "Circle" icon to match the size of the area to be covered. These icons are found in the Drawing toolbar. Click on the appropriate icon and move the mouse pointer to a location on the screen. In

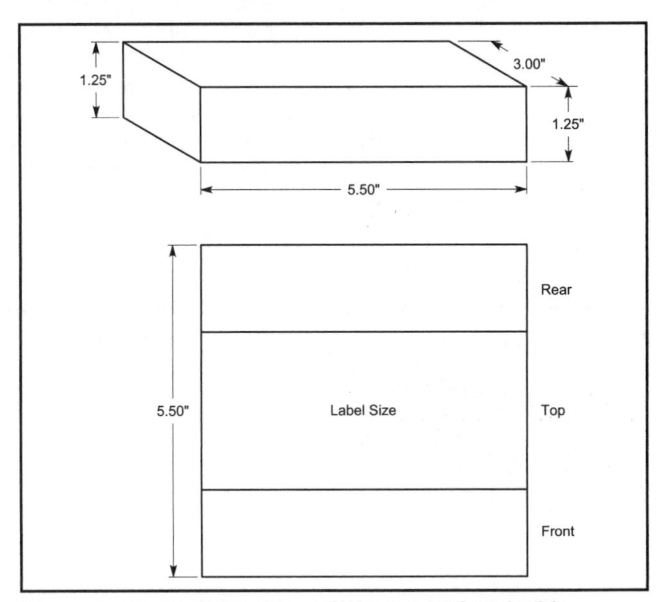

Figure 1—A sample label for a CW keyer project. In this example, the label covers 3 sides of the enclosure.

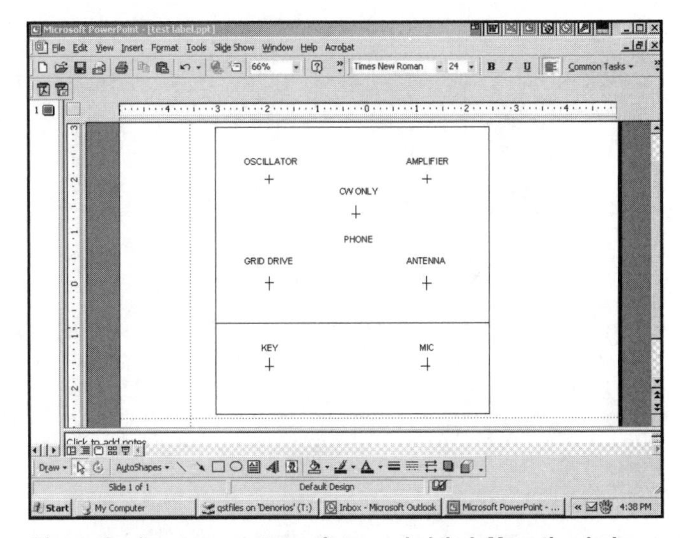

Figure 2—A screen capture of a sample label. Mounting hole locations can be clearly seen.

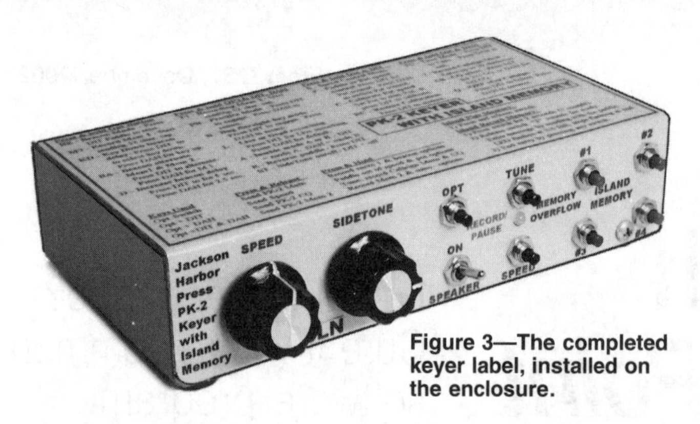

Figure 3—The completed keyer label, installed on the enclosure.

this example, a box for a label size of 5.50 inches by 5.50 inches will be made.

When the initial box is created, a default color will appear within the box. Click the "Fill Color" down arrow on the Drawing toolbar and click on "No Fill" window to remove the color.

2. Measure the locations of all holes and identify them by creating a location for the center of each hole. Use the "Line" drawing tool or simply create a small square or diamond at the hole locations. Save the file.

3. Add labeling (by creating a text box and then typing within that box) around the areas identified for mounting components. Save the file.

4. Print a copy of the label and cut it out to match the panel. Make corrections as needed to align the mounting holes and to center any information on the holes. Repeat this until correct. A screen capture of a sample label that was made using this technique can be seen in Figure 2.

Applying the Double-sided Tape and Clear Cover to the Label

1. If the hole location marks are small enough, leave them to aid in label positioning. If they won't be covered by the panel holes, remove them. Print the final label and cut it to size.

2. With the label cut out, apply the double-sided rug tape to the backside of the label. Multiple pieces will be required, so be careful when you place the tape side by side. Trim off any excess around the edges of the paper label.

3. After the tape is applied to the rear of the label, the clear cover is applied. Cut the clear cover larger than the actual label, overlapping about 1 inch. Apply the cover carefully to avoid air bubbles and trim off the excess at the edges. The label is now ready to apply to the panel/box.

Applying the Label to the Panel/Box

1. On the front bottom edge of the label, fold back about a half inch of paper backing from the double-sided tape. Align the bottom edge and sides of the label to the bottom and side of panel. This is best accomplished by laying the panel on a flat surface and the edge of the label on the same surface. Gently apply pressure with your fingers to the front of the label to stick it to the panel. Now remove the paper backing on the tape and slowly rub the label on the panel. Repeat this for each section of tape until the label is completely attached to the panel. The completed keyer label can be seen in Figure 3.

2. With the label now attached to the panel, cut out the holes for the components. Use a hobby knife to smoothly remove the label material in the holes.

Congratulations… your panel/box label is complete. It will take a bit of practice to use the software efficiently, but the hard parts—panel layout, hole positioning and label font sizes and styles—will now be easier to design. Try it on your next project!

Bill Sepulveda, K5LN, was first licensed in 1963 as WN4PVW. He has worked for a major computer manufacturer in central Texas and he has a background in Manufacturing Engineering. K5LN is an avid CW and QRP operator. He can be reached at **K5LN@aol.com**.

By Bob Kavanagh, VE3OSZ

From *QST*, December 2003

Front Panel Layout—Another Approach

Here's one more way to make that homebrew project shine.

The technique described by K5LN for laying out and printing labels for front panels (B. Sepulveda, K5LN, "Panel Layout with Microsoft *PowerPoint,*" *QST*, Dec 2002, p 61) made me think I should share with other hams my somewhat different approach to this problem.

My technique involves making a computer-printed paper cover for the front panel of a chassis. The paper cover is then covered by a sheet of Lexan [This is a polycarbonate plastic and is very resistant to scratching, impact and cracking damage. Other acrylic plastics (Plexiglas, for example) can be used, but they may not be as damage resistant.—*Ed.*] The fastening nuts for the potentiometers, switches and jacks secure the paper cover and the Lexan sheet to the front panel. No glue is needed.

Having decided upon the layout and labeling of the front panel controls, I produce a corresponding layout on a computer. I use Lotus *Freelance Graphics*, but other similar programs can be used. The layout is created to ensure that the paper sheet matches the desired control positions and labels on the panel. This is after laser printing and trimming the sheet to size. The font can be chosen to suit your taste. I place small circles (or crosses) at the locations where the holes for controls are to be drilled. The layout can be printed on paper of any suitable color and weight (thickness). It is a good idea to print several copies of the final layout.

The next step is to cut a piece of Lexan to correspond to the size of the front panel. I use 1/16 inch sheets. Then, using one of the printed layouts as a template clamped between the Lexan and the panel, drill the Lexan and the panel at the same time with the required holes for the controls.

Finally, using a fresh copy of the paper layout (in case the first copy has been damaged during the drilling process) with the layout and the Lexan clamped in place, I pierce holes in the paper at the appropriate locations. This requires some care in order to not tear the paper while removing the excess paper from the holes. Controls with knobs will cover up any imperfections in the paper holes, but special care must be taken with holes like those for LEDs. The controls are then added and everything is fastened together with the control nuts.

Figure 1 shows two of my recent projects that have been packaged using this technique. The method works best for boxes that have an overhang at the front sides and top, as shown in the picture. The overhang conceals the edge of the Lexan sheet.

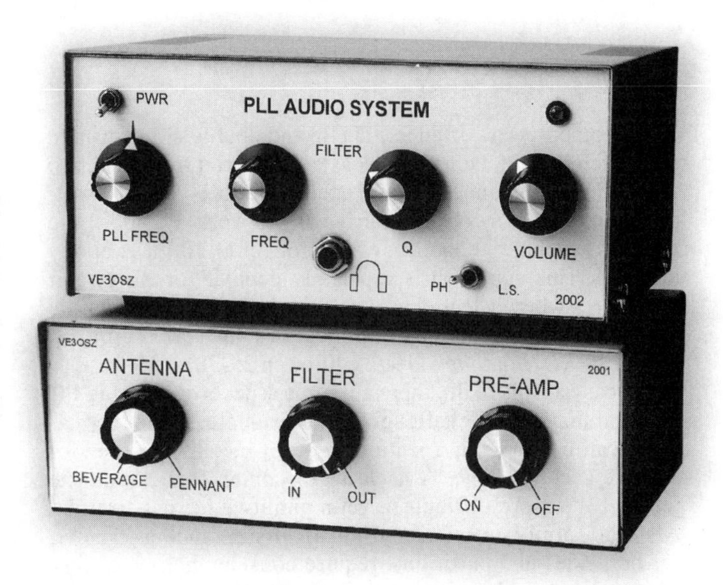

Figure 1—Two examples of completed projects that use VE3OSZ's front panel finishing technique.

The result is a professional looking front panel with lettering that is protected and will never wear off.

Bob Kavanagh, VE3OSZ, was first licensed as VE1YW in 1951 and since then he has held VE3AQO, VE1AXT and his present call. Bob has a PhD in electrical engineering and has taught at the University of Toronto and the University of New Brunswick. He was also a Director General at the Natural Sciences and Engineering Research Council of Canada. Bob's Amateur Radio interests include DXing (especially on 160 meters), contesting, experimenting with circuits, antennas and propagation. He is an authorized Amateur Radio examiner for Canada's licensing authority, Industry Canada. He can be contacted at 849 Maryland Ave, Ottawa, ON K2C 0H9, Canada or at **ve3osz@rac.ca**.

By Bob Kopski, K3NHI

From *QST*, September 2003

An Easier Way to Build PC Board Enclosures

Can't find a chassis or cabinet for your latest project? K3NHI shows you how to build effective and attractive enclosures...easily, inexpensively and quickly.

Homebrewers often use single and double-sided printed circuit (PC) board stock as raw material from which to build custom electronic enclosures. It is readily available, not too expensive and fairly easy to work. Popular base materials include phenolic or the stronger G-10 glass epoxy material. This approach is especially popular for RF circuit enclosures where a leak-free enclosure is desired. One classic example is the step attenuator project in some earlier editions of *The ARRL Handbook* (1992 edition, p 25-38).

As easy as it is to do, there are some aspects of working PC material that can be challenging. For example, while the material can be cut with a stationary sheet metal sheer, few of us have access to one. A hacksaw or coping saw can do the job, but it's more difficult to get a quality edge cut straight. Also, uniformly cut pieces of identical size, such as desired for box sides and partitions, require considerable measuring and cutting care. I decided there had to be a way to ease these tasks, and I set about to find one.

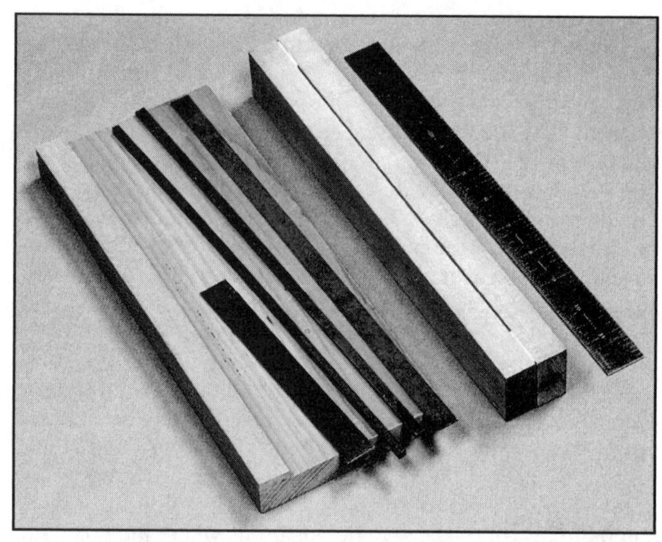

Figure 1—These are some of the simple fixture parts needed to build functional, good looking enclosures from printed-circuit material. A few dollars' worth of wood and metal are all that are needed.

Simple Wood Fixtures and Other Tools

After some thought and trial and error, it seemed clear that only a few very simple tools were needed to do a really good job. Figure 1 shows it all. The items needed include two simple wooden fixtures, some straight edge metal pieces and a hobby knife. Not shown is some sandpaper for final touches on the cut edges.

The first wood fixture needed is a "cutting jig." Mine is simply a piece of white pine about 4$^1/_2$ inches wide and 1 foot long. The exact dimensions are not critical but the wood should be warp free. I glued an equal length of $^1/_4$×$^3/_4$ inch pine trim material along one edge, forming a full length "dog." This becomes a simple resting edge during material cutting and it must be a straight piece. The fixture itself can be seen in Figure 2 and the technique for scoring the work is shown in Figure 3.

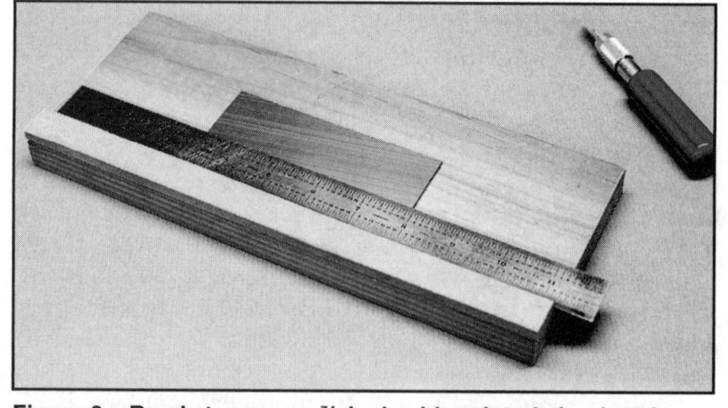

Figure 2—Ready to score a $^7/_8$ inch wide printed circuit strip. Here a metal scale gauges the width; the simple cutting jig assures accurate, uniform dimensions piece to piece.

I call the second fixture the "break tool." Mine consists of two pieces of 1×2×12 inch maple stock separated but fixed to each other at the ends by $^1/_{16}$ inch thick double-sided foam tape. My break tool began life in the local hardware store as a 1×2×24 inch piece of quality maple, which I selected for

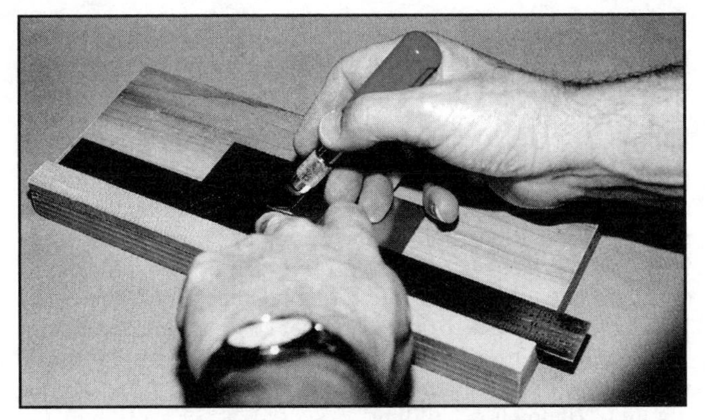

Figure 3—Keep everything snugly in place and against fixture dog, keep fingertips in back of the gauge edge, then score the material several times on both sides with a hobby knife.

Figure 4—Scored piece of printed-circuit board material held in 1/16 inch break fixture slot. Phenolic based board snaps right off with a little push; G-10 type material requires some back and forth "push-pull" motion.

straightness and then cut in half. It cost less than $2 and can be seen in Figure 4.

It's important to use a hard wood for the break tool (not pine). While I chose maple, I think birch or oak would do the job just as well. It's also important when taping the pieces together to make sure they are exactly even along the long edges. Choose a good quality double-stick foam tape similar to the 3-M brand—easy to get in most hardware and variety stores. One-sixteenth thick tape is just the right size to work 1/16 inch PC board material.

Cutting Gauges

Begin your project by knowing what it is you want to assemble and the dimensions of the PC board pieces needed to do it. It helps to have a dimensioned sketch. It's also best to choose dimensions that equal readily available metal straight edges like those shown.

Examples of the latter include metal drafting scales, which are readily available in a variety of widths and lengths. An even larger variety of metal straight edge material can be found in hobby and craft stores. These stores usually have a "Metal Center" display stand holding many different brass sheet strips and rectangular sections. If the exact width you seek is not available, simply "parallel" narrower pieces and "build up" to the desired dimension. For example, a 1/8 inch square tube and a 1/2 inch wide flat strip "add up" to a width of 5/8 inch. These chosen strips become "cutting gauges"; this will become clearer later.

Cutting and Tools

I have used a variety of hobby knives for this task, although I tend to favor the huskier types like the one shown in Figures 2 and 3. Whatever your choice, be sure to have some spare blades on hand. "Box cutters" with snap-off blades may be a good choice too, although I've not tried these.

Now the fun part begins. Since most enclosures have all sides and may have internal partitions that require the same width material, it makes sense to cut lengths of this particular width first. Simply place the PC board material on the cutting jig snugly against the dog, lay the selected metal gauge on top of the work and against the dog, and draw the knife along the gauge edge scoring the board material several times. Flip the work over, and do the same thing on the other side.

It's important to cut completely through the copper, and to have at least some scoring of the base material on both sides of the PC board. Some extra slices are recommended for material like G-10, since this is a bit tougher to work with. Notice that using the width gauge ensures that the score marks on both sides of the work are located exactly opposite each other.

Now insert the scored board material in the break tool. It should slide in snugly between the two wood pieces. Locate the score marks exactly at the edges of the wood rails, and snug the whole works in a bench vise as shown in Figure 4.

At this point all you need do is push on the material outside the wood rails and snap—off comes the exposed piece of board. Depending on the size of the work, you may want to use a spare piece of wood as a "push piece" to get uniform pressure along the work edge. Also, that "snap" is actually the case for phenolic based PC material. Tough stuff like G-10 will break less dramatically and usually needs some "back and forth" push and pull on the exposed piece. Both materials will come apart with nice straight edges with only a little sanding needed to de-burr and smooth the final edges. I prefer to use a sanding block—sandpaper tacked to a wood block—to get the best edges. This method allows repeatable, accurate widths to be cut and requires only a little care in the overall task to achieve.

Cutting the resulting PC strips to length is done in much the same way. Place the just-cut strips in end-to-dog fashion, lay in place a gauge of width equal to the desired finished piece length and score both sides of the work as described above. Depending on the exact dimensions of all involved, it may be helpful to use a small drafting triangle to ensure squareness. Just place one edge of the triangle against the dog, nest the PC strip against both the dog and the other triangle edge, and then place the gauge. Here again, similarly cut PC pieces come out uniformly dimensioned—very nice when it comes to accurate, good looking final box assembly. The enclosure top and bottoms are handled the same way.

While obvious, I want to include this advisory anyway: Hold everything snugly while doing the scoring—no slipping allowed. *Be absolutely sure* your fingertips do not overhang the gauge edge! (Guess how I found out that this is a good idea?)

Figure 5—Use one hand to hold the printed circuit board pieces and small wood block in the cutting jig. Two 1/4 inch lengths of solder are strategically placed at the board intersection. Use the other hand to bring an iron in and tack the pieces together.

Figure 5 shows an additional trick for achieving accurate box assembly. Using the cutting jig and an additional small, squarely cut wood block, place the enclosure base plate against the dog

Figure 6—Two examples of simple printed circuit board structures made as described herein. The temporarily uncovered box is a 50 Ω power splitter. The "U" shaped assembly is a component test jig.

and put the small block on top also against the dog. Place a PC board sidepiece on the previously marked board material base against both the dog and the block. Adjust all for proper relative position and hold firmly. Using tweezers or similar, drop some precut 1/4 inch lengths of solder right at the interface edge of the PC board pieces. Now bring to bear a suitable iron to spot solder or "tack" the pieces. (Yes, it requires only two hands!) When this cools, the work can be separated from the wood tooling and the corner soldering completed. [It should be easy to "drag" a bead of solder (while feeding solder) along the joint with the soldering iron tip when the work temperature is proper. Allow a sufficient, but not excessive, time for heating. It's a good idea to use flux remover to clean the joint after soldering. As with anything else, "practice makes perfect" and the technique is easily learned. The editor has built many enclosures this way and can say that it works well!—*Ed.*]

While these procedures may seem intrinsically simple, intuitive and inexpensive, you'll get better results with a little practice. I suggest trying this on some PC scraps before you undertake a "real" project. Of course, all the normal advisories and precautions apply. For example, I like to drill any holes in the pieces before assembling them. Also, shiny and clean copper solders a whole lot better than any other kind. Similarly, a clean and tinned, proper temperature, adequate mass soldering tip makes the solder seams look truly professional. Two examples of finished enclosures can be seen in Figure 6. Happy box building!

All photos were taken by the author.

Bob Kopski, K3NHI, has been licensed since 1959. Having a strong interest in aeromodeling and electronics, he has homebrewed extensively in the radio control field and flies R/C aircraft routinely on 6 meters. He has also operated both fixed and mobile on that band. Bob writes at length on the R/C modeling field for a model aviation magazine and his modeling and R/C interests date back to the early 1950s with a specialty in electrically powered R/C models. He has published many articles covering model aircraft design and electronics. Bob is the author of "An Advanced VHF Wattmeter," which appeared in the May/June 2002 issue of QEX. *He's a retired senior design engineer and has BSEE and BSEP degrees from Lehigh University. You can reach him at 25 W End Dr, Lancaster, PA 19446.*

From *QST*, April 2000

By Charlie Hansen, N0TT

A Tool for Winding Small Toroidal Cores

This simple tool and winding approach will help you snake thin wires through tiny iron and ferrite cores.

Have you ever tried to wind thin wire onto a very small toroidal core? It can be a tedious, time-consuming and sometimes frustrating job! Recently, I needed to wind several RF chokes of fairly large inductance on some FT-37-43 cores. You know the ones I mean: You need a magnifying glass and a good light source to work with them! For the job at hand, I used hair-fine #34 wire and intended to put as many close-wound turns on the core as I could. (That is, up to the classic 30· wedge of open core to avoid capacitance effects at the ends.[1]) After struggling to wind a few turns using a bent paper clip and fishing the wire through the core with each turn, I knew I had to come up with a better way!

The Old Shuttle

I thought about using the shuttle shown in Figure 1.[2] But because the hole in the FT-37-43 core is so small, I needed a *very* slender piece of plastic, stiff paper or other material to get through the hole. Also, I had used that shuttle many times before only to have the core slip from my fingers and fall to the floor. (Of course, it *had* to unwind itself on the way down!) I decided I needed a different approach. After some experimenting, here's what I came up with.

A New Shuttle Experiment

The shuttle I designed is shown in Figure 2 and the accompanying photograph. No specialized tools or parts are needed to reproduce it. It's made of two lengths of small-diameter brass tubing available at hobby shops and hardware stores just about everywhere. You'll need a 2$^3/_4$-inch length of $^1/_{16}$-inch-OD tubing and a 1$^1/_4$-inch length of $^3/_{32}$-inch-OD tubing. The incremental sizes of brass tubing available telescope together for a perfect slide fit. Other materials used in this shuttle are common items likely found your junk box: a short length of plastic insulation and a 1$^5/_8$-inch length of heat-shrink tubing to fit the $^1/_{16}$-inch-OD brass tubing.

Cutting the Tubing

To cut the brass tubing to size, secure it in a vise. Pad the vise jaws to prevent deforming the tubing. Keep the area of the tubing being cut close to the vise jaws. Using a sharp edge of a small triangular file, make a groove around the tubing at the proper length dimension, then simply snap off the piece. This action leaves a slightly burred end on the tubing. Use the file to square off the tubing ends, then finish the edges with fine sandpaper. (You could use a hacksaw with a fine-toothed blade to cut the tubing, but if the blade catches, it might bend the tubing.)

It's important to remove all the burrs from the tubing. This reduces the chance of nicking the insulation of the wire being wound on the toroid. Also, the larger-diameter tube (the bobbin) must rotate freely on the smaller-diameter tube (the shaft). To remove small burrs on the inside of the smaller-diameter tubing, I slightly flattened the point of a safety pin with a hammer to produce a small reamer. To ream the ID of the bobbin, I used a small drill bit. Both reamers can simply be twisted by hand.

The Wire Guide

Strip a short piece of thick plastic insulation from a piece of wire (about #16) and force the insulation onto one end of the shaft to act as a wire guide (see Figure 2). The insulation should extend about $^1/_{16}$ inch from the end of the shaft. If you cut the wire-guide insulation with wire cutters, the cut end of the tube won't be perfectly square. That's okay because the uneven areas tend to catch the wire and prevent it from wrapping around the guide as the wire is being pulled from the bobbin. The wire guide also provides enough friction to tension the wire while winding the toroid. Normally the wire makes less than one full spiraling turn around the wire guide as it is pulled from the bobbin, through the shaft and out the heat-shrink tubing tail.

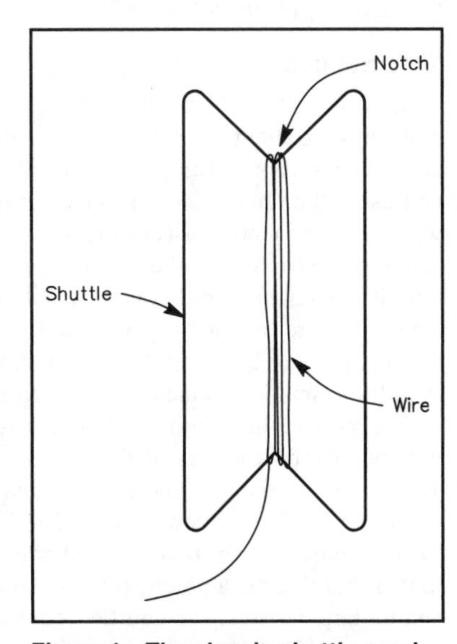

Figure 1—The classic shuttle used for winding toroidal cores. This shuttle, described in *QST* and *The ARRL Handbook* (see Note 2), is usually cut from stiff paper or thin plastic material.

[1]Notes appear on page 1-30.

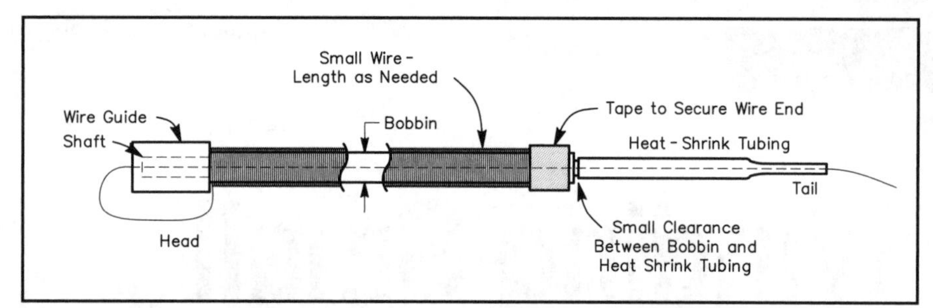

Figure 2—Mechanical details of a shuttle designed for winding small-diameter toroidal cores. The shaft is made of a 2³/₄-inch length of ¹/₁₆-inch OD brass tubing. A 1¹/₄-inch length of 3/32-inch OD brass tubing is used for the bobbin. In the drawing, one end of the wire (the right-hand end next to the heat-shrink tubing) is held to the bobbin with a piece of tape. The wire is then loaded onto the bobbin, passing the free end over the left-hand wire guide, through the tubing and out the heat-shrink tail (right).

Table 1
Approximate Wire Capacity of a Closely-Wound One-Inch Area of the Shuttle's Bobbin

Wire Size (#)	*Length (Inches)*
26	17
28	22
30	27
32	34
34	42
36	53
38	65
40	84

The #16-wire insulation I used for the wire guide in my prototype is adequate for smaller-diameter wires, ie, those smaller than #33. For larger wire diameters, I use thick-walled Teflon tubing. This tubing is mechanically stiff and damage resistant. If needed, you can use several tubing layers to increase the wall thickness. The thicker plastic tubing works better than thin-walled insulation because it creates a larger bending radius for the wire as it is pulled through. Caution! If the wire cuts its way through the end of the wire guide to the brass tubing, the wire's insulation will likely be damaged.

Slip the bobbin onto the shaft and place the heat-shrink tubing tail at the opposite end of the shaft to hold the bobbin in place. Cut the heat-shrink tubing about ³/₈ inch longer than the end of the shaft to allow its use as a threading guide. The tail also helps protect the insulation from damage as the wire is dispensed. Leave a small gap between the bobbin and the heat-shrink tubing to allow the bobbin to turn freely. I didn't apply any lubricant between the bobbin and the shaft, but you could use a single drop of lightweight machine oil at that location. Avoid using too much oil as it might migrate to the outside of the bobbin and prevent the tape (used in the next step) from adhering.

Wire Handling

Before loading the bobbin with wire, cut some short strips of masking tape about ¹/₈-inch wide. Tape one wire end to the bobbin at the heat-shrink-tail end. Load the bobbin with a rolling motion, and using the thumb and forefinger of one hand, guide the wire with the other hand into a closely wound coil. When the bobbin is full, slip the free end of the wire over and into the wire guide, and push it through the shuttle shaft and out the heat-shrink tubing tail. Hair-fine wire (about #40) is normally difficult to push through the shaft. I use a length of #30 wire with one end bent into a hook to pull the smaller-diameter wire through the shaft.

For its size, the bobbin holds a surprising amount of wire. I've wrapped over 40 inches of the #34 wire on it in a single layer. Table 1 shows the nominal bobbin capacity for a closely wound single layer of a given wire size. The lengths given allow for a ¹/₄-inch-wide area for taping. The bobbin could be made slightly longer, but there is a practical limit. To accommodate more wire and/or larger cores, the shuttle could also be made of larger-diameter tubing. I've used wire sizes as large as #26 with the tubing sizes specified. Larger-diameter wire is more difficult to work with because of the small radius at the tip of the wire guide. With larger wire diameters, it helps to push the wire on the bobbin toward the guide. As more wire is dispensed, the tape securing the other end of the wire can be removed to allow the remaining wire to move closer to the guide.

Using the Shuttle

Holding a miniature core in your hand and simultaneously wrapping wire around the core can be difficult. Instead, use a small, smooth-jaw vise, lining the jaws with a layer of tape to hold the core by a small section of its sidewalls. The tape cushions the brittle core and provides some friction to keep the core in place. Clamp the core at a corner of the vise. That leaves most of the core material and the hole exposed to accept the wire. Tighten the vise just enough to keep the core from slipping. About halfway through the winding, I usually rotate the core to make the winding more visible, but never clamp the wire that's wound on the core.

To use the shuttle, slip the tail of the shuttle through the toroid and pull out about 1¹/₂ inches of wire. Secure the wire end to the top of the vise jaws with some tape. This keeps the wire end out of the way and allows the first few turns to be tensioned. Pull out about four more inches of wire; pass the shuttle around the core, then through the center of the core. With two turns of wire on the core, continue to wind, pulling more wire from the bobbin as needed.

When close-winding toroids (especially with very small-diameter wire), I first space-wind the wire for a few turns while keeping tension on the wire, then push the turns together. This helps avoid any crossovers. The coiled wire has a tendency to spring around the core, so I use a drop of Superglue at the start of the winding and the end. I apply the glue to the core next to the wire and allow capillary action to carry the glue into the winding; it dries in a few seconds. In lieu of glue, I sometimes use a long, narrow strip of cloth tape around the core, pressing it into place as winding progresses.

This simple and inexpensive shuttle makes winding small toroids easy and fast! Add it to *your* workshop arsenal!

Notes

[1]See Figure 25.34 in the Circuit Construction Chapter of *The 2000 ARRL Handbook*, p 25.23—*Ed.*

[2]See Harold Muensterman, N9DEO, "Toroid-Coil-Winding Aids," Hints and Kinks, *QST*, Nov 1984, p 55, and *The 1984 ARRL Handbook*, Chapter 2, p 2-31, Figure 57.—*Ed.*

8655 Hwy D
Napoleon, MO 64074; n0tt@arrl.net

Photo by Joe Bottiglieri, AA1GW

QST

Chapter 2

Transceivers

By Dave Benson, K1SWL

From *QST*, April 2003

The RockMite—A Simple Transceiver for 40 or 20 Meters

There's something wondrous about an effective rig that's little bigger than a PL-259 connector. This minimalist ½ W transceiver has several sophisticated features—built-in keyer, PIC µcontroller, one-button control —and it fits in an Altoids tin.

It All Started Innocently Enough...

It's refreshing to take a step back and pursue a project that promises fun without an investment in complexity. In June 2002, I'd been eagerly awaiting several gatherings with amateur friends. Upon discussing our plans for these events, colleague Doug Hendricks, KI6DS, and I agreed that nothing could be more rewarding than giving away a kitted construction project at these get-togethers.

The challenge was to design a compact and simple transceiver, which could be kitted inexpensively, yet wouldn't wind up in the scrap bin after one or two frustrating contacts. There's something fundamentally appealing about using a minimalist radio on the air successfully, so it's not surprising that the topic has been visited before. Wes Hayward, W7ZOI's Mountaineer[1] stands out as an example of this genre, as does the work of the late Doug DeMaw and the others who followed. DeMaw's Tuna-Tin transmitter[2] has retained a faithful following with builders for more than a quarter century.

When designing for maximum simplicity consistent with good signal quality, one is drawn almost inexorably to the crystal-controlled transceiver. This article describes the *RockMite*, named naturally enough for its crystal control and its small size. The printed-circuit board measures 2.0 × 2.5 inches and fits in the Altoids tin that is beloved by the QRP community as an enclosure. It uses the familiar direct conversion (D-C) receiver scheme.

Figure 1 shows a simplified block diagram of the direct-conversion transceiver. There isn't much to it—an oscillator and a mixer convert received signals directly to audio and an amplifier boosts that audio to usable levels. On transmit, the same oscillator serves as the transmitter frequency source, and only gain and keying stages are needed to bring the oscillator signal up to levels usable for making CW contacts.

Several crucial details are missing from this oversimplified picture, however. The operator who calls "CQ" with a crystal-controlled D-C rig will most likely get replies on zero-beat with his signal and without some means of shifting frequency (offset) between transmit and receive, will copy only low-frequency thumps. Additionally, the joy of sending CW will be somewhat tempered by the lack of a sidetone circuit to monitor one's own sending.

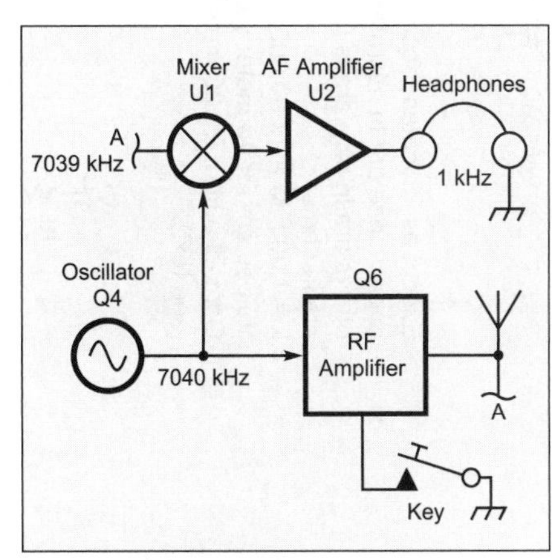

Figure 1—The 40 meter RockMite transceiver simplified block diagram.

I couldn't resist adding a new wrinkle to the rock-bottom-radio saga. By using an 8-pin PIC microcontroller, it becomes possible to add an iambic keyer along with other functions. This can be done with minimum cost and with little printed circuit board acreage. Having made the decision to use a controller chip, a spare pin on that IC was dedicated to providing a 700 Hz sidetone during key-down conditions. The controller also supplies a TR control signal and a shift signal. This shift signal merely provides a dc voltage level to a varicap (tuning) diode to pull the crystal oscillator frequency between transmit and receive. "Heck, I don't need a computer to do that!" True enough, but the TR offset is reversible, as described later, so that the RockMite offers two possible operating frequencies. This function has traditionally been done with a double-pole switch, but it's easier and cheaper to perform that function in firmware.

There is one other noteworthy trick employed in the RockMite. Builders of simple receivers for 40 meters have all

[1]Notes appear on page 2-4.

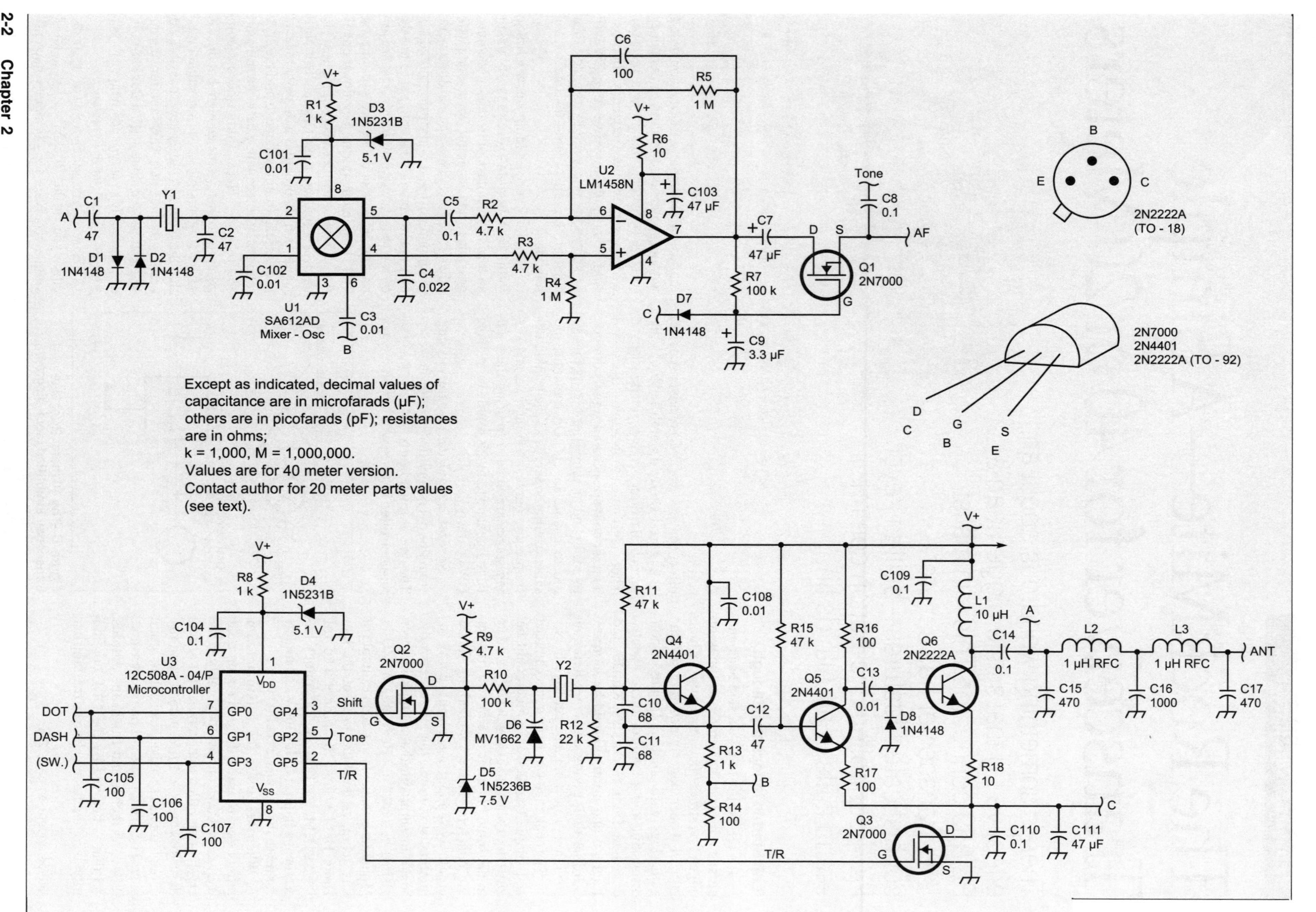

Except as indicated, decimal values of
capacitance are in microfarads (µF);
others are in picofarads (pF); resistances
are in ohms;
k = 1,000, M = 1,000,000.
Values are for 40 meter version.
Contact author for 20 meter parts values
(see text).
U1
SA612AD
Mixer - Osc
U2
LM1458N
U3
12C508A - 04/P
Microcontroller
VDD
VSS
GP0
GP1
GP2
GP3
GP4
GP5
DOT
DASH
(SW.)
Shift
Tone
T/R
AF
ANT
V+
A
B
C
C1
47
C2
47
C3
0.01
C4
0.022
C5
0.1
C6
100
C7
47 µF
C8
0.1
C9
3.3 µF
C10
68
C11
68
C12
47
C13
0.01
C14
0.1
C15
470
C16
1000
C17
470
C101
0.01
C102
0.01
C103
47 µF
C104
0.1
C105
100
C106
100
C107
100
C108
0.01
C109
0.1
C110
0.1
C111
47 µF
R1
1 k
R2
4.7 k
R3
4.7 k
R4
1 M
R5
1 M
R6
10
R7
100 k
R8
1 k
R9
4.7 k
R10
100 k
R11
47 k
R12
22 k
R13
1 k
R14
100
R15
47 k
R16
100
R17
100
R18
10
D1
1N4148
D2
1N4148
D3
1N5231B
5.1 V
D4
1N5231B
5.1 V
D5
1N5236B
7.5 V
D6
MV1662
D7
1N4148
D8
1N4148
Y1
Y2
L1
10 µH
L2
1 µH RFC
L3
1 µH RFC
Q1
2N7000
Q2
2N7000
Q3
2N7000
Q4
2N4401
Q5
2N4401
Q6
2N2222A
2N2222A
(TO - 18)
2N7000
2N4401
2N2222A (TO - 92)
E
B
C
D
G
S

experienced the joys of listening to shortwave broadcasts mixed in with their CW. For most simple gear, the high levels of broadcast RF cause intermodulation distortion (IMD). This can be mitigated by the use of more robust (higher-current consumption and complexity) receiver front ends. Another approach is to ensure that the broadcast signal levels reaching the receiver mixer are attenuated enough to avoid their IMD effects. If you're interested in only a small segment of an amateur band, a sharply tuned (narrow) band-pass filter may be used to good effect to accomplish that. The RockMite uses this approach by utilizing a second crystal at the operating frequency at the receiver front end. The performance improvement with the added crystal is significant.

Figure 2—The 40 meter RockMite schematic. All resistors are 5%, ¼ W, carbon composition. Part numbers are Mouser unless otherwise specified. Mouser Electronics, 1000 North Main St, Mansfield, TX 76063, tel 800-346-6873; local 817-804-3888; fax 817-804-3899; www.mouser.com/; sales@mouser.com. Hosfelt Electronics, 2700 Sunset Blvd, Steubenville, OH 43952-1158, tel 888-264-6464; fax 800-524-5414; e-mail tonia@hosfelt.com; www.hosfeltelectronics.com. Avnet, www.avnet.com/.

C1, C2, C12—47 pF NPO disk capacitor, 5% (140-50N5-470J).
C3, C13, C101, C102, C108—0.01 µF disk capacitor (140-50Z5-103M).
C4—0.022 µF monolithic capacitor (80-C322C223K5R).
C5, C8, C14, C104, C109, C110—0.1 µF monolithic capacitor (80-C322C104M5U).
C6, C105-107—100 pF disk capacitor (140-50S5-101J).
C7, C103, C111—47 µF, 25 V electrolytic capacitor (140-XRL25V47).
C9—3.3 µF, 50 V electrolytic capacitor (140-XRL50V3.3).
C10, C11—68 pF NPO disk capacitor, 5% (140-100N5-680J).
C15, C17—470 pF disk capacitor, 5% (140-50S5-471J).
C16—0.001 µF (1000 pF) C0G monolithic capacitor 5% (80-C322C102J1G).
D1, D2, D7, D8—1N4148 diode (625-1N4148).
D3, D4—1N5231B Zener diode, 5.1 V, 0.5 W (625-1N5231B).
D5—1N5236B Zener diode, 7.5 V 0.5 W (625-1N5236B).
D6—MV1662 varicap diode (Hosfelt MV1662).
L1—10 µH RF choke, 10% tolerance (434-22-100).
L2, L3—1 µH RF choke, 10% tolerance (434-22-1R0).
Q1, Q2, Q3—2N7000 FET (512-2N7000).
Q4, Q5—2N4401 transistor (512-2N4401).
Q6—2N2222A transistor (610-2N2222A).
R1, R8, R13—1 kΩ.
R2, R3, R9—4.7 kΩ.
R4, R5—1 MΩ.
R6, R18—10 Ω.
R7, R10—100 kΩ.
R11, R15—47 kΩ.
R12—22 kΩ.
R14, R16, R17—100 Ω.
U1—SA612AD mixer/oscillator IC (Avnet).
U2—LM1458N dual op-amp IC (512-LM1458N).
U3—12C508A-04/P microcontroller IC (Digikey, Mouser—see Note 5).
Y1, Y2—7.040 MHz (40 meters), 20 pF load, HC49/U crystal (see Note 6) (14.060 MHz for 20 meters).

Walk This Way, Please...

The RockMite schematic is shown in Figure 2. Local oscillator Q4 is a crystal-controlled Colpitts oscillator and runs continuously. Its operating frequency is determined by crystal Y2 and the surrounding components. Diode D6 is a varicap (tuning) diode and it furnishes a voltage-dependent capacitance. This effect is used to pull the crystal oscillator frequency about 700 Hz between transmit and receive to provide a beat-frequency offset. The voltage applied to D6 through resistor R10 is 0 V with Q2 turned on (conducting) or it is the rated Zener voltage of D5 with Q2 off.

A sample of the local oscillator signal is coupled to the base of Q5. Q5 provides no voltage gain but instead serves to improve key-up isolation between the local oscillator and the antenna. This ensures that the key-up energy to the antenna (back-wave) is negligible. Equally important, the lowered signal level at the antenna terminal prevents blocking effects from desensitizing the receiver.

The output of the buffer stage is coupled via C13 to the power amplifier stage, Q6. I'll concede that term (power) is being stretched a bit to encompass a 2N2222A transistor. Diode D8 provides a clamp function, making it easier to drive the base of Q6. Transistor Q6 runs Class C, is driven hard and, in theory, has only conducting and nonconducting states for high efficiency. The waveform at Q6's collector would ideally be a square wave. In practice, there's considerable waveform distortion at that signal point and, in any case, it's nothing you'd want to apply directly to an antenna.

Capacitor C14 couples this waveform to the output harmonic filter, which comprises L2 and L3 and C15, C16 and C17. In an effort to save space and reduce construction complexity, sub-miniature epoxy-molded RF chokes were used instead of the traditional toroids. For the frequencies and power levels encountered in the RockMite, performance appears adequate—loss and self-heating were not significant. The signal at the antenna has a maximum harmonic content of –34 dBc (–33 dBc for 20 meters) and it complies with FCC requirements for spectral purity. Power output is about 500 mW with a 13 V dc supply. Incidentally, it will work at lower supply voltages—one intrepid experimenter reports making contacts with 30 mW of output power at a supply voltage of 6.8 V.

The receiver is continuously connected to the antenna through coupling capacitor C1. Diodes D1 and D2 limit the key-down voltage swing appearing at the receiver front-end to safe values. The presence of Y1 at the receiver front-end may seem somewhat startling, but it serves as a narrow band-pass filter to keep RF energy from frequencies far removed from the operating frequency to a minimum. The SA612 mixer, which does the conversion from RF to audio, needs all the help it can get.

Readers may recognize the circuit as an adaptation of a Roy Lewallen, W7EL, circuit—a widely used series-LC TR switch. The inductance in this circuit is being furnished by crystal Y1 at a frequency slightly off its series-resonant point. Perhaps less obviously, capacitor C2 forms an L network in combination with a portion of the crystal motional inductance. It's impedance step-up; there's about 10 dB of voltage gain prior to the mixer input (U1, pin 2). The values of C1 and C2 were twiddled empirically to yield a 6 dB bandwidth of about 2 kHz and to straddle the two operating frequencies fairly evenly. Receiver filter response is –35 dB at 7100 kHz and up. Although this value of ultimate rejection is unacceptably poor for typical crystal filters, here it needs to be only good enough to yield significant improvement in IMD performance.

The mixer IC, U1, converts the received signal from the operating frequency to audio; that signal appearing at pins 4 and 5 of U1. C4 provides some low-pass filtering to cut unwanted audio hiss. U2 is a garden-variety dual op-amp (one-half is unused) configured for a gain of about 200 (46 dB). This boosts the mixer's output audio to headphone-usable levels. Capacitor C6 provides an additional pole of audio low-pass filtering.

Transistor Q1 provides a simple mute function to reduce the amount of key-down thump. It disconnects the audio amplifier from the headphones whenever the rig is keyed. The large (transmitted) signal appearing at the receiver during key-down yields a dc offset at the mixer output, which is amplified to a large transient by the audio amplifier. The muting isn't perfect but it's a lot less fatiguing than none at all. Key-up recovery time is set by C9—this value may be reduced if you prefer quicker QSK (break-in).

U3 is a 12C508A microcontroller device and has been custom programmed to provide iambic keyer (Mode B) and fre-

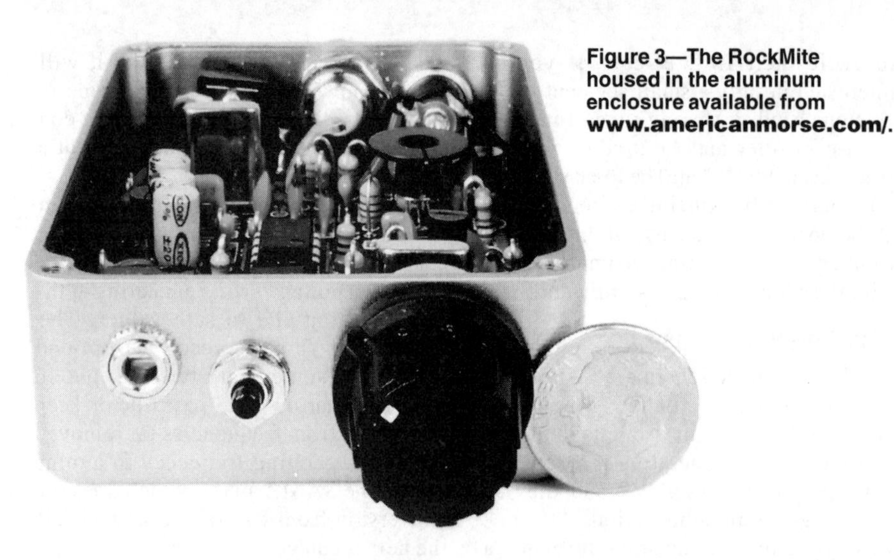

Figure 3—The RockMite housed in the aluminum enclosure available from www.americanmorse.com/.

quency shift functions. U3 pins 6 and 7 are typically connected to a pair of paddle inputs to provide the keyer functions. If a straight key is your cup of tea, ground one of those two inputs during rig power up and the RockMite will use the other input for the straight key. (This also facilitates the use of a more capable external keyer.)

There are two operator controls on the RockMite and they're both implemented via a push-button switch closure, in order to ground controller pin 4. The two functions are discriminated by the duration of the switch closure, to wit:

A brief (< 250 ms) closure on the switch reverses the offset to provide a second operating frequency. When you wish to work another station, use this function to select the *higher* of the two pitches on a received signal. Note that the pitch at the converse setting is a measure of how close to zero-beat you are; ideally it would be just a low-frequency thump. If the two selections yield a high pitch and a still higher pitch, you probably won't be able to work the other station.

A longer closure on the switch input puts the keyer in a speed-adjustment mode. The RockMite outputs a Morse code S to acknowledge entry into this mode. Tapping (or holding) the dot paddle speeds up the keyer, the same operation on the dash paddle slows it down. The default (power-up) speed is approximately 16 WPM and the speed range is about 5 to 40 WPM. If no dot/dash inputs are received after about 1 second, the RockMite outputs a lower-frequency tone and reverts to normal operation. The Morse S and subsequent tones are not transmitted on the air.

The idea of a transceiver whose only control is a pushbutton switch probably flies in the face of recent trends in transceiver design. If you feel the need to "manage" your radio, resistor R5 may be replaced with a 1 MΩ audio taper potentiometer (wiper and one end-terminal used) to serve as a volume control. Figure 3 shows the RockMite packaged in a commercially available enclosure.

Does It Really Work?

The receiver is direct-conversion, so the audio you hear is busier than what's typically found in a big rig. There's some audio low-pass filtering, but it still doesn't have the sharp roll-off characteristics prevalent with crystal IF filtering. Because the D-C receiver receives both sidebands equally well, there are twice as many signals as you'd expect of a more capable receiver. Once you get the hang of selecting which of the two operating frequencies to call someone on, the operation is pretty straightforward.

Other Frequencies

The RockMite became practical because of the economics of purchasing crystals in large quantities. Changing the RockMite frequency is a matter of replacing the two (identical) crystals with frequencies of your choosing.[3] If you change bands, however, the output harmonic filter and C10/C11 must be scaled accordingly and the value of Zener diode D5 may need to be changed to adjust the TR offset. Modification information is maintained at the URL listed later in this article.

Did I Mention "Innocently"?

This project started out as a party favor and indeed, it was initially dubbed "a wireless code practice oscillator"—somewhat tongue-in-cheek. Once the first samples were available, it became clear that the RockMite was a usable radio. Seabury Lyon, AA1MY, wrote:

> That first night thriller of real operating with the RockMite was on 7/11/02 on 40m; G3JFC, Brian; 559 him, 339 me. That set the tone and high expectations for many fun evenings later on when the "starter batch" of 'Mites began to multiply. In early August, KD1JV and NØRC floated the idea of a "Mite Nite." Starting on 8/7/02 I worked between eight and 22 stations per night, proving that…once you get used to it, the 'Mite really worked!

At present, the distance record with the 20 meter version is held by Jerry Brown, N4EO, with a contact from his home in Columbia, Tennessee, to Dunedin, New Zealand—a distance of 8485 miles.

Much of this success can be attributed to the QRP community's use of 7040 kHz (and 14,060 kHz for 20 meters) as a watering hole. Many QRPers monitor those frequencies when they're in the shack and your chances of success with a "CQ" are surprisingly good. That 40 meter frequency (7040 kHz) is used by QRPers here in North America (it's 7030 kHz in Europe) and 14,060 kHz is used universally on 20 meters. Needless to say, the RockMite has caught on like wildfire! As of this writing, about 1000 have already been assembled by builders of all ages.[4] There's even a great Web site maintained by Rod Cerkoney, NØRC—a gallery of construction pictures, modification information, links and related topics. See **www.qsl.net/n0rc/rm/**.

Acknowledgments

A special thanks to Doug Hendricks, KI6DS, for his material support with this project and to Rod Cerkoney, NØRC, for his enthusiastic Web site support. Thanks also to Steve Weber, KD1JV, for design suggestions during the development phase.

Notes

[1]W. Hayward, W7ZOI, "An Ultra Portable CW Station—The 'Mountaineer,'" *QST*, Aug 1972, p 23.

[2]D. DeMaw, W1FB, "Build the Tuna-Tin 2," *QST*, May 1976, p 14.

[3]Custom frequencies are available from International Crystal Mfg; **www.icmfg.com/**. Specify HC-49/U, 20 pF loading, 0.01% frequency tolerance.

[4]A complete kit of parts including two 7040 or 14,060 kHz crystals, PC board, all on-board parts, ICs and instructions is available from the author for $27 including shipping (US), $30 elsewhere. Specify 40 or 20 meters, and send the order to Dave Benson, K1SWL, 32 Mountain Rd, Colchester, CT 06415. The 12C508A supplied with the kit is preprogrammed. Educator/Scouting inquiries are welcomed. A European (7030 kHz) version is available from **www.qrpproject.de**.

[5]The 12C508A as available commercially is a blank (unprogrammed) device. Programmed ICs alone are available from the author for $5, including shipping. RockMite object code is available to individuals, for non-commercial use only, upon request to the author.

[6]7040 kHz or 14,060 kHz crystals (two required) are available from Doug Hendricks, KI6DS, 862 Frank Ave, Dos Palos, CA 93620, for $3 each, including shipping.

Dave Benson, K1SWL, is a frequent contributor to QST, *a QRP fan and an inveterate builder and designer. He can be reached at* **dave@smallwonderlabs.com**.

From *QST*, December 2005

The HiMite—A RockMite Transceiver for the Higher Bands

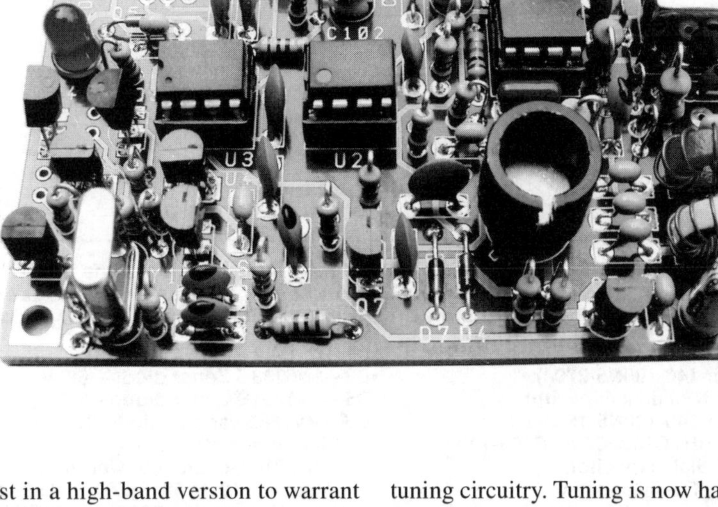

Dave, K1SWL, provides a redesign of the popular RockMite low power transceiver that can work on the higher HF bands, has more output power— and provides RIT capability.

Dave Benson, K1SWL

When the ARRL published my original "RockMite" article,[1] I had no idea just how popular it would prove to be. Picture, if you will, the sorcerer's-apprentice scene from "Fantasia"—it depicts perfectly the tidal wave of enthusiasm for that project! As of this writing, there are at least 6000 RockMites out there, reassuring proof that construction projects remain an enduring part of Amateur Radio.

As that project took on a life of its own, I'd receive occasional questions and comments about extending the tuning range of the RockMite, or about putting it on the higher HF bands. In general, I discouraged the idea of simply replacing the existing crystal oscillator with a variable-frequency oscillator (VFO). Two complications arose from this prospective change—the need to maintain the desired transmit-receive frequency offset, and the rather narrow passband established by the receiver front-end crystal filter.

Having said that, I did sense enough interest in a high-band version to warrant further design work. The *HiMite*, to stretch a name somewhat, echoes the simple architecture of the RockMite's direct conversion (D-C) receiver. The most significant changes to the design include those necessary for adequate RF output power on the high bands, and those required by the conversion from fixed frequency to variable crystal oscillator (VXO) tuning. This article describes just the 15 meter version of the HiMite. Component values for 20, 17, 12 and 10 meters differ primarily in the band-pass and low-pass filtering circuitry. The details for the other bands may be found on my Web site.[2]

Ch-Ch-Ch-Changes

The most notable change with this incarnation of the 'Mite is in the varicap diode (D6) control circuitry. The RockMite stepped between two discrete tuning voltages to yield two operating frequencies 700 Hz apart. This time around, this fixed tuned scheme is replaced with a potentiometer controlled tuning voltage. A three terminal voltage regulator (U4) provides a stable 8 V supply to the tuning circuitry. This ensures that a battery supply isn't the source of chirp, or frequency shift, as it approaches a nearly discharged state.

As seen in the schematic (Figure 1), there's a good deal more complexity to the tuning circuitry. Tuning is now handled by the TUNE control, a potentiometer with a wiper voltage range from 0 to 8 V over its rotation. Q3 and Q4 are added components, and provide a receiver incremental tune (RIT) function for the HiMite. Just as with the RockMite, a means of shifting frequencies between key-up and key-down is required. I evaluated a number of RIT schemes, and indeed had considered the idea for the RockMite. I was somewhat disappointed in the lack of consistency in RIT offset at various portions of the tuning range, and had not incorporated RIT in that previous design. Because of the somewhat limited tuning range of a VXO, I wanted to avoid further restricting its coverage by reserving part of its voltage tuning range for RIT. The RIT function is achieved by turning Q3 and Q4 off (non-conducting) during key down and thereby removing any contribution to the tuning voltage on transmit. During receive, the RIT control is active and does influence the tuning voltage.

Components R12 and R13 provide a *weighted summing* function to establish a primary tuning control for the tune function and a more limited RIT range. Lower values of R12 will provide additional RIT range. As a concession to conventional RIT operation, the entire range of the tuning control is restricted to 0 to 8 V, resulting in minimal action at each

[1]Notes appear on page 2-7.

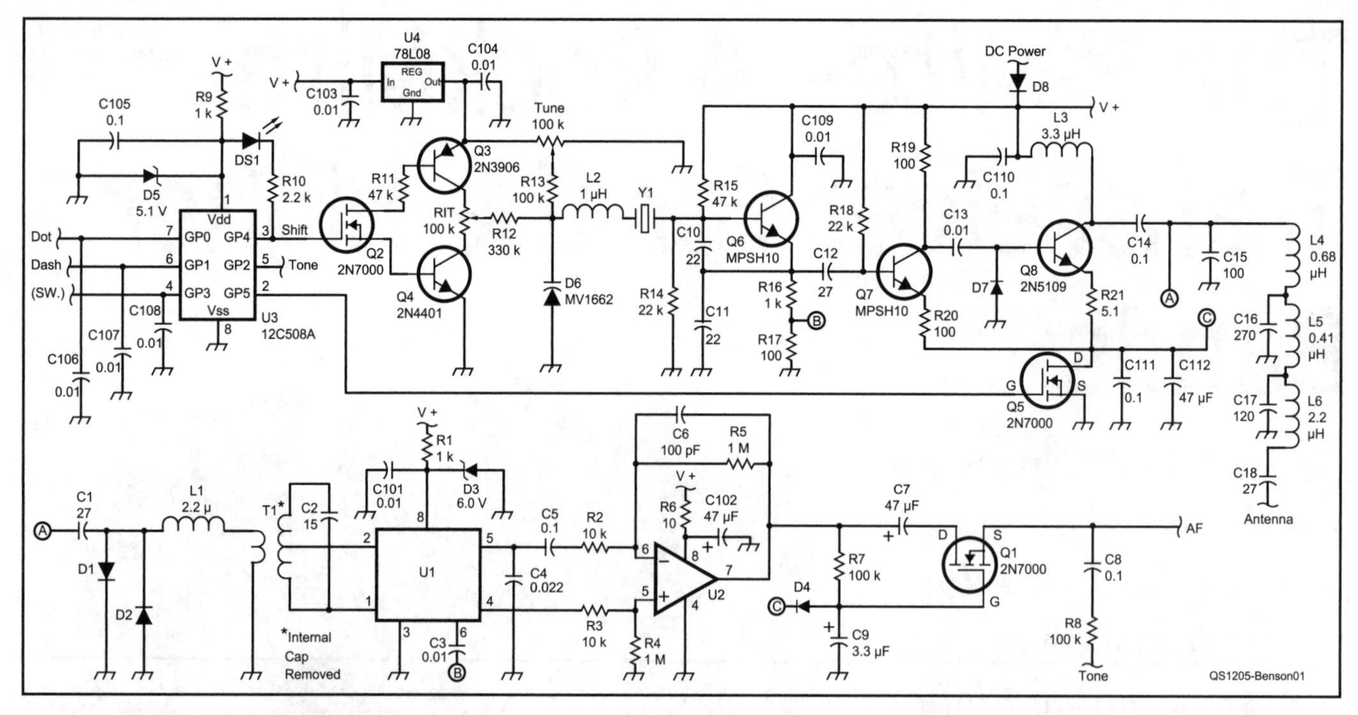

Figure 1—Schematic diagram and parts list for the 15 meter version of the HiMite HF transceiver. Most components are stocked by major distributors such as Digi-Key (www.digikey.com), Mouser (www.mouser.com) and Ocean State Electronics (www.oselectronics.com). Resistors are 5% 1/4 W carbon composition. Capacitors are in pF and 5% unless otherwise noted.

C1, C12, C18—27 pF NPO disk capacitor (Mouser 140-100N5-270J).
C2—15 pF NPO disk capacitor (Mouser 140-100N5-150J-TB).
C3, C13, C101, C103, C104, C106-109—0.01 µF disk capacitor.
C4—0.022 µF poly capacitor.
C5, C8, C14, C105, C110, C111—0.1 µF monolithic capacitor.
C6—100 pF disk capacitor.
C7, C102, C112—47 µF electrolytic capacitor.
C9—3.3 µF electrolytic capacitor.
C10, C11—22 pF NPO disk capacitor (Mouser 140-50N5-220J).
C15—100 pF COG capacitor.
C16—270 pF COG capacitor.
C17—120 pF COG capacitor.
D1, D2, D4, D7—1N4148 diode.
D3—1N5233B Zener diode—6.0 V, 1/2 W.
D5—1N5231B Zener diode—5.1 V, 1/2 W.
D6—MV1662 varicap diode (Hosfelt Electronics MV1662).
D8—1N5818 Schottky power diode (Mouser 512-1N5818). TO-5/TO-39 heat sink (Mouser 532-322505B00, Digi-Key 294-1105).
DS1—T1-3/4 LED (Mouser 606-CMD5053).
L1, L6—2.2 µH RF choke, 10% tolerance.
L2—1 µH RF choke, 10% tolerance.
L3—3.3 µH RF choke, 10% tolerance.
L4—0.68 µH, wind 15 turns #26 magnet wire on T37-6 (yellow) toroid core (Amidon or Palomar T37-6).
L5—0.41 µH, wind 10 turns #26 magnet wire on T37-2 (red) toroid core (Amidon or Palomar T37-2).
Q1, Q2, Q5—2N7000 transistor.
Q3—2N3906 transistor.
Q4—2N4401 transistor.
Q6, Q7—MPSH10 transistor.
Q8—2N5109 transistor.
T1—10.7 MHz IF transformer with internal capacitor removed (Mouser 42IF123).
U1—SA612AN mixer-oscillator IC (Avnet Electronics, Future Electronics).
U2—LM1458N (Mouser 512-LM1458N).
U3—12C508A-04/P microcontroller (Digi-Key, Mouser—see Note 5).
U4—78L08- 8 V 100 mA regulator (Mouser 511-L78L08ACZ).
Y1—21.060 MHz, HC49/U crystal (see Note 6).

end of the control rotation. Elsewhere within the tuning range, the RIT control may be set on either side of "zero beat" (more on this later). In practice, I saw a 0.8 kHz "upward" RIT range at one extreme and 2.5 kHz of "downward" RIT at the other. Intermediate tuning voltages, of course, yielded RIT travel in both directions.

How do I Know if the RIT's On?

There's an LED visible in the photo on the upper left corner of the HiMite circuit board. During key down, the LED is on in time with the transmitted signal. As with the RockMite, a brief ground closure at the switch input (U3 pin 4) toggles between RIT on (receive) and RIT off (transmit). A longer ground closure activates the keyer speed control sequence. This functions exactly as it did with the RockMite. Whether the LED is on or off, RF is transmitted only on key down. Physical placement of the LED is non-critical. Although shown mounted on the board, it may be mounted remotely via wire leads without adverse effect.

Transmitter Changes

Because the HiMite must work at higher frequency, a change in the semiconductor lineup is needed. The garden-variety high-speed switches used for the local oscillator and buffer have been replaced with a higher speed device, the MPSH10. This transistor has a much higher gain-bandwidth product than its predecessors. The power amplifier (PA) device itself has been replaced with the heftier 2N5109, with a gain-bandwidth product of 1200 MHz.

I took one other step to improve output power for the high bands. The RockMite uses a 50 Ω bilaterally designed low-pass filter. For a 1/2 W amplifier stage, there's a noticeable mismatch to the PA collector impedance, and a resultant poor efficiency. When I modified the HiMite low-pass filter for a collector impedance of 200 Ω, output power nearly tripled to the 500 mW level. Consequently, with the improved efficiency the HiMite, the PA runs cooler than that of its RockMite sibling.

The transmitter low-pass filter uses toroidal cores for L4 and L5. In addition to the beneficial self-shielding properties of toroidal windings, off-the-shelf RF choke and capacitor combinations weren't handy for all of the upper bands. There's also an important extra element to the filtering—a band-pass filter comprising L6 and C18. Since the RockMite article was published, FCC spectral purity requirements have become more rigorous.[3] The new requirement of spurious signals 43 dB below the carrier requires additional filtering.

Since the crystal filtered front end of the

RockMite has been eliminated, I wanted additional receive out of band rejection as well and thus applied additional band-pass filtering to both paths. The resulting filter combination complies with current FCC requirements—spurious and harmonic content was measured as –52 dBc or less.

The combination of L6 and C18 is series resonant at the frequency of interest, 21 MHz. A caution about the selection of component values for this filter is in order. At higher values of component reactance (resulting in higher Q), component tolerances become critical to the extent that the two component reactances may no longer cancel. Additionally, at higher Q values, allowable current ratings for the inductor may be exceeded. I used a Q of 5 as a compromise between band-pass performance and filter loss. This value did not materially affect the available output power.

Receiver Changes

The primary change to the receiver is in the front end filtering. As I described above, the narrow passband of the RockMite proved an obstacle to VFO operation on the low bands. The situation did not improve much on the higher bands. Try as I might, in no case did the front end bandwidth exceed more than 3 to 4 kHz—too little for effective VXO coverage. I reverted to the more conventional tuned LC front end circuit, with the several series LC tuned circuits providing an additional measure of front-end selectivity.

The AF gain of the HiMite has been cut in half compared to its predecessor. The RockMite gain proved too high for some conditions involving weak batteries or control wiring proximity and was found to howl occasionally. This reduction in gain should be helpful—there's still enough signal to drive a good pair of headphones.

I received a number of comments about the sidetone level—a number of folks thought it a tad too energetic. R8 has been added in series with the controller AF output to reduce the level somewhat. Its value can be readily changed to suit your preferences. Power diode D8 was also added to the board layout—this protects the circuitry in the event of a reverse polarity power-supply hookup.

Frequency coverage on the 15 meter HiMite is 14 kHz on my sample unit. This coverage ranged from 21.026 to 21.040 MHz with the values shown on the schematic. This coverage may be shifted by changing the value of L2. Higher frequency coverage is achieved by reducing or shorting L2, and by replacing it with a series capacitor. I was able to cover well above the 21.060 MHz low power calling frequency. I'll encourage experimentation here with one caution. At high- enough values of L2 (or small enough values of a replacement series capacitor), the local oscillator will become somewhat unpredictable. There may be discontinuous jumps in tuning coverage or low output power. That's a sign that you've taken the process too far, and that it's time to return to a previous component value. Thevalues of C10 and C11 may also beadjusted somewhat to effect further frequency shift.

Operating the HiMite

Tuning in a received signal is fairly self-evident, but the need to zero-beat a station you want to call introduces a new wrinkle. A tap of the switch input turns on the LED and shifts to the transmit frequency. Zero-beat the station using the TUNING control and then tap the switch again to turn RIT back on. Adjust RIT as needed for a comfortable signal pitch on the station you're copying. For portions of the tuning range, you'll be able to hear the station on either side of zero-beat. This is normal for a direct-conversion receiver, and you can use RIT to choose whichever sideband yields less interference.

Band openings on the upper HF bands tend toward brief and infrequent during the current sunspot minimum. Conditions vary by time of day and seasonally—it's a case of patience and luck!

Notes

[1]D. Benson, K1SWL, "The RockMite—A Simple Transceiver for 40 or 20 meters," *QST*, Apr 2003, pp 35-38.

[2]Schematics/parts lists for other bands are at **smallwonderlabs.com/HiMite_docs.htm**.

[3]Component values to retrofit this modification to the existing RockMites are available from the author on request. For retrofit parts, send a self-addressed, stamped envelope with two units of postage to the author and specify 40 or 20 meters.

[4]A kit of parts including crystal, PC board, all on-board parts, ICs, two potentiometers, and instructions is available from the author for $32 including shipping (US), $35 elsewhere. Specify frequency: 14040, 18096, 21060, 24906 or 29060 kHz and send to Dave Benson, K1SWL, 32 Mountain Rd, Colchester CT 06415. The 12C508A supplied with the kit is pre-programmed.

[5]The 12C508A as available commercially is a blank (unprogrammed) device. Programmed ICs alone are available from the author for $5, including shipping. *RockMite/HiMite* code (.hex file) is available from the author on request.

[6]Crystals for other frequencies are available from **expandedspectrumsystems.com** or **AF4K.com**.

Dave Benson, K1SWL, has been licensed since 1967. An electrical engineer by profession, he worked in the aerospace industry for 20 years before pursuing small business aspirations. Dave is a frequent contributor to QST, *with a number of QRP oriented construction articles to his credit. A strong believer in "doing-it-yourself," he's currently designing and will be building an energy-efficient home in New Hampshire in anticipation of his wife's retirement. He can be reached at* **nn1g@earthlink.net**.

By Wes Hayward, W7ZOI, and Terry White, K7TAU

The Micromountaineer Revisited

Although this easy-to-build transceiver was initially designed for use on 10 meters, you can move it to other bands as well. As a bonus, you get a power meter, too!

An inside view of one of the Micromountaineers enclosed in a die-cast aluminum box.

In the early 1970s, we described a pair of small, crystal-controlled QRP stations.[1, 2] Both use direct conversion (D-C) receivers in the 40-meter band with a transmitter output of half a watt. The first (see Note 1) uses a tunable receiver, while the other, dubbed the Micromountaineer (see Note 2), uses the transmitter crystal oscillator as the receiver LO. We reasoned that there is little need to be able to receive on frequencies far removed from those of the transmitter. While a generation has brought much more sophistication to the QRP operator, the premise remains valid.

Although QRP operation has grown to become a major subculture within Amateur Radio, the activity is still largely confined to bands where commercial equipment predominates. Only the high-end commercial QRP boxes or 100-W transceivers with reduced power operate on the 10-meter band. This updated version of the Micromountaineer is aimed primarily at the 10-meter band where excellent propagation allows international communications with low power and simple antennas, but the rig can be built for use on other bands as well. (We're including component values for a 40-meter version of the rig, too.) The original Micromountaineer theme is retained: a $^1/_2$-W transmitter and a D-C receiver share a crystal oscillator. Electronic TR switching and *almost incremental tuning*, AIT, (explained later) have been added to enhance performance while retaining a primitive simplicity.

Circuit Details

Transmitter Section

The heart of the transceiver is the crystal-controlled oscillator, Q2 (Figure 1). The circuit uses third-overtone mode 28-MHz crystals and is tuned to frequency with T1 and C4. (See T1 detail in Figure 2.) Q1 is an electronic switch that shifts the oscillator frequency by about 1 kHz. (Large frequency shifts are available in fundamental-mode oscillators, but are more difficult

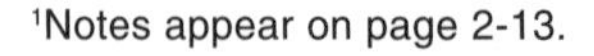

[1]Notes appear on page 2-13.

From *QST*, July 2000

This Micromountaineer is housed in an LMB enclosure. Take care to remove paint on the inside of these enclosures to ensure contact with grounded components.

to obtain in overtone oscillators.) Without this shift, you might send CQ, get a reply *on-frequency* and never hear a beat note!

The oscillator output, nearly 10 mW, is extracted from T1 and applied to a power splitter consisting of three 51-Ω resistors. One output provides receiver LO injection, while the other is applied to Q3, the transmitter driver. Q3 is keyed through Q4, a PNP switch and shaping integrator that prevents key clicks. Driver output is extracted via ferrite transformer T2 (shown in greater detail in Figure 2). This transformer uses a binocular balun core in which *one wire turn* constitutes a complete pass through *both* holes.[3] This transformer has the primary exiting one end with the secondary at the other end.

The power amplifier (Q5 and Q6 in parallel) uses a pair of modest and inexpensive 2N3904s. Emitter degeneration forces the transistors to equally share current and provides thermal stability. A Zener diode, D5, prevents severe stress on the transistors during momentary operation without a load. Experimenters might want to try using other transistors in the PA stage. A single 2N4427 we tried worked well, as did four 2N2222As in parallel, both PAs producing over 1 W output. The 2N3904 pair is normally operated at about 1/2 W output, a level at which the transistors are not thermally abused, even without a heat sink. L3 and L4, with C13, 14, and 15, form a low-pass filter doubling as an impedance-matching network.[4] The result is a cool, robust amplifier with an efficiency of over 50%. Measured second-harmonic output is 58 dB below the desired output, easily meeting FCC 2002 spectral-purity requirements.[5] To obtain maximum output, the turns on L3 are spread or compressed as required. If you decide to try an alternative PA, you might need to alter the output-network component values to obtain maximum output with reasonable efficiency.

Receiver Section

The receiver is a variation on the familiar Neophyte popularized by John Dillon, WA3RNC.[6] Mixer U2, an NE602 (an obsolete part, but still available—*Ed.*) or NE612 mixer, serves as a product detector followed by U3, an LM386D audio amplifier. The detector is biased at 5 V from U1, a 78L05 regulator. To provide receiver muting and a simple way of injecting a sidetone oscillator, the receiver is modified slightly from the original circuit. MOSFETs Q8 and Q9 are turned on during transmit intervals, shorting the audio from the detector. (The MOSFETs have a very low *on* resistance that is unavailable from a bipolar transistor with modest base current.) The input to U2 is tuned, but with an unbalanced input. This produced a large dc offset during transmit intervals until Q7 was added to enhance receiver muting. Replacing L5 with a balanced transformer will also reduce dc offset.

U2 uses very little current. Although this is a great advantage for portable applications, the low current results in severely degraded dynamic range. Enterprising experimenters can expect a large-signal performance improvement of up to 20-dB using diode-ring-based designs.[7]

A sidetone oscillator (Q10 and Q11) is keyed with a PNP switch, Q12, to produce a signal that is fed to U3 via R32. Sidetone level may be adjusted by changing the value of R33. Oscillator pitch can be decreased by increasing the values of R39 and R41.

Two PNP switches generate additional timing voltages when the key is pressed. Q16 mutes the receiver for a mute time determined by the values of C35 and R25. Q15 causes the +12R line to go low while the +12T line goes high. These signals are both present at toggle switch S1. Operating S1 allows you to apply the frequency offset to the crystal oscillator during either transmit or receive, affording some ability

A close-up of a portion of the prototype Mountaineer. Yes, it works just fine!

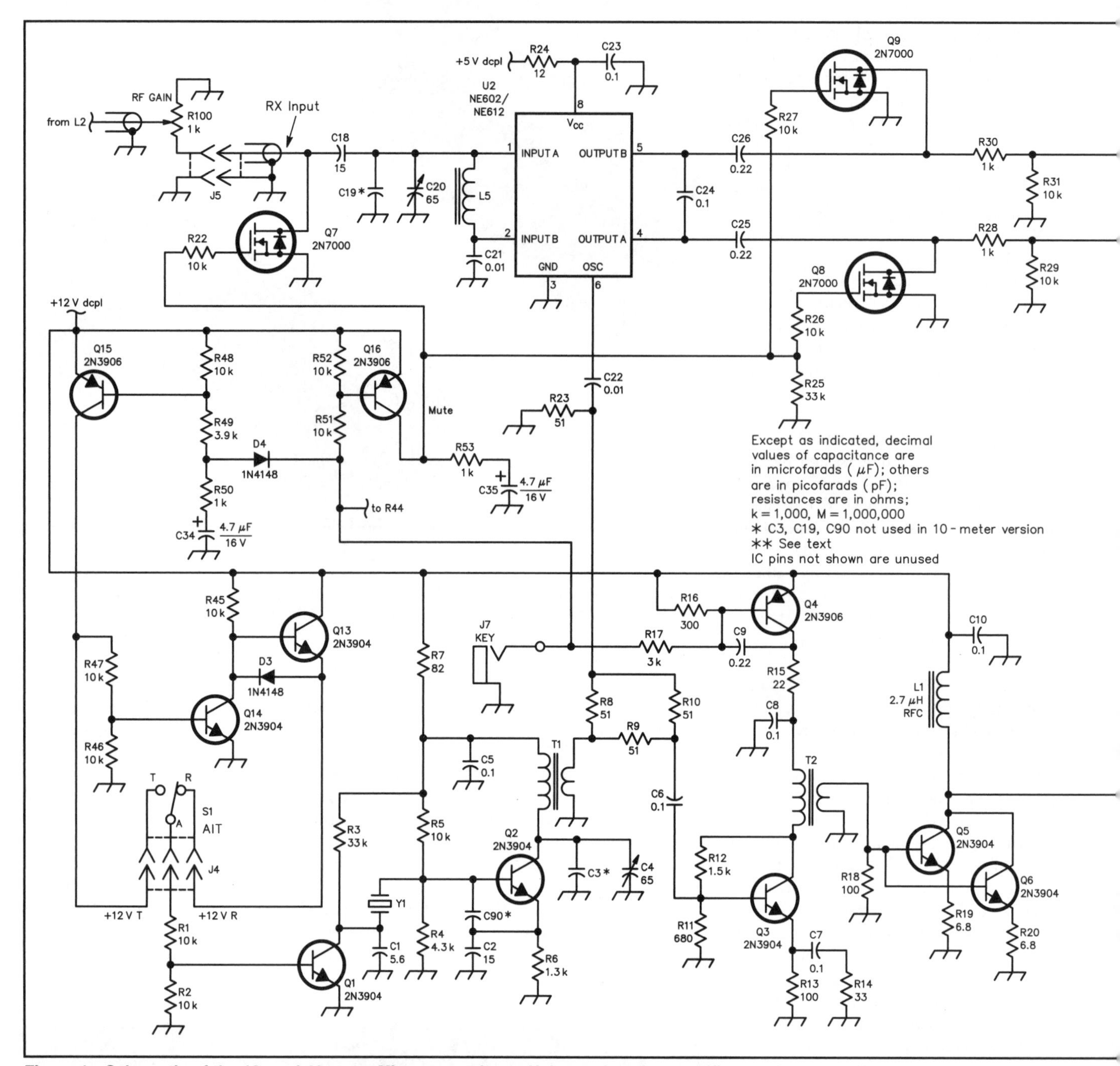

Figure 1—Schematic of the 10- and 40-meter Micromountaineer. Unless otherwise specified, resistors are ¼-W, 5%-tolerance carbon-composition or film units. Fixed-value capacitors should be 5% tolerance; either ceramic (NP0/C0G) or mica are suitable. Equivalent parts can be substituted; n.c. indicates no connection. The parts list identifies band-specific components. In addition to the PC-board/component collection in Note 11, parts are available from several sources: Mouser Electronics, 958 N Main St, Mansfield, TX 76063-4827; tel 800-346-6873, 817-483-4422, fax 817-483-0931; sales@mouser.com; http://www.mouser.com; Digi-Key Corp, 701 Brooks Ave S, Thief River Falls, MN 56701-0677; tel 800-344-4539, 218-681-6674, fax 218-681-3380; http://www.digikey.com; RadioShack and others.

to dodge QRM. It's almost like having receiver incremental tuning (RIT) in a more-refined radio, hence the term AIT: *almost incremental tuning*. We have used the scheme in several simple VFO-controlled transceivers proved in severe portable situations.[8] Timing of the frequency toggle is controlled by the values of R49 and C34. You can reduce the frequency shift (about 1 kHz in our transceivers), by increasing the value of C1.

TR switching is handled by D1 and D2.

The front panel of the dual-range QRP power meter.

C12 and L2 form a series-tuned 28-MHz circuit that routes antenna signals from the transmitter to the **GAIN** control and the receiver. When the transmitting key is pressed (creating a strong signal), the diodes conduct, keeping the received-signal level low enough to prevent damage to mixer U2. Gain is controlled by R100, a panel-mounted potentiometer at the receiver input. Although a trimmer capacitor is used at C12 in the 10-meter transceiver, a fixed-value capacitor is employed in the 40-meter version.

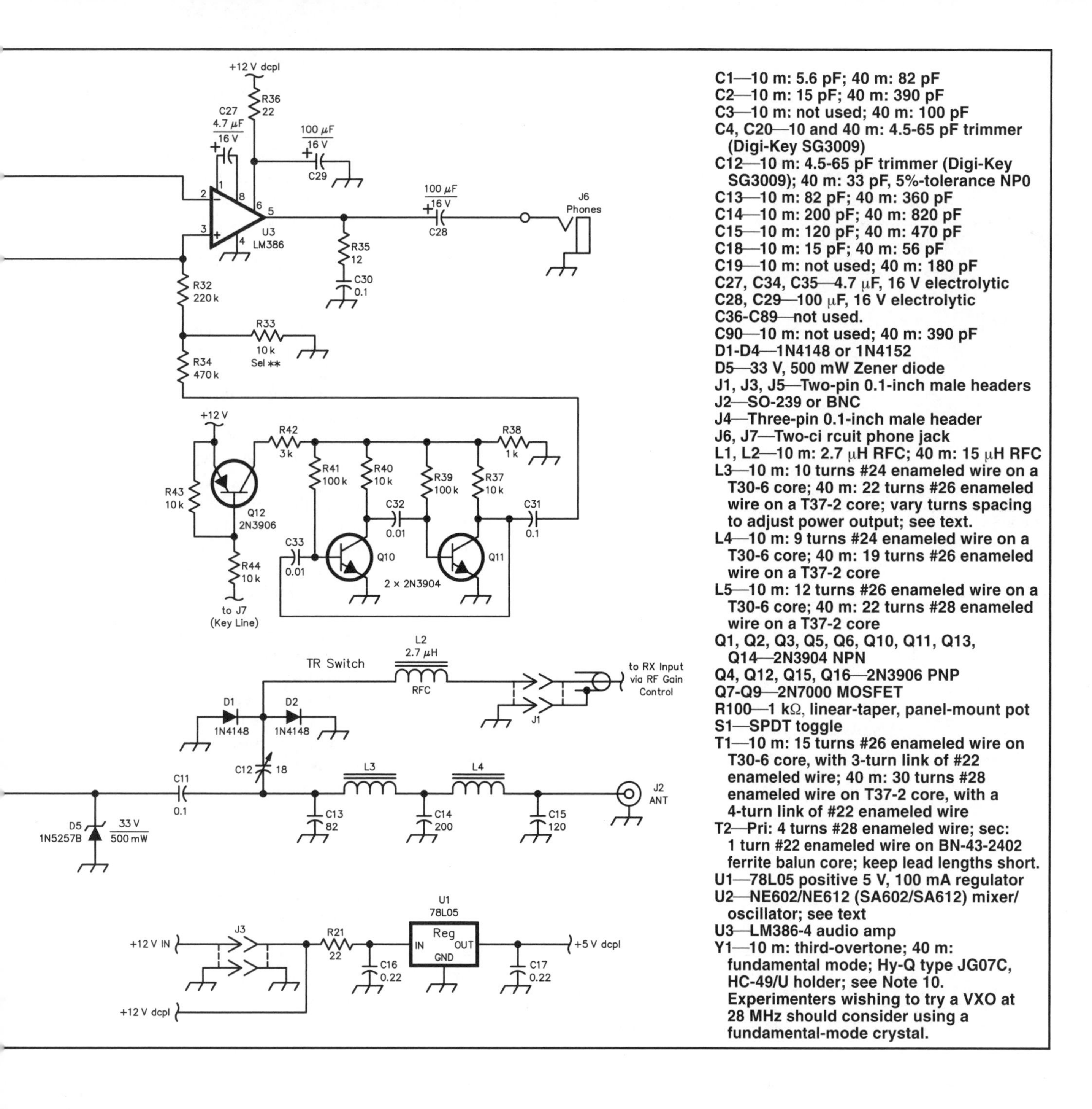

A Simple Power Meter

Figure 3 is a schematic of a simple dual-range power meter that you can use to test this rig (or other similar power sources). Consider first the 1-W range. D200 rectifies the peak RF voltage appearing across the 50-Ω, 1-W load formed by the parallel combination of R200 and R201. The resulting dc voltage is applied to a voltmeter (formed by R202 and the meter) having a 10-V full-scale reading.

After we built the single-range power meter, we noted that the dc voltage across the meter movement was small. Investigation showed that the meter had an internal resistance of only 100 Ω, a typical value for

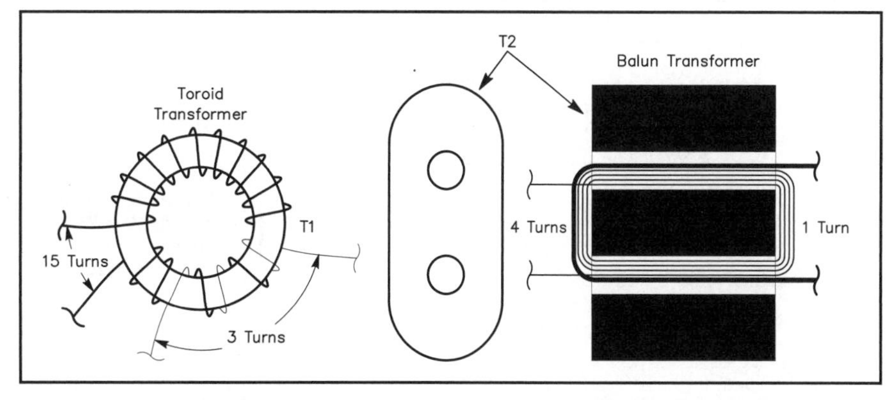

Figure 2—Winding details for the two transformers used in the 10-meter Micromountaineer. Each pass through the core center counts as a turn in a traditional toroid, T1 above. But each turn must pass through *both holes* of a balun core (see Note 3).

As this inside view of the power meter shows, it's simplicity itself. The components mounted on the rear panel of the box serve functions unrelated to that of the power meter.

inexpensive 1-mA movements. This allowed us to add a second detector and RF load to form a second, more-sensitive (50-mW) range. We calibrated this against the 1-W meter using the transmitter and a step attenuator, resulting in the calibration chart of Figure 4. This curve may be used directly for approximate measurements.[9]

This power meter includes 50-Ω terminations, acting as loads for whatever source is applied. This design differs from popular in-line power meters used by QRP operators that require an external load.

Building the Transceiver

There are probably as many ways to build this rig as there are experimenters. For our first version, we used "ugly construction."[10] But there are sure to be many prospective builders who want a PC-board version of the transceiver.[11] In any case, we strongly recommend that you build and test the rig as you build it *one stage at a time.*

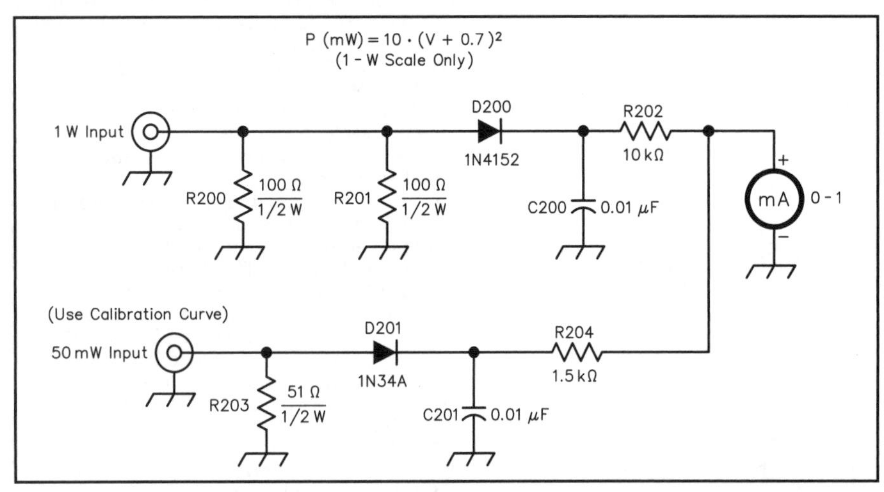

Figure 3—The transmitter can be tested with this simple power meter. When the meter reads 0.5 mA with RF applied to the 1-W input port, the indicated peak RF voltage is 5 V and the power is 325 mW, calculated with the formula shown. D200 is a 1N4152, 1N4148, or 1N914, while D201 is a more-sensitive 1N34A germanium diode.

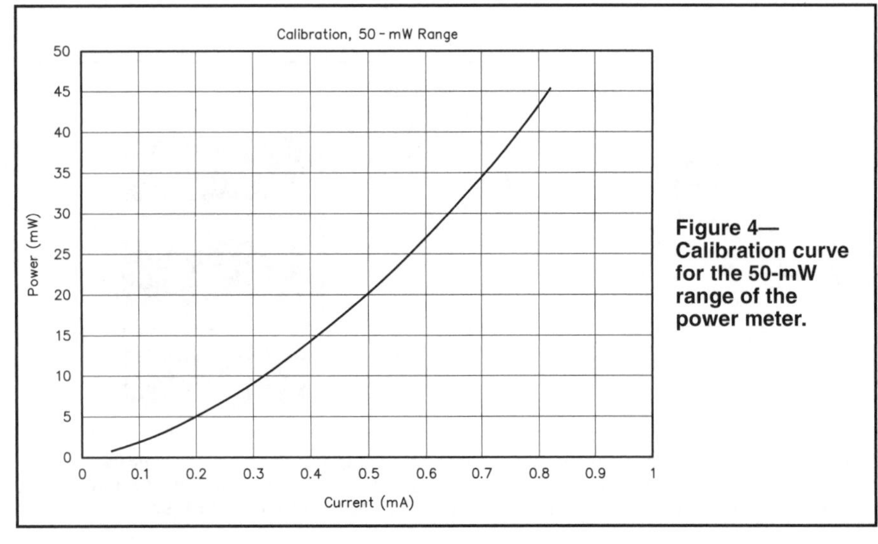

Figure 4—Calibration curve for the 50-mW range of the power meter.

Transmitter Section

Begin by building the crystal oscillator, including the frequency-controlling switch, Q1. Apply power and confirm oscillator operation by listening for its signal in a receiver tuned to the crystal frequency. To test the oscillator, attach the T1 link to a short piece of 50-Ω coaxial cable connected to the power meter. Adjust C4 for maximum output. Using the sensitive (50-mW) range of the power meter, the output from T1 should measure about 10 mW into the 50Ω load. Only after this measurement is made and confirmed is the transformer secondary (link) attached to the next stage.[12]

Next, build the driver stage (Q3), including the keying switch (Q4) and confirm operation. Start with a 15-Ω resistor for R14, knowing that its value will be adjusted later. Confirm this stage's operation by attaching the output of T2 to a 50-Ω coaxial cable connected to the power meter. The indicated power should be about 35 mW. After testing, the T2 output is routed to the PA transistor bases.

Assemble the PA and output network. Then attach the output to the power meter's 1-W range, apply dc power and press the key. Adjust the circuit for maximum power output by squeezing or compressing the turns on L3. We found it useful to measure the inductor values prior to installing them. For this, we used a small LC meter available from Almost All Digital Electronics.[13]

Receiver Section

Wire the audio amplifier (U3) and test it by listening for a slight hiss in headphones plugged into J6. Touching one of the input leads with your finger or a screwdriver should produce some hum or perhaps even an AM broadcast signal. After building and testing the regulator circuit, U1, assemble the detector (U2) and attach an antenna. Even if it's late at night and the band is dead, you should still be able to hear background noise. Adjust C20 for maximum noise output. If you've got a signal generator, by all means, use it. Signal levels of 0.1 µV are easily copied with this receiver.

If you build this receiver on a PC board that's sitting on a table or workbench, you might observe considerable hum in the receiver. Don't be concerned—this tunable hum will go away once the board is installed in a metal box. With the ugly breadboard unit, the hum was barely detectable, but the hum was considerably louder when testing the PC-board version.

Other Circuits

Once the basic transceiver parts are working, you can start to add the "frills." Install the muting transistors (Q8 and Q9) and related switch, Q16. Make sure you have a 50-Ω load attached to the transmitter output. (Don't operate the transmitter without a proper load attached.) Then apply power and press the key. You should hear the receiver noise drop to nearly nothing. Add and test the AIT-related parts, followed by the sidetone components. Finally, install the TR diodes and L2/C12 at the PA. Route a coaxial cable between the receiver input and the TR output and adjust C12 for maximum receiver output.

You're now ready to put the transceiver in a box. Avoid using a small enclosure that makes component access difficult. Once the rig is in the box, you might want to adjust R14 (discussed earlier) to maintain the power output at around 500 to 650 mW. For the 10-meter transceiver, we ended up with a value of 33 Ω for R14; on 40 meters, R14 is 47 Ω. It's also wise to anticipate changing R33's value for sidetone-level adjustment.

Putting the Rig on the Air

Now it's time to see what the rig can do! As always, a lot depends on the antenna used. We used a modest dipole at 30 feet. When we first put the rig on the air, 10 meters was open, but relatively quiet. After calling CQ a few times, a station 2000 miles away answered. Following a 20-minute chat, we ended the QSO by getting his e-mail address so we could QSL with a digital photo of the rig. The following day was even more productive, netting contacts from Vermont to Alaska.

Good operating practice calls for *listening first* to be sure that the frequency is not in use before calling CQ. If an answering station returns your call right on the same frequency, you'll hear the signal shortly after releasing the key, and at a pitch equal to the offset. By switching in the AIT, you can pick the best place to answer any other station that might have responded.

Concluding Thoughts

Although *any* QRP activity can be great fun, 10-meter CW QRP is about as good as it gets! You can work the world with a watt or less, even with a modest antenna. But the good band conditions won't be with us forever, so seize the moment!

Aggressive experimenters will want to consider expanding the performance of this transceiver. The rig can, of course, be put on bands other than 10 and 40 meters. The next refinement of interest is more flexible frequency control. Probably the simplest way to accomplish this is with a VFO. VFO stability is not as easy to obtain at 28 MHz as it is on lower bands, so a heterodyne approach may be required. In any case, use care to maintain spectral purity.

Builders of the original Micromountaineer often converted the oscillator to a VXO. This is considerably more difficult to do with overtone crystal-oscillator circuits than it is with fundamental crystal-oscillator circuits. It's also challenging to implement an offset circuit with uniform frequency shift in a VXO. But let's see what *you* can do!

Notes

[1]Wes Hayward, W7ZOI, and Terry White, K7TAU "The Mountaineer—An Ultraportable CW Station," *QST*, Aug, 1972, pp 23-26.

[2]Wes Hayward, W7ZOI, "The Micromountaineer," *QST*, Aug, 1973, pp 11-13 and 45. Also see Wes Hayward, W7ZOI, and Doug DeMaw, W1FB, *Solid-State Design for the Radio Amateur* (Newington: ARRL, 1977), p 219.

[3]Experiments easily demonstrate that there is minimum coupling from one hole to the next. A single winding through one hole produced a signal 16 dB lower in a single winding in the other hole. When both windings were in the same hole, the loss was only 4 dB. Measurements were at 28 MHz with 50-Ω terminations.

[4]The network was originally designed as a 31-MHz, 0.1-dB ripple Chebyshev low-pass filter with 50 Ω terminations at each end. L3 and C13 were then modified using Smith Chart analysis to provide a 28-MHz impedance of 100+*j*0 Ω at the PA collectors.

[5]Rick Campbell, KK7B, "Unwanted Emissions Comments," Technical Correspondence, *QST*, Jun 1998, pp 61-62.

[6]John Dillon, WA3RNC, "The Neophyte Receiver," *QST*, Feb 1988, pp 14-18.

[7]Rick Campbell, KK7B, "High-Performance Direct-Conversion Receivers," *QST*, Aug, 1992, pp 19-28.

[8]Read more about portable operation beyond the traditional mobile situations at the Web site for the Adventure Radio Society, **www.natworld.com/ars** and look at the online magazine, *The ARS Sojourner*.

[9]RadioShack sells a 0-15 V dc meter (RS 22-410), which is a 1-mA meter movement equipped with an external 15 kΩ resistor. The resistor is not needed in this application.

[10]This method is a point-to-point wiring scheme using circuit board scraps serving as a ground foil. Most components are supported by other components if they are not themselves soldered to the ground foil. Additional mechanical supports can be added in the form of dummy resistors of high value. See Roger Hayward, KA7EXM and Wes Hayward, W7ZOI, "The 'Ugly Weekender'," *QST*, Aug, 1981, pp 18-21. Also, visit our Web page at **www.teleport.com/~w7zoi/bboard.html**.

[11]A double-sided, plated-through-hole PC board and a component collection (but *not a kit*) are available from Kanga USA. See their Web site for price and availability information: **www.bright.net/~kanga/kanga**. The crystal used in this circuit is a third-overtone type in an HC-49 package. We recommend Hy-Q type JG07C from Hy-Q International, 1438 Cox Ave, Erlanger, KY 41018-3166; tel 606-283-5000, fax 606-283-0883; e-mail **sales@hyqusa.com**, **http://www.hyqusa.com**. The 10-meter QRP calling frequency is 28.060 MHz. A crystal lower in the band might be more productive for DX enthusiasts.

[12]This is a substitutional measurement, which is typical of RF studies. In contrast, most measurements in analog electronics are in situ (in place) measurements where probes are attached to functional systems.

[13]Almost All Digital Electronics 1412 Elm St SE, Auburn, WA 98092; tel 253-351-9316, fax 253-931-1940; e-mail **neil@aade.com**; **www.aade.com.**

Wes Hayward, W7ZOI, was first licensed in 1955 while in high school. A career in electron-device physics and circuit design took him to companies in the western states. Wes is now semi-retired, devoting his time to writing and research, with a smattering of backcountry hiking. You can contact Wes at 7700 SW Danielle Ave, Beaverton, OR 97008; **w7zoi@teleport.com**.

Terry White, K7TAU, taught himself Morse code and theory and received his Novice license, KN7TAU, in 1962. Employment has taken him worldwide, operating from New Delhi, India (VU2TAU) and Fairbanks, Alaska (KL7IAK). In 1992, he joined TriQuint Semiconductor in Hillsboro, Oregon, and is part of the advanced-receiver development group. Terry enjoys homebrewing his radios and test equipment with a twist on craftsmanship to each project. You can contact Terry at 9480 S Gribble Rd, Canby, OR 97013; **twhite@TQS.com**.

By Dan Metzger, K8JWR

From *QST*, December 2002

Build the "No Excuses" QRP Transceiver

With this simple 40-meter transmitter-receiver combo, your dream of operating a "build-it-yourself" ham station will finally become a reality. No exotic parts are required—so warm up your soldering iron!

"Say… you don't happen to have a few FT-37-43 toroids in your junk box, do you? Or maybe a T-50-6? No? …I didn't think so. How about an MC3362 integrated circuit or a 2N3553 RF power transistor? Well…don't feel bad, almost no one else does, either. I guess my low-power transceiver project will just have to wait until I can find the parts…"

I've been through that scenario half a dozen times in the past decade, and it was always a list of special parts that kept me from building my own solid-state ham station. I was sure that even if I ordered all those special parts, a crucial one would be out of stock or I'd blow one up and have to wait two weeks for the replacement to arrive.

Well, potential homebrewers, there are no more excuses; this QRP transceiver uses no—I repeat, no—special parts. There's not a toroid in it. The 12 transistors are all garden-variety 350 mW types and the only integrated circuit is the dirt-common LM386 audio amplifier. The hardest part to scrounge is a transmitter crystal for your favorite patch of 40 meters.

Still not convinced? Here's a list of features to help nudge you off the fence.

- The transmitter and receiver are completely separate. One can be built first and then the other, so the whole project doesn't have to be tackled at once. (The transmitter can be started first. It's easier—and is something to brag about on the air.)
- Power output is an ample 2.5-3 W—about half an S unit below the QRP full-gallon of 5 W.
- The transmitter is VXO (variable crystal oscillator) controlled, to give crystal-controlled frequency stability and the ability to change frequency by a few kHz to get away from interference.
- The receiver uses a superhet design with a crystal filter and offers true single-signal selectivity. (Most "simple" homebrew receivers use regenerative or direct-conversion circuits, which effectively double the perceived amount of interference on the band because they produce a signal on both sides of zero beat.)
- The receiver has audio-derived automatic gain control (AGC) to save the ears when a strong station comes on.
- A frequency counter can be connected to display both transmit and receive frequencies. I used the UniCounter[1] on these examples and it fit right into the case.
- All of the circuits can be built on an unetched copper-clad board using ground-plane construction; you can procure a ready-made board[2] or etch a board using the available pattern and parts-placement guide.[3]

If that's not enough persuasion, I'm not sure what is! Figure 1 shows the two versions I built: a ground plane on the left and a printed-circuit version on the right. Now let's take a look at the innards.

Transmitter Circuit

See Figure 2. Q1 is a crystal oscillator that feeds buffer Q2 and drivers Q3 and Q4. Q2, Q3 and Q4 are untuned switching transistors that switch class C final amplifiers Q5 and Q6 from saturation to cutoff as quickly as possible. The finals are able to put out about 3 W of RF while dissipating only 0.5 W each under continuous key-

[1]Notes appear on page 2-25.

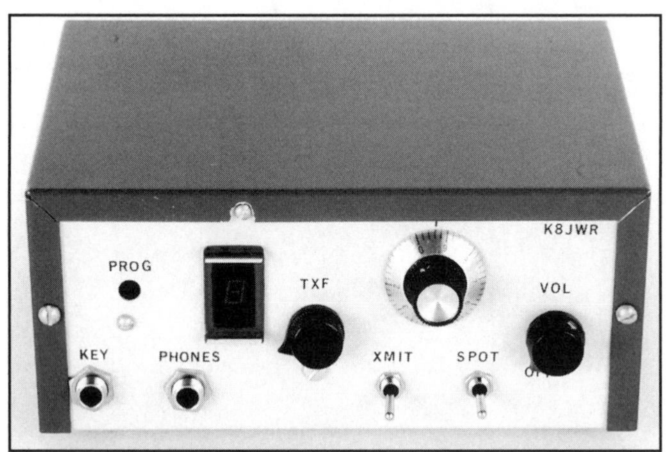

Figure 1—Two versions of the "No Excuses" QRP transceiver. The "ground-plane" version is at the left; the printed circuit version, at right.

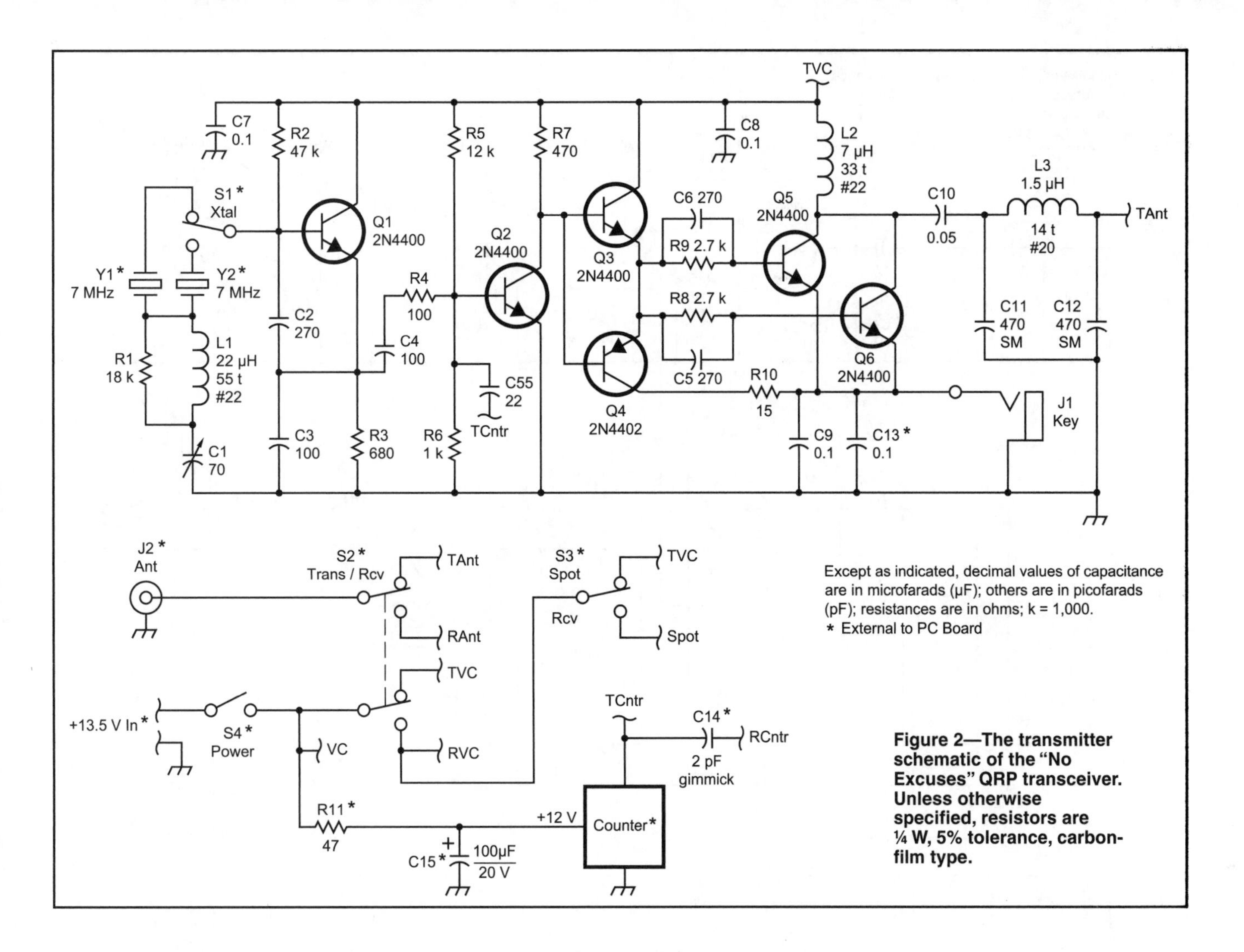

Figure 2—The transmitter schematic of the "No Excuses" QRP transceiver. Unless otherwise specified, resistors are ¼ W, 5% tolerance, carbon-film type.

down condition—which is quite safe if a ½ × ¾" metal plate is super-glued to the flat face of each transistor as a heat sink. The transistors will survive a brief key-down period while feeding a badly matched or disconnected antenna, but I wouldn't make a habit of doing that!

The oscillator and buffer run continuously during transmit, while the drivers and final amplifiers are keyed. The emitter current of Q5 and Q6 consists of square pulses that are rich in harmonics. If this current were flowing in the key leads it would produce considerable RFI, so an additional 0.1 µF bypass capacitor is soldered right across the key jack (J1). In stubborn cases, an RF choke (10 turns of #22 enameled wire wound on a 1 MΩ, 1 W resistor), in series with the key lead may be needed. The Q2-Q6 signal chain is direct-coupled, so it's important that voltage divider R5/R6 biases Q2 *on* in the event of oscillator failure. This keeps Q5 and Q6 *off* so they don't draw excessive current.

The oscillator frequency can be "pulled" below the marked crystal frequency by adjusting capacitor C1 (70 pF) toward maximum capacitance. The amount of "pull" varies from about 1 to 2 kHz to more than 10 kHz, depending on the characteristics of your crystal and whether or not you elect to use L1 and R1 (see below). The crystals from Ocean State Electronics can be offset about 8 kHz if that circuit is used.

Coil L1 (22 µH) is rather critical, and some experimenting may be necessary to find the right value. It was found that a value of 20 µH cut the tuning range in half, while a value of 24 µH put the oscillator in a free-run mode, well outside the ham band. It was also found that slug-tuned coils and coils wound with finer #28 wire did not work well. If experimentation hassles are to be avoided, L1 and R1 can be omitted altogether, replacing them with a short circuit (jumper). The QSY range of the transmitter will be reduced to 1 or 2 kHz, but oscillator stability won't be a problem.

Transmitter Tune-Up

There is nothing to tune in this transmitter save for the aforementioned fiddling with L1 to get the greatest frequency range while maintaining a stable signal. It is worth mentioning that L3 must be wound with #20 wire (or larger) and that C11 and C12 must be silver mica types. If smaller wire is used or ceramic caps are substituted, quite a bit of RF will be lost at the output. It may also be worthwhile to try different transistors at Q1 through Q4. It was possible to raise the RF output by 0.5 W by swapping transistors here and there. If a transistor checker is available (now on many digital volt-ohmmeters,) try to match the betas of Q5 and Q6, in order to equalize the current gain of the parallel final amplifier transistors. These transistors can also be selected for maximum output.

Also, if a 60 MHz oscilloscope is available, observe the waveform at the antenna. It should be a clean, symmetrical sine wave. If it's lopsided, there is second or third harmonic content present. Try minor adjustments to L3, C11 and C12. If a ragged-looking trace is seen, higher order harmonics and RFI are present. Try to improve the grounding or power supply bypassing or use a better-regulated power supply.

Receiver Circuit

There's nothing fancy to understand here (see Figure 3). Two fixed-tuned cir-

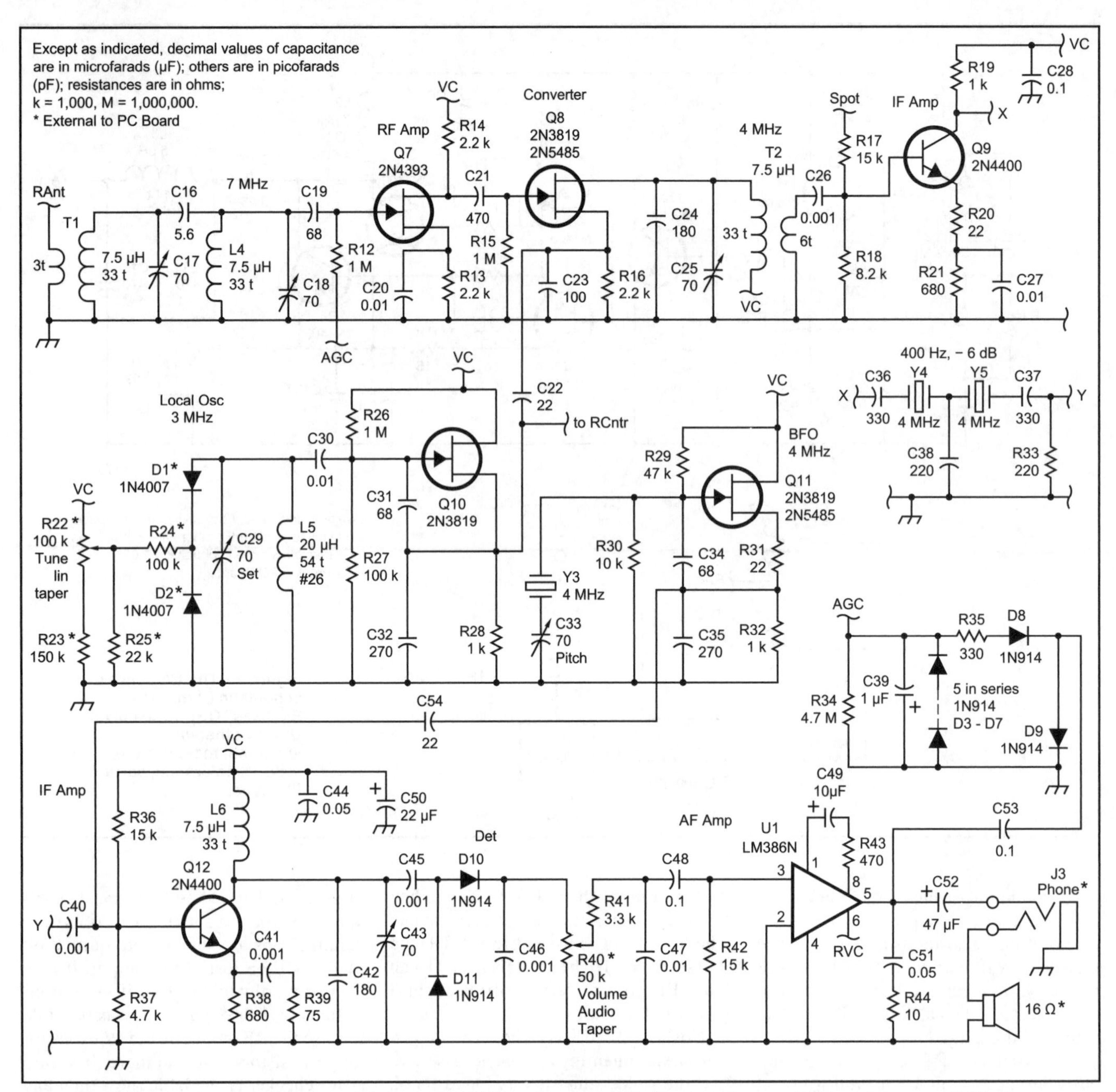

Figure 3—The receiver schematic of the "No Excuses" QRP transceiver. Unless otherwise specified, resistors are ¼ W, 5% tolerance, carbon-film type. Coils are wound of #22 wire, except L5, as noted.

cuits limit the passband of RF amplifier Q7 to signals near 7 MHz. The RF amplifier provides a maximum gain of 5 or 10. The amplified 7 MHz signal is mixed with a 2-V P-P signal at 3 MHz from local oscillator Q10, which produces a 4 MHz difference signal at the output of converter Q8. The 4 MHz IF signal is amplified by Q9, limited to a 400 Hz bandwidth by crystal filter Y4-Y5 and amplified further by Q12.

The beat frequency oscillator (BFO) stage, Q11, runs about 1 kHz above 4 MHz and its signal is combined with the 4 MHz signal in Q12. The nonlinearity of detector diodes D10-D11 causes an audio difference signal of about 1 kHz to appear at the top of volume control R40. The audio is amplified by the LM386 IC and is fed to headphones or a small speaker. A portion of the audio is rectified by diodes D8-D9 to produce a negative dc voltage that biases the gate of RF amplifier Q7. Audio levels above a few volts peak-to-peak, which would be hard on the ears, produce enough negative bias to drive Q7 into cutoff, reducing its gain to one or less, thus keeping the volume below the "pain" level. All receiver circuits except the audio amplifier are powered continuously to minimize frequency drift.

Choosing Receiver Components

Transistors Q7, Q8 and Q11 can be just about any N-channel junction FETs, provided that the drain current drops to near zero with a gate voltage of –2 V or less. If the AGC loop is to keep the volume at a pleasant level, be sure to use a FET for Q7 that cuts off close to –1 V.

Figure 4 shows a simple test circuit for determining each FET's cutoff voltage. Many different FETs are labeled "2N3819," and some have gate turnoff voltages of –5 V or more. (Conversely, some may be suitable for this project; hence the usefulness of the tester.) The tester won't be necessary if some 2N4393s are used, as these already have the required low gate cutoff voltage and experimentation probably won't be necessary.

Table 1
Parts List

For part numbers in parentheses, OS = Ocean State Electronics, 6 Industrial Dr, Westerly, RI 02891; voice 800-866-6626, fax 401-596-3590; **www.oselectronics.com**.

C1, C17, C18, C25, C29, C33, C43—7-70 pF ceramic trimmer (OS TC-777).
C2, C5, C6, C32, C35—270 pF disc ceramic (OS CD270-1).
C3, C4, C23—100 pF disc ceramic (OS CD100-1).
C7, C8, C9, C13, C28, C48, C53—0.1 µF monolithic (OS CM104).
C10, C44, C51—0.05 µF ceramic disc (OS CD05-5).
C11, C12—470 pF silver mica (OS CSM470).
C14—Gimmick capacitor made of twisted wire.
C15—100 µF, 25 V electrolytic (OS CEA100-25).
C16—5.6 pF disc ceramic (OS CD5.6-1K).
C19, C31, C34—68 pF disc ceramic (OS CD68-1).
C20, C27, C30, C47—0.01 µF disc ceramic (OS CD01-5).
C21—470 pF disc ceramic (OS CD470-1).
C22, C54, C55—22 pF disc ceramic (OS CD22-1).
C24, C42—180 pF disc ceramic (OS CD180-1).
C26, C40, C41, C45, C46—0.001 µF disc ceramic (OS CD001-5).
C36, C37—330 pF disc ceramic (OS CD330-1).
C38—220 pF disc ceramic (OS CD220-1).
C39—1 µF, 50 V electrolytic, axial (OS CEA1-50).
C49—10 µF, 25 V electrolytic, axial (OS CEA10-25).
C50—22 µF, 25 V electrolytic, radial (OS CEA22-25).
C52—47 µF, 25 V electrolytic, axial (OS CEA47-25).
D1, D2—1N4007 (OS 1N4007).
D3-D11—1N914 (OS 1N914).
J1, J3—Key and earphone jacks, ¼-inch (OS 30-412). Bend to keep circuit open or use smaller 3.5-mm size (OS 30-702J).
J2—Antenna chassis connector, SO-239 style (OS 25-5630) or smaller BNC style (OS 27-8460) or RCA-type phono style (OS 30-428).
J4—Binding posts for 12 V power (OS 90-799, black; OS 90-800, red).
L1—22 µH (55 turns #22 enameled wire wound on a "New Home" type sewing machine bobbin).
L2, L4, L6—7.5 µH (32 turns #22 enameled wire on sewing machine bobbin).
L3—1.5 µH (14 turns #20 enameled wire on sewing machine bobbin).
L5—20 µH (54 turns #22 enameled wire on sewing machine bobbin).
Q1, Q2, Q3, Q5, Q6, Q9, Q12—2N4400 (OS 2N4400). or 2N4124 or similar NPN (see text).
Q4—2N4402 (OS 2N4402), or 2N4126 or similar PNP.
Q7—2N4393 (OS 2N4393), or similar N-channel JFET with 1 V gate cutoff voltage.
Q8, Q11—2N5485 (OS 2N5485), or selected 2N3819 or similar N-channel JFET with –2 V gate cutoff voltage (see text).
Q10—2N3819 or MPF105 (OS 2N3819).
R7—470 Ω, ½-W carbon film resistor.
R22—100 kΩ linear potentiometer (OS P100K).
R40—50 kΩ audio taper potentiometer w/switch (OS PAS50K).
S1, S3—Min SPDT switch (OS 10002).
S2—Min DPDT switch (OS 10012).
S4—SPST switch, part of R40.
T1—7.5 µH (32 turns #22 enameled wire wound on a "New Home" sewing machine bobbin, with 3-turn primary wound over top).
T2—7.5 µH (32 turns #22 enameled wire on sewing machine bobbin, with 6-turn secondary wound over top).
U1—LM386N audio amplifier IC (OS LM386N).
Y1—Crystal, 7.040 MHz (OS CY7040).
Y2—Crystal, 7.030 MHz (OS CY7030).
Y3-Y5—Crystal, 4.000 MHz (OS CY4).

Miscellaneous
Transistor sockets (12), (OS T3400); IC socket (1), (OS LP8); knobs (3); crystal socket (FT-243 type), OS CS243 (HC6/U type) (OS CS6U); Octal socket (OS STM8). 3" × 7" × 5" (HWD) chassis box for the printed circuit version (or larger for the ground-plane version).

The three receiver crystals should all be of the same type, from the same manufacturer and production lot. Crystals with different characteristics will probably not work well together, even if they are all marked 4.000 MHz.

The receiver is tuned by varying the reverse bias voltage on a pair of ordinary rectifier diodes; this varies their junction capacitance. Common 1N4004s will work, but it was found that the 1 kV PIV 1N4007s were more temperature stable. ("varicap" diodes were tried, too, but the 1N4007s worked just as well). Some experimenting may be necessary to find a pair of diodes that give the desired tuning range. It is important that the tuning potentiometer, R22, has a resistance curve that varies smoothly or the tuned frequency will jump erratically. Test this potentiometer on a digital volt-ohmmeter by picking a few random resistance values and see how difficult it is to repeatedly set the pot to exactly those values on the meter. It is possible to eliminate the diode tuning scheme altogether, by lifting diode D1's anode lead or by not installing R22 through R25, D1 and D2. An air-variable capacitor of about 50 pF can then be wired across C32 in order to tune the receiver.

Coil Winding

All of the coils for this project are wound on plastic bobbins made to hold the lower thread in "New Home" brand sewing machines. These bobbins are readily available from most sewing stores for about 50 cents each. The bobbins have holes in the sides that you can feed the wire ends through to keep the coil from unwinding. There are no tapped coils in this project, no bifilar windings and nothing else to confuse the builder.

Two of the receiver coils require the winding of a few turns over the main winding to make a transformer, and for that two more holes will have to be drilled to let the extra wire leads out, but that's about as complicated as it gets. The coils can be wound by hand, but a simple coil winder will make things neater and the winding easier. The coil winder used to wind these coils consists of a $^3/_{16}$ × 4-inch bolt with a crank made from a piece of coat hanger soldered to the bolt head, one nut to cinch the bobbin and a second nut (held in a vise) to hold the bolt and coil steady as the crank is turned. The coil winder in use can be seen in the photograph (Figure 5). Figure 6 shows the number of turns to wind for any value of inductance up to 35 µH and is useful if you experiment with L1 and the other coils.

Receiver Alignment

Adjust C29 for a 3 MHz signal from the local oscillator using a frequency counter at the source of Q10 or a general-coverage receiver tuned to 3 MHz. Then check for a 4 MHz signal from BFO Q11 in the same manner. Now, as trimmers C17, C18, C25 and C43 are adjusted, some signals, or at least some band noise, should be audible. Adjust these trimmers for maximum volume. An antenna should be connected for these adjustments.

If self-oscillation (a loud "screech") is heard, increase the value of R20 to 33 or 47 Ω to reduce the gain of stage Q9. Tune in a CW signal for maximum volume, regardless of pitch. Now adjust trimmer C33 for the desired pitch. The signal on the other side of zero beat should be barely audible if the circuit has been adjusted correctly.

At night, short wave AM broadcast stations may be heard in the back-

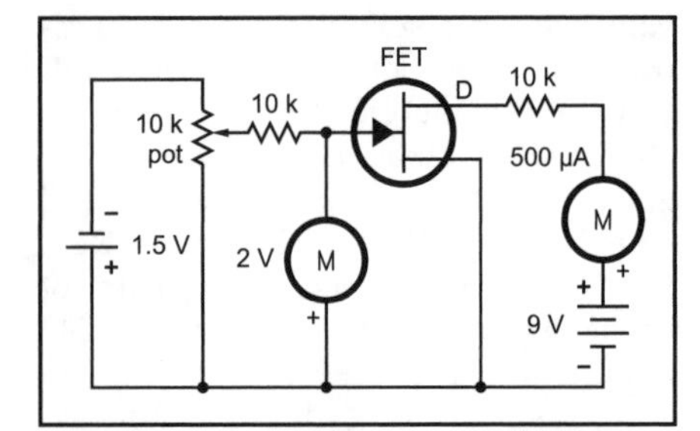

Figure 4—FET test circuit for selecting FETs (may not be necessary; see text). Adjust the potentiometer for a drain current of 10 µA and read the gate cutoff voltage on the voltmeter.

ground. This occurs when two extremely strong stations are exactly 4 MHz apart; their signals mix and get into the IF. This type of interference can be minimized by careful adjustment of C17 and C18.

It was possible to get a 0-100 calibrated knob to read the received frequency to within 2 kHz over the range 7010 to 7055 kHz by using the following procedure:

1) Temporarily replace R23 and R25 with 200 kΩ potentiometers. Set them to mid-range.

2) Adjust R22 for maximum voltage (highest frequency). This will occur 30 to 50 degrees before full clockwise rotation. Set the knob to read the maximum desired frequency (7070 kHz in this case).

3) Adjust R22 for minimum voltage (not necessarily fully counterclockwise) and adjust R23 so the receiver picks up the lowest desired frequency (7000 kHz in this case).

4) Set the R22 dial for mid-range (7035 kHz in this case) and adjust R25 so the receiver picks up that frequency.

5) Repeat steps three and four until the calibration is satisfactory. Remove the potentiometers; measure their values and solder in the appropriate fixed resistors for R23 and R25.

Frequency Readout

When the SPOT switch is thrown, the transmitter's oscillator and driver are on and the receiver's sensitivity is greatly reduced by removing the bias from IF amplifier Q9. This allows the transmitter signal to be heard at a comfortable volume. Learning how to adjust the transmitter frequency for the pitch that "zero-beats" with a desired station will come quickly. A frequency counter isn't absolutely necessary, but it is convenient.

On receive the counter will read 3040 for 7040 kHz, but that shouldn't present a problem. The receiver oscillator signal is coupled to the counter by bringing a piece of hookup wire from the source leg of Q10 close to the counter input wire, or twisting the wires together for a turn, if necessary. This forms a "gimmick" capacitor of 1-2 pF. On transmit, the 22 pF from the base of Q2 couples a signal to the counter that overpowers the weakly coupled receiver input and the counter reads the transmitted frequency.

Put the "No Excuses" on 30

Here are the circuit changes necessary to put the "No-Excuses" Transceiver on 30 meters. Please note the parts revisions in Table A. The same circuit board can be used; it will accommodate all of the necessary changes.

To eliminate the varactor circuit in order to save a few components and revert to a more traditional tuning method, either lift the anode of diode D1 (the side that connects to the gate of Q10) or remove R22, R23, R24, R25, D1 and D2. The receiver is then tuned with a 40 pF air-variable capacitor in parallel with C32. Diode tuning can still be used if the builder desires, however.

There are several powerful shortwave broadcast stations just below the 10 MHz band. The 2 element crystal filter in the IF stage has an off-frequency feedthrough that is only 32 dB down from the passband and those AM signals may come crashing through. Quite a bit of relief can be obtained by adding a low-Q series-resonant circuit at the crystal filter output. The filter, comprised of a 6.8 µH inductor in series with a 150 pF capacitor, can be installed across R33. It does not impair normal 4 MHz signals fed through the crystal elements, but it greatly attenuates the off-frequency signals coupled through the crystal holder capacitance.

The transmitter output is 2.5 W on 30 meters. I worked 32 states in 6 weeks with a dipole 25 feet up—and winter propagation hadn't even started yet. So go ahead, get on a new band!

Table A

Parts Needed for 30-Meter Version

C2—180 pF disc ceramic.
C3—68 pF disc ceramic.
C5, C6—100 pF disc ceramic.
C11, C12—330 pF silver mica.
C16—3 pF disc ceramic.
L3—1.0 µH, 11 turns #18 enameled wire on "New Home" sewing bobbin.
L4—5.2 µH, 28 turns #22 enameled wire on bobbin.
L5—5.2 µH, 28 turns #22 enameled wire on bobbin (oscillator will then be at 6.1 MHz).
R1, L1—Replace with a single 100-Ω resistor. The frequency shift will then be +4 to +5 kHz.
T1—5.2 µH, 28 turns #22 enameled wire on bobbin, 2-turn secondary wound over top.
Y1, Y2—Use 10.1 MHz crystals (OS CY1010) (OS CY1011) (OS CY1015) or (OS CY1012).

Construction Notes

The first version of this radio was actually built on a couple of solderless breadboards. The second, a "ground-plane" version, was built on an unetched piece of copper-clad board, 4½ × 6½-inch in size, and housed in a 3×7×5-inch (HWD) aluminum box. See Figure 7. (It all fit, but a bigger box would have been better.)

The photographs show how chips of copper-clad board about 3/16-inch square are super-glued to the copper surface to form the connecting points. The unetched board forms a "ground plane" beneath the circuit and aids in decoupling adjacent stages. This version uses an air variable

Figure 5—Winding the coils using the author's winding fixture (see text). A third hand is helpful!

for both transmitter and receiver frequency control, with a vernier drive on the receiver tuning capacitor. It also uses a front-mounted crystal socket, with the UniCounter mounted on the back panel with a bundle of wires connecting the front-mounted display to the counter electronics. This version works well, but some of the parts are hard to find and some of the wiring is a little difficult. The third version uses an etched circuit board available from FAR Circuits,[4] and fits easily into the 7×5×3-inch box (Figure 8). Circuit board artwork is available for those wishing to "roll their own" board. This transceiver uses an easy-to-find trimmer capacitor for transmitter frequency control; a piece of 1/8-inch diameter rod is soldered to its screwdriver slot so a knob can be used. It also uses a potentiometer with common diodes (1N4007), which are used as "varicaps" to tune the receiver. It dispenses with the vernier drive and the air-variable capacitor. The pin arrangement of the transistors on the circuit board is always E-B-C (S-G-D for FETs), reading clockwise when viewed from the bottom.

The receiver frequency tends to drift 100 Hz or so if air currents pass over the diodes, so keep its box closed. If a small air variable is available, you may want to consider using it rather than the diodes

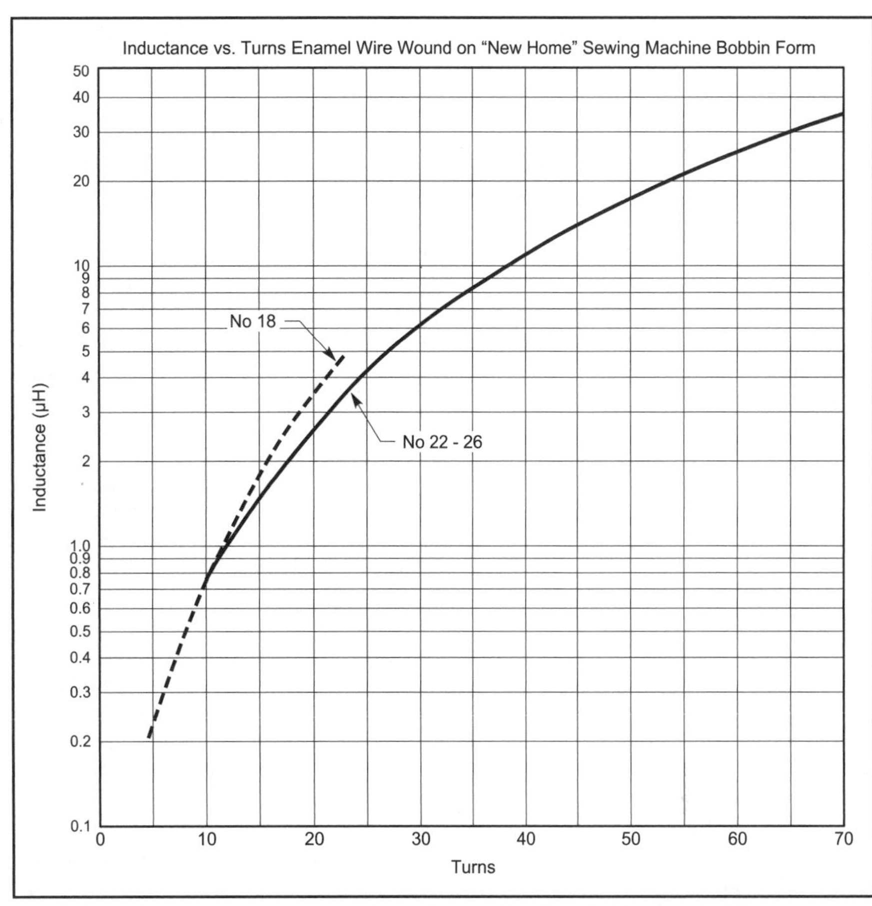

Figure 6—Inductance versus turns, for coils used in the "No Excuses" QRP transceiver.

Figure 7—Inside the "ground-plane" version. A small speaker can be seen mounted to the cover, as well as the "remoted" display for the UniCounter (see text) and the air-variable capacitors.

Figure 8—Inside the printed circuit version. An octal tube socket accommodates two crystals; these are switch selected. The UniCounter can be seen mounted on the front panel.

to control the receiver's frequency the old-fashioned way. The final version uses an octal tube socket that holds two old-style FT-243 crystals and a switch to quickly change between them. HC6/U crystals with a suitable socket can be used. If the modern HC-18 type crystals with wire leads are used they can be soldered in without a socket.

The antennas and keys at this station are set up for use with tube-type radios, so a large SO-239 antenna connector and ¼-inch key and headphone jacks were used to stay compatible with the rest of that equip-ment. A smaller BNC or a less costly RCA-type phono connector for the antenna can be used, as well as the smaller 3.5-mm jacks for the key and headphones. They are listed in the parts list—see Table 1.

Modifications

This transceiver was run from a well-regulated 13.5 V dc supply so there was no need to re-regulate the voltage to the receiver frequency control pot, R22. If mobile power, battery power or an unstable dc source is used, a 10 V Zener diode or a 78L08 regulator to "clamp" that voltage should be employed.

Several hams have suggested using easier to find 3.58 MHz crystals instead of the 4 MHz crystals. That wouldn't present a problem; however, the receiver frequency would not read out nicely on the counter. The UniCounter has a programmable offset that would solve that problem, but then the counter wouldn't read right while transmitting.

Forty meters can be called the "7/24 band." Except during a rare solar storm it's open for business day and night, summer and winter, during all 11 years of the solar cycle. It's doubtful that can be said of any other HF ham band. Still, some folks have asked if the "No Excuses" rig could be put on other bands.

For 80 meters you'd have to add some turns to the RF coils to move the tuned circuits to 3.5 MHz and use fewer turns on L5 to move the oscillator to 7.5 MHz. For 30 meters the local oscillator should run at 6.1 MHz and the RF coils should have fewer turns to cover 10.1 MHz. The sidebar provides details for building a 30-meter version of this project.

Considering my experiences with "varicap" tuning, it would be preferable to use an air variable capacitor in the local oscillator for operation on frequencies other than 7 MHz. You shouldn't attempt operating at 14 MHz or higher with this design, as that would require frequency doubling in the transmitter and a rather different circuit. It would also be pushing the limits of these common garden-variety transistors.

On the Air

How does the "No Excuses" rig perform on the air? The receiver is quite sensitive if you use a proper antenna. Most anything can be heard on the receiver that a modern, solid-state transceiver can hear. It doesn't do that well with short, random-length wires and other nonresonant antennas, so use a proper antenna and cut it to frequency.

Overall, the receiver tuning is more critical and the QRM filtering isn't quite as good as a modern commercial receiver. The transmitter gets consistent praise for having a clean, stable signal with good keying characteristics.

If you are new to low-power operation you may experience some frustration when the station calling CQ always seems to answer someone else with a stronger signal. One just has to learn to "tail-end" ongoing QSOs or call on an open spot near the 40 meter QRP calling frequency of 7040 kHz.

This little transceiver has afforded many hour-long ragchews and 599 reports from stations within a 500-mile radius, using a dipole antenna at 25 feet. Contacts with the West Coast (from near Detroit) tend to last about 10 minutes and draw 459 reports, but they are not rare. And four or five times, very late on a cold winter night, I've worked Western Europe on the low end of 40 meters.

I hope your interest has been sparked. Whatever you do, don't wait to get started. Build this simple—yet useful—circuit to put your own homebrew station on the air. You'll be surprised with its capabilities!

Notes

[1]R. Stone, KA3J, "The UniCounter—A Multi-purpose Frequency Counter/Electronic Dial," *QST*, Dec 2000, pp 33-37.

[2]A printed-circuit board is available from FAR Circuits, 18N640 Field Ct, Dundee, IL 60118-9269; tel 847-836-9148. Price: $12.

[3]A circuit board pattern and a parts placement guide are available on the *ARRLWeb* at **www.arrl.org/qst-binaries/NoExcusesXcvr1202.pdf**.

[4]See Note 2.

Dan Metzger, K8JWR, was licensed as a high-school freshman in 1958 and has held the same call sign ever since. He holds a bachelor's degree in electrical engineering and a master's degree in education from the University of Toledo (Ohio). He has held engineering positions at Magnavox, Toledo Scales and Owens-Illinois, and for the past 35 years has been teaching electronics at Monroe County Community College (Michigan).

Dan has written six textbooks, including Electronics Pocket Handbook, Electronics For Your Future, *and* Microcomputer Electronics. *His construction articles have appeared in* Electronics World, Radio-Electronics *and* Popular Electronics *magazines. As this article implies, he is active primarily on 40-meter CW. Dan can be reached at 6960 Streamview Dr, Lambertville, MI 48144,* **dmetzger@monroe.lib.mi.us**.

By Dave Benson, NN1G, and George Heron, N2APB

From *QST*, March 2001

The Warbler—A Simple PSK31 Transceiver for 80 Meters

Small and inexpensive, this transceiver is packed with fun!

JOE BOTTIGLIERI, AA1GW

There's no doubt that PSK31 has taken the Amateur Radio community by storm! In fact, *tidal wave* might be a more fitting description! In this Internet age, the enjoyment and satisfaction of using your computer and an HF transceiver to communicate using this reliable and low-bandwidth digital mode goes beyond words. PSK31 has been rekindling the interest and excitement in hams of all ages, and is drawing new amateurs into the ranks because of its simplicity and the appeal of modern technology. Now, the low-cost entry and high success rate for those trying PSK31 for the first time has been enhanced by Dave (NN1G) Benson's inexpensive PSK31-ready transceivers, the latest of which is described here.[1, 2] When used in conjunction with innovative PC software such as *DigiPan*, hams can have solid contacts on any HF band.[3]

Even more astounding, PSK31 seems to be providing the means for a rebirth of an old way of communicating for us hams. We're not referring to the data modulation/demodulation techniques of SSB. Nor are we alluding to this mode's ability to pack dozens of active QSOs simultaneously into the same bandwidth as a single SSB QSO. What we're talking about is the *real use* of the spectrum. PSK31 is providing a way for hams of all ages to gather with record ease and efficiency around new watering holes to communicate as friends and club members.

Warbler Opens 80 Meters to Low-Cost PSK31

The PSK-80 is the newest PSK-capable transceiver design of Dave Benson, NN1G. The New Jersey QRP Club, whose members are kitting the rig, dubbed it the "Warbler." This very low-cost 80-meter transceiver provides a way for friends, club members, schoolmates and ham relatives located within a 200-mile (or greater) radius to have solid, enjoyable, lively contacts on a regular basis. The natural propagation characteristics of 80 meters offers PSKers a way to have regular roundtable QSOs and club get-togethers on the air during the evening hours. You've probably heard of (and may have participated in) CW or SSB nets for traffic handling, weather tracking, used-equipment auctions and so on. The same net activities are now taking place using PSK31, building on the same strengths of this digital mode.

Pockets of 80-meter PSK31 activity have been springing up with increasing frequency throughout the country, due in great part to the popularity of the Warbler. Hams in Denver led by Rod Cerkoney, NØRC, have started some Rocky Mountain Warbler group-build sessions to help others get on the air with this mode. QRPers in northern California, led by Bill Jones, KD7S, and Doug Hendricks, KI6DS, started a Sunday evening weekly "ragchew" session called the Western Warbliers. New Jersey QRPers are on the air nightly with their Warblers and every Sunday night with a club meeting. QRPers in Atlanta are starting their own group-build of the 80-meter kit. Veteran PSKer Ken Hopper, N9VV, in Chicago, is one of the biggest on-the-air promoters of PSK31. We, the authors of this article, can be found most evenings operating around 3580.5 kHz.

Warbler Activity

The map in Figure 1 shows the distribution of current 80-meter Warbler PSK31 activity throughout the country. The red circles indicate a 200-mile radius of solid contacts. As you can see, strong areas of PSK31 activity are in northern California, the Northeast, Chicago and Atlanta. Canada is also quickly coming on as a strong PSK31 player on 80 meters. The areas of heaviest overlap offer the highest density of PSK31 activity, hence the greatest possibility of success for newcomers to this mode. We know that there's been some success in attracting new blood to HF: Marc Ziegler, W6ZZZ, of Los Gatos, California, reported making his first-ever HF contact using a Warbler!

The 200-mile radius of solid copy 80-meter propagation, though, gets bigger during the winter months. KD7S in northern California has reported increasingly better contacts with Derry, VE7QK, of BC, Canada. Phil Wheeler, W7OX, in California, has been in regular contact with a station in Utah. Doug Hendricks, KI6DS, in Dos Palos, California, and co-leader of the immensely popular NorCal QRP Club, reports "I worked Bill, KD7S, in Sanger; Dave, AB5PC, in Fresno; Ben, NW7DX, near Seattle, and Ron, K7UV, in Brigham City, Utah." Phil, W7OX, in

[1]Notes appear on page 2-25.

Los Angeles, reports partial copy (including a complete call sign) of NN1G's signal from Connecticut.

Although the map represents only 80-meter PSK activity with the Warbler, there is an increasing amount of non-Warbler PSK31 activity springing up as well. It seems that many PSK31 operators using other hardware and software equipment up on the higher bands are moving down to play with the Warblers on 80 meters. These higher-band PSKers are seeing the proliferation of Warblers as fertile new territory for ragchews, contests, experiments and propagation-favorable local communications. At any given time during the evening, we see QSOs in progress outside the Warbler passband, showing us that Warblers are facilitating a growth in 80-meter activity.

Local Communication Opens Again!

Remember when you had to go to your monthly radio club meeting to hear all the

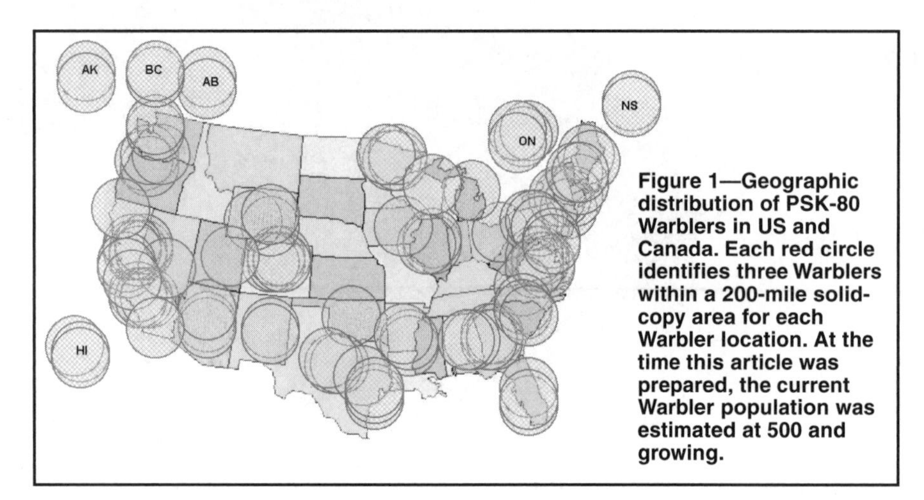

Figure 1—Geographic distribution of PSK-80 Warblers in US and Canada. Each red circle identifies three Warblers within a 200-mile solid-copy area for each Warbler location. At the time this article was prepared, the current Warbler population was estimated at 500 and growing.

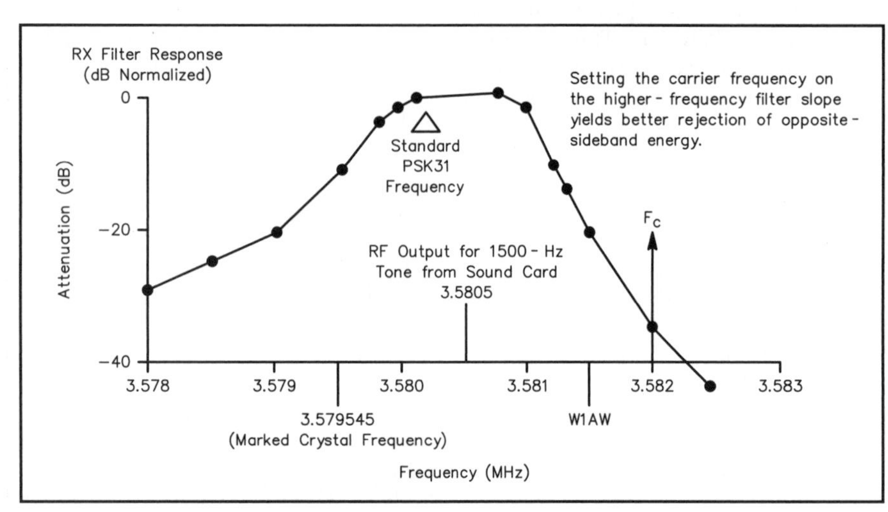

Figure 2—Crystal-filter passband response.

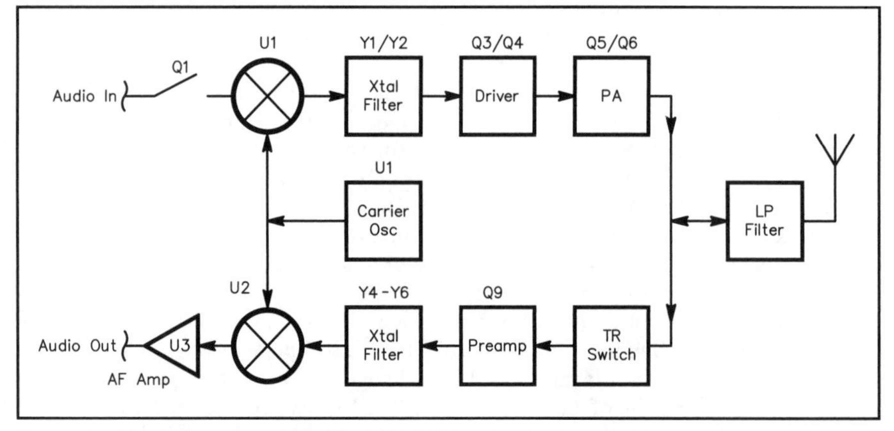

Figure 3—Block diagram of the Warbler D-C transceiver.

Figure 4—Schematic of the Warbler 80-meter D-C transceiver. Unless otherwise specified, resistors are 1/4-W, 5%-tolerance carbon-composition or metal-film units. For part numbers in parentheses, DK = Digi-Key Corp, 701 Brooks Ave S, Thief River Falls, MN 56701-0677; tel 800-344-4539, 218-681-6674, fax 218-681-3380; www.digikey.com; ME = Mouser Electronics, 958 N Main St, Mansfield, TX 76063-4827; tel 800-346-6873, 817-483-4422, fax 817-483-0931; sales@mouser.com; www.mouser.com; RF = RF Parts, 435 S Pacific St, San Marcos, CA 92069; tel 888-744-1943, 760-744-1943; www.rfparts.com; order@rfparts.com. Equivalent parts can be substituted; n.c. indicates no connection.

C1, C2, C23, C24—1 µF, 50 V electrolytic, radial leads (ME 140-XRL50V1.0)
C3—4-20 pF trimmer (DK SG20015)
C4—47 pF disc, 5% NP0/C0G (ME 140-50N5-470J)
C5—68 pF disc, 5% NP0/C0G (ME 140-100N5-680J)
C6-C8, C15-C18—33 pF disc, 5% NP0/C0G (ME 140-50N5-330J)
C9, C14, C101-C106, C108-110, C112—0.01 µF disc (ME 140-50Z5-103M)
C10—330 pF disc (ME 140-50S5-331J)
C11, C12—0.001 µF NP0/C0G monolithic (ME 581-UEC102J1)
C13—100 pF disc ceramic,5%
C20—0.022 µF monolithic (DK P4953)
C21, C107—0.1 µF monolithic (DK P4924)
C22—150 pF disc (ME 140-50S5-151J)
C111, C113—47 µF, 25 V electrolytic, radial leads (ME 140-XRL25V47)
D1, D2—7.5 V, 500 mW Zener, 1N5236B (DK 1N5236BDICT)
D3-D5—1N4148 (DK 1N4148DICT)
D6—1N4001 (DK 1N4001DICT)
J1, J2—3.5-mm 3-circuit jack, PC board mount (ME 161-3501)
J3—DB9, PC board mount (Jameco 104951)
J4—Dc power jack, 2.1×5.5 mm, PC board mount (ME 163-5004)
J5—BNC female, PC board mount (Jameco 146510)
L1—6.8-µH RF choke (ME 43LS686)
L2—23 turns #24 solid, insulated wire on a T37-2 core
L3—22-µH RF choke (ME 43LS225)
Q1, Q7—2N7000 N-channel enhancement-mode FET (DK 2N7000)
Q2-Q4—2N4401 NPN (DK 2N4401)
Q8—2N3906 PNP (DK 2N3906)
Q5, Q6—2SC2166 or 2SC2078 NPN RF power (RF)
T1—4 trifilar turns #24 solid insulated wire on an FT37-43 core
T2—Pri: 4 bifilar turns #24 solid insulated wire; sec: 8 turns, #24 enameled wire on an FT37-43 core
U1, U2—SA612A double-balanced mixer/oscillator
U3—LM1458N or MC4558N dual op amp (DK LM1458N)
U4—LM393N dual differential comparator (DK LM393N)
Y1-Y6—3.579-MHz crystal, series-resonant, HC-49/U holder (DK X011)
Misc: P1—2.1/ 5.5 mm power plug, heat sinks (DK HS106)

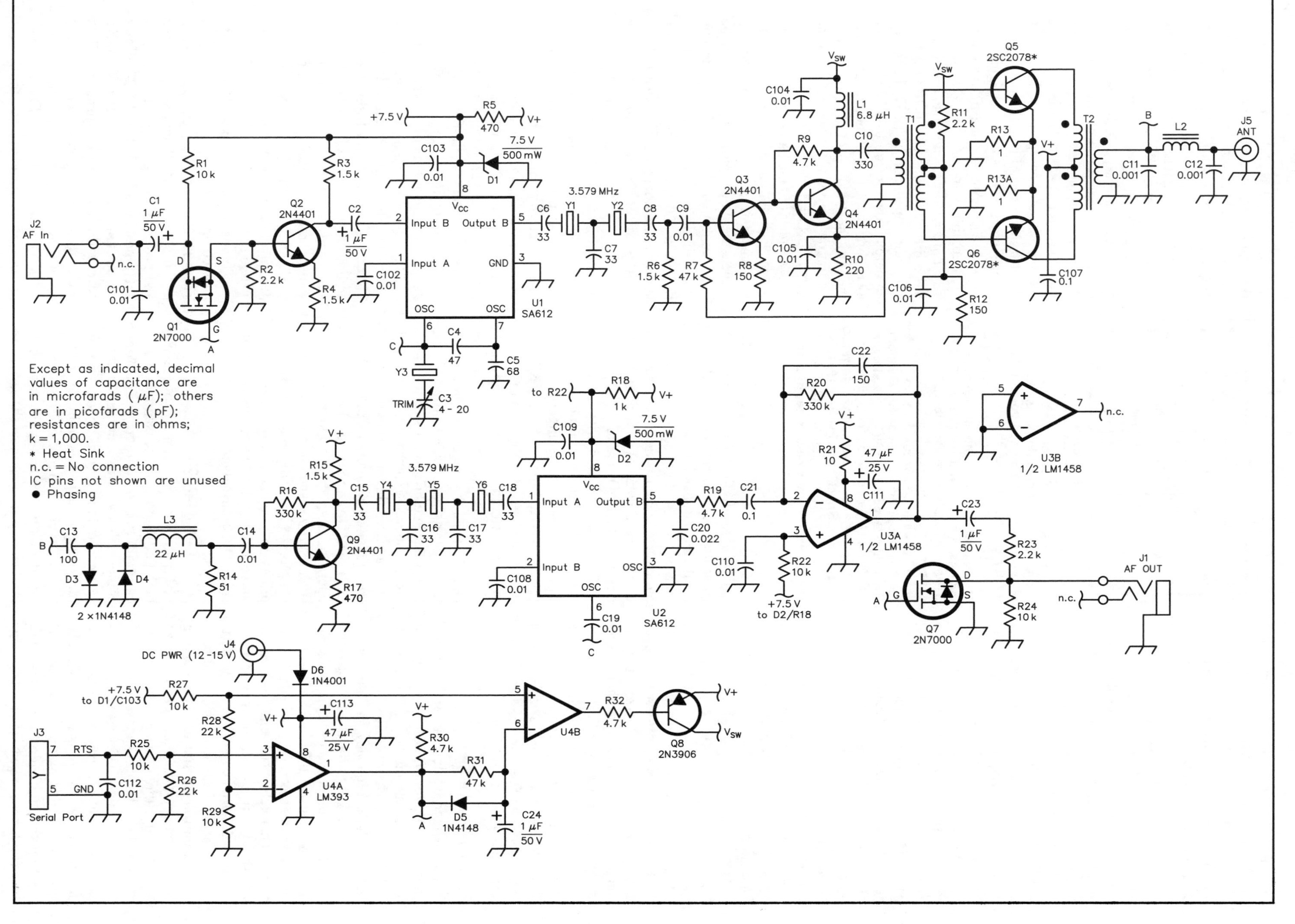

Except as indicated, decimal values of capacitance are in microfarads (μF); others are in picofarads (pF); resistances are in ohms; k = 1,000.
* Heat Sink
n.c. = No connection
IC pins not shown are unused
● Phasing
J2 AF In
J5 ANT
J1 AF OUT
J3 Serial Port
J4 DC PWR (12-15 V)
U1 SA612
U2 SA612
U3A 1/2 LM1458
U3B 1/2 LM1458
U4A LM393
U4B
Q1 2N7000
Q2 2N4401
Q3 2N4401
Q4 2N4401
Q5 2SC2078*
Q6 2SC2078*
Q7 2N7000
Q8 2N3906
Q9 2N4401
3.579 MHz

latest info about new rigs, swap meets and things? The Internet now supplants a lot of that need, but the scale is so wide that you might as easily chat with someone in Spain about some parts you need, as opposed to someone in the next town. In some cases, this is fine, but the camaraderie of local club members can better be achieved through PSK31's local communications capabilities, and that's just what many folks are doing!

Perhaps leading the pack are the Western Warbliers in California. This group had a head start because kits were first distributed at a symposium hosted by the NorCal QRP club in October 2000. These operators are maximizing that 200-mile radius of solid communications to help bring others into the fold, to help find parts and just have some regular ragchews at QRP levels. They're contributing application notes, tips, techniques and circuit improvements for other PSKers around the country. The Internet, of course, has enabled this sharing of information, and these notes are maintained by the New Jersey QRP at their project Web site.[4]

Along with the Western Warbliers, the New Jersey QRP club members have shown that it doesn't take a full-blown 80-meter antenna farm to get out with this mode. Sure, "the bigger the better" usually applies, but reduced-size antennas can put you on 80 meters without requiring an acre-size backyard![5] The commercially available verticals also combine effective operation with a low footprint. Dave, NN1G, uses a dipole about 15 feet high for all the 80-meter PSK work he's done to date. Not bad!

So, What's a Warbler?

Last winter, while preparing for a talk on PSK31, Dave was casting about for a low-parts-count means of handling a PSK31 signal. He noted that the PSK31 watering-hole frequency on 80 meters is at 3580.15 kHz, darn close to the color-burst frequency of 3579.545 kHz. After an intensive thirty minutes of cut-and-paste engineering, a schematic was born. Remarkably, this early schematic withstood further evaluations and refinements without much growth in the parts count. The hardware design started with an evaluation of simple filters using color-burst crystals. Figure 2 shows an example of a three-crystal filter and its measured passband response.

The asymmetric skirt response is typical of a crystal ladder (Cohn) filter—the upper-frequency slope is steeper. We take advantage of this by setting the carrier/BFO on the high side of the passband. This yields better rejection of W1AW signals and results in LSB operation. The filter uses series-resonant crystals. As a result, the passband is actually *above* the marked crystal frequency. The BFO is pulled to the high side of the passband using a small value of capacitance in series with the BFO crystal.

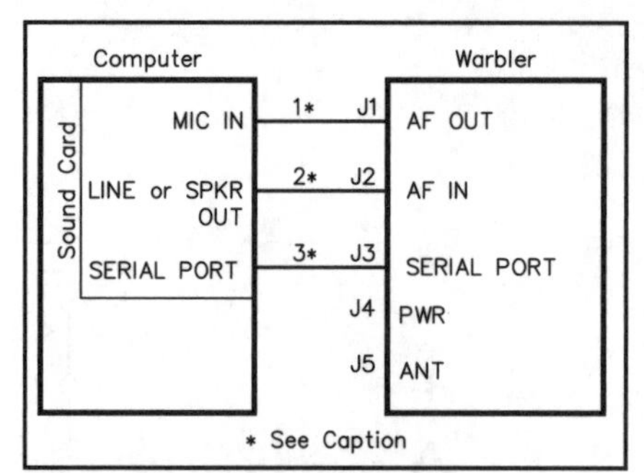

Figure 5—Pictorial of the Warbler/computer interconnections. Use interconnecting cables that suit your equipment requirements. As shown here, three-conductor cables are used at 1 and 2; they are equipped with 3.5-mm stereo connectors at each end (RadioShack 42-2387). Serial port cable 3 has female DB9 connectors at each end (RadioShack 26-117 or Jameco 25700). Jameco Electron-ics, 1355 Shoreway Rd, Belmont, CA 94002; tel 650-592-8097, domestic fax, 800-237-6948, international fax, 650-592-2503; info@jameco.com; www.jameco.com; RadioShack.com, PO Box 1981, Fort Worth, TX 76101-1981; tel 800-843-7422; fax 800-813-0087; www.radioshack.com.

So what do we do with a filter that works right at the operating frequency? Consider the Neophyte direct-conversion (D-C) receiver,[6] the epitome of simplicity: It consists of nothing more than a product detector/oscillator and an AF amp. Add a crystal filter to its front end, and it's still a D-C receiver, but its selectivity and its resistance to (out-of-passband) intermodulation distortion (IMD) are considerably improved. On transmit, adding such a filter to the output of a balanced modulator alters its output from a DSB signal to an SSB signal—right at the operating frequency. A block diagram of such a setup is shown in Figure 3. Pretty simple, eh? Naturally, reducing this simplicity to practice always seems to involve adding a few components, but it's still a D-C transceiver.

Circuit Description

Figure 4 is the schematic of the Warbler. Let's start our discussion of the circuit with the transmitter. Audio from a computer's sound-card output (LINE OUT) is connected to J2. Q1 is conducting during transmit and passes audio and dc bias to Q2. Several hundred millivolts of audio are applied to mixer U1 to generate DSB energy at a (suppressed) carrier frequency of 3582 kHz. Y1 and Y2 and C6 through C8 remove the bulk of the unwanted sideband energy. Q3 and Q4 amplify the remaining SSB signal. The output of Q4 includes an L network (L1 and C10) that matches the driver output impedance to the PA, Q5 and Q6.

Things begin to look a little different around the PA—a push-pull stage. The two halves of the PA show equal gains on their respective half cycles of conduction; this balance pays off in rejecting second-harmonic energy. A trifilar-wound input transformer (T1) splits the driving signal into two out-of-phase signals fed to Q5 and Q6. Another multifilar winding (T2) combines the Q5 and Q6 outputs. T2's third winding is done separately and has a different turns count. It's set for a collector impedance of 12 Ω. In theory, an output power of up to 6 W should be available from this stage. In practice, though, the IR drop of R13/R13A and saturation effects of Q5 and Q6 limit the output to 4 or 5 W PEP.

As a result of the balance provided by the push-pull configuration and the improvement in second-harmonic performance, the output harmonic filter can be considerably simplified. In this design, it's a single-section network. Thanks to the 15 to 20 dB of second-harmonic suppression inherent in the push-pull PA configuration, the minimum harmonic rejection for this design is 33 dB, compliant with current FCC regulations.

C13, D3 and D4 and L3 act as a TR switch and provide a measure of front-end selectivity for the receiver. Q9, a preamplifier stage, provides a gain of 10 dB. In addition to providing gain, Q9 offers a flat 1.5-kΩ source impedance to the crystal filter. The initial Warbler design lacked this stage and the filter passband shape was poor. Y4 through Y6 and the associated capacitors deliver the passband selectivity shown in Figure 2. Mixer U2 converts the filter output to audio, where it's low-pass filtered and amplified by U3. R23, R24 and Q7 provide a muting function to prevent feedback during transmit; this subject is described in more detail in a recent *QST* article (see Note 1).

Comparators U4A and U4B and related components provide TR sequencing. The RTS signal of the computer's serial port is a negative voltage during receive and positive during transmit. A turn-on delay produced by R31 and C24 allows the dc voltages around Q2 to stabilize before the transmitter turns on. This minimizes the transient energy emitted during the transition to transmit. Catch diode D5 serves a similar function during the return to receive by shutting off

The Considerate Operator's Guide to 3580 kHz

The Warbler operates over a fixed 1-kHz slice of 80 meters: 3580 to 3581 kHz. PSK31 users aren't the only inhabitants of this portion of the band. Most notably, the Glowbugs, a community using simple gear and experimenting with crystal-controlled CW rigs, uses and monitors 3578 kHz. PSK users should adhere to the published band plan for data operation (3580 kHz and up) to minimize interference to other users of the frequency. If your transceiver is set to LSB and the dial is set much below 3582 kHz, the chances for inadvertent interference are good.—*Dave Benson, NN1G*

the transmitter bias as soon as possible.

Hookup and Alignment

The Warbler connects to your computer as shown in Figure 5. Sources for the interconnecting cables are shown. Necessary cables are available at most electronics retail outlets and other stores.

Setup

You need software to use the transceiver. If you don't already have it, download and run *DigiPan 1.5*, the most recent version (see Note 3). Once *DigiPan* is running, initialize the frequency display to 3582 kHz and select **LSB**, which places 3582 kHz at the right edge of the display. Connect an antenna and dc power (12 to 15 V) to J5 and J4, respectively. Adjust the sound-card microphone volume-control slider (**CONFIGURE | WATERFALL DRIVE** in *DigiPan 1.5*). Set the level with this control to yield blue-to-yellow speckles on the screen. This should yield a band covering approximately one-third of the computer screen width when properly adjusted.

Adjustment

There's only one adjustment on the transceiver board—trimmer cap C3. The ARRL was kind enough to furnish a calibration marker to adjust these rigs. During many of the afternoon and evening hours, W1AW is transmitting on 3581.5 kHz.[7] If you're located east of the Mississippi, you should have little trouble spotting W1AW's CW transmissions onscreen. Using a small screwdriver, simply adjust C3 until W1AW's signal is lined up under the 3581.5 tick mark of the *DigiPan* frequency display. Lacking W1AW's signal, adjust C3 to center the brightest portion of the display screen in the range of 3580.0 to 3581.0 kHz.

If you live close enough to W1AW so that its signal causes spurious traces on the display, and reducing the sound-card's microphone slide-control setting to cure this effect causes PSK31 signals to disappear into the noise, try this approach: Set the *DigiPan* start frequency to 3581.5 kHz and adjust C3 so that W1AW's signal is zero beat at the right extreme of the display. This takes advantage of the low-frequency rolloff characteristics of the receiver's audio amplifier to knock the signal down to manageable levels. Dave, NN1G, lives about two miles from W1AW and its signals are *very* strong there. Once this adjustment was performed though, Dave could copy PSK31 signals without difficulty.

Transmit Adjustment

In *DigiPan*, select **Mode** and click on **Tune**. This places the transceiver in transmit mode with a 100% duty cycle. Click on the speaker icon in *Window*'s tray and advance the volume slider until the transmitter output power is set at 3 W. Although the Warbler's PA stage can be driven harder for more output, the additional power comes at the expense of poorer IMD performance. If you don't have a wattmeter, you can effectively accomplish the job using a 50-Ω resistive load and peak-voltage detector.[8]

Operation

Clicking your computer's mouse cursor over the typical "railroad-track" PSK31 signal should cause text to begin appearing in *DigiPan*'s upper text window. Clicking on **T/R** in the *DigiPan* menu switches to transmit and your typed text in the lower window streams out on the air.

Do It, Use it, Enjoy PSK31!

Just when it seemed to some that the flames of excitement in ham radio were dwindling to smoldering embers, along has come a new mode of communications to stoke us up again. Overwhelming evidence is showing us that folks all over the country are having *tons* of fun building and operating PSK radio equipment such as the Warbler!

QRPers and high-power operators alike are pulling others into this digital aspect of the hobby by conducting coordinated construction and instruction classes, forming statewide nets on 80 meters for club and special-interest support groups, and plain old ragchewing. There isn't a night that goes by here on the East Coast without having up to a half dozen QSOs going on at once throughout the evening hours.

The fun doesn't stop here! A number of experimenters are using DSP evaluation kits instead of the computer/sound-card approach, so we may see PSK31 terminals that cut the tether to the PC. This will enable an even more portable and lower-cost operation for PSK31.

Start enjoying PSK31! Get a local PSK31 ragchew net going in your state. Put on a demo for the local high school science class showing how much fun can be had communicating *without* using the Internet. Get a PSK31 transceiver group-build going with your ham club. No matter how you approach it, do it, use it and have fun with PSK31!

Acknowledgements

Thanks to the New Jersey QRP Club, the Western Warbliers and many others for their enthusiastic support and contributions to this activity.

Notes

1 Howard "Skip" Teller, KH6TY, and Dave Benson, NN1G, "A Panoramic Transceiving System for PSK31," *QST*, Jun 2000, pp 31-37.

2 The New Jersey QRP Club offers a complete kit of parts including a PC board, all on-board components and assembly instructions. Price: $45, including shipping in the US and Canada; foreign orders add $5. Make your check or money order payable to George Heron, N2APB. Send your order to George Heron, N2APB, 2419 Feather Mae Ct, Forest Hill, MD 21050. Please allow two to four weeks for delivery. All sales proceeds benefit club-sponsored public activities.

3 *DigiPan* is available for free from **members.home.com/hteller/digipan/**. The current version is 1.5. Links to additional software products may be found at **psk31.com**.

4 Warbler project updates and errata are maintained at **www.njqrp.org/warbler/kitnotes.html**.

5 Loaded verticals such as the PM-1 offered by Vernon Wright, W6MMA, is one example of a suitable compact antenna; **www.superantennas.com**. See also Robert Johns, W3JIP, "A Ground-Coupled Portable Antenna," *QST*, Jan 2001, pp 28-32.

6 John Dillon, WA3RNC, "The Neophyte Receiver," *QST*, Feb 1988, pp 14-18.

7 See the W1AW Operating Schedule in this issue.

8 Chuck Hutchinson, K8CH, ed, *The ARRL 2001 Handbook* (Newington: ARRL, 2000), p 26-11.

Dave Benson, NN1G, is well known to QST *readers. His life has been captured on the installment plan in prior issues of this magazine. You can contact Dave at 80 E Robbins Ave, Newington, CT 06111;* **nn1g@arrl.net**.

George Heron, N2APB, plays a lead role in the New Jersey QRP Club and has been active in the QRP community throughout the last decade. He organizes the annual Atlanticon QRP Forum for the NJQRP and edits and publishes the club's quarterly journal QRP Homebrewer. *An inveterate homebrewer by nature, with strengths in software and digital design, N2APB's latest project is the design of a PC-less, single-board controller for portable operation using PSK31. Contact George at 2419 Feather Mae Ct, Forest Hill, MD 21050;* **n2apb@amsat.org**.

Chapter 3

Transmitters

By Ed Hare, W1RFI

From *QST*, March 2000

The Tuna Tin 2 Today

Ham radio lost its kick? Go QRP with this weekend project! Worked All States with a 40-meter half-watter? **You betcha!**

In the 1970s, the late Doug DeMaw, W1CER/W1FB, ARRL Technical Editor, was one of several Headquarters staff who published homebrew projects, many with a QRP twist. One of those was a simple, two-transistor 40-meter transmitter that used a tuna can as the chassis. Dubbed the "Tuna Tin 2," it was a popular project, introducing many hams to homebrewing and QRP. A series of events, some quite amazing, have come together to keep the magic alive—the original Tuna Tin 2, built in the ARRL Lab, is still on the air and articles, Web pages and kits are available for this famous rig. Some have dubbed the Tuna Tin 2 revival as "Tuna Tin 2 mania"—an apt term to describe the fun that people are still having with this simple little weekend project.

This article has been edited from the original, written by DeMaw and published in the May 1976 QST. You can download a copy in Adobe PDF format from the ARRL Members-Only Web site at: **http://www.arrl.org/members-only/extra/features/1999/0615/1/tt2.pdf**. Some of the original parts are no longer available, so modern components have been substituted, using values that were featured in a column in QRP with W6TOY on the ARRL Web Extra. I think that Doug would have been pleased to see just how popular that little rig still is, almost a quarter century after he first designed it and built it in the ARRL Lab.— *Ed Hare, W1RFI, ARRL Laboratory Supervisor*

The original Tuna Tin 2

Workshop weekenders, take heart. Not all building projects are complex, time consuming and costly. The TunaTin 2 is meant as a short-term, gotogether-easy assembly for the ham with a yen to tinker. Inspiration for this item came during a food shopping assignment. While staring at all of the metal food containers, recollections of those days when amateurs prided themselves for utilizing cake and bread tins as chassis came to the fore. Lots of good equipment was built on make-do foundations, and it didn't look ugly. But during recent years a trend has developed toward commercial gear with its status appeal, and the workshop activities of many have become the lesser part of amateur radio. While the 1-kW rigs keep the watt-hour meters recording at high speed, the soldering irons grow colder and more corroded.

A tuna fish can for a chassis? Why not? After a few hours of construction, 350 milliwatts of RF were being directed toward the antenna, and QSOs were taking place.

Maybe you've developed a jaded appetite for operating (but not for tuna). The workshop offers a trail to adventure and achievement, and perhaps that's the elixir you've been needing. Well, Merlin the Magician and Charlie the Tuna would probably commend you if they could, for they'd know you were back to the part of amateur radio that once this whole game was about—creativity and learning!

Parts Rundown

Of course, a tunafish can is not essential as a foundation unit for this QRP rig. Any 6 1/2-ounce food container will be okay. For that matter, a sardine can may be used by those who prefer a rectangular format. Anyone for a Sardine-2? Or, how about a "Pineapple Pair?" Most 6 1/2-ounce cans measure 3 1/4 inches in OD, so that's the mark to shoot for. Be sure to eat, or at least remove the contents before starting your project!

Although the original project used all RadioShack parts, some of the parts are no longer stocked. The 2N2222A transistor is

Kits and Boards

While the original Tuna Tin 2 can be built from scratch, surprisingly, printed-circuit boards and kits are still available.

The September 16, 1999 *QRP with W6TOY* column in the *ARRL Web Extra* featured a modern version of the Tuna Tin 2[1]. FAR Circuits can supply the printed circuit for W6TOY's version (not built on a tuna tin) as well as the original design PC board.[2]

Those who want to buy everything all in one place can buy a complete kit, including PC board from the NJ-QRP Club[3]. Send a check for $12 postpaid to George Heron, N2APB, New Jersey QRP Club, 2419 Feather Mae Ct, Forest Hill, MD 21050. Doug Hendricks, KI6DS also designed a version of the Tuna Tin 2, for the Northern California QRP Club (NorCal)[4].

JOE BOTTIGLIERI, AA1GW

W6TOY's version of the Tuna Tin 2 design—without the tuna can.

[1]See: **http://www.arrl.org/members-only/extra/features/1999/09/16/1/**.

[2] FAR Circuits, 18N640 Field Ct, Dundee, IL 60118-9269, tel 847-836-9148; **http://www.cl.ais.net/farcir/**

[3]NJ-QRP Club, contact: George Heron, N2APB, 2419 Feather Mae Ct, Forest Hill, MD 21050; **n2apb@amsat.org**; **http://www.njqrp.org/**. NJ-QRP has a section of their Web site devoted to the Tuna Tin 2 revival. See **http://www.njqrp.org/tuna/tuna.html**.

[4]Northern California-QRP Club (NorCal), 3241 Eastwood Rd, Sacramento, CA 95821; tel 916-487-3580; **jparker@fix.net**; **http://www.fix.net/NorCal.html**. Like the NJ-QRP Club, NorCal also has a Tuna Tin 2 revival page at: **http://www.fix.net/~jparker/norcal/tunatin2/tunatin.htm**.

widely available. The original coils have been replaced with inductors wound on toroidal cores. Printed circuit boards are available from several sources and the NJ QRP Club is offering a complete kit of parts. (See the sidebar "Kits and Boards".)

The tiny send-receive toggle switch is a mite expensive. The builder may want to substitute a low-cost miniature slide switch in its place. A small bag of phono jacks was purchased also, as those connectors are entirely adequate for low-power RF work.

Finding a crystal socket may be a minor problem, although many of the companies that sell crystals can also supply sockets (you can locate a number of crystal manufacturers and distributors on the ARRL TISFIND database at **http://www.arrl.org/tis/tisfind.html**). Fundamental crystals are used in the transmitter, cut for a 30-pF load capacitance. Surplus FT-243 crystals will work fine, too, provided the appropriate socket is used. If only one operating frequency will be used, the crystal can be soldered to the circuit board permanently. Estimated maximum cost for this project, exclusive of the crystal, power supply and tunafish, is under $20. The cost estimate is based on brand new components throughout, inclusive of the

The Tuna Tin 2 on the Road

Those who've read our on-line publication, the *ARRL Web Extra,* probably saw the article that appeared in the June 15th edition titled "The Tuna Tin 2 Revival." This article told an incredible tale of how the original Tuna Tin 2 was lost from the ARRL Lab and was found years later in a box of junk under a fleamarket table in Boxboro, Massachusetts. The Tuna Tin 2 was refurbished by Bruce Muscolino, W6TOY, and put back on the air by me on June 4, 1999. Since that time, over 400 hams have had the pleasure of working the original Tuna Tin 2, some using their own Tuna Tin 2 rigs built in the 70s (or built anew from the available kits).

California Dreamin'

After making about a hundred contacts from home, I was asked to attend an IEEE meeting in Long Beach, California. My sister, Bev, lives in the area, so I planned a week-long visit. I tossed the Tuna Tin 2 and a G5RV into my suitcase, hoping to give a few West Coast hams a chance to make a contact with the original.

After all the hugs and kisses, I explained to my sister what I was up to. She grinned, remembering the wild days of my youth, climbing trees to string wires all over our property, back when I was WN1CYF. As I looked over the site, though, I was not too hopeful; about the best I thought I could do would be to try a random wire around the balcony, maybe risking a run over to a small tree or two. I looked roofward and sighed, "Gee, it would be nice to get an antenna up on the roof." She made a quick call to Debbie, the building manager and close friend, who winced painfully and said, "Don't fall off!" and, in a classic Schultz accent, "I know nothing!"

We took the G5RV up to the emergency roof access, walked boldly out, and I proceeded to string the antenna up while Bev stood guard. I got the antenna up, dropped the feedline past the upstairs apartment balcony and hoped for the best.

Sure enough, the "antenna police" were on alert—the tenant right below us heard the noise and wondered what was going on. Just as we got back to the apartment, the phone rang; it was Debbie. She told us of the complaint, told us the excuse she gave and wished us luck.

With Bev watching with great interest, I hooked up the Heath HW-8 I used as a receiver, hooked up the Tuna Tin 2, the code key and antenna tuner, and gave the band a fast listen. Signals were booming in. On June 19, I worked my first contact with the Tuna Tin 2 from the West Coast, W6PRL/QRP. Every evening, after a day of offshore fishing, Bev and I expected to find that the antenna police had confiscated the wire, but somehow, it stayed up the whole week. By the end of the week, 45 new stations were in the Tuna Tin 2 log!

Among the Monsoons and ScQRPions

I was then asked if I would be willing to attend the ARRL Arizona State Convention at Ft Tuthill. That is an annual pilgrimage for many a QRPer; how lucky could I get? I agreed, but warned the ARRL Division Director that I might spend a bit more time away from the ARRL booth than usual. In the meantime, I casually asked Joe Carcia, NJ1Q, the W1AW station manager, if he could arrange for W1AW/7/QRP to be used at the convention. After some consultation with Dave Sumner, a new QRP "first" was in the works. In the meantime, the Arizona ScQRPions[1], an Arizona QRP club, asked me if I would give a presentation at the QRP forum they sponsor at Ft Tuthill every year. I agreed, but with one condition—they had to be willing to host W1AW/7/QRP at their booth. I would have loved to be a fly on the wall as that e-mail was read!

A *great* time was had by all, but W1AW/7/QRP did not go off without a hitch. An operator error (mine) damaged the receiver (the binaural receiver, designed by Rick Campbell). The local QRPers came through, though, and several receivers were made available to the operation to finish the day. Even worse, later in the day, it looked like all was lost! During a quick test of the Tuna Tin 2, one of the resistors emitted a puff of smoke, and the power went to 0 W. I had just blown up the original Tuna Tin 2!

I did a quick troubleshooting job and identified that the output transistor had short-circuited. Special thanks go to Niel Skousen, WA7SSA, who dug into his portable junkbox. (Niel is a real ham's ham! How many hams do you know who bring their junkbox to a hamfest?) He quickly located a 2N2222A. I handed him the Tuna Tin 2 and asked him if he would mind installing it. After that W1AW/7/QRP was back on the air.

After the convention, using a borrowed receiver, I took the Tuna Tin 2 on a whirlwind tour of Arizona, although I only got to operate two

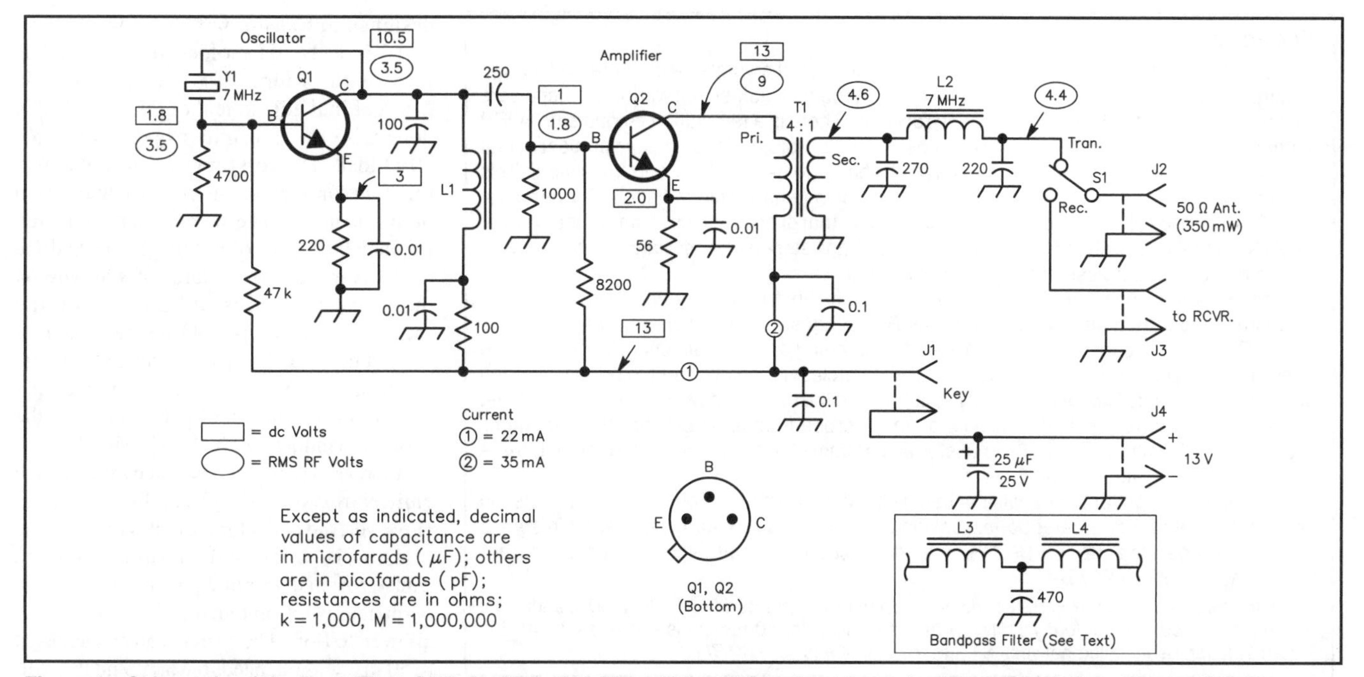

Figure 1—Schematic of the Tuna Tin 2 QRP rig. Note that the polarized capacitor shown in the schematic is an electrolytic.

J1—Single-hole-mount phono jack. Must be insulated from ground. Mounts on printed circuit board.
J2, J3, J4—Single-hole-mount phono jack. Mount on tuna tin chassis.
L1—22 μH molded inductor
L2—19 turns of #26 wire on a T-37-2 toroidal core
L3, L4—21 turns of #24 wire on a T-37-6 toroidal core
Q1, Q2—2N2222A or equivalent NPN transistor.
S1—Antenna changeover switch. Miniature SPDT toggle (see text).
T1—4:1 broadband transformer. 16 turns of #26 wire on the primary, 8 turns of #26 wire on the secondary, on an FT-37-43 toroidal core.
Y1—Fundamental crystal, 7 MHz.

nights from a campsite in Williams. I had brought along my DK9SQ[2] 33-foot portable fiberglass mast, so my antenna went up and down quickly. (Let me tell you, this is one great product. I literally put up my 40 meter inverted **V** in 5 minutes, 33 feet in the air. Taking it down was even faster.) It was monsoon season in Arizona and it rained each night. Despite the downpours, I doggedly squeezed in operating time in between thunderstorms, and added a few new ones to the log.

Hanging Out in the Park

Just two weeks later, I was off to Golden, Colorado for the Colorado State Convention (during which I got to show off the Tuna Tin 2 to the Colorado QRP Club[3]) and the trusty Tuna Tin 2 and portable mast came along with me. I scoped out the hotel area—no good. The noise level from the high-tension lines was just too high. The convention was held in a small park, so after the confab ended I walked a mile back to the hotel, loaded up the Tuna Tin 2, batteries, key, antenna and mast, and trekked back to the park. Fifteen minutes later, the antenna was standing proud and tall, and I made my first CQ. A security guard stopped by, and fearing the worst, I explained what I was doing. "Okay," she said, and drove away. A few minutes later I had a nice surprise—Rod Cerkoney, N0RC, showed up to operate with me!

The Tuna Tin 2 came back home, and I got it ready for the QRP Extravaganza Weekend (my name for it) on Halloween, with the QRP-ARCI/ARRL "Black Cat" party and the NorCal Zombie Shuffle operating event. You can read that tale in Rich Arland's "QRP Power" column in this issue.

Are We Having Fun Yet?

Did I have fun? Do you need to ask? I guess I was just in the right place at the right time, and have been privileged to be the center of all this Tuna Tin 2 activity. What is important to me, though, is that the magic that DeMaw created in the ARRL Lab still lives. It has, in fact, it has taken on a life of its own.

The Tuna Tin 2 will be on the air on 40 meters a lot over the rest of the winter, spring and summer. You'll hear it from W1RFI, from W1AW, and possibly some other station locations. I do have one more "special event" in the works, but I am sworn to secrecy. The Tuna Tin 2 will play a part in it. I won't tell you what call it will use, but I will say that you will know it when you hear it. And when you do, you will know that the magic is still alive.

I hope that lots of hams build some of the various Tuna Tin 2 replicas, and that they get a chance to work the original. I will do my best to keep it on the air. I am sure that Doug DeMaw would approve.—*W1RFI*

[1]See the Arizona ScQRPions site on the Web at: **http://www.extremezone.com/~ki7mn/sqrppage.htm.**
[2]The DK9SQ mast is available for $99 plus $5 shipping and handling from Kanga US, 3521 Spring Lake Dr, Findlay, OH 45840; tel 419-423-4604; **kanga@bright.net**; **http://www.bright.net/~kanga/kanga/**.
[3]Colorado QRP Club, PO Box 371883, Denver CO 80237-1883; **rschneid@ix.netcom.com**; **http://www.cqc.org/**.

Ed Hare, W1RFI, operating the TT2 from his sister's apartment in Los Angeles.

TT2 Performance

Keying quality with this rig was good with several kinds of crystals tried. There was no sign of chirp. Without shaping, the keying is fairly hard (good for weak-signal work), but there were no objectionable clicks heard in the station receiver. There is a temptation among some QRP experimenters to settle for a one-transistor oscillator type of rig. For academic purposes, that kind of circuit is great. But, for on-the-air use, it's better to have at least two transistors. This isolates the oscillator from the antenna, thereby reducing harmonic radiation. Furthermore, the efficiency of oscillators is considerably lower than that of an amplifier. Many of the "yoopy" QRP CW signals on our bands are products of one-transistor crystal oscillators. Signal quality should be good, regardless of the power level used.

The voltages shown in Figure 1 will be helpful in troubleshooting this rig. All dc measurements were made with a VTVM. The RF voltages were measured with an RF probe and a VTVM, The values may vary somewhat, depending on the exact characteristics of the transistors chosen. The points marked 1 and 2 (in circles) can be opened to permit insertion of a dc milliammeter. This will be useful in determining the dc input power level for each stage. Power output can be checked by means of an RF probe from J2 to ground. Measurements should be made with a 51- or 56-Ω resistor as a dummy load. For 350 mW of output, there should be 4.4 V_{rms} across the 56-Ω resistor.

Operating voltage for the transmitter can be obtained from nine Penlite cells connected in series (13.5 volts). For greater power reserve one can use size C or D cells wired in series. A small ac-operated 12- or 13-V regulated dc supply is suitable also, especially for home-station work.—*W1FB*

[Although this rig met all the Part 97 surious emission requirements when built in 1976, additional filtering is needed to meet today's rules. A bandpass filter for 40 meters is shown as an inset in Figure 1. It can be installed between S1 and the antenna jack.—*W1RFI*]

left-over parts from the assortments. Depending on how shrewd he is at the bargaining game, a flea-market denizen can probably put this unit together for a few bucks.

Circuit Details

A look at Figure 1 will indicate that there's nobody at home, so to speak, in the two-stage circuit. A Pierce type of crystal oscillator is used at Q1. Its output tickles the base of Q2 (lightly) with a few mW of drive power, causing Q2 to develop approximately 450 mW of dc input power as it is driven into the Class C mode. Power output was measured as 350 mW (1/3 W), indicating an amplifier efficiency of 70%.

The collector circuit of Q1 is not tuned to resonance at 40 meters. L1 acts as an RF choke, and the 100-pF capacitor from the collector to ground is for feedback purposes only. Resonance is actually just below the 80-meter band. The choke value is not critical and could be as high in inductance as 1 mH, although the lower values will aid stability.

The collector impedance of Q2 is approximately 250 Ω at the power level specified. Therefore, T1 is used to step the value down to around 60 Ω (4:1 transformation) so that the pi network will contain practical values of L and C. The pi network is designed for low Q (loaded Q of 1) to assure ample bandwidth on 40 meters. This will eliminate the need for tuning controls. Since a pi network is a low-pass filter, harmonic energy is low at the transmitter output. The pi network is designed to transform 60 to 50 Ω.

Ll is a 22-μH molded inductor. L2 is made with 19 turns of #26 wire on a T-37-2 core. Final adjustment of this coil (L2) is done with the transmitter operating into a 50-Ω load. The coil turns are moved closer together or farther apart until maximum output is noted. The wire is then cemented in place by means of hobby glue or Q dope

T1 is made with 16 turns of #26 wire on the primary, 8 turns of #26 wire on the secondary, on an FT-37-43 ferrite core. This is good material for making broadband transformers, as very few wire turns are required for a specified amount of inductance, and the Q of the winding will be low (desirable).

Increased power can be had by making the emitter resistor of Q2 smaller in value. However, the collector current will rise if the resistor is decreased in value, and the transistor just might "go out for lunch," permanently, if too much collector current is allowed to flow. The current can be increased to 50 mA without need to worry, and this will elevate the power output to roughly 400 mW.

Construction Notes

The PC board can be cut to circular form by means of a nibbling tool or coping saw. It should be made so it just clears the inner diameter of the lip that crowns the container. The can is prepared by cutting the closed end so that 1/8 inch of metal remains all the way around the rim. This will provide a shelf for the circuit board to rest on. After checkout is completed, the board can be soldered to the shelf at four points to hold it in place. The opposite end of the can is open.

Summary Comments

Skeptics may ch□ortle with scorn and amusement at the pioneer outlook of QRP enthusiasts. Their lack of familiarity with low-power operating may be the basis for their disdain. Those who have worked at micropower levels know that Worked All States is possible on 40 meters with less than a watt of RF energy. From the writer's location in Connecticut, all call areas of the USA have been worked at the 1/4-W power plateau. It was done with only a 40-meter coax-fed dipole, sloping to ground at approximately 45° from a steel tower. Signal reports ranged from RST 449 to RST 589, depending on conditions. Of course, there were many RST 599 reports too, but they were the exception rather than the rule. The first QSO with this rig came when Al, K4DAS, of Miami answered the writer's "CQ" at 2320 UTC on 7014 kHz. An RST 569 was received, and a 20-minute ragchew ensued. The copy at K4DAS was "solid."

If you've never tried QRP before, the first step is easy. Just contact the QRP Amateur Radio Club International (QRP-ARCI), 848 Valbrook Court, Lilburn, GA 30047-4280; **http://www.qrparci.org/.** QST

Fishy Excitement at the Meriden ARC

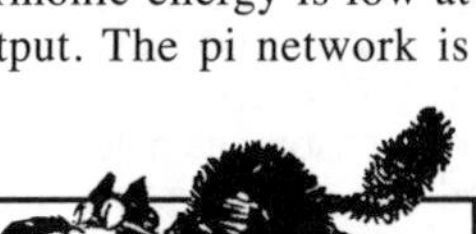

Renewed interest in the Tuna Tin 2 transceiver prompted the Meriden (Connecticut) Amateur Radio Club to build these classics as a club project. Bob Stephens, KB1CIW and Jamie Toole, N1RU secured components for 20 kits. Tim Mik, WY1U, supplied 20 cat food cans, cleaned and stripped of labels. (We had to assume that each can had, in fact, contained tuna flavor cat food. We didn't want to stray too far from the original design!) Tim also brought along his original Tuna Tin 2, which he had built as a newly licensed teenager over 20 years ago.

Several of the more experienced members were quite helpful in assisting those less knowledgeable in the arcane arts of schematic reading and toroid winding. Counting the number of turns, especially on the transformer, is not quite the simple task that it seems at first. Other tips on soldering and building in general were freely passed on from the veterans.

Honors for the first contact went to MARC president Bill Wawrzeniak, W1KKF. After finishing his rig, he brought it home, connected an antenna and almost immediately made contact with a California ham. With his new Tuna Tin 2, WY1U worked Ed Hare, W1RFI, operating the W1AW special event at ARRL HQ on Halloween. Most of the other kits were completed and put on the air over the next several weeks.

Building the Tuna Tin 2 is a terrific activity for any club. It can be completed in one or two evenings. The circuit is simple enough to provide an excellent springboard for education in electronic and RF theory without getting bogged down in too many esoteric topics. Building the kit is a great way to learn or sharpen construction skills. And, of course, there's no substitute for the pride and satisfaction of telling the station at the other end of the QSO, "RIG HR IS HMBRW TT2".—*John Bee, N1GNV,* QST *Advertising Manager*

CATS BY GIL, W1CJD

By Steve Johnston, WD8DAS

From *QST*, January 2003

The Two Tube Tuna Tin Transmitter (T5)

The tuna tin emerges once again as a classic two-tube transmitter. This easy-to-build, diminutive blowtorch can pump out 8 W with a comforting glow that transistors can't match.

My homebrew circuits are rarely original. I generally spend a couple of weeks looking through my old magazines, books and online sources, asking questions of friends and experts, combine the good ideas I find, then, maybe, start building. My design ends up being a hybrid of other peoples' projects, fine-tuned by my troubleshooting efforts. The projects used for ideas were probably a merging of yet other projects, fine-tuned by other troubleshooting efforts, and so on. A form of evolution, I suppose.

After the success of my three-tube *Secret Dream* breadboard 100 W transmitter[1] and five-tube HF receiver,[2] I was ready for my next challenge—QRP.

I noticed a posting on an e-mail list offering a super bargain: a carton of 100, new-in-the-box, 1985-vintage, type 5763 (a beam power pentode) tubes at a great price—25 cents each! From my reading and experience working on old commercial rigs, I knew the 5763 to be a very useful tube for building and repairing receivers and transmitters. It was long a popular choice in *ARRL Handbook* construction projects. I couldn't pass up the deal.

Reference Material

The Internet is a great stimulus for Amateur Radio projects. I get extra satisfaction in the application of a modern medium to the needs of vintage radio builders and restorers. Two of the most useful online places to swap ideas on glow-in-the-dark radios are the *Glowbugs*[3] and *Boatanchor*[4] mailing lists.

While waiting for the tubes to arrive, I researched existing designs for something that would be fun to build. My favorite books to consult on tube projects include:

1962 *RCA TT-5 Transmitting Tube Manual*

1975 *RCA RC-30 Receiving Tube Manual*

ARRL Handbook of 1945, '50, '59, '66, '71 and '75

Radio Magazine/Editors and Engineers *Radio Handbook* from 1938-1975

Understanding Amateur Radio, ARRL, 1971.

The latter is one of my most cherished books on Amateur Radio. When I was starting as a Novice I read *Understanding Amateur Radio* again and again.

On-line vacuum tube specifications can be found at **www.hereford.ampr.org/cgi-bin/tube**. Back issues of magazines such as *CQ, QST, 73* and *Ham Radio* can be very useful, too. A contemporary magazine for tube project enthusiasts is the excellent *Electric Radio*.[5] My research turned up some great designs for transmitters using the 5763 and similar pentodes.[6-9]

A Web search yielded excellent ideas from the modern ARRL *Technical Information Service*. The *HF Transmitters & Receivers* Web page, **www.arrl.org/tis/info/tranrcvr.html** is a good example. I found a great design, The *Novice Special Transmitter*, from *The Radio Amateur's Handbook*, 1971, pp 181-183 or **www.arrl.org/members-only/tis/info/pdf/71hb181.pdf** (ARRL Members only). This 15 W, 80 and 40 meter CW transmitter was intended for the Novice builder using a 6C4 oscillator and a 5763 power amplifier stage.

The ARRL *Technical Information Service* has some great projects from past ARRL publications. The service also includes lists of appropriate sources for parts: air core inductors, air variable capacitors, surplus components, transformers and vacuum tubes.

The Design

After consideration of all these sources, combined with my growing collection of notes, I decided on the design shown in Figure 1.

My chassis selection was a major turning point in the project. Looking at the small parts count of this transmitter, just an oscillator and power amplifier, I knew I could keep the size to a minimum. The relatively low voltages and currents from the transmitter meant the parts could be small, many on the same scale as solid-state QRP projects. When my eyes fell upon my version of the original Tuna Tin 2 transmitter on the shelf, I was struck by an idea—wouldn't it be wonderful to also build this rig into a tuna fish can! It would be in line with one of the finest traditions of homebrewing—the Doug DeMaw, W1FB, Tuna Tin rigs.[10] That comparison can be seen in Figure 2.

I emptied a tuna can in the kitchen

[1]Notes appear on page 3-8.

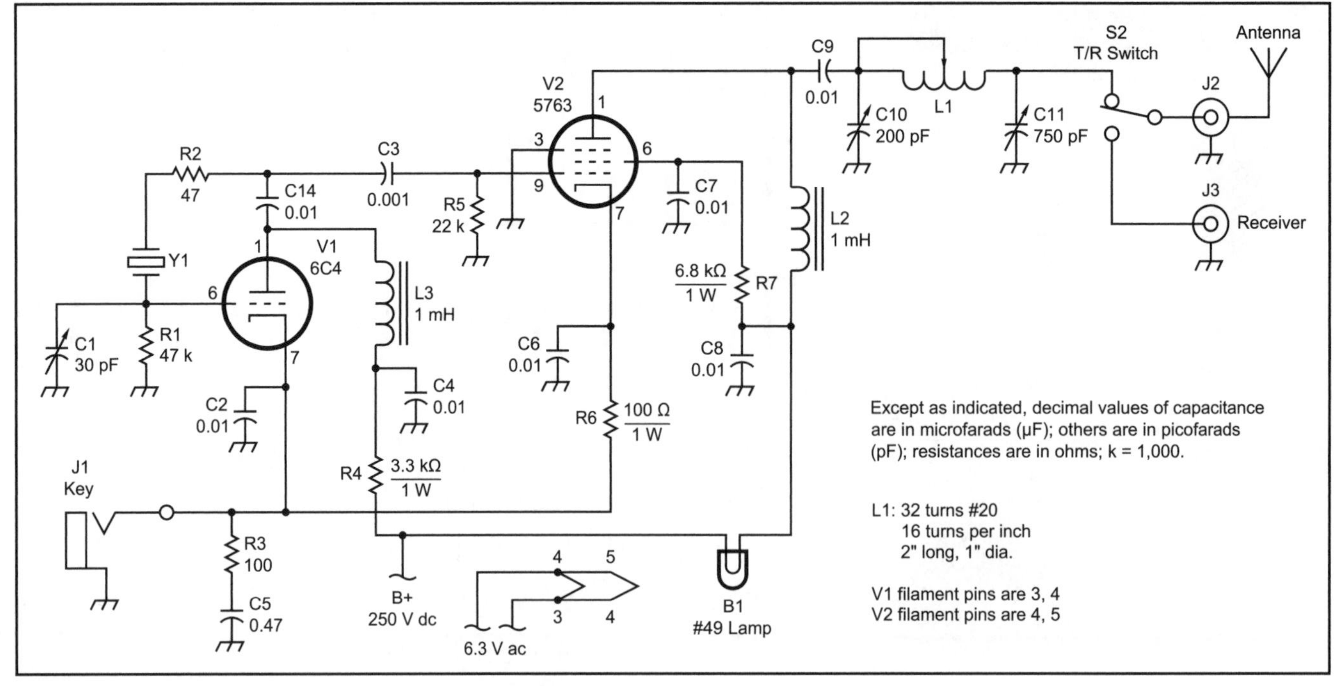

Figure 1—The T5 schematic. C10 is a 200 pF air variable capacitor; C11 is a 750 pF compression trimmer capacitor. All other capacitors are ceramic disc type, 500 V minimum working voltage. L1 is constructed on a 1-inch diameter form (see parts list). All resistors are ¼ W except where noted. All capacitors are ceramic disc type, 400 V dc minimum, unless otherwise specified. (B&W can be contacted at www.bwantennas.com/ama/mini.ama.htm.)

B1—#49 panel lamp held in rubber grommet.
B2—Neon lamp with current limit resistor (47 kΩ or 68 kΩ, ¼-W) or 120 V ac neon lamp with built-in resistor.
C1—30 pF compression trimmer.
C2, C4, C6, C7, C8, C9, C14—0.01 μF.
C3—0.001 μF.
C5—0.47 μF tubular.
C10—200 pF air variable.
C11—750 pF compression trimmer.
C12, C13—22 μF, 250 V dc minimum.
D1, D2—1N4007 silicon diode.
F1—Fuse, 0.4 a, 250 V dc, cartridge type.
F2—Fuse, 1.5 a, 120 V ac, cartridge type.
J1—Key jack, ¼″ or ⅛″.
J2, J3—SO-239 or BNC or RCA type coaxial antenna jacks.
L1—32 turns #20 wire, 16 turns per inch, 2″ long, 1″ diameter. (A B&W miniductor, type 3015, may be used, if available.)
L2, L3—1 mH RF choke.
R1—47 kΩ.
R2—47 Ω.
R3—100 Ω.
R4—3.3 kΩ, 1-W.
R5—22 kΩ.
R6—100 Ω, 1-W.
R7—6.8 kΩ, 1-W.
R8—470 kΩ, ½-W.
S1—SPST mini-toggle switch.
S2—SPDT mini-toggle switch.
T1, T2—120 V primary / 12.6 V CT secondary, 3 A; see note 11. (RadioShack 273-1511).
V1—6C4 tube.
V2—5763 tube.
Y1—Crystal (fundamental frequency; see text).

Misc
7-pin miniature tube socket.
9-pin miniature tube socket.
Crystal socket.
Rubber grommets.
AC power cable and plug.
4 conductor dc power cable.
Cylindrical fuse-holders, (2) panel-mount.
4-pin Molex style plug and socket.
Tuna tin and mini-loaf tin (see text).

(and devoured a couple of tuna sandwiches!) as I pondered the possibilities. A few trial fits of the parts showed the smallest standard size of tuna tin would be too small for easy construction. But the next larger size, the 12-ounce version, would be perfect. Besides, that's how I usually build things—unusual housings and recycled parts. They don't call it a *junk* box for nothing. The resulting transmitter is very compact, 4×4 inches, and it's *cute* (as my wife put it).

Tube Safety Habits

Although this project is considerably safer than my previous breadboard transmitter, there are still some serious safety issues the builder/operator must keep in mind. Hams who have only built solid-state projects may not have developed "tube safety habits." This is no reason to shy away from building a tube project; just be aware of the risks and treat the equipment with due respect:

1. Keep in mind even the medium voltages used to power this rig can be lethal.
2. Lay out the circuit so exposed, live conductors are inside the chassis.
3. Use insulated wire and spaghetti tubing to minimize junctions with exposed voltage.
4. Don't work on the equipment when tired or feeling stressed.
5. Turn off the power supply *every* time a circuit change is needed and discharge the power supply filter capacitors.
6. Turn off and unplug the power supply if the equipment is to be left unattended.
7. Keep children, pets and visitors clear of the energized rig.
8. When testing and adjusting the op-

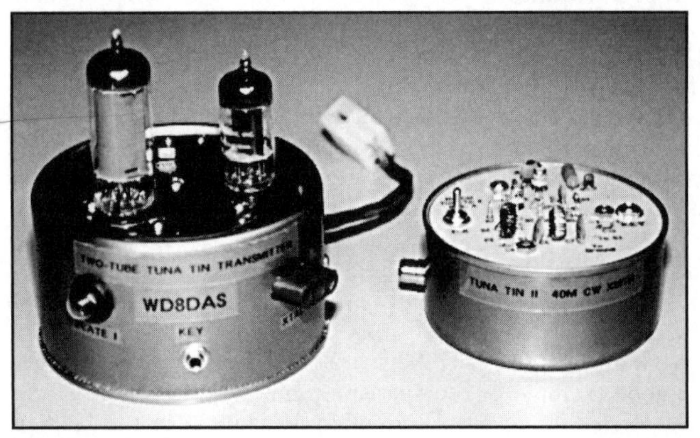

Figure 2—A side-by-side comparison: The T5 is on the left and the classic W1FB design, the Tuna Tin 2, is on the right

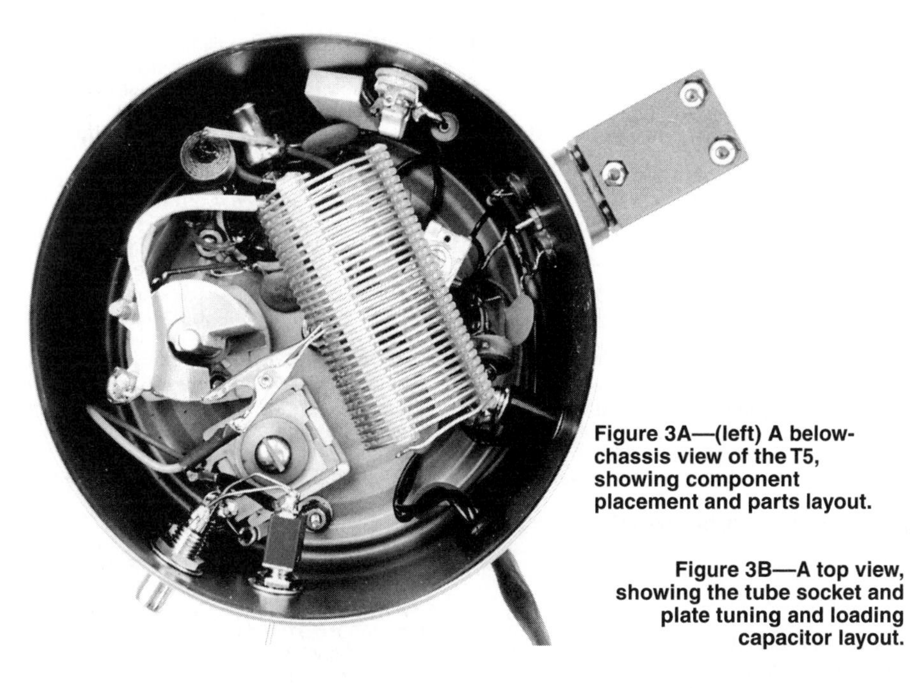

Figure 3A—(left) A below-chassis view of the T5, showing component placement and parts layout.

Figure 3B—A top view, showing the tube socket and plate tuning and loading capacitor layout.

erating transmitter, keep one hand in a pocket.

This final rule provides some back-up protection; in case of accidental contact with the circuit, it reduces the possibility of heart-stopping current flow across the chest. Remember, these voltages can be lethal... so switch to safety! An additional caution—there is high-voltage across the key terminals of this transmitter, so be careful.

Construction

Construction used normal hand tools. The metal of the tuna cans is incredibly easy to work; I found it was easy to make holes for the jacks, tie strips, crystal socket and tube sockets. I chose the older FT-243 style crystals since I already had a few in the shack, but other types can be used. Select the socket style to suit your collection, or install multiple sockets in parallel to allow various types to be used (but only plug in one at a time!). A below-chassis view of the completed transmitter will give you some idea of the component layout. It can be seen in Figure 3A. A top view of the transmitter appears in Figure 3B.

I initially powered the prototype transmitter's plates and filaments from an old EICO variable regulated bench supply I bought at a hamfest, but I later built the matching power supply, whose schematic is shown in Figure 4.[11] For its chassis I used a small baking pan, about the size suitable for a tiny loaf of bread. This maintained the food container theme. It is shown in Figure 5, next to the transmitter.

On the Air

The initial tests were conducted into a 50 Ω power resistor and wattmeter. It worked! The power amplifier was stable, showing no tendency to oscillate. The oscillator ran loud and clear and did not stall under any condition of tuning. Listening on a nearby receiver, I heard no key clicks. After adjusting the trimmer capacitor in the oscillator grid circuit, the CW note sounded near perfect and one of the best I've heard from a simple homebrew rig.

The warmth (ouch!) of the resistor RF load indicated significant power output. Moving the output to a wattmeter and a conventional 50 Ω dummy load revealed about 8 W out at the optimum settings of the 5763's pi-network (coil and capacitors). The scope showed a smooth waveform low in harmonic energy.

I discovered a bonus—the transmitter works well on four bands: 80, 40, 30, and

Sources of Classic Parts

HF crystals in FT-243 holders can be a challenge to locate. Used and surplus crystals can occasionally appear at hamfests, but ones on useful amateur frequencies seem rare. Places to buy new, cut-to-frequency, FT-243 type (same pin spacing) crystals include the following:

- JAN Crystals, PO Box 60017, Fort Myers, FL 33906-6017, 800-526-9825.
- PR Crystals, 2735 Avenue A, Council Bluffs, IA 51501, 712-323-7539.

Many hams bemoan the difficulty in locating vacuum tubes, but in some respects it has never been easier. In recent years a number of excellent on-line or mail order tube dealers have appeared. Examples include:

- Electron Tube Enterprises, PO Box 8311, Essex, VT 05451-8311, 802-879-1844; **members.aol.com/etetubes/**.
- ERSC, 1599 SW 30th Ave, Unit 4, Boynton Beach, FL 33426, 561-737-8044; **home.att.net/~esrc/esrcmain.html**.
- Radio Electric Supply, PO Box 1939, Melrose, FL 32666, 352-475-1950; **www.vacuumtubes.net/npage/indexc.html**.

The source of the 5763 tubes mentioned in the text was a newsgroup posting: "New, old-stock 5763 tubes in packs of 100—only $25!" **stevie@foothill.net**.

Additionally, for tubes, sockets, variable capacitors, connectors and other vintage electronic components, look at these resources (there are others):

- Antique Electronic Supply, 6221 S Maple Ave, Tempe, AZ 85283, 480-820-5411; **www.tubesandmore.com**.
- Dan's Small Parts and Kits, PO Box 3634, Missoula, MT 59806-3634, 406-258-2782; **www.fix.net/dans.html**.
- Ocean State Electronics, 800-866-6626; **www.oceanstateelectronics.com**.
- Fair Radio Sales, PO Box 1105, Lima, OH 45802, 419-227-6573; **www.fairradio.com**.
- Play Things of Past, 9511 Sunrise Blvd, #J-23, Cleveland, OH 44133, 216-251-3714; **www.oldradioparts.com**.

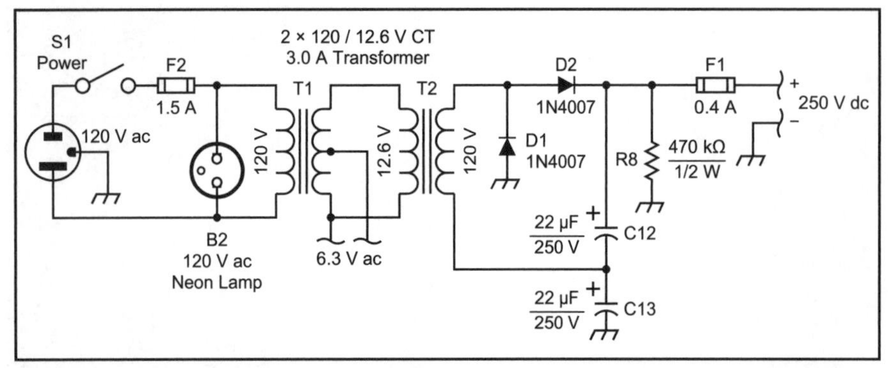

Figure 4—The T5 power supply schematic is a good example of a voltage doubler circuit. This project is meant to rely on your junk box, but some parts may have to be purchased. The transformers shown use 12.6 V CT, 3 A windings, but 6.3 V transformers can also be used (see note 11). The neon indicator lamp, B2, must have an internal current limiting resistor. The common NE-2 neon lamp does *not*. Be sure to observe polarity on C12 and C13 and make sure to include R8, the 470 kΩ, ½ W bleeder resistor—it's a required safety item.

Figure 5—The completed project: Mini-Loaf power supply next to the T5.

20 meters. Using fundamental crystals the rig tuned beautifully on all frequencies with output ranging from 5 to 10 W on the various bands. Do not use crystals on a lower band to try to work a higher band on harmonics—this rig won't. If a tuned stage was added to peak the oscillator on a harmonic it could work, but not with the simple RF choke used in my design. The pi-network components might be less than perfectly configured for 80-meter operation, but I found that the transmitter had a decent RF power output level with no strain obvious on the pentode.

The first on-air test was a snap. I tuned around the low end of 40 meters and didn't hear anything exciting, so I plugged in a crystal and called CQ. Unlike my previous homebrew solid-state transmitters such as the flea-power Tuna Tin 2, the Two Tube Tuna Tin Transmitter has significant RF output. Unless the other operator was using a "brick" for a receiver I should be heard. Bob, N6WG, answered my first CQ and we had a solid contact. The first QSO on a new homebrew rig is always satisfying.

Discussions with friends on the mailing lists have resulted in further 5763 projects. I sent some tubes to my fellow NERTs (the proud "New Era Repeater Technocrats"—a gang of old friends who keep in touch on the air and on the Internet) and we're planning a building contest. We'll bring our finished rigs to the traditional NERT-gathering at the Dayton Hamvention and compare our work. Some decided to build one-tube power oscillator transmitters, while others are going for more elaborate, multistage rigs—including an AM transmitter and a high-level SSB design. We'll compete for the "Most Unusual, Non-Conformist Construction Technique" award, the "Built All From Used Parts" award and the "It Could Never Work" award.

Conclusion

I am proud of the Two Tube Tuna Tin Transmitter. It's my retro-tribute to Doug DeMaw's original Tuna Tin design and it generates considerably more RF output. Not bad for some 25-cent tubes and a few evenings' work!

Notes

[1]S. Johnston, WD8DAS, "The Secret Dream Transmitter," *CQ Magazine,* Feb 2001, p 70. Also shown at **www.qsl.net/wd8das/dream.html**.

[2]Five-tube HF receiver; **www.qsl.net/wd8das**.

[3]The *Glowbugs* email mailing list. To subscribe see **www.home.cfl.rr.com/happysurfer/glowbugs.htm**. To post, **glowbugs@piobaire.mines.uidaho.edu**. File archive at **www.mines.uidaho.edu/ftp/pub/Glowbugs**.

[4]The *Boatanchors* e-mail mailing list. To subscribe see **www.qth.net**. To post, **boatanchors@qth.net**.

[5]*Electric Radio,* 14643 County Rd G, Cortez, CO 81321-9575—monthly magazine.

[6]D. Ishmael, WA6VVL, "A Two-Tube 6AG7 80/40M CW Transmitter," *Electric Radio,* Dec 1993.

[7]L. McCoy, W1ICP, "Novice 80 and 40 Meter One Tube Rig," *QST,* Nov 1953, p 28.

[8]V. Chambers, W1JEQ, "A Two-Band Miniature Mobile Transmitter," *QST,* Sep 1952, p 11.

[9]Two-tube 5763 transmitter shown in the 1st edition (1955) of the ARRL *Mobile Manual for Radio Amateurs,* pp 92-95.

[10]Solid State Tuna Tin 2 articles from *QST*: D. DeMaw, W1CER, "Build A Tuna-Tin 2," May 1976, p 14; F. Stevens, WA5LIE, "Tuna-Tin 2" (Strays), Feb 1977, p 36; R. Arland, K7SZ, "Of Tuna Tins, Black Cats and Zombies," Mar 2000, p 91; E. Hare, W1RFI, "The Tuna Tin 2 Today," Mar 2000, p 37; "Up Front in *QST*—Another Tuna Tin Aficionado," Aug 2000, p 21.

[11]Power supply transformers T1 and T2 are shown as 12.6 V, 3 A units with a center-tapped secondary, as these are readily available. You can improvise, however. A 12.6 V CT, 1.4 A transformer for T2 can be used, while a 2.4 A transformer will be needed for T1, if you can find these in your junk-box. If available, you can use two 6.3 V filament transformers without a center-tap, provided they can supply the required current. You would need a 3.8 A, 6.3 V unit for T1 and a 2.8 A, 6.3 V unit for T2. T2 has to supply about 16 W to the 5763 power amplifier, assuming a plate efficiency of about 60% and about ½ W for the oscillator. Transformer T1 has to supply the 1 A filament load for both tubes, plus the current required for T2. Bear this in mind when picking transformers for this project, as these transformers will affect the input regulation of the power supply and hence the output voltage under load.—*Ed.*

Steve Johnston, WD8DAS, has been a builder, repairer and restorer of electronic equipment nearly all his life. He started taking apart radios as a youngster and became a ham at age 13. He has been active in school radio clubs, including the University of Akron station W8UPD in Ohio and, in later days, the international NERT radio club. A Broadcast Engineer for the past 20 years, Steve is Director of Engineering and Operations for Boise State Radio; a 20-station, 5-network, public radio system in the Northwest. In the rare hours not spent with radio he pursues his other avocation as a writer and historian. He has been a ham for 26 years and lives in Boise, Idaho with his wife, Christy, and two children, Kaitlin and Noah. You can contact the author at **sbjohnston@aol.com**.

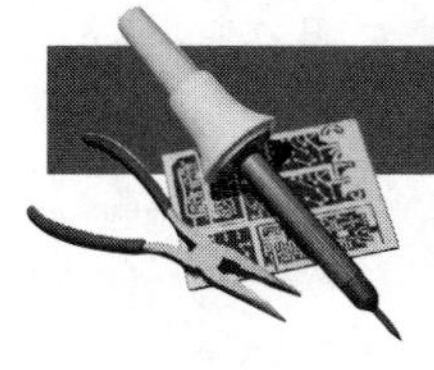

HINTS & KINKS

From *QST*, January 2003

ALTERNATIVE PARTS FOR THE TWO TUBE TUNA TIN TRANSMITTER (T5)

◊ I enjoyed reading the January (2003) *QST* article about a vacuum-tube version of Doug DeMaw's Tuna Tin transmitter.[1] Steve Johnston, WD8DAS, was lucky enough to come across a "lifetime" supply of 5763 tubes at a reasonable price. For others interested in building the transmitter, there are readily available substitutes for the 5763. In the mid-1960s, transmitter construction projects in *QST* and *The ARRL Handbook* started using the 6GK6 as a low-power driver stage for a pair of 6146 amplifier tubes. Table 1 compares characteristics of the 5763 and 6GK6.

Other possible substitutes are the 12BY7A and the 6CL6. These two substitutes have lesser maximum plate dissipations and will have less output power than the 5763 or the 6GK6. Don't forget that the substitutes all have different pin arrangements.

If the T5 is to be used on frequencies above 80 meters, the values of the two RF chokes in the plate circuits (L2 and L3) should be reduced. A 1 mH choke has a self-resonant frequency of around 6 MHz, which is suitable for 80 meters, but may have series resonances on the higher amateur bands. For operation from 80 meters to 20 meters, choke values of 120 µH would keep self-resonance above 14 MHz. The only change needed is to increase the inductance of the π-network coil slightly to account for the parallel inductance of the plate choke. If single-band operation is planned, the values of the two plate chokes can be selected for the specific band. Good values to avoid self-resonance problems are: 40 meters, 220-270 µH; 30 meters, 150-220 µH; 20 meters, 100-120 µH. *—Mal Crawford, K1MC, 19 Ellison Rd, Lexington, MA 02421;* **Malcolm_Crawford@Raytheon.com**

[1]S. Johnston, WD8DAS, "The Two Tube Tuna Tin Transmitter (T5)," *QST*, Jan 2003, pp 39-42.

Table 1
Tube Characteristics

Tube	*5763*	*6GK6*	*12BY7A*	*6CL6*
Maximum Plate Voltage (V)	350	330	330	300
Maximum Screen Voltage (V)	250	330	190	300
Filament Current (A)	0.76	0.76	0.60	0.65
Grid Capacitance (pF)	9.5	10.0	10.2	11.0
Plate Capacitance (pF)	4.5	7.0	3.5	5.5
Grid-Plate Capacitance (pF)	0.3	0.14	0.063	0.12
Plate Dissipation (W)	13.5	13.2	6.5	7.5
Transconductance (mS)	—	11.3	11.0	11.0

Table 2
Tube Base Connections

	Pin Number			
Tube	*5763*	*6GK6*	*12BY7A*	*6CL6*
EIA Base Code	9K	9GK	9BF	9BV
cathode	7	1	1	1
control grid	8, 9	2	2	2, 9
screen grid	6	8	8	3, 8
suppressor grid	3	3, 9	3, 9	7
plate	1	7	7	6
heater 1	4	4	4	4
heater 2	5	5	5	5
heater center tap	—	—	6	—

From *QST*, November 2001

By Erik Westgard, NY9D

Updating the W1FB 80-Meter "Sardine Sender"

Ingenuity is still the mother of invention. Follow this tale of determination and use NY9D's results to build your own 80-meter QRP transmitter.

Not long after acquiring a nice stock of RF parts from some old VCRs and television sets, I started wondering if it was possible to build a QRP (low power) rig with the parts they could provide, and possibly a few from the nearest RadioShack store. Each TV and VCR you take apart will reward you with transistors, capacitors, RF chokes and a color-burst crystal, which is in the 80-meter CW band. With that crystal and a common 2N2222 transistor you can build an oscillator. The difficulty lies in adding a power amplifier to that tiny transmitter so you can make some headway on 80 meters, which is not an easy place for milliwatt power.

In many published RF amplifier designs that use bipolar transistors, you have to contend with the impedance mismatch between oscillator outputs and RF amplifier inputs. In QRP construction books, there are schematics for transmitters with broadband matching transformers. Depending on the band, you can take the right toroid core, wind the correct number of turns for the primary, and use the square of the turns for calculating the impedance of the secondary.

A Toroid Alternative?

Would it be possible to accomplish the matching without a toroid? (Bear in mind that I wanted to keep the entire project as simple as possible without resorting to mail-order shopping.) I recalled that the classic 40-meter Tuna Tin 2 by Doug DeMaw, W1FB, worked its impedance-matching magic with 10-µH RF chokes. I soon discovered, however, that RadioShack no longer carried 10-µH chokes. These chokes weren't available in my VCR/TV scavenger assortment, either. Scaling Doug's design for 80 meters looked complicated as well.

Some more digging in my article archives revealed the 80-meter W1FB Sardine Sender transmitter (see Figure 1). This was exactly what I needed. All of the parts came from RadioShack, except the all-important 10-µH chokes, which were also used for the broadband transformer. Back to the books this time. How to adapt the currently available 100 µH Radio Shack 276-102 choke to be the broadband transformer?

There is a lot buried in the W1FB books the League publishes. In the original 1986 *QRP Notebook* (now out of print) there is a good discussion of broadband transformers and how to use the "A_L factor" to wind toroids. With that information

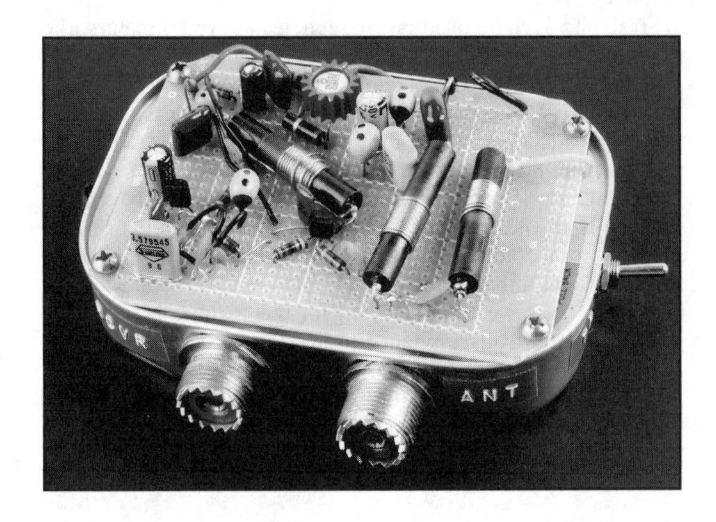

and the permeability factor, you can calculate the right number of turns for a given inductance. There was no mention of how to do to this with rods instead of toroids, or what to do without the A_L factor.

An e-mail response from **Radioshack.com** provided the permeability (220) for the core used in the 100 µH RadioShack 276-102 choke. No A_L was available. A helpful break occurred at the Midwinter Madness Hamfest in St Paul, Minnesota. One of the vendors was selling a Doug DeMaw book I had not seen before—*Ferromagnetic Core Design and Application Handbook* published by MFJ. On page 42 there was a critical bit of information: "It is difficult if not impossible to construct a set of A_L factors for rods and bars." This is because the location of windings on the bar or rod and the spacing of the turns had a big impact on the inductance. The identical number of turns spaced differently or on a different place on the rod, say at the end, might cause the inductance to change.

Doug provided, as usual, a hint for getting out of the dilemma. In a November 1974 *QST* article on building a 160-meter transmitter, he says it is okay to experiment your way out of design problems you can't solve by mathematics and theory, using "empirical effort," as he called it. So how do you measure inductance down to at least one decimal place?

I thought the answer could be found with an old Heathkit

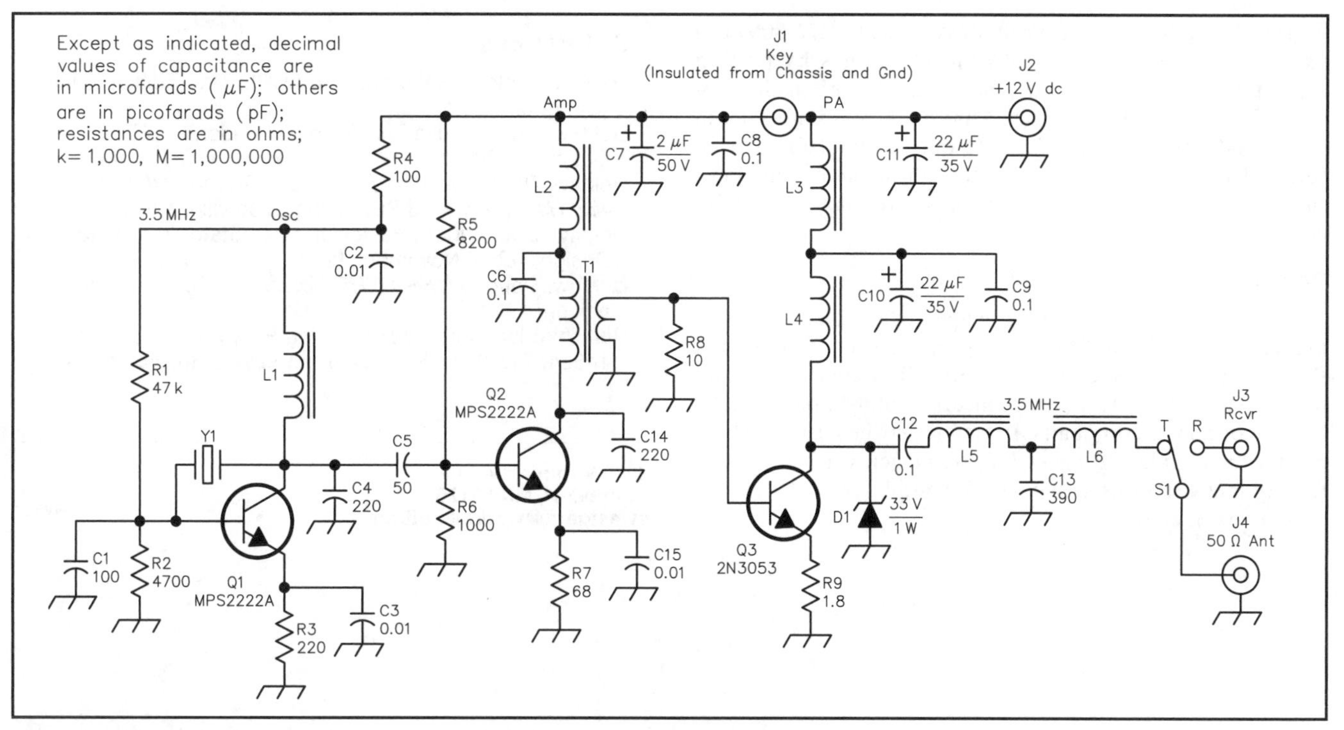

Figure 1—Schematic diagram of the updated W1FB Sardine Sender transmitter.

C1—100 pF ceramic.
C2, C3, C15—0.01 µF ceramic.
C4—220 pF ceramic.
C5—50 pF ceramic.
C6, C8, C9, C12—0.1 µF ceramic.
C7—2 µF electrolytic, 50 V.
C10, C11—22 µF electrolytic, 35 V.
C13—390 pF ceramic.
C14—220 pF ceramic.
D1—33-V, 1-W zener diode.
J1-4—Single-hole phono jacks.
L1—100-µH choke (RadioShack 273-102).
L2-L4—10-µH choke. Unwind the RadioShack 273-102 to 15 turns close-wound near the center of the choke for about 10.6 µH. Or use some 10-µH chokes (brown/black dots with black on the side) from an old TV or VCR.
L5—12-µH choke. Unwind all but 16 turns from a RadioShack 273-102 choke for 11.8 µH.
L6—8.9 µH choke. Unwind all but 14 turns from a RadioShack 273-102 choke for 8.9 µH.
Q1, Q2—MPS2222A transistors.
Q3—2N3053 transistor, heat sinked.
R1—47 kΩ.
R2—4.7 kΩ.
R3—220 Ω.
R4—100 Ω.
R5—8.2 kΩ.
R6—1 kΩ.
R7—68 Ω.
R8—10 Ω.
R9—1.8 Ω.
S1—SPDT toggle switch.
T1—Broadband impedance-matching transformer. See Figure 2 and caption.
Y1—Color burst crystal, 3.579545 MHz, or other 80-meter frequency of your choice.

IB-5281 impedance bridge that I picked up for $20 at an earlier hamfest. However, this one only provided a rough guess at the inductance of the RadioShack choke as I removed the windings. I tried for a few minutes to calculate how many turns I would need, but gave up. My saving grace appeared in the latest Ramsey Electronics catalog. That's where I found a new gadget—the $99 DMM 240 LCR meter, which could measure inductance down into the microhenry range. Sure, you can get multimeters that measure capacitance and inductance, but you need the smallest possible range for QRP designs where coils and chokes had microhenry values. The DMM 240 fit the bill.

Doug thoughtfully provided the µH values for most of the coils and windings in the Sardine Sender. I just took a stock of 100-µH RadioShack chokes and started unwinding and measuring. Once I reached my target, I scraped and soldered the wire end back on and I was done. The broadband transformer is shown in Figure 2. The test leads cause the readings to be a little high, but you can use the contact set on the body of the meter for greater accuracy.

The Sardine Sender Lives Again

The rest was pretty easy. I used a RadioShack universal board, which is a little larger than a sardine can. I saved board space by using some 10- and 100-µH chokes from scrap VCRs. I was a little nervous about the 10-µH VCR chokes—they were tiny and used fine wire.

For the first time I took the often-given kit-building advice and tried the oscillator stage first—it was fine. The big test was the transformer, output stage and filter coils. These were fine, too, but the resulting signal sounded grungy on my receiver. The power output was right on—slightly more than 1 W. I did some poking around, and shortened up some connections. On a hunch I tried my larger station power supply; the grunge was gone!

Figure 2—A close-up view of T1, the broadband transformer. I created my version by using a RadioShack 273-102 choke, unwound to 15 turns (10.6 µH) for the primary. Save the wire! The secondary windings consist of two turns of the removed wire.

The RadioShack disk capacitor assortment (272-809) is a useful resource. You can make up odd values by putting capacitors in series or parallel, such as the 390 pF made from a 56 pF and a 330 pF in parallel. Almost all the resistors are in stock, and you can make a 1.8-Ω out of two 1-Ω parts in series. I used mostly 1/4-W resistors throughout, but don't substitute wirewound resistors as these are made from wire coils, which are inductive.

Conclusion

So there you have it—a classic updated, with all parts still available from RadioShack. It is interesting to note that RadioShack is still stocking the 2N3053 RF transistor after all these years. My only caveat concerns color-burst crystals. Beware of poor quality units. In fact, it may be best to order a crystal for popular 80-meter CW frequencies. I had trouble finding many stations active on 3.579545 MHz.

Above all, enjoy!

References

DeMaw, Doug, *QRP Notebook*, ARRL, First Edition, 1986. (Out of print)

DeMaw, Doug, "Build This Sardine Sender," *QST*, October 1978.

DeMaw, Doug, *Ferromagnetic Core Design and Application Handbook*, MFJ Publishing, Starkville, MS

DeMaw, Doug, "More Basics on Solid-State Transmitter Design," *QST*, November 1974.

DeMaw, Doug, *W1FB's QRP Notebook*, ARRL, second edition, 1999.

Ramsey Electronics, 793 Canning Parkway, Victor, NY 14564; 716-924-4560; **www.ramseyelectronics.com/**.

3990 Virginia Avenue
Shoreview, MN 55126
ewestgard@worldnet.att.net

By Lew Smith, N7KSB

From *QST*, March 2000

A Simple 10-Meter QRP Transmitter

Take advantage of this 10-meter/QRP combo to get more miles per watt!

Now that the sunspots are back, 10 meters has again become a QRP paradise! Worldwide DX can be easily worked with this "homebrew" QRP transmitter and a simple antenna. It uses only 23 electronic parts, yet puts out nearly 4 W of good-sounding CW on 10 meters.

Circuit Description

The circuit uses a 74AC240 octal inverter logic IC as a combination oscillator and driver.[1] One inverter is configured as a classical Pierce crystal oscillator. R1 improves oscillator start-up and C3 "rubberizes" the crystal, allowing a degree of frequency change. To prevent chirp, the oscillator runs continuously during transmission. Four of the 74AC240's inverters are wired in parallel to make a very simple driver.

An inexpensive VN88AF power MOSFET is used as a keyed final amplifier. On 10 meters, this device is much easier to drive than the more-popular IRF510 used in many lower-frequency transmitters. A TIP115 keying transistor and a 7805 5-V regulator complete the lineup.[2]

To keep things simple, I use a TR switch to switch the antenna and mute the receiver. I mute my homebrew receiver by removing power from the audio stage. Other receivers may require a different arrangement. The **SPOT** switch, S2, allows frequency adjustment without causing interference.

Special Parts

The oscillator *will not operate properly with overtone crystals* that are commonly used above 25 MHz. Use a *fundamental-mode* crystal with a 20-pF load capacitance; the vendor should acknowledge that this type has been shipped. With the crystal specified to operate with a 20-pF load capacitance, the minimum transmitter frequency will correspond to the value marked on the crystal can. The crystal's maximum frequency will be about 17 kHz higher.A

This inside view of the 10-meter QRP transmitter shows hobby-shop brass put to good use. To the left is the homemade tuning capacitor, C3 of Figure 1. An angled brass shield cuts across the IC (U1 of Figure 1) that lies legs up on a piece of sheet brass; the shield separates the tuning capacitor from the rest of the transmitter. Mounted on the vertical brass strip in the background are Q1 (top) and Q2 (bottom). L1, L2 and L3 are at right angles to each other. Voltage regulator U2 lies between S1 (foreground) and S2 near the rear of the enclosure. Small pieces of PC board, glued to the sheet brass, provide for component-lead isolation and interconnection.

A close-up view of the homemade tuning capacitor, C3 of Figure 1. The 1/4-20 carriage bolt passes through the front-panel-mounted T nut and is capped with a tuning knob. The bolt's head rests against a piece of piece of Fiberglas PC-board material (sans copper) epoxied to a flexible flap of brass acting as the rotor plate of C3. A small section of copper foil on the Fiberglas board behind the flap is C3's stator plate (see Figure 2). The short length of bare wire from C3's stator connects directly to one pin of the crystal socket.

[1]Notes appear on page 3-16.

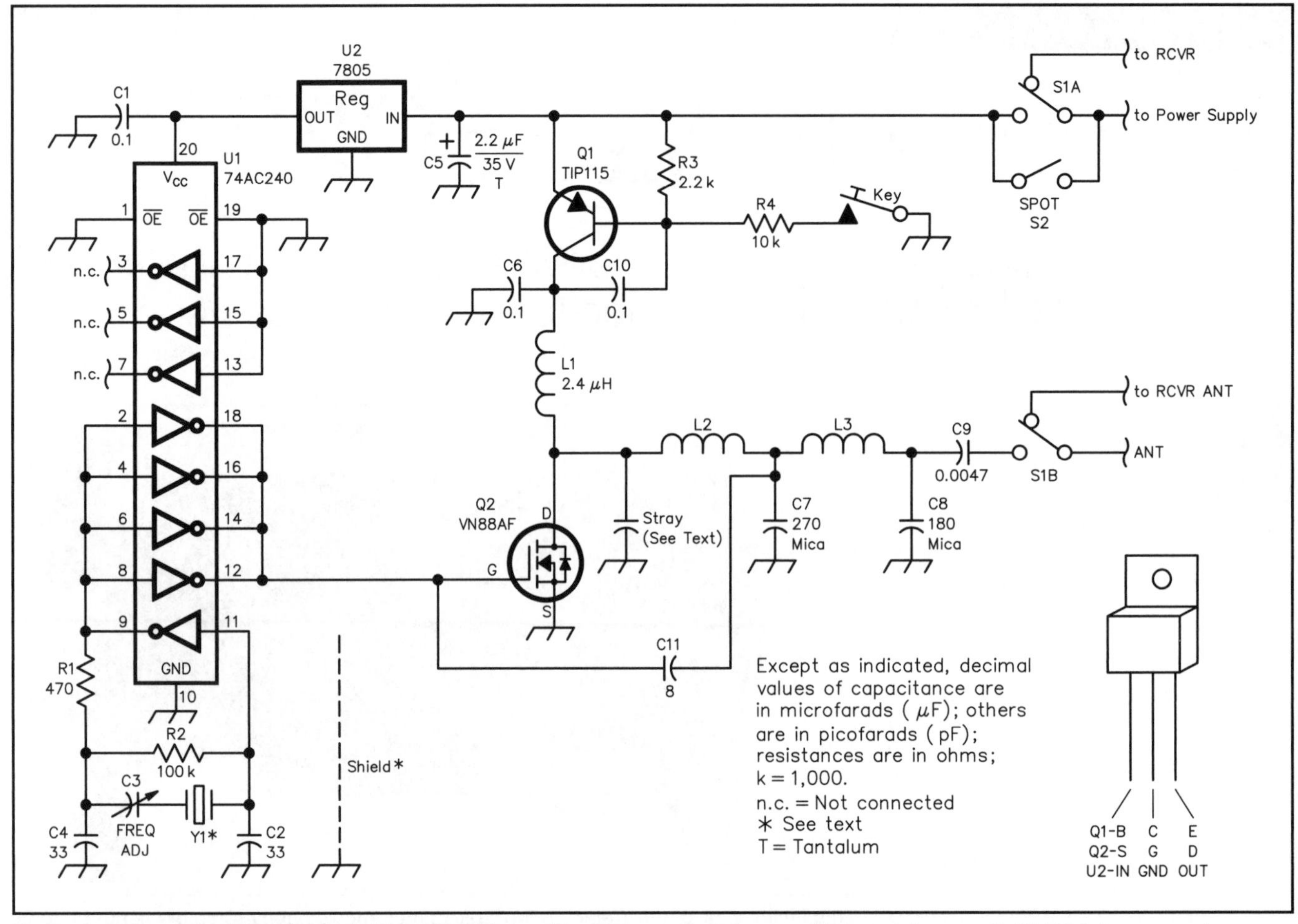

Figure 1—Schematic of the Simple 10-Meter QRP Transmitter. Unless otherwise specified, resistors are 1/8-W, 5%-tolerance carbon-composition or film units. MO part numbers in parentheses are Mouser (Mouser Electronics, 958 N Main St, Mansfield, TX 76063-4827; tel 800-346-6873, 817-483-4422, fax 817-483-0931; sales@mouser.com; http://www.mouser.com; RS part numbers are RadioShack; the 900-series numbers (not available in stores) identify parts available by Web-site ordering at http://www.radioshack.com; tel 800 THE SHACK). Equivalent parts can be substituted; n.c. indicates no connection.

C1, C6, C10—0.1 µF, 50 V monolithic ceramic (RS 272-109)
C2, C4—33 pF, 50 V ceramic (from RS 276-806 Picofarad 50-Pack)
C3—Homemade 2- to 60-pF air-dielectric variable capacitor; see Figure 2.
C5—2.2 µF, 35 V tantalum (RS 900-2172)
C7—270 pF, 300 V or greater, mica (Mouser 5982-15-500V270)
C8—180 pF, 300 V or greater, mica (Mouser 5982-15-500V180)
C9—0.0047 µF, 2 kV ceramic (RS 900-7214)
C11—8 pF, 50 V ceramic (selected from RS-276-806 Picofarad 50-Pack), or try two 4.7 pF, 50 V (RS 272-120) in parallel
L1—2.4 µH, 1.5-A RF choke (RS 900-4834)
L2—11 turns #18 enameled wire, 1/4-inch ID, 1/2-inch long
L3—10 turns #18 enameled wire, 1/4-inch ID, 9/16-inch long
Q1—TIP115 PNP Darlington power transistor (MO 511-TIP115)
Q2—VN88AFD power MOSFET (RS 900-5544)
S1—DPDT toggle
S2—SPST toggle
U1—74AC240N octal 3-state inverting buffer (RS 900-3626)
U2—7805 5-V, 1-A positive regulator, TO-220 case (RS 276-1770)
Y1—28,010-kHz fundamental-mode crystal in HC-6/U case, 20-pF load capacitance; (JAN Crystals, 2341 Crystal Dr, PO Box 06017, Ft Myers, FL 33906-6017; tel 800-JAN XTAL, 941-936-2397, fax 941-936-3750; International Crystal Mfg Co, 10 N Lee, PO Box 26330, Oklahoma City, OK 73126-0330; tel 800-725-1426, 405-236-3741, fax 800-322-9426). See text.
Misc: Enclosure (2 1/8×3×5 1/4-inch [RS 270-238]),hardware and a crystal holder

28,010-kHz crystal covers about 75% of the DX portion of the band.[3]

C3, the homebrew compression capacitor shown in Figure 2 and photographs, gives better resolution and tuning range than can be obtained from a conventional air-dielectric variable capacitor. A bent piece of 0.016-inch-thick brass becomes the equivalent of a capacitor's rotor plate.[4] A bit of copper foil mounted on a section of PC-board material acts as the capacitor's stator. (The PC-board's foil can be cut with a sharp knife and the unwanted foil removed after heating it with a soldering iron.) Connect the capacitor's stator to the crystal.

Use a 0.0047-µF, 1-kV disk-ceramic output-coupling capacitor at C9. The rather large antenna current may destroy physically smaller capacitors (especially monolithic ceramic capacitors). Use mica capacitors at C7 and C8.

Construction

A 2-inch wide, 0.016-inch-thick brass strip is used as a ground plane for this circuit. Small pieces of PC-board material epoxied to the ground plane act as solder lands for several components. This construction approach results in much lower stray inductance, better heat sinking, reduced construction time and, perhaps, better appearance than possible with standard PC-board construction.[5]

The 74AC240 and 7805 ICs, switches and crystal socket are mounted on the brass ground plane. The final amplifier and keying transistor are mounted on another 2×2-inch piece of 0.016-inch-thick brass that is soldered at right angles to the main ground plane. This configuration makes efficient use of the space in the 2 1/8×3×

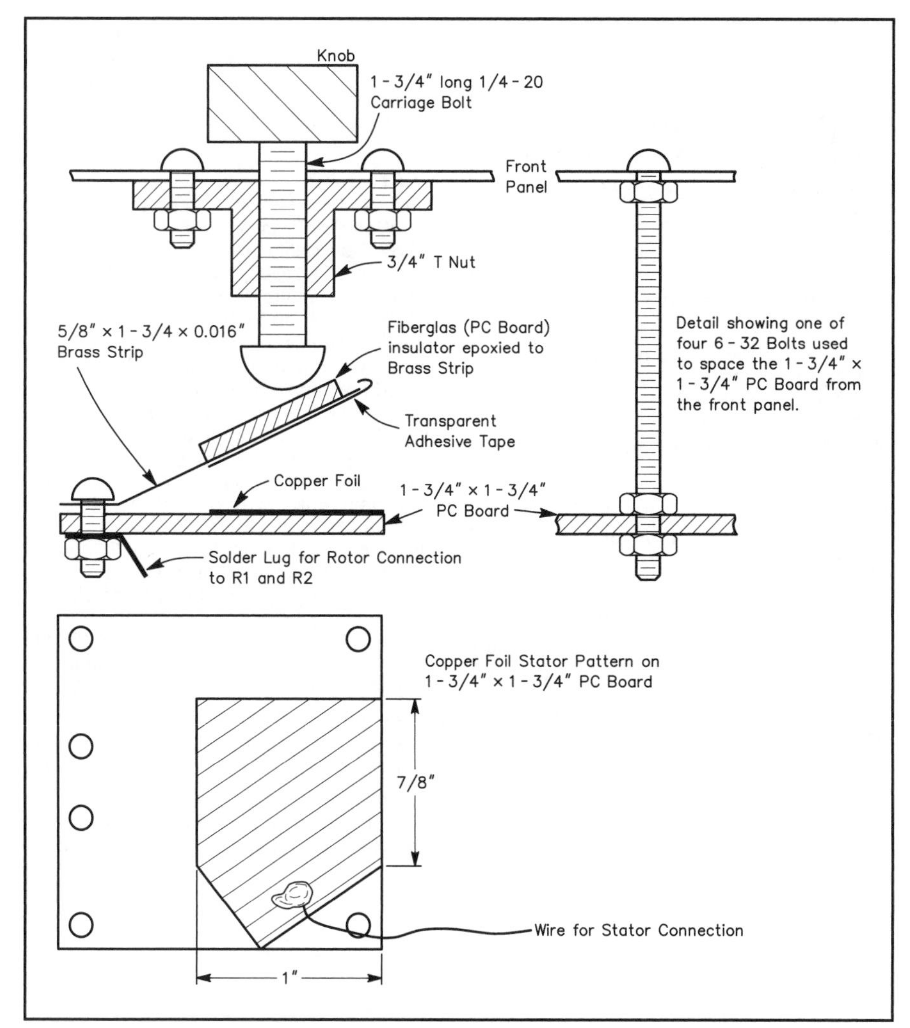

Figure 2—Mechanical assembly of C3; see text and accompanying photographs.

Here's another view of C3, this time from the top.

On the rear panel of the 10-meter transmitter, four phono jacks provide connection to the power supply (PWR), receiver (RCVR PWR and RCVR ANT) and the station antenna (ANT). A 1/4-inch KEY jack is beneath the phono connectors. The receiver/transmit (R/T) and SPOT toggle switches are at the rear of the top panel, with the crystal socket toward the enclosure's front.

Ready to go for 10-meter QRP!

5 1/4-inch aluminum box (Radio Shack 270-238) used to house the transmitter.

Holes drilled in the top surface of the aluminum box align with matching holes in the ground plane keeping the ground-plane in contact with the aluminum box for efficient heat flow and good grounding. The homebrew capacitor is mounted on the front lip of the box; the input and output connectors are mounted on the box's rear lip.

U1, the 74AC240, is mounted "dead bug" style (ie, on its back with its legs pointing up). This minimizes several very critical lead lengths. The pin-10 ground lead and the leads of the bypass capacitor at pin 20 must be as short as possible. U1's unused pins (3, 5 and 7) are folded onto the IC's belly; the grounded pins (1, 10, 13, 15, 17 and 19) are bent downward and soldered to the ground plane. Pins 2, 4, 6, 8 and 9 are strapped together, as are pins 12, 14, 16 and 18.

Satisfactory heat-sinking is obtained by bolting Q1, Q2 and U2 to the brass groundplane. Because the tabs of Q1 and Q2 are not at ground potential, mica insulators and nylon shoulder washers (RadioShack 276-1373, TO-220 mounting hardware) are needed. Because the mica insulator forms part of the capacitance used in the output filter, using a different heat-sinking technique will require output-circuit component-value changes.[6] The 7805 voltage regulator, U2, does not require a mica insulator.

Most of the components are wired point-to-point. Five 3/8×3/8-inch pieces of PC-board material epoxied to the groundplane act as solder lands for the coils, one end of R4, and the junction of R1, R2, C2, and C3. The coils are mounted at right angles to each other to minimize coupling.

Power Supply

Although this transmitter can be powered by a standard 13.8-V supply, best performance requires 24 V. The simple, well-filtered (but unregulated) supply shown in Figure 3 is ideal. Physically separate the power supply from the transmitter to prevent pin 11 of U1 from picking up 60-Hz hum.

Troubleshooting

This transmitter is easy to troubleshoot. It draws roughly 60 mA key up and 200 to 300 mA (depending on the supply voltage) with the key down. Check that the 7805 output is +5 V and that the collector voltage of Q1 (the TIP115) rises to about 1 V less than the supply when the key is closed. I measured 3.7 W of RF output with a 24-V

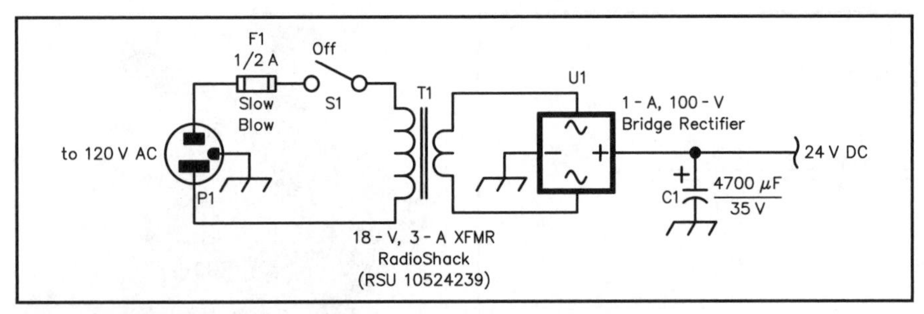

Figure 3—A simple 24-V dc power supply for the transmitter.

C1—4700-µF, 35-V electrolytic capacitor (RS 272-1022)
F1—0.5-A slow-blow fuse (RS 270-1018)
S1—SPST toggle
T1—Transformer, 120-V pri; 18-V, 3-A sec (RSU 10524239)
U1—1-A, 100-PIV bridge rectifier (RS 276-1171 [4 A])
Misc: Enclosure, hardware, fuse holder

supply and 1.5 W with a 13.8-V supply.

Initially, I detected a slight chirp at the high end of the tuning range. This was caused by stray capacitive coupling from the output circuit (Q2, the coils, filter capacitors and TR switch) to the junction of the tuning capacitor and the crystal. A grounded 2×2-inch brass shield between the output circuitry and the tuning capacitor eliminates the chirp.

C11 aids measurably in reducing backwave.[7] Without C11, the backwave is about 40 dB down during key up. By connecting an 8-pF capacitor (C11) between the MOSFET gate and the ungrounded end of C7, the backwave is suppressed another 13 dB, to 53 dB down.

The MOSFET final amplifier operates as a symmetrical square-wave current source. This results in very low even-order harmonics. With a 24-V supply, the second harmonic measures –58 dBc.

Results

This transmitter was fun to build and even *more* fun to operate! The first 10 contacts I had while using this transmitter and a roof-mounted groundplane antenna nabbed these prefixes: three LUs, two Ws, ZL, T22 (Tuvalu), VK, VE and FG. I'm pleased!

Notes

[1]Lew Smith, N7KSB, "An Experimental 1/2-Watt CW Transmitter," Hints and Kinks, *QST*, Nov 1994, p 84.

[2]Lew Smith, N7KSB, "An Easy-to-Build, 15-Watt Transmitter," *Hambrew Magazine*, Spring 1994, pp 9-13.

[3]Most DX seems to be in the pirate-free window between 28,008 and 28,030 kHz.

[4]I used brass packaged by K&S Engineering and sold by hardware and hobby stores.

[5]No PC board is available for this project.

[6]Stray capacitance at the MOSFET drain is estimated at 90 pF and is mostly related to the heat-sink insulator and the MOSFET output capacitance.

[7]Backwave is key-up low-level RF output caused by an oscillator signal feeding through a keyed, unneutralized amplifier.

Lew Smith, N7KSB, was first licensed in 1947 at age 12. After receiving a BSEE and MSEE from MIT in 1959, he spent 33 years designing analog and analog-to-digital circuits. Lew is now retired and enjoys hiking and paragliding in addition to ham radio. He likes to chase CW DX with a variety of homebrew rigs. You can contact Lew Smith, N7KSB, 4176 N Soldier Trail, Tucson, AZ 85749.

By Charles Kitchin, N1TEV

From *QST*, February 1998

A Simple CW Transmitter for 80 and 40 Meters

If you've never used a MOSFET-powered transmitter before, give this one a try!

Here's a simple CW transmitter for the 80 and 40 meter bands. It's a modern day equivalent of the classic two-tube MOPA (master oscillator, power amplifier) transmitter used by so many hams for decades—usually their first transmitter and usually home-brewed. Building this equivalent is much simpler and probably less expensive. The two-stage transmitter uses a low-cost, high-speed op amp in the crystal oscillator circuit and a modern HEXFET device (an N-channel, enhancement-mode power MOSFET) for the output amplifier. Unlike some other solid-state designs, this little gem requires few components and can be built by most hams in an evening or two. On 80 meters, it provides a power *output* of 5 to 9 W into a 50 Ω load, depending on Q1's drain voltage (the trans-mitter's input power is about double the output). ARRL Lab measurements indicate power outputs of 6 W on 80 meters with a 30 V supply, and 1 W on 40 meters using a 25 V supply.

Using a HEXFET (rather than a bipolar transistor) for the output amplifier provides a much higher load impedance for the oscillator. That eliminates the need for a step-down transformer generally required between bipolar transistor stages, and allows the use of simple RC coupling instead. In turn, RC coupling greatly improves circuit stability, making this a simple project to build and get working. With a typical breakdown voltage of 500 V, and built-in Zener diode overvoltage protection, the HEXFET output stage can operate without a load and under conditions of high SWR, *without damage*. Another advantage is that this circuit can use a commonly available 99 cent microprocessor crystal for operation on the 80 meter Novice band.

Oscillator

An Analog Devices AD811AN op amp (U1) is used in the oscillator circuit. This low-cost op amp has a current-output capability of 100 mA, which allows it to drive the typical 350 pF input capacitance of Q1's gate. U1 is biased to mid supply by resistors R1 and R2. Oscillation occurs because of positive feedback applied from U1's output through the crystal (Y1) and R2 and R3 to the positive op amp terminal. By using R3, rather than a direct connection to the positive op amp terminal, the positive feedback level is reduced (as is the crystal current), so that a low-cost microprocessor crystal can be used without noticeable chirp. (Be sure to use 40-meter crystals for operation on 40 meters; doubling in the final using an 80-meter crystal at Y1 isn't efficient.)

Dc stability is maintained by a negative feedback loop using R4 and C2. C2 rolls off the higher frequencies while operating the op amp at a dc gain of one. Note that the transmitter's power supply is directly keyed and that C1, the 33-μF power-supply bypass capacitor, is located *ahead* of the key. If C1 is connected behind the key—to the *circuit side* of the key—the oscillator turns on too slowly and chirps. A 0.01 μF capacitor (C3), right at pin 7 of U1, maintains circuit stability and helps smooth out key clicks.

U1's output impedance is about 30 Ω at 3.6 MHz. R5 provides only a dc return for Q1's gate; it is U1's low-impedance output that permits adequate drive levels to Q1 at 3.5 and 7 MHz.

Power Amplifier

Q1's characteristics greatly simplify the transmitter design and provide a high power-output level when using a battery power source. The high impedance input of Q1 makes it much easier to drive than a bipolar transistor. Also, the HEXFET's output impedance is higher than that of a bipolar transistor, so component values for a pi-network output circuit are reasonable.

Q1's drain connects through a 100 μH RF choke to the dc supply. The output signal is capacitively coupled to a pi-network consisting of L1, C7, C8, and C9. L1 is close-wound on a plastic film can, using standard 20 gauge insulated hook-up wire.

The output network is tuned to resonance using C7 (**TUNE**), a 500 pF variable capacitor. S1, **BAND**, switches output inductances for the 40 and 80 meter bands. S3, **LOADING**, selects none (center) or one of two additional output capacitors (C10 or C11) to permit optimum matching to the antenna. The **OPERATE/TUNE** switch (S2) can switch in a #44

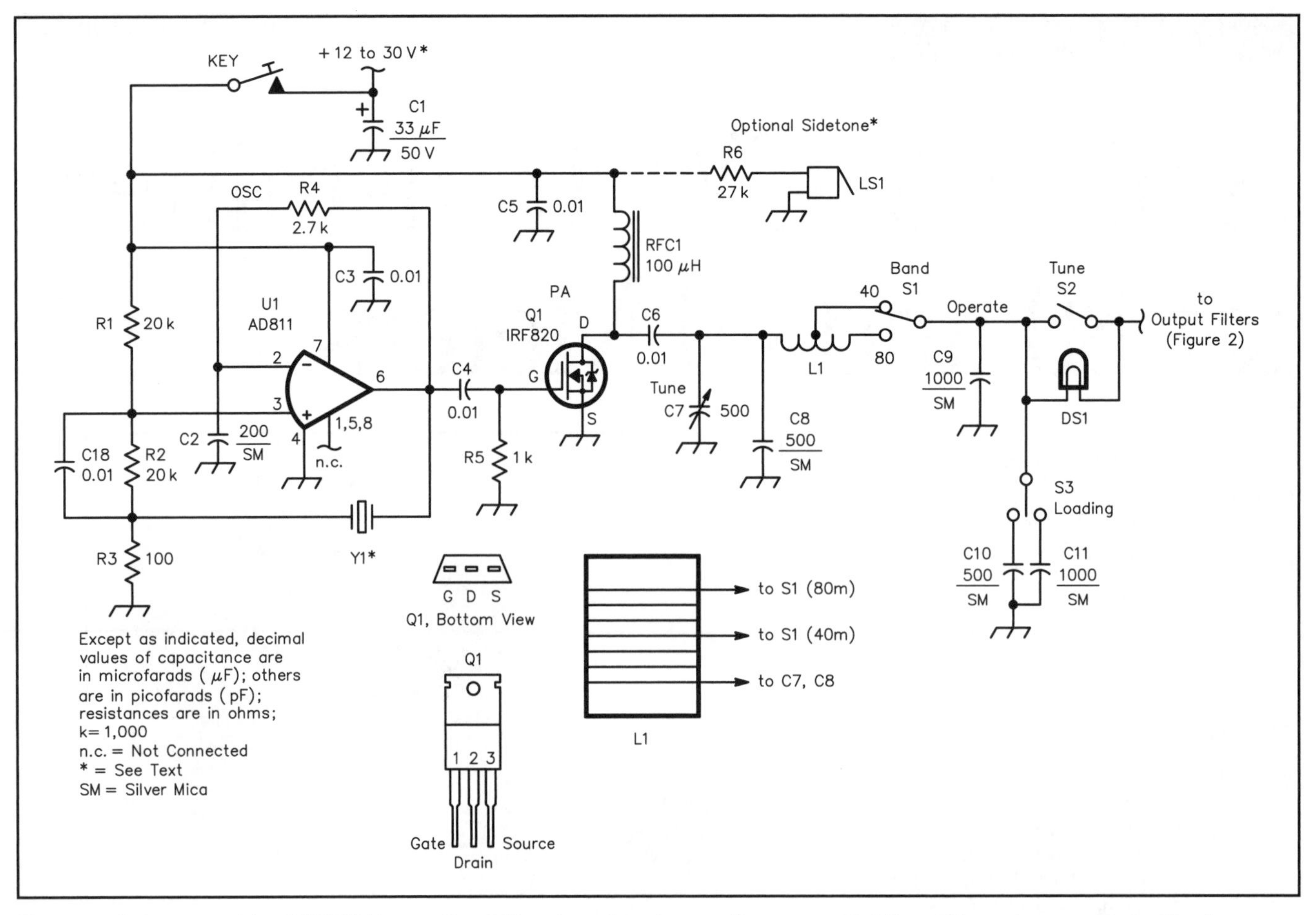

Figure 1—Schematic of the MOSFET transmitter. DK part numbers in parentheses are Digi-Key, RS numbers are Radio Shack; equivalent parts can be substituted. Unless otherwise specified, resistors are 1/4 W, 5% tolerance carbon-composition units.

C1—33 μF, 50 V electrolytic or tantalum
C2—200 pF mica capacitor
C3-C6, C18—0.01 μF disc ceramic
C7—500 pF air-dielectric variable capacitor; connect in parallel sections of a discarded AM radio air-dielectric variable capacitor to achieve a total capacitance of about 500 pF; (Fair Radio Sales Co, Inc, 1016 E Eureka St, PO Box 1105, Lima OH 45802, tel 419-223-2196; 419-227-6573; fax 419-227-1313; e-mail **fairadio@alpha.wcoil.com**; **http://alpha.wcoil.com/~fairadio/**. Antique Electronics Supply, 6221 S Maple Ave, PO Box 27468, Tempe AZ 85285-7468, tel 602-820-5411; fax 602,820-4643.)
C8 ,C10—500 pF silver mica
C9, C11—1000 pF silver mica
DS1—#44 lamp (RS 272-1108A; lamp holder RS 272-355)
L1—See text.
LS1—Piezo buzzer (RS 273-060; DK P9914; Digi-Key Corp, 701 Brooks Ave S, Thief River Falls, MN 56701-0677 tel 800-344-4539, 218-681-6674; fax 218-681-3380 **http://www.digikey.com**)
Q1—International Rectifier IRF820 HEXFET (Digi-Key IRF820, or Active Electronics 26036; Active Electronics [a division of Future Electronics] 811 Cummings Park, Woburn, MA 01801; tel 617-932-0050, fax 617-933-8884)
RFC1—100 μH (RS 273-102C)
S1—SPDT toggle (RS 275-613)
S2—SPST toggle (RS 275-612)
S3—SPDT center-off toggle (RS 275-325)
U1—AD811AN op amp; available from Newark Electronics (check your telephone book for a branch near you). Main office: 4801 N Ravenswood Ave, Chicago, IL 06040-4496; tel 312-784-5100; fax: 312-907-5217, and Allied Electronics, 7410 Pebble Dr, Fort Worth, TX 76118, tel 800-433-5700.
Y1—3.6864 MHz microprocessor crystal; see text (Digi-Key X080; Active Electronics 68010). For conventional 80 and 40 meter crystals, contact Peterson Radio at 2735 Ave A, Council Bluffs, IA, 51501, tel 712-323-7539; closed Fridays.
Misc: Crystal socket, binding posts for battery/power supply connections (RS 274-662); hook-up wire (RS 278-1219), hardware, heat sink, knobs, enclosure, PC boards (see Note 2).

lamp during tune-up to provide a visual indication of output current to the antenna.

The simple output network is, unfortunately, insufficient to reduce transmitted harmonic levels sufficiently to pass FCC specifications. That obstacle is readily overcome by adding some low-pass filtering to the output of the transmitter (see Figure 2).[1]

Construction

Take steps to avoid static discharges (use a grounded wrist strap) and keep Q1 inside its protective foam as long as possible. Whether you build the transmitter using "dead-bug," perfboard or PC board[2] methods, first wire and test the oscillator circuit, then build and test the output stage. When building the output stage, install all other components first; install Q1 last.

Component layout and lead dress for any HF RF circuit is important. Keep all leads, especially ground wires, as short as possible. Locate power supply bypass capacitor C3 physically as close as possible to U1 pin 7, using a short ground lead. Use noninductive carbon-composition resistors in RF circuits (not wire-wound types). By housing the transmitter in a small metal box, the box will also serve as the groundplane. Alternatively, a 5 1/2×8×3/4-inch piece of wood (poplar or pine are soft) can be used to hold a piece of copper clad PC board screwed (or glued) to it to serve as a groundplane. U1 and Q1 can be wired to a small breadboard mounted above the groundplane using short spacers. All ground leads should be as short as possible and be soldered to the copper groundplane.

L1 is wound on a standard 1 1/4-inch OD 35 mm plastic film can or pill bottle coil form. L1 consists of 8 turns of #20 hook-up wire with a tap at the center. Before winding the coil around the form, drill two small holes at the beginning of the winding. Feed one end

[1]Notes appear on page3-18.

Substituting Parts

Being inveterate experimenters (or cheapskates!), some builders might decide to make substitutions for the AD811, U1, and the IRF820 MOSFET, Q1. Don't do so without some forethought! The AD811 was carefully chosen because it has a beefy 100-mA output stage and a very low output impedance at the operating frequencies chosen. Other op amps may work, but they're likely to be expensive ones.

If you're thinking of using a JFET or bipolar transistor oscillator to drive the IRF820, you may have difficulty in getting the oscillator started, or it may not provide enough drive to Q1.

Providing a substitute for the IRF820 is another story. I selected the '820 because it has a fairly low input capacitance of 350 pF. Other HEXFETs, such as the IRF830 and '840 have *very high* input capacitances on the order of 800 to 1500 pF, so you'll never be able to drive them properly without regressing to the use of a step-down transformer between the oscillator and amplifier. Other MOSFETs that do have low values of C_{in} can't handle as much power, have lower break-down voltages and are *not* Zener-diode protected. So, if you're thinking of substituting parts, do so carefully.—*Charles Kitchin, N1TEV*

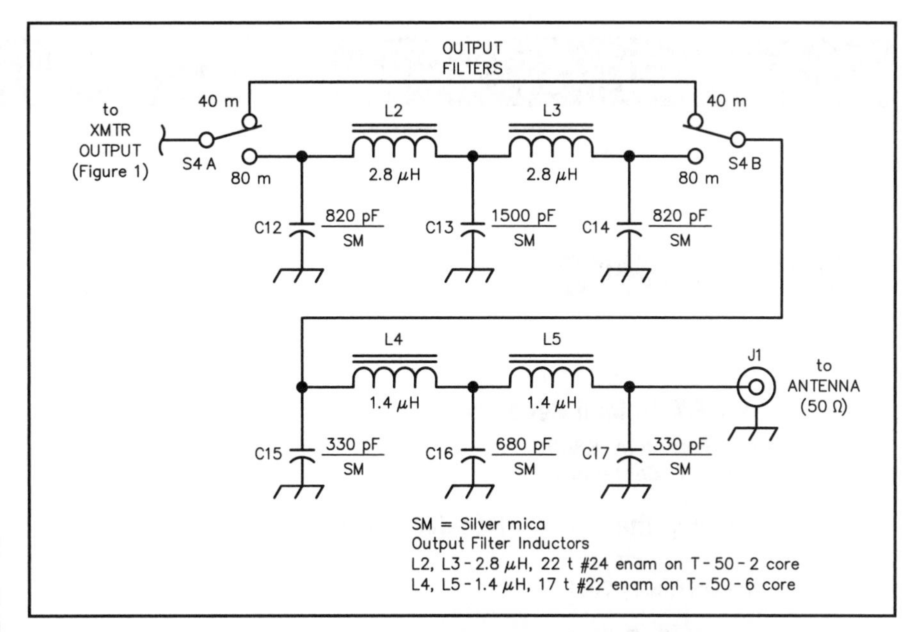

Figure 2—These output filters ensure the transmitter's spectral purity meets FCC requirements. Both filters are in the circuit for 80-meter operation. On 40 meters, the first filter is bypassed. All capacitors are silver mica units.

J1—SO-239 (Digi-Key ARFX1005; RS 278-201)
L2, L3—2.8 µH, 22 turns #24 enameled wire on a T-50-2 powdered-iron toroidal core
L4, L5—1.4 µH, 17 turns #24 enameled wire on a T-50-6 powdered-iron toroidal core
S4—DPDT toggle (RS 275-614)

of the wire into the form through the first hole, then out again through the second. Knot the wire where it enters the form—this will hold the wire securely in place. Then wind the wire tightly onto the form so that there are no spaces between the turns. You may need to push the turns together as you wind.

After you have wound four turns, make the tap by removing some insulation from the wire and soldering a short length of wire to that point. When the winding is finished, hold the wire in place while you drill two more holes in the form at the end of the winding. Feed the wire through these holes as before. Tie a knot at the wire end to hold the coil in place and keep it from unwinding.

With a machine screw and nut, attach a one-inch square TO-220 heat sink to Q1. J1 can be a standard SO-239 connector. Connect the key to the circuit using a standard headphone jack, or simply hard-wire it. S1, S2 and S3 are standard subminiature toggle switches. Use short leads between the switches and associated components. You may want to add a crystal socket if you plan to operate on more than one frequency.

Testing and Operation

Series-connected 6-V lantern batteries are ideal for powering this transmitter—they are low cost and deliver many ampere-hours of energy. A 24 V pack of four lantern batteries should deliver 10 to 20 hours transmitting time. Use duct tape to hold the batteries together.

To test the transmitter, attach the batteries (or other power supply—start with a 24 V, or lower, supply), plug in the crystal and place S2 in the **TUNE** position. Connect a 50 Ω resistor or dummy antenna to J1. (You can use four 200 Ω, 1 W resistors connected in parallel.) With S3 in the center position, close the key and adjust C7 for maximum output as indicated by DS1's glow. Then, move S3 first to the left position, then to the right, retuning C7 each time. The correct position is whichever provides the greatest brilliance on the lamp. Using a 24 V supply, the lamp should be lit to near full brilliance on 3.6 MHz and about half that at 7.1 MHz.

To avoid overheating the op amp or Q1, don't keep the key closed any longer than necessary during tune-up. After the best impedance match has been found, move S2 to the **OPERATE** position (to bypass the lamp and provide full power output). Once testing is complete, remove the resistive load, connect the output to a 50 Ω antenna and repeat this procedure.

The optional piezo buzzer shown in Figure 1 provides sidetone for monitoring your keying. It's a good idea to check your keying quality using a nearby receiver with its antenna disconnected.

Summary

Apply no more than 30 V to this transmitter's dc-input line because the AD811 may be destroyed. Battery operation is more forgiving as the batteries' internal resistance reduces the voltage applied to the transmitter. Use of a "hefty" power supply may cause the AD811 to overheat and that can cause "chirp" (a signal-frequency change during keying). If you want more power output, increase the voltage applied to Q1's drain by breaking the connection to Q1's power-supply line and inserting additional series-connected batteries (or a suitable power supply) to increase the supply voltage to Q1 *only*. If you do this, provide Q1 with a larger heat sink to dissipate the additional heat. Some experimentation with the output matching components may also be needed.

Notes

[1] During ARRL Lab tests for spectral purity, we found the transmitter required more filtering than that provided by the existing output network. The second harmonic was only −22 dBc rather than the −32 dBc required by the FCC. All other harmonics were within legal limits.

[2] PC boards for the transmitter and filters are available from FAR Circuits, 18N640 Field Ct, Dundee, IL 60118-9269, tel 847-836-9148 (voice and fax). Price: $6 plus $1.50 shipping for up to four boards. Visa and MasterCard accepted with a $3 service charge.

Charles (Chuck) Kitchin, N1TEV, is a hardware applications engineer at Analog Devices Semiconductor Division in Wilmington, Massachusetts. His main responsibilities include customer applications support and writing technical publications such as application notes and data sheets. He has published over 50 technical articles and two applications booklets.

Chuck graduated with an ASET from Wentworth Institute in Boston. Afterwards, he continued studying electrical engineering at the University of Lowell's evening division.

Chuck has been an avid radio builder and shortwave listener since childhood, and a licensed radio amateur (Technician Plus) for three years. His other hobbies include astronomy, brewing beer, and oil painting. You can contact Chuck at Analog Devices, MS 126. 804 Woburn St, Wilmington, MA 01887; tel 781-937-1665; fax 781-937-2019; e-mail **charles.kitchin@analog.com**.

Chapter 4

Receivers

By Steve Bornstein, K8IDN

From *QST*, September 1997

The MRX-40 Mini Receiver

Here's a 40-meter receiver you can build in a single evening!

After exploring very low power (QRPp) communication by building a 40 meter Micronaut CW transmitter,[1] I took on the challenge of constructing a tiny 40 meter companion receiver. Not only did I think the receiver would complement the Micronaut, I thought it might also have potential as a kit project for my hometown group, the CQRP (Columbus, Ohio, QRP) Club. The final push to action came from the discovery that there were 93 licensed amateurs in my neighborhood ZIP code. Visions of a local mini-milliwatt net flashed through my imagination!

The result is the MRX-40, a 40 meter CW receiver barely larger than a half dollar. You don't need to have a Micronaut transmitter to use the MRX-40. This receiver can be paired with *any* 40 meter transmitter—low power or otherwise.

Design Details

The main design objectives for the MRX-40 were small size and simplicity. I arbitrarily decided to limit the size of the printed

[1]Notes appear on page 4-2.

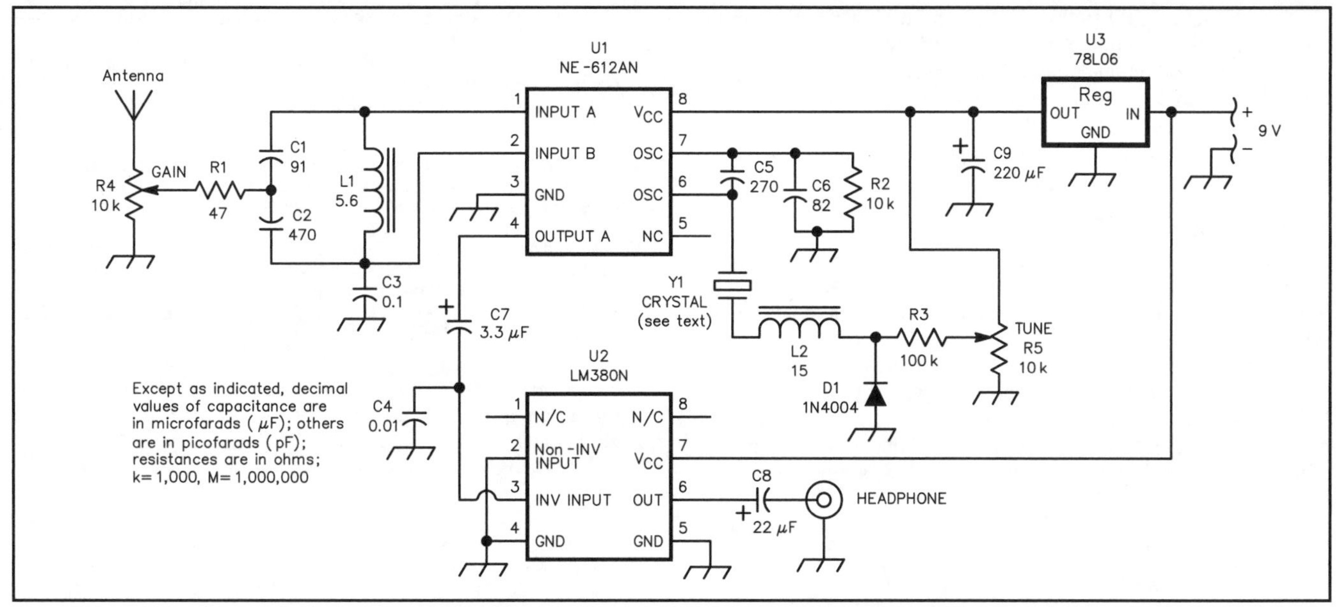

Figure 1—Schematic of the MRX-40 receiver. Equivalent parts can be substituted. With the exceptions noted below, all parts are available from Mouser Electronics, 958 N Main St, Mansfield, TX 76063-4827; tel 800-346-6873.

C1—91 pF ceramic disc capacitor (Mouser 140-CD50S2-091J)
C2—470 pF ceramic disc capacitor (Mouser 140-CD50P2-471K)
C3—0.1 µF monolithic capacitor (Mouser 581-UDZ104K1)
C4—0.01 µF monolithic capacitor (Mouser 581-UEZ103K1)
C5—270 pF monolithic capacitor (Mouser 581-UEC271J1)
C6—82 pF monolithic capacitor (Mouser 581-UEC820J1)
C7—3.3 µF electrolytic capacitor (Mouser 208-50V3.3)
C8—22 µF electrolytic capacitor (Mouser 208-50V22)
C9—220 µF electrolytic capacitor (Mouser 208-10V220)
D1—1N4004 (Mouser 592-1N4004A)
L1—5.6 µH molded choke (Mouser 43LS566)
L2—15 µH molded choke (Mouser 43LS155)
R1—47 Ω, 1/4 W resistor (Mouser 30BJ250-47)
R2—10 kΩ, 1/4 W resistor (Mouser 30BJ250-10K)
R3—100 kΩ, 1/4 W resistor (Mouser 30BJ250-100K)
R4, R5—10 kΩ potentiometers (Mouser 317-2091-10K)
U1—NE-612AN (Dan's Small Parts, Box 3634, Missoula, MT 59806; tel 406-258-2782; http://www.fix.net/dans.html)
U2—LM-380N-8 (Dan's Small Parts; see U1)
U3—78L06ACZ voltage regulator (Mouser 511-78L06ACZ)
Y1—Crystals in HC49U holders for 7040 or 7122 kHz are available for $3 each from Doug Hendricks, KI6DS, 862 Frank Ave, Dos Palos, CA 93620.

From left to right, the Micronaut transmitter, MRX-40 receiver and an equally small key (made by DK7UD).

circuit board (PCB) to 1×2 inches. To accomplish that goal, I used miniature molded chokes and other small components.

The simplicity is in the circuit (see Figure 1). The MRX-40 is a crystal-controlled direct-conversion receiver consisting of an NE-612AN oscillator/mixer chip followed by an LM-380N audio amplifier.

By using a 1N4004 diode as a varicap, the receive frequency can be shifted about 1.5 kHz above or below the crystal frequency. This tuning technique eliminates the need for bulky variable capacitors. The CQRP Club kit[2] includes a crystal for 7040 kHz, the 40 meter QRP frequency. If you're a Novice or Technician Plus, you'll want to substitute a crystal for 7122 kHz. See the parts list in Figure 1 for crystal sources.

The voltage supply to the NE-612 and tuning circuit is regulated by a 78L06 so that the 8 V limit of the NE-612 is not exceeded. On the other hand, the full battery voltage is applied to the LM-380. The audio output is more than ample for Walkman-style headphones. Instead of an audio **VOLUME** control, the MRX-40 uses a **GAIN** control at the antenna input to accomplish the same purpose.

Etching Your Own PCB

The MRX-40 can be built on a piece of perforated board, but you can also opt to etch your own printed circuit board for a neater appearance. The etching template and overlay are available from ARRL Headquarters.[3]

The board layout for the CQRP Club kit was created on a computer using *Easytrax* and *Easyplot* software.[4] Once the circuit board was designed and printed on paper, I transferred the pattern to TEC-200 film[5] using a copy machine. If you've never used TEC-200, you'll find that it is quite handy for single-board production. The image placed on the film is transferred to the circuit board using a clothes iron. The board is then etched in the conventional manner.

The *Easyplot* software can also produce the files necessary for multiboard production on a Gerber Plotter. For our project we produced four files: bottom layer, solder mask, overlay, and drill plot. With these files, a PCB production facility can produce as many boards as you desire. (If you walk into a PCB plant with just a schematic and ask them to do the layout they charge by the hole. For this project board [with its 62 holes], the charge for layout alone would have been about $240!)

Construction

There is really nothing unusual about the construction of the MRX-40. You will be working in a very small area, so a hobby vise is recommended to hold the board steady when soldering. A good set of eyes also helps, as does a 60 W iron with a small tip. I find it best to mount the smallest components first. Sockets for ICs are optional. An enclosure can be made for the MRX-40 from circuit boards soldered together, or anything else you have available.

Operation

Connect a 9 V battery to the receiver and check the voltage at pin 8 of U3. It should be about 6 V. If you can check the current drain from the battery, you should find something in the range of 16 to 17 mA.

Assuming that the voltage and current measurements are normal, you should be home free. No alignment is necessary. You can check the local oscillator function by listening to it with another 40 meter receiver. Now fire up your 40 meter CW transmitter and enjoy!

Notes

[1]Micronaut transmitter kits are available from Dave Ingram, K4TWJ, 4941 Scenic View Dr, Birmingham, AL 35210. They are $15 (without crystal), plus $2 shipping and handling.

[2]MRX-40 receiver kits are available for $18 from Steve Bornstein, K8IDN, 475 East North Broadway, Columbus, OH 43214. The kit contains all parts, PCB with mask and overlay, jacks, controls, and a step-by-step manual.

[3]A PC-board template package is available from the ARRL, at a cost of $2 for members, or $4 for nonmembers. Send your request for the BORNSTEIN MRX-40 TEMPLATE along with a business-size SASE to the Technical Department Secretary, 225 Main St, Newington, CT 06111-1494.

[4]You'll find a demo version of *Easyplot* software on the World Wide Web at **http://www.tol.mmb.com/E147**. *EasyTrax* freeware can be downloaded at **http://www.protel.com/download.htm**.

[5]TEC-200 film is available from Meadow Lake Corp, 25 Blanchard Dr, Box 497, Northport, NY 11768.

475 East North Broadway
Columbus, OH 43214
e-mail saborns@freenet.columbus.oh.us

Rescaling the MRX-40 Receiver for 80 Meters

Last October I found myself on a flight to Tyndall Air Force Base in Florida to meet my new grandson, Kyle Charles Stanfield II, born Sept 27, 2000, to my daughter, Gwen, KB4UNT, and her husband, Kyle, KF4TIV. Before leaving, I put a request on the QRP-L e-mail reflector (**qrp-l@lehigh.edu**) for ideas or designs for a companion receiver kit to match the NoGANaut 80-meter transmitter. Several folks replied, including Mike Boatright, KO4WX, who said he'd meet me at the Atlanta airport during my layover and show me his newly completed design.

After a two-hour flight, I met Mike in Atlanta. He took me into the Delta Air Lines Crown Room, where we had a great time talking QRP. Mike pulled out his new receiver, a redesign of the MRX-40 ("The MRX-40 Mini Receiver" by Steve Bornstein, K8IDN, *QST* September 1997, page 59), for use on 80 meters. Mike furnished me with a basic parts kit, a PC board and a set of modification instructions. I decided to build the receiver while on vacation in Florida.

Subbing Parts

Figure 1 shows the modified schematic of the MRX-80 receiver, set up for operation on the NoGA (North Georgia QRP Club) net frequency, 3686.4 kHz. Mike substituted the NE612 (U1) used on the MRX-40 with a NE602 since the latter has 4 dB more conversion gain. The LM380 (U2) was also changed to a LM386, which is a more easily obtainable part. Although the LM386 has less output power, it is adequate for use with headphones. C10 was included in the audio stage to increase the LM386's gain to approximately 200 (46 dB). Finally, the 6-V regulator used in the 40 meter version was replaced with a 78L08 (8-V) regulator. The 78L08 is about top end for the mixer chip, but it yields an additional 1dB of conversion gain in the NE602.

In order to make the MRX-40 work on 80 meters, the LC ratios had to be rescaled. C1 worked out to 172 pF. By paralleling one 150 pF with a 22 pF capacitor, the desired value is achieved. Mike thinks a 180 pF would also work

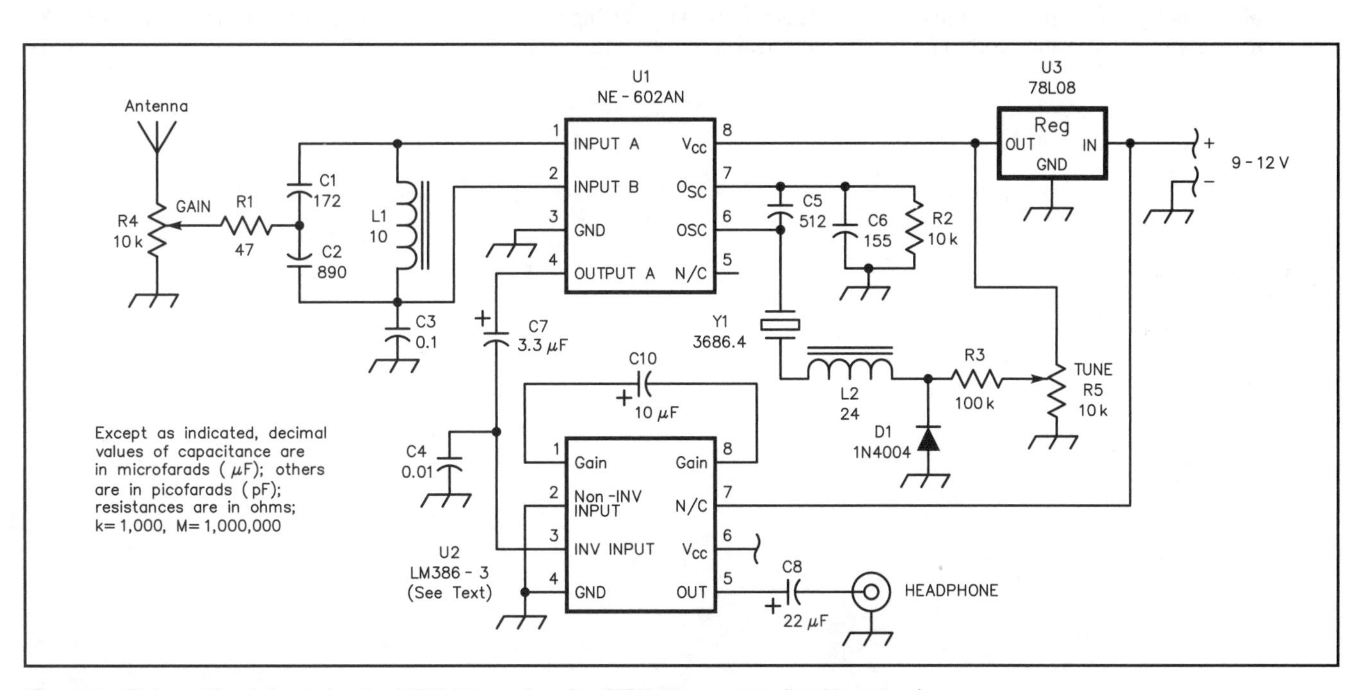

Figure 1—Schematic diagram for the MRX-80 receiver (an MRX-40 converted to 80 meters).

Component Changes:

U1, the NE612, was replaced with an NE602. Either component will work in this circuit, however, the NE602 has 4 dB more conversion gain than the NE612.

U2, the LM380, was replaced with an LM386. The LM386 is a more easily obtainable part (available at most Radio-Shack stores) than the LM380N-8. However, it does have less power output. The LM386-3 provides up to 700 mW of power, more than adequate for headphone use. Note that C10 was added to the circuit to boost the LM386 gain to 46 dB (200).

U3, the 78L06 was replaced with a 78L08. Either component will work in this circuit, but running the NE602 at 8 V yields an extra 1 dB of conversion gain.

RF Component Scalings:

C1—scaled to 172 pF (X_c = 250 Ω); not a standard value, but can be made using a 150 pF and a 22 pF (or 20 pF) in parallel. A 180 pF would probably work also.

C2—scaled to 890 pF (X_c = 48 Ω); not a standard value, but can be made using an 820 pF and a 68 pF (or 82 pF) in parallel.

C5—scaled to 512 pF (X_c = 84 Ω); not a standard value, but 470 pF works just fine as a replacement.

C6—scaled to 155 pF (X_c = 277 Ω); not a standard value, but 150 pF is close enough.

L1—scaled to 10.6 μH (X_l = 246 Ω); can use standard value 10 μH molded inductor

L2—scaled to 28 μH (X_l = 660 Ω); can use standard value 27 μH or 33 μH molded inductor.

Rich Arland, K7SZ ♦ 25 Amherst Ave, Wilkes Barre, PA 18702 ♦ k7sz@arrl.org well. Same with C2, which scaled to 890 pF. By paralleling an 820 pF with a 68 pF you can get very close to the 890 pF capacitance. C5 and C6 were rescaled to 512 pF and 155 pF respectively. A 470 pF works for C5 while 150 pF works for C6. L1 and L2 are rescaled to 10.6 μH and 28 μH respectively. Standard-value 10 μH and 27 μH molded inductors work just fine for these two inductors.

Mike's PC board is what I call "Ugly Manhattan Style" construction. Everything is soldered to the circuit pads *etched on top of the PC board.* This saves drilling holes and, in the case of standard "Manhattan Style" construction, gluing the small pieces of PC material to the main PC board groundplane. I'd never tried this style construction before and was pleasantly surprised at how easily and rapidly the circuit went together. Total construction time was around one hour. I clipped a 9-V battery to the circuit, plugged in the headphones and wire antenna and was rewarded with CW near the 3686.4 kHz operating frequency. There are virtually no alignment steps after the receiver is built and working.

Figure 2—My version of the MRX-80.

Just plug everything in and listen to 80 meters.

Great Little Receiver

How well does this little receiver work? Considering the simplicity of design and minimal parts count, I am amazed! Granted there is only a 500-600 Hz swing using the tune control, which severely limits the tunable reception, but for coupling to a NoGANaut transmitter or KnightLite SMiTe transmitter (also on 3686.4 kHz) the MRX-80 is just the ticket. There is no AF filtering to speak of, so you hear a lot of stuff around the operating frequency. It would not be hard to include an active AF filter on output of the receiver to help limit the AF bandwidth. Figure 2 shows my MRX-80 receiver.

Mike's idea of rescaling the MRX-40 into an 80-meter receiver was brilliant. The overall simplicity of design and common parts means that most people can build one on a shoestring budget. While the MRX-80 is not on par with a superhet design, it certainly will work well for the intended purpose, which is 80 meter net operations. You could pair the MRX-80 with any 80-meter transmitter. Those folks who modified their Tuna-Tin II transmitters for operation on 80 now have a companion receiver.

This is a fun project. My thanks to Mike Boatright, KO4WX, for sharing his redesign of the MRX-40 with the readers of "QRP Power." This receiver should be dead simple to duplicate using perfboard or Manhattan Style construction. Parts are easily obtainable and there's not much to go wrong. Should you desire a one-night project, why not try the MRX-80 receiver? QST

A Cascade Regenerative Receiver

Extreme selectivity—here's a receiver with two *regenerative stages.*

By Bill Young, WD5HOH

A reference to a cascade regenerative receiver can be found on page 78, Fig 47 of *Vacuum Tubes in Wireless Communication* originally published in 1918.[1] The tickler coils in the 1918 circuit are located at or near circuit ground as they are in my cascade regenerative receiver and in the regenerative superheterodyne receiver recently published in *QEX*.[2] I used two circuits that I have used in other receivers that seemed amenable to being connected in cascade. I wanted to try a receiver design that offered the gain and selectivity of a regenerative receiver incorporating two tuned circuits without any of the complications associated with heterodyning. This cascade regenerative receiver tunes from just above 3 MHz to just above 5 MHz. The selectivity, in my opinion, justifies the effort expended to build and operate the extra regenerative stage. One must learn to turn one knob only about half as far as usual and then turn the second knob about the same distance to avoid tuning past stations. In retrospect, I would say that anyone building a cascade regenerative receiver should buy the best vernier drives available.

I'm sure some readers will wonder why I have written an article about a regenerative receiver in 2003 when other experimenters are working on software defined receivers and other advanced concepts. Interest in regenerative receivers hasn't gone away. Every now and then, one appears in print. I have written this article because other experimenters might like to build this or a similar circuit

Nothing in a regenerative receiver should move with respect to anything else except variable capacitor rotors, potentiometer wipers and switch contacts. Even those items can cause problems. Conducting paths should not move or vibrate. Everything must be tied down. Components that exhibit any tendency to move when held by their leads alone should be held down with hot glue. The largest possible wire gage should used. Use lacing cord liberally. Where two or more wires run together for any distance (assuming it's electrically permissible for them to do so) they should be tied together. It need not be pretty; but it cannot move. I have made this receiver as mechanically and electrically stable as I could, and since it's primarily used to tune AM signals with the RF amplifiers at high gain but not oscillating, it's very stable and free of microphonics.

There is a common-source, untuned RF stage between the antenna and the first regenerative RF stage. This serves—as similar stages have for over 80 years—to isolate the regenerative circuits from the antenna, and it presents high impedance to the electrically short wire antenna. The high impedance that the antenna sees makes receiver performance somewhat independent of antenna length. The "gimmick" coupling capacitor allows surprisingly high signal strength on strong signals and, so far, has eliminated any interference from nearby strong stations.

There are two regenerative RF amplifier stages ahead of the bridge detector. Coupling between the first and second stages and the second and third stages is accomplished by 1.8 pF to 10 pF trimmer capacitors. These coupling capacitors should be adjusted for the best possible regeneration control for both stages.

The first regenerative stage design was worked out experimentally with a breadboard circuit. The 560 Ω drain load was determined by temporarily installing a resistance decade box as a drain load and varying the drain load resistance until regeneration occurred.

Each of the two JFET regenerative stages has a tickler winding in series with the drain load at the bottom or "cold" end of the drain load. This circuit was worked out experimentally with "breadboard" receivers, and it results in better, more positive control of regeneration in my opinion. The apparent reason for this is that changing the regeneration controls in this circuit does not change the drain load impedance very much. My article "A Mathematical Model for Regenerative RF Amplifiers,"[3] is a brief discussion of the relationship between regeneration control and drain load impedance.

[1]Notes appear on page 4-9.

343 Forest Lake Dr
Seabrook, TX 77586
blyoung@hal-pc.org

Also, each tickler winding is positioned directly over the tuned circuit winding. This configuration results in better control of regeneration, but I don't really know why. I've tried the conventional configuration with the tickler winding separated from the tuned circuit winding by a small gap, but regeneration control was rough and erratic that way. I suspect that this behavior is unique to this circuit. Other experimenters apparently get better results with a separated tickler winding and their "leak" detector circuits, but this circuit works best with the tickler winding over the center of the tuned winding. Furthermore, the ratio of tickler turns to tuned-winding turns was determined by trial and error in earlier breadboard circuits, but once determined, seems to remain about the same over a frequency range from below 2 MHz to above 5 MHz, as long as the regeneration control resistances remain about the same.

Each 140 pF tuning capacitor is mounted to the aluminum chassis by its rear mounting lug alone. This makes tuning a little smoother by allowing some motion of the capacitor. The first regenerative stage (second stage) has a "floating rotor" FINE tuning capacitor which consists of a moving aluminum plate or rotor held by a nylon screw and turned by a small vernier drive (see Fig 1). The rotor swings in and out of the space between a grounded plate screwed to the chassis and an insulated plate held at a fixed distance from the grounded plate by nylon screws and spacers. The rotor or moving plate is shaped to have a greater perimeter with less area. The capacitance of the floating rotor capacitor appears to depend more on the "fringe effect" at the edges of the plates and less on the area of the plates. This concept has come from experience with an earlier receiver that incorporated an earlier version of the floating rotor capacitor (see Note 2). The observed change in capacitance is much less than can be accounted for by applying the usual expression for the capacitance of parallel plates with air dielectric. Further experimentation is called for, but the floating rotor capacitor as it exists is very useful when connected in parallel with the tuning

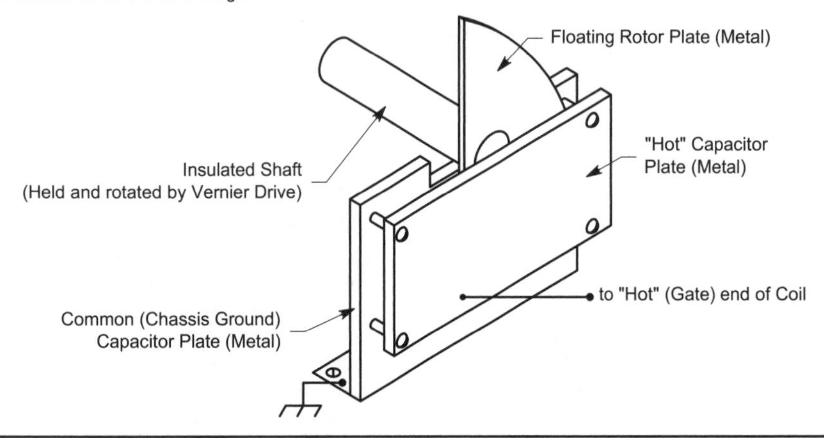

Fig 1—The floating-rotor capacitor.

Tuning and Transformer Resonance

I suspect that self-resonance of a bifilar coupling transformer driving a diode bridge detector results in increased "smoothness" of regeneration control in receivers of this type.

A 90-turn bifilar transformer similar but not identical to T3 exhibited a "dip" or resonance very near 17 MHz. This dip was present with power on and with power off. Regeneration control was smoother near 17 MHz than anywhere else within its tuning range, although the 19-meter band above 15 MHz was almost as good. The 22 meter band just below 14 MHz was noticeably worse, as was the 13 meter band just below 22 MHz. This result, if it's correct seems to agree with the information published in my earlier *QEX* article (see Note 3). Self-resonance would result in increased impedance and improved control. I further suspect that if self-resonance was occurring in the receiver referred to above (not the present cascade regenerative receiver), the only reason the stage wasn't oscillating uncontrollably was the diode-bridge load on the secondary of the bifilar transformer.

The most obvious application of the self-resonant bifilar transformer is in a regenerative intermediate amplifier stage as part of a superheterodyne receiver. The gate tuned circuit and the bifilar transformer could be made resonant at a chosen frequency and left there. Designing such a receiver may not be straightforward, though. Arranging for optimum coupling between a doubly or triply balance mixer and the regenerative IF stage will require some thought.

It may be possible to establish self-resonance of the bifilar transformer at the upper end of the tuning range of a regenerative receiver and switch small, fixed capacitors across the primary of the bifilar transformer to extend resonance across the rest of the tuning range. I'm sure, however, that there's a limit to how far the Q of the bifilar transformer can be increased without loss of regeneration control.

Fig 2—A schematic of the Cascade Regenerative Receiver. The Very Fine regeneration controls were added after the photo in Fig 3 was taken. The MPS2222A stage is an audio preamplifier circuit from the *1992 ARRL Handbook*. Unless otherwise specified, use 1/4 W, 5%-tolerance carbon composition or film resistors. You can contact Mouser at 958 N Main St, Mansfield, TX 76063; tel 817-483-4422; fax 817-483-0931; e-mail sales@mouser.com; Web www.mouser.com.

B1—Rayovac NM1604, 150 mAh 8.4 V (for 9 V applications, rechargeable).
C1—"Gimmick" capacitor, 4 turns of insulated solid hook-up wire.
C2, C5—1.8-10 pF ceramic trimmer capacitor (Mouser 242-1810).
C3—140 pF variable capacitor.
C4—Floating-rotor capacitor, see text and Fig 1.
C5—1 µF 25 V, low leakage capacitor.
FB—Fair-Rite EMI shield bead (Mouser 623-2643000101).
S1—SPST ganged with the fine regeneration pot in the third stage.
T1—Transformer 22 t primary 3/4 inch long on a 7/8 inch PEX form, with a 4 t secondary wound over the primary's cold end. The lead to the potentiometers is RG-174 coax. Shield the transformer by covering it with a small metal food can, such as that for Mandarin oranges.
T2—Transformer 22 t primary 3/4 inch long on a 7/8 inch PEX form, with a 4 t secondary wound over the primary's cold end. The lead to the potentiometers is a twisted pair. Shield the transformer by covering it with a small metal food can, such as that for Mandarin oranges.
T3—Bifilar transformer, 90 t 24 AWG enameled wire twisted, and then wound over two inches on a 3/8 inch diameter PEX form. T3 is mounted inside the aluminum chassis box.

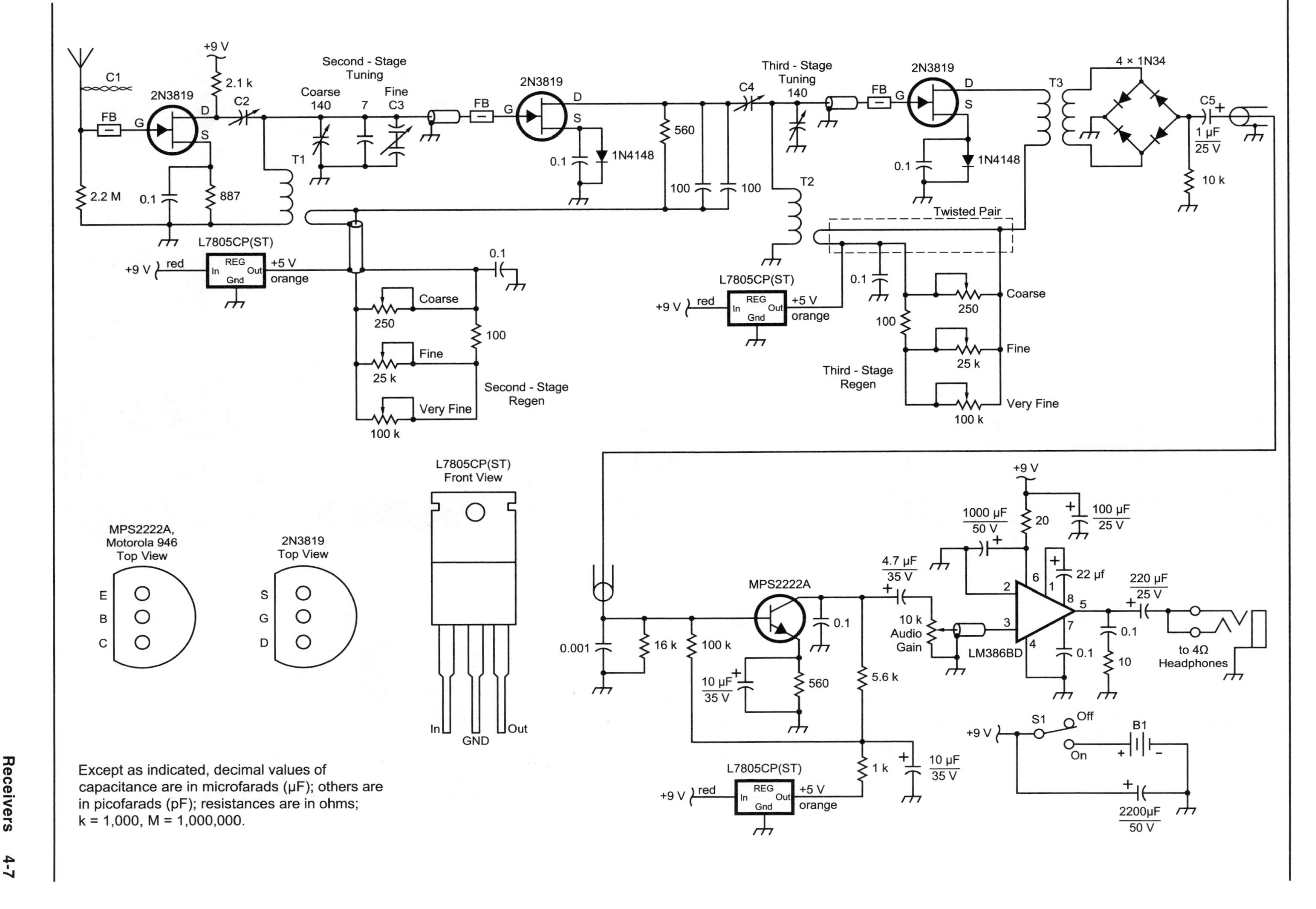
C1
2N3819
FB
G
D
S
+9 V
2.1 k
C2
2.2 M
0.1
887
T1
Second - Stage
Tuning
Coarse
140
7
Fine
C3
2N3819
0.1
1N4148
560
100
100
C4
Third - Stage
Tuning
140
T2
2N3819
1N4148
T3
4 × 1N34
C5
1 µF
25 V
10 k
Twisted Pair
L7805CP(ST)
+9 V
red
REG
In
Out
Gnd
+5 V
orange
Coarse
250
Fine
25 k
Very Fine
100 k
Second - Stage
Regen
Third - Stage
Regen
L7805CP(ST)
Front View
In
GND
Out
MPS2222A,
Motorola 946
Top View
E
B
C
2N3819
Top View
S
G
D
MPS2222A
0.001
16 k
100 k
10 µF
35 V
560
5.6 k
1 k
4.7 µF
35 V
10 k
Audio
Gain
LM386BD
1000 µF
50 V
20
100 µF
25 V
22 µf
220 µF
25 V
10
to 4Ω
Headphones
S1
Off
On
B1
2200µF
50 V
Except as indicated, decimal values of capacitance are in microfarads (µF); others are in picofarads (pF); resistances are in ohms; k = 1,000, M = 1,000,000.

capacitor of a regenerative stage. It enables very accurate tuning of AM signals. This accurate tuning when used together with precise control of regeneration made possible by the combination of COARSE, FINE and VERY FINE regeneration control potentiometers results in improved sensitivity and selectivity. An increment of one small division of the FINE tuning control can result in a noticeable change in signal level, so the FINE tuning control does serve a useful purpose.

Some builders of regenerative receivers advocate a single-point ground. The circuit board for this receiver has a buss wire "fence" around the edge of the board. The "fence" does not form a closed loop. All ground connections are made to this wire which is then connected to a single ground lug held to the aluminum chassis by a machine screw and nut. There are several other connections at some distance from the board made in the same way with a lug and a machine screw. The receiver performs well, so I assume the grounding scheme is adequate.

I have attempted to evaluate potential RF coil form materials by placing them, one at a time, in the field of an RF coil connected to an oscillating RF regenerative amplifier stage. I set the tuning for an audible beat note and then placed a small sample of the material in the field of the coil. I have tried cardboard, glass, wood, PVC, acrylic, PEX (cross-linked polyethylene), the much recommended black-plastic film container and a plastic "pill bottle." Each of these changed the frequency of oscillation substantially, some more than others. Each was suspended in the field of the coil at the end of a length of waxed nylon lacing cord. The length of lacing cord was shown to affect the frequency very little by itself. I decided by a process of repeated comparison that the PEX would be the material of choice. It doesn't crack as easily as acrylic, and it's easier to cut and drill. The two coils are mounted above the chassis deck on square pieces of Vector board, and each coil is enclosed in a cylindrical metal shield can (formerly filled with Mandarin orange slices).

I have been reminded recently that conventional regenerative detectors tend to exhibit capture effect, where a strong signal close in frequency to a weak signal takes over the receiver from the weak signal. That does not appear to happen with this circuit. Strong signals are audible "behind" or "under" weak signals, but there is no capture.

Each of the regenerative RF stages incorporates a silicon 1N4148 diode in the source circuit of the 2N3819 JFET. This idea was suggested in an unpublished private correspondence with Charles Kitchen, N1TEV. It is something he said he had thought of but not tried. I tried it and have been much impressed with the performance it contributes. It makes the receiver much more capable of tuning AM signals below oscillation and somewhat less capable of tuning CW signals above oscillation. This trade off is fine with me. I have been trying to do this for years with partial success. The diode seems to delay the onset of oscillation allowing higher gain below oscillation.

The cascade regenerative receiver was originally constructed as a single board receiver. All of the transistors, diodes and the integrated circuits were mounted on a single perforated board. The board is mounted about an inch below the underside of the chassis deck on six nylon standoff insulators. I have now added a single MPS2222A junction-transistor audio preamplifier taken from page 28-5 of the *1992 ARRL Handbook* and mounted on a small board adjacent to the main board. I removed the transformer-coupled 2N3819 audio amplifier circuit originally placed between the bridge detector and the LM386 because it contributed too little gain and caused an annoying audible "quench" frequency. The MPS2222A stage works well, and I'm confident that the receiver can be built as a single-board receiver. There's plenty of room on the main board for the MPS2222A stage.

The LM386 final audio stage is conventional except possibly for the 2200 µF capacitor across the 9 V dc supply, which is necessary to prevent "motorboating" or instability.

The cascade regenerative receiver can be operated as follows: Start by setting both main tuning capacitors to about 80 on their logging scales. With FINE and VERY FINE regeneration controls fully clockwise adjust the COARSE control so that the stage is just oscillating. Do this at reduced audio gain. Then turn the FINE control counter-clockwise so that oscillation just stops and then go back clockwise with the FINE control until the stage is just oscillating. Now, turn the VERY FINE control counter-clockwise until the stage stops oscillating, and use the VERY FINE control to adjust stage gain just short of oscillation. Turn the AUDIO GAIN pot clockwise to increase audio gain. This procedure can be followed with each of the two regenerative RF stages. Some careful tuning will help as these adjustments are made. It may be necessary to repeat some of the adjustments until the receiver has stabilized after several minutes, but it will stabilize and can be maintained at high gain for extended listening.

Now, turn each of the two main tuning knobs to discover which one causes oscillation to stop when tuned in the direction you want to tune. Oscillation can then be started again by turning the other main tuning knob in the same direction. Proceed to tune incrementally this way (turning one main tuning knob, then, the other) until an interesting signal is heard.

When you have acquired a signal adjust the FINE regeneration controls appropriately depending on what you're trying to receive (AM, CW, RTTY and so on). For AM signals, back

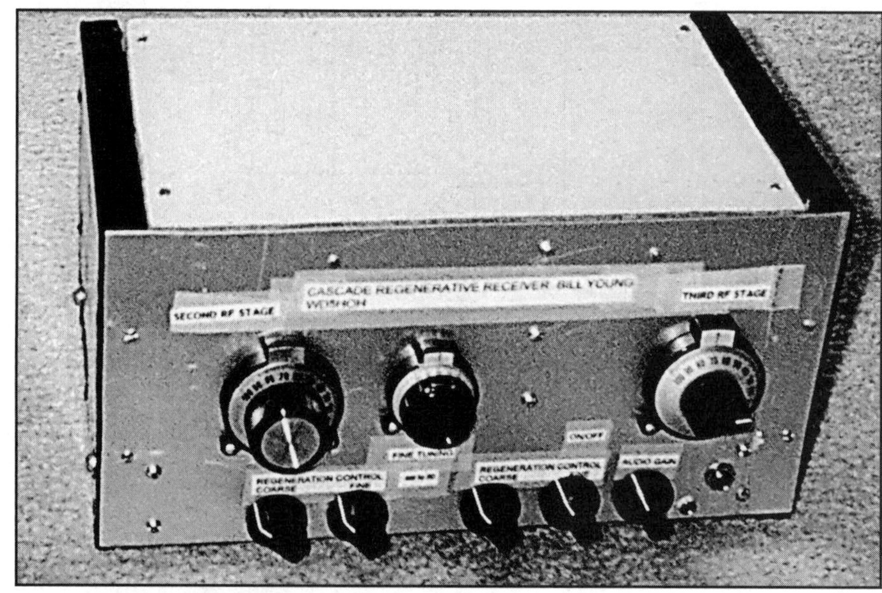

Fig 3—Front view of receiver.

off the FINE controls until you have a clear AM signal, and then advance each VERY FINE control in turn for maximum signal level. Work with the two FINE regeneration controls and the VERY FINE tuning control to tune exactly to the center of an AM signal at maximum gain. Increasing the VERY FINE regeneration controls beyond a certain point results in distortion and the onset of "fringe howl." It's easy with this selective receiver to tune past a signal without hearing it.

If you have reached the conclusion that this receiver is laborious to tune, you're right. It is. Now you know one reason why the superhet came to be dominant. If you need to switch quickly and repeatedly between several air-to-ground frequencies, you really can't do it with a receiver like this. That's one of the things a crystal controlled superhet could do well.

On the other hand it's fun to tune in a weak signal, peak both RF stages to that signal and then pull it up out of the "mud" using the two VERY FINE regeneration controls without either stage going into oscillation. A disadvantage of this receiver is the presence of the "skirts" of powerful shortwave broadcast signals tens of kilohertz away from their carrier frequency. However, most of the weaker signals will increase in signal strength as they are tuned in and as the regeneration is increased enough to overcome this interference. At my location, this has been a problem only around 5 MHz.

There are a couple of things about the performance of this receiver that are puzzling:

1. Even if both regenerative stages are oscillating and are not tuned to the same frequency there is no audible heterodyne.
2. There is some interaction between the two regenerative RF stages at high gain even though their coils are shielded and they are both voltage regulated.

This receiver was powered at first by disposable 9 V alkaline batteries, but I am now using a Rayovac NM 1604 NiMH rechargeable battery with good results. This battery powers this receiver for about three hours between charges. The internal impedance is low enough for good performance until the battery voltage drops off the "plateau" and needs recharging. The transition from plateau to a weak battery is abrupt. It suddenly becomes necessary to increase regeneration to maintain gain, but gain decays faster than regeneration is increased. At this point switch the receiver off, and recharge the battery.

You will probably notice that the ON/OFF switch is built into one of the FINE regeneration controls rather than the audio gain potentiometer as is usual. This choice is a result of parts availability only. If you can find a 10 kΩ pot with a switch, use it.

Bill is retired following a 36-year career as a project engineer and manager with NASA in the biomedical-hardware area. He was first licensed as KN5DNM in about 1953 and has been WD5HOH (General class) since about 1980. He holds a BSEE from the University of Texas (1961) and an MS in environmental management from the University of Houston (1981) at Clear Lake.

Notes

[1]E. Bucher, *Vacuum Tubes in Wireless Communication,* (New York: Wireless Press, 1918).

[2]B. Young, WD5HOH, "A Homebrew Regenerative Superheterodyne Receiver," *QEX*, May/Jun 2002, pp 26-35.

[3]W. Young, WD5HOH, "A Mathematical Model for Regenerative RF Amplifiers," Jul/Aug 2001, pp 53-54.

By Charles Kitchin, N1TEV

From *QST*, December 1997

An Ultra-Simple Receiver For 6 Meters

Here is a simple VHF receiver you can build without any special components or test equipment.

This receiver uses superregeneration for high sensitivity and low parts count. It can receive both FM and AM modulated signals. This design differs from previous superregenerative circuits because it uses a "quench waveform" control to allow the reception of narrow-band FM. Receiver sensitivity is around 1 μV. Builders can easily modify the radio to operate over a wide band of VHF. It is inexpensive (about $20), can be built quite compactly and powered from a 9 or 12 V battery.

The performance of this rig does not equal that of modern commercial transceivers, but you can build it yourself and be monitoring all types of local communications in a few hours. This includes 6 meters and the adjacent frequencies. With easy modifications, you can receive police, snowplows, fire stations, telephone paging, maintenance crews, etc on VHF. This receiver is also useful for low-power wireless data links. As with any regenerative set, you will need practice and patience in learning to adjust the receiver's controls for best performance.

Regenerative Receivers

Regenerative receivers use a special type of detector that is essentially a user-controlled oscillator. In a straight regenerative circuit, the input signal couples to the detector, and some of the output signal is fed back to its input, in phase. This repeatedly amplifies the input signal. The result is very high gain in a single stage. If we allow the feedback to go past the point of oscillation, the circuit's gain stops increasing and starts decreasing, as most of the transistor's energy works to maintain the oscillation. Some type of regeneration control is necessary, so that you can keep the feedback at a point just short of oscillation. Using this technique, a single transistor or JFET can achieve circuit gains of 20,000 easily.

The superregenerative circuit uses an oscillating regenerative detector that automatically stops or "quenches" the oscillations periodically. This allows the input signal to build up to the oscillation point repeatedly, providing single-stage gains close to 1 million, even at UHF. These detectors can use two approaches for the required quenching: Either a separate lower-frequency oscillator supplies the quenching signal (*separately quenched circuitry*), or a single JFET can produce both oscillations (a *self-quenched circuit*), as shown here.

A Superregenerative Receiver for 49 to 55 MHz

The circuit shown in Figure 1 consists of an RF stage, a superregenerative detector and an audio amplifier. The common-gate RF stage, Q1, provides RF gain and helps prevent the receiver from radiating its signal out the antenna.

The detector, Q2, operates as a grounded-gate oscillator. C4 applies in-phase feedback between the JFET's source and drain. RFC2 raises Q2's source above ground (at RF) enough for oscillation to take place.

R3 provides bias for the JFET and, together with an RC network, provides the necessary quenching oscillations. The time constant set by C8A, C8B, R7 and bias resistor R3 is deliberately made long enough so that the dc-bias level across R3 increases until it inhibits the oscillating detector. The bias voltage then discharges through the network until the bias is low enough for oscillations to start again. This creates the necessary quenching action that produces the superregenerative effect.

The received signal from the RF stage couples to the detector through a small "gimmick" capacitor made by twisting together two one-inch-long pieces of #20 AWG insulated hook-up wire. You can also use a 1 or 2 pF mica capacitor in place of the gimmick.

The detector's operating voltage is set by the 10 kΩ **REGENERATION CONTROL**. This control affects both sensitivity and selectivity. Because the detector is a modulated oscillator, it generates a double-sideband signal. Increasing regeneration (more voltage applied to the detector) increases sensitivity but also generates greater sidebands that reduce selectivity (the sidebands interfere with a narrow-band signal).

The **QUENCH-WAVEFORM-ADJUST** potentiometer, R7, adds a small resistance in series with C8 that changes the quench waveform from its normal sawtooth shape to a sine wave. A sine wave is a much

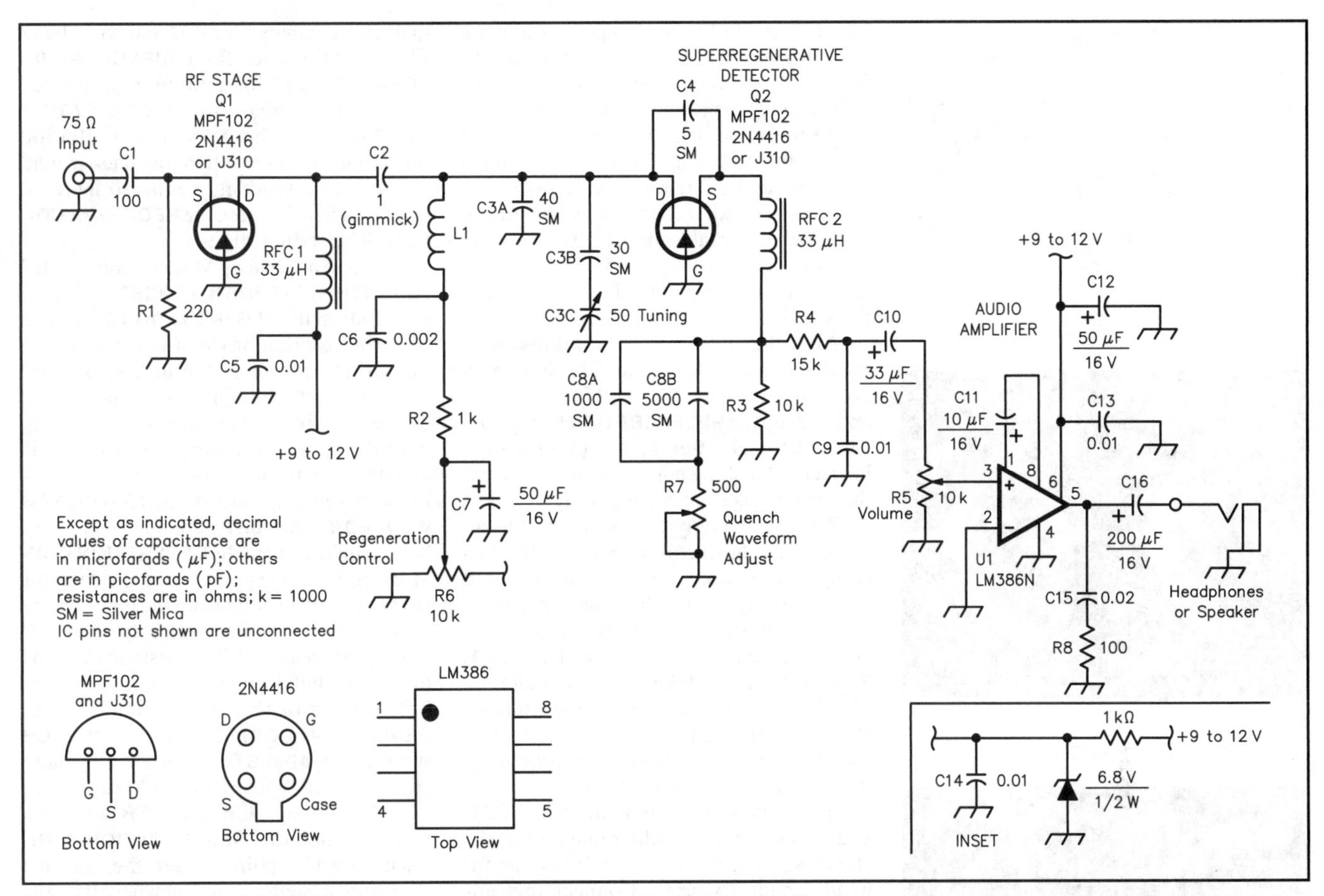

Figure 1—Schematic of the VHF receiver. Unless otherwise specified, use 1/4 W, 5%-tolerance carbon composition or film resistors. Equivalent parts may be substituted. Digi-Key Corp, 701 Brooks Ave S, PO Box 677, Thief River Falls, MN 56701-0677; tel 800-344-4539 (800-DIGI-KEY), fax 218-681-3380; URL **http://www.digikey.com/**

C2—Gimmick capacitor (see text)
C3C—50 pF variable
C7, C12—50 μF, 16 V electrolytic
C10—33 μF, 16 V electrolytic
C11—10 μF, 16 V electrolytic
C16—200 μF, 16 V electrolytic
L1—7 turns, (air-core) #14 AWG solid copper wire space wound 3/4 inch long on a 0.25 inch form (a pencil).
R6—10 kΩ, 10-turn pot
RFC1, RFC2—33 μH (Digi-Key M7330-ND)

cleaner waveform (with fewer harmonics) than a sawtooth, so the sidebands are smaller and selectivity is much better. The oscilloscope photos of Figure 2 show the quenched RF envelope of the receiver with and without R7.

A simple low-pass filter (R4 and C9) removes the quench voltage from the detector's audio output. The output of the detector drives an LM386 audio-amplifier IC.

The receiver can be connected to a discone or other 75 Ω antenna via coax cable, or you can use TV twinlead cable to make a folded-dipole antenna. For a 6 meter dipole, use a nine-foot length of 300 Ω twinlead for the antenna. Solder the two wires at each end of the twinlead together, then cut one of the two twinlead wires in the center of its length. Solder the transmission line, a second piece of twinlead, to the cut ends at that point (solder two places). A good antenna greatly increases the number of narrow-band stations you can receive.

Construction

Stray circuit capacitances and multiple ground paths can prevent the detector from oscillating. It is vitally important that the detector's tuning coil (L1) be located away from other conductive objects—particularly chassis ground, the bottom and sides of the equipment box and any other metal object.

Avoid mounting the tuning coil on a printed circuit board: This loads the detector so that it fails to oscillate properly, if at all. A hand-wired universal breadboard works fine as long as the detector coil mounts well above it, or you can just use a piece of copper-clad board and some terminal strips (ie, solder lugs). Suspend the components above the board on the lugs, or you can use the parts that have grounded leads as standoffs to hold the other components above the board. (Some call this "dead bug" or "ugly" construction.)

Put the completed circuit inside a small box or use a block of wood and a piece of metal for the front panel. If you plan to place the entire receiver inside a closed metal box, build the circuit outside the box first, and be sure it oscillates properly before placing it inside.

It is very important to mount the **TUNING** capacitor, C3C, directly onto the board and pass its shaft through an oversized hole in the front panel: Avoid mounting it directly to a metal front panel. If the capacitor's frame contacts both the front panel and the ground plane, it creates a multiple ground path (ground loop), which usually prevents the detector from oscillating. Mount all other controls directly on the front panel and connect them to the board using the shortest leads possible. Use shielded wire to connect the **VOLUME** and **QUENCH WAVEFORM ADJUST** controls to the PC board. You can connect the **REGENERATION CONTROL** to the circuitry with a twisted pair of wire leads. Connect C13 directly to U1, pin 8.

I recommend a 10-turn potentiometer for the **REGENERATION CONTROL** and a reduction drive for the **TUNING** capacitor. These make the receiver much easier to operate.

Always build receiver circuits backwards. Start with the audio stage. Build the circuitry from the speaker to the **VOLUME** control. Then test the stage by turning the

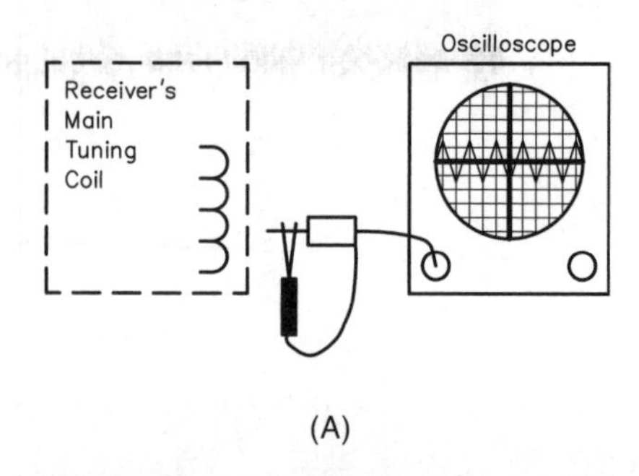

(A)

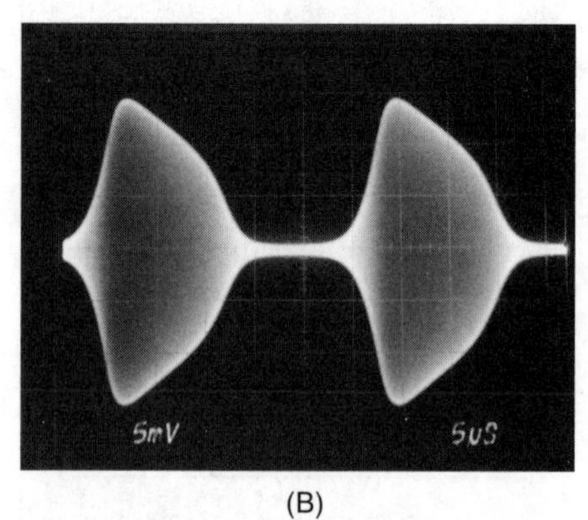

(B)

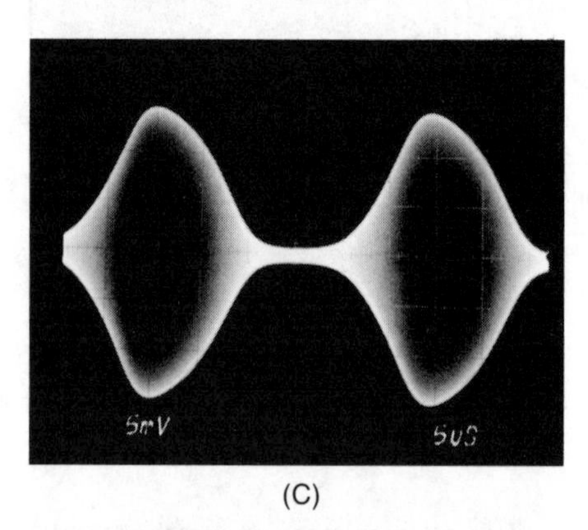

(C)

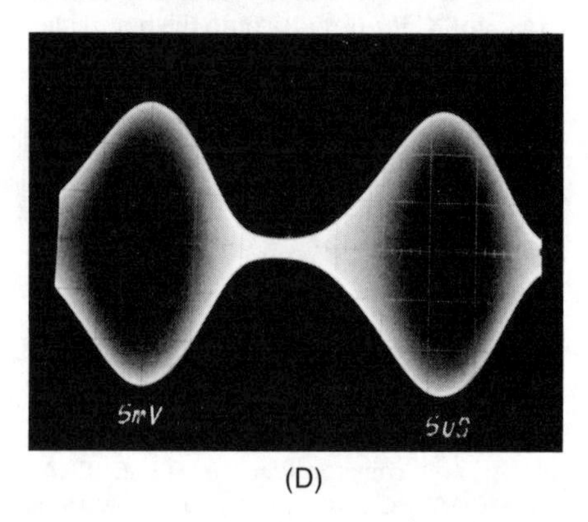

(D)

Figure 2—The effect of **QUENCH WAVEFORM ADJUST** control on the shape of the detector's oscillation waveform. A shows the test arrangement. The oscilloscope was coupled to the receiver by connecting the probe's ground clip to its tip and placing the probe tip near the receiver's main tuning coil. B, C and D show the waveforms with R7 set to 0, 250 and 500 Ω, respectively.

VOLUME control to midrange and placing your finger on the wiper (listen for a buzz). (In this test, your body serves as an antenna for the radio to pick up the noise from surrounding ac wiring. If you have no ac power, the test won't work. Then you'll need an audio signal source.—*Ed.*) If there's no sound, recheck the wiring or use a voltmeter to troubleshoot the problem. Be sure the supply voltage is present and that the voltage on pin 5 of the LM386 is half of the supply voltage.

After the audio stage is working, wire the detector, but leave out C2. Now, with no load on the detector, set R7 to midrange and turn-up the **REGENERATION CONTROL**, R6, until oscillation starts. (You should hear a loud rushing noise that indicates that the detector is superregenerating.)

RFC2 is the only component in the receiver that is at all critical. Since individual component layouts (and RF chokes) will vary, you may need to do some experimentation to get the detector oscillating properly. With a 5 pF value for C4 and the 33 µH RF choke specified (Digi-Key part number M7330-ND), the detector should oscillate strongly. If it doesn't, check the wiring very carefully. If the wiring is okay, try changing the value of the RF choke.

Here's how to do it with an RS 273-102C RFC: First, unsolder one end of the choke winding from its lead. Remove (unwind) about 15 turns. Connect this unwound end to Q2's source and the other end to the junction of R3 and R4. Switch the radio on. Does it oscillate? If not, unwind more turns with the set operating, a few at a time, until there's a strong oscillation. Then, kill the power. Bend the coil's free end over to its lead, solder it in place and cut-off the extra wire. Connect the lead to Q2 and do a final test.

Miscellaneous

For optimum sensitivity from this receiver, use a fresh battery. A 9 V transistor radio battery is fine for portable use. Two series-connected 6 V lantern batteries will operate this receiver for many months.

You can expand or reduce the receiver's tuning range by varying the values of C3A and C3B. C3B sets the total tuning range, so you can use different values of tuning capacitor other than the 50 pF specified. Make C3A's value greater to lower the tuning range. Likewise, you can compress or expand the turns on the main tuning coil for the same effect.

Operation

For the best performance, this receiver needs to have its regeneration level reset every time its tuning changes. The **REGENERATION CONTROL** changes the voltage that powers the detector. Higher detector voltages increase sensitivity but they broaden the selectivity.

In these self-quenched circuits, the **REGENERATION CONTROL** also varies the quench frequency. For AM and wide-band FM reception, set R7 (**QUENCH WAVEFORM ADJUST**) for minimum resistance and simply increase the **REGENERATION CONTROL** past the detector's oscillation threshold to a point where the background (mush) noise suddenly begins to increase rapidly. Then decrease the **REGENERATION CONTROL** setting slightly.

For narrow-band FM reception, set R7 (**QUENCH WAVEFORM ADJUST**) at midscale, adjust the **REGENERATION CONTROL** for strong oscillation (high sensitivity) and tune in the carrier of the desired station. After tuning to the center frequency of the carrier, decrease the regeneration level until the audio level increases sharply. (If you decrease the level too much, the detector will squeal.) Adjusting R7 (**QUENCH WAVEFORM ADJUST**) creates a narrow-band window on the **REGENERATION CONTROL** between the point where the detector first begins to oscillate and the point where (narrow-band) audio begins to drop off rapidly. Increasing R7's resistance widens this region but decreases detector sensitivity. Because of their interaction, the **REGENERATION CONTROL** and the **QUENCH WAVEFORM ADJUST** control need repeated adjustment for narrow-band FM reception.

You can copy CW and SSB with this receiver. Set the **REGENERATION CONTROL** to a low point, where the detector stops superregenerating, but where it is still oscillating. The receiver now operates as a straight regenerative set.

You can easily convert this receiver to operate on other bands. For 2 meters, make the following component changes: Omit C8A and C3A, change C3B to approximately 15 pF, change C3C to a 25 pF variable capacitor, change C4 to 2 pF, and change RFC1 and RFC2 to 15 µH; L1 is 3 turns, 1 inch long. Add a 1 kΩ resistor and 6.8 V Zener diode before the **REGENERATION CONTROL** (see Figure 1 inset) for increased stability on the higher bands, but that's not needed at 6 meters.

Charles Kitchin is a hardware applications engineer at Analog Devices Semiconductor Division in Wilmington, Massachusetts, where he has been employed for the past 21 years. His main responsibilities include customer applications support and writing technical publications such as application notes and data sheets. He has published over 50 technical articles and two applications booklets. Chuck graduated with an ASET from Wentworth Institute in Boston, and afterward spent many years studying electrical engineering at the University of Lowell's evening division. Chuck has been an avid radio builder and shortwave listener since childhood and a licensed radio amateur (Tech Plus) for two years. His other hobbies include astronomy, beer brewing and oil painting. You can reach Chuck at 804 Woburn St, Wilmington, MA 01887; tel 781-937-1665, fax 781-937-2019; e-mail **Charles.Kitchin@analog.com.**

By Dan Wissell, N1BYT

From *QST*, August 2001

The WBR Receiver

Build this simple receiver and "bridge" the gap between regenerative and direct-conversion receivers!

Despite the well-known drawbacks of regenerative receivers, the elegance and simplicity of the regenerative detector is still appealing. I'm always looking for a better way to implement Armstrong's brilliant design, and with the introduction of the Optically Coupled Regenerative Receiver (OCR),[1] the major problems of the regenerative detector were overcome. The OCR receiver demonstrated the potential of this nearly 90-year-old design to hold its own as a simple all-mode receiver. The design is still quite complex, however, and relies on expensive, hard-to-find, electro-optical components with limited bandwidths.

The key to a simple regenerative receiver design is coupling the antenna directly to the oscillating detector. Anyone who has ever tried to couple an antenna directly to a regenerative detector has found that signals from dc to daylight show up everywhere on the tuning dial—and all at the same time! Overcoming this problem by isolating the antenna from the detector causes the design to become as complex, or more so, than that of a simple direct-conversion receiver. Because of this complexity factor, regenerative detectors have largely given way to other simpler circuits.

The future of Major Armstrong's namesake may be more open-ended, however, because a simple and effective solution to the coupling problem has been found. The method of coupling the antenna to the tank circuit described below is reminiscent of a Wheatstone Bridge circuit, and thus the receiver name, "Wheatstone Bridge Regenerative (WBR) Receiver." I'm reluctant to claim that this is a "new" detector design, even though an extensive search hasn't yielded anything similar. But with nearly 90 years of use, I'm sure every method of detector-antenna coupling has been tried at one time or another!

This WBR may very well be the ultimate simple, *high-performance* regenerative receiver. As an added plus, the design virtually eliminates the negative aspects of regenerative receivers such as antenna radiation, frequency pulling, microphonics and hand capacitance effects. A printed circuit board is available to speed construction of this project.[2]

[1]Notes appear on page 4-16.

Design Overview

The schematic of the WBR Receiver is shown in Figure 1. The basic circuitry is the same as that used in the OCR Receiver. The two most significant differences are the removal of the optocoupler from the oscillator, and the oscillator tank circuit configuration.

The highly stable Colpitts oscillator and infinite-impedance detector have been retained in this design. The major difference is that the oscillation is now controlled by directly varying the base current of Q1 (with R5 and related components) instead of using photons from the optocoupler LED.

The tank circuit is comprised of inductor L1, capacitors C7 and C8, along with tuning diode D1. This circuit is redrawn in Figure 2 to highlight the unusual component arrangement. C7, C8 and D1 have been omitted in Figure 2 for clarity. As shown, Figure 2 represents a classic Wheatstone Bridge circuit. Inductor L1 is center-tapped, with equal inductance on both sides of the tap. Capacitor C1 represents the oscillator and detector load capacitance. Balancing capacitor C2 is selected to match the value of capacitor C1. In this ideal case, the bridge is balanced and there is no voltage present at the center tap. The full oscillator voltage appears at nodes V1 and V2. Because no voltage is present at the center tap, it could be grounded—*or an antenna could be directly connected to this point without impacting the oscillator signal.*

In this design, the antenna is coupled to the center tap of L1 through an impedance represented by Z1. This is simply a one-inch length of wire connected to ground. The antenna is connected at the midpoint of Z1. This provides a low-impedance connection point for the antenna, as well as providing a dc ground for detector Q2 and tuning diode D1.

In practice, the bridge can't be perfectly balanced because the oscillator load capacitance changes as the level of regeneration is changed. Despite that, this arrangement still yields a *significant reduction* in oscillator voltage at the center tap of L1. Voltage measurements taken at 7 MHz show that the voltage present at the center tap of L1 is about 46 dB less than at nodes V1 and V2. The practical impact of this is good antenna isolation. When monitoring the oscillator signal from a WBR on a communications receiver, the WBR antenna can be removed and reconnected *with no perceptible change in the audio beat note from the communications receiver!* It turns out that if the oscillator coupling capacitance and the balancing

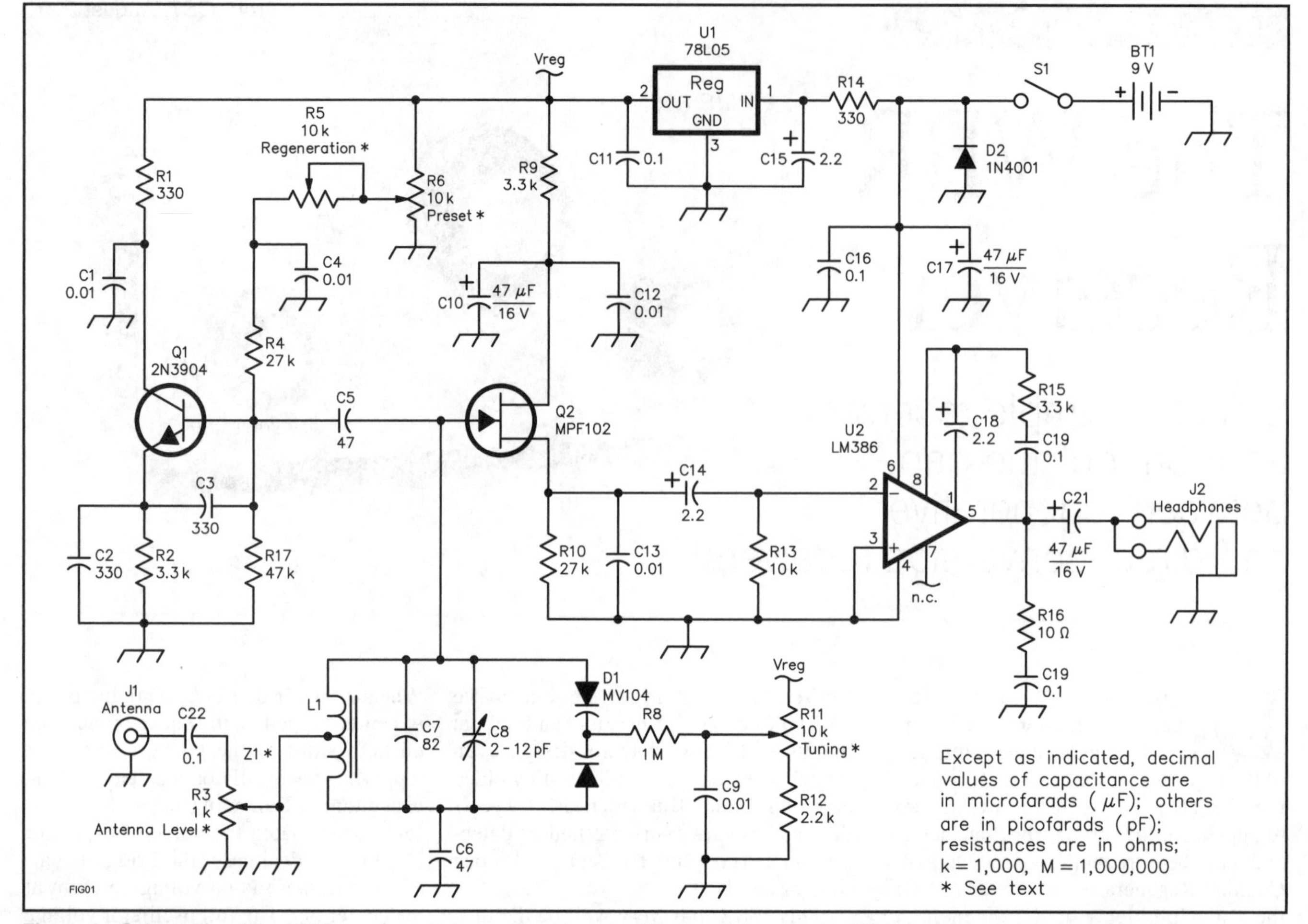

Figure 1—Schematic of the WBR receiver. Unless otherwise specified, resistors are 1/4-W 5% tolerance carbon composition.

C2, C3—330 pF, 5% NPO
C5, C6—47 pF, 5% NPO
C7—82 pF, 5% NPO
C8—2-12 pF NPO
C1, C4, C9, C12, C13, C19, C22—0.01 µF
C11, C16, C20—0.1 µF
C10, C17, C21—47-µF, 16-V electrolytic
C14, C15, C18—2.2 µF, 16 V electrolytic
D1—MV104
D2—1N4001
J2—Three-conductor phone jack, 1/8 inch.
L1—Approximately 3.7 µH: 28 turns of #22, center tapped, on T-68-6 core (yellow).
Q1—2N3904
Q2—MPF102
R1, R14—330 Ω
R2, R9, R15—3.3 kΩ
R3—1 kΩ linear-taper potentiometer. Panel mount.
R4, R10—27 kΩ
R5—10 kΩ linear-taper potentiometer. Panel mount.
R6—10 kΩ linear-taper potentiometer. Panel or PWB mount.
R7—47 kΩ
R8—1 MΩ
R11—10 kΩ, 10-turn potentiometer. Digi-Key # 3590S-1-103-ND.
R12—2.2 kΩ
R13—10 kΩ
R16—10 Ω
S1—SPST
U1—78L05
U2—LM386

capacitor (C5 and C6 in Figure 1) are matched, a good balance can be obtained. If the oscillator design is changed, the balancing capacitor may have to be made variable to null the circuit.

Diode D1, a voltage-variable capacitor (VVC), is used to tune the oscillator. A low-cost plastic 10-turn potentiometer is used as the main tuning control (R11). Resistor R12 is used to set the lower voltage limit at D1 to about 0.9 V, below which the capacitance change of D1 is quite small. Regulator U1 is used to provide a stable voltage source for D1, Q1 and Q2. Regeneration is controlled by R5, a single-turn, panel-mounted potentiometer. R6 is used as a "preset" for R5 and allows for smooth regeneration control.

To keep the overall design simple, only a single stage of audio amplification is used (U2). This provides adequate volume for headphone operation when using a simple 40-meter dipole antenna. Reducing the signal level applied to the detector via R3 controls the headphone volume. A 9-V battery supplies power for the WBR. A well-filtered bench supply in the 8- to 13.8-V range may also be used. Diode D2 is added as a safety measure. The receiver works well with dipole or random wire antennas and an earth ground.

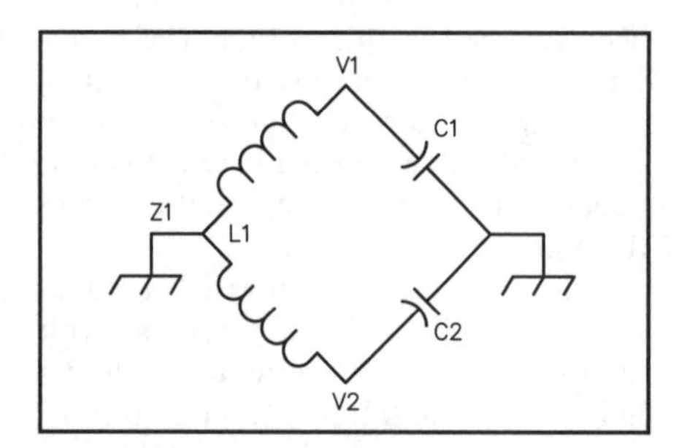

Figure 2— The tank circuit is comprised of inductor L1, capacitors C7 and C8, along with tuning diode D1.

Constructing a 40-Meter WBR Receiver

The caption of Figure 1 contains the parts list for a 40-meter version of the receiver. The parts are available from a variety of suppliers. With the given values, the receiver will tune the entire 40-meter band.

A printed-circuit board is available and contains most of the components, but the circuit is quite simple and lends itself to "dead-bug"-style construction on a bare copper PC board. The only critical part is the oscillator circuit. Note that

NP0 capacitors are used to enhance frequency stability. Short, direct leads should be used in this part of the receiver. Make the circuit as mechanically robust as possible to improve stability. I typically use high-value resistors as standoff supports for signal components. For power-related parts of the circuit I often use ceramic capacitors in the range of 0.01 to 0.1 µF as supports. This also adds additional power-supply bypassing.

Inductor L1 is wound with #22 enamel wire. It's easy to create the center tap by using two separate windings. Start with two 15-inch lengths of wire. Wind the first 14 turns of L1, remembering that one pass through the center of the core is one "full" turn. Leave about 1 inch of wire for connections. The winding should fill about 40% of the core. Add the second 14 turns, as above, winding in the same direction as the first winding. The second winding should start next to one end of the first winding. Again, leave about 1 inch of wire for connections. Connect the end of the first winding to the start of the second to create the center tap.

As mentioned previously, Z1 is a one-inch length of #20 solid copper wire connected from the center tap of L1 to ground. The antenna connection is made at about the midpoint of Z1. While it's tempting to increase the amount of impedance at Z1, it's not a good idea because of the potential for detector overload, especially at 5 to 15 MHz, where strong AM stations dominate.

The regeneration preset control (R6) can be a small PWB-style unit or a standard panel-mount type. Because it's a "set and forget" control, it may be placed in any convenient location.

The hardware used for antenna connector J1 and switch S1 may be whatever the builder prefers. A fully enclosed case isn't required for good operation. My prototype WBR receivers are built as open breadboards and work well.

I have kept the complexity of the WBR design at a minimum to encourage builders to give it a try. For those wishing to add loudspeaker operation, however, or to increase the sensitivity of the receiver, I would recommend adding the audio preamp and volume control used in the OCR II receiver.[3] As presented, the basic oscillator will work up to about 18 MHz with D1 and C7 removed. The upper frequency is limited by the combination of capacitors C5 and C6. If desired, the frequency-dependent portion of the design (C7 and C8) can be scaled for other frequencies of interest in the lower HF region. The tuning voltage applied to D1 will need to be adjusted to provide the desired tuning range at other frequency ranges.

A rear view of the WBR receiver.

Checkout and Operation

Carefully check your work before applying power for the first time. Once everything has been checked, plug in headphones at J2 and apply power. Advance the regeneration control (R5) to about 75% of its maximum setting. Adjust the regeneration preset (R6) until a distinct increase in background noise is heard. This indicates that Q1 is oscillating and that the audio section is working. Varying the regeneration control should produce a smooth transition when going into and out of oscillation. The oscillator can now be set to the correct operating frequency. Set the main tuning (R5) potentiometer to its minimum setting. With the regeneration control set to the point of oscillation, adjust C8 while listening for the signal on your station receiver or a communications receiver set for CW reception at 7.00 MHz. You will probably need to connect a short wire from the antenna connector on your station receiver and place it near the WBR to receive a signal. Once the frequency has been set, connect an antenna to J1 and you're all set! If a station or communication receiver isn't available, connect an antenna and adjust C8 until the CW portion of the band is found. Continue setting C8 until the lower edge of CW subband can be determined. This is best done in the evening when there is plenty of CW activity.

Using the WBR receiver will take some practice if you've never used a regenerative receiver before. Maximum sensitivity is obtained in the area just before oscillation (for AM reception) and just at oscillation (for CW). For SSB reception, the best operating point is usually found at a point that's just a bit past the setting required for CW reception. You will get the "feel" of the receiver quickly. The interaction of the regeneration, gain and selectivity controls will become apparent.

Summary

The Wheatstone Bridge Regenerative Receiver works as well as its predecessor, the OCR. It has the added advantages of greater bandwidth, increased simplic-

This ultra-tight close-up illustrates the so-called "dead-bug construction" that the author used in this version of the WBR. As an alternative, a circuit board is available from FAR circuits.[2]

ity and a much lower cost. It virtually eliminates the negative aspects of the regenerative receivers that came before it. This simple receiver is well suited for beginners who would like to build a simple all-mode shortwave receiver. QRP ops and home-brewers in general will also be interested in the WBR. Given that the antenna is quite isolated from the oscillator, the WBR can be used as a simple receiver for transmitter-receiver operation. It could easily be paired with a simple crystal-controlled transmitter, creating a small, portable "trans-receiver."

Notes

[1]Daniel Wissell, N1BYT, "The OCR Receiver," Jun 1998 *QST*, pp 35-38.

[2]Circuit boards are available for $4 (plus $1.50 shipping and handling) from FAR Circuits, 18N640 Field Ct, Dundee, IL 60118; tel 847-836-9148; **www.cl.ais.net/farcir/**.

[3]Daniel Wissell, N1BYT, "The OCR II Receiver," Sep 2000 *QST*, pp 35-38.

You can contact the author at 7 Notre Dame Rd, Acton, MA 01720-2108; **n1byt@arrl.net**.

By Dan Wissell, N1BYT

From *QST*, September 2000

The OCR II Receiver

Here's the radio a number of readers have been asking for: A simple, all-mode shortwave receiver based on the combination of the popular SLR and OCR receiver designs.

Since the introduction of the SLR (shielded loop receiver)[1] and OCR (optically coupled regenerative)[2] receiver designs in *QST*, I have been gratified by the overwhelmingly positive response from builders of these receivers. Many builders have asked the same question: "How can I convert the receiver to cover a wider frequency range?" Independently converting the SLR or the OCR to cover a broader frequency range poses design challenges. Being a simple D-C design, it's easy to make the SLR cover a broader frequency range, but this receiver is not suitable for good AM shortwave reception. On the other hand, the OCR is by design an all-mode receiver, but it's quite difficult to make it cover a broader frequency range. To answer the question, I combined the SLR and OCR designs to produce an all-mode multiband (ie, 3.5 to 8.5 MHz) shortwave receiver that I'll describe here.

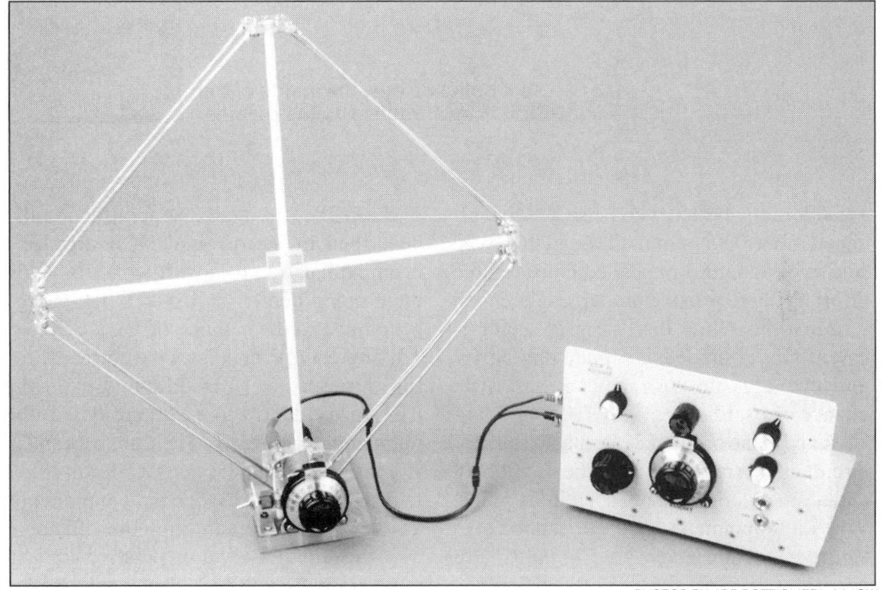

PHOTOS BY JOE BOTTIGLIERI, AA1GW

The challenge I faced was designing a receiver that retains the qualities of both earlier unique designs. For the SLR, these qualities include its sensitivity and ability to use a small loop antenna to reduce local noise pickup. The OCR offers the extraordinary performance of the optically isolated regenerative detector, providing all-mode operation. The receiver presented here answers the design challenge, yet contains about the same number of components as used in either *one* of the other receivers. Because I've provided a means of using simple random-wire antennas *and* a tuned-loop antenna, I've dubbed this receiver the OCR II. A PC board and kit of parts are available to speed construction.[3] I encourage you to review the previous two *QST* articles to gain a greater insight into the evolution of this design (see Notes 1 and 2).

The Receiver Circuit

Overview

Refer to Figure 1. The OCR II is basically a single-conversion receiver with a 455-kHz IF. An incoming 3.5- to 8.5-MHz signal is converted to the IF, amplified and presented to the detector, which is an OCR operating at a fixed frequency of 455 kHz. An audio preamplifier and a headphone amplifier follow the OCR. This approach is similar to that employed by simple receiver designs of the 1950s and 1960s that use oscillating (regenerative) detectors at a fixed IF. There is, however, no comparison between the performance of those earlier detectors and the better OCR!

Description

As in the SLR design, the receiver's converter employs an SA602 mixer, U1. L1 and **BANDSET** capacitor C9 control U1's internal oscillator frequency. Tuning diode D1 provides bandspread. The oscillator tunes between about 3 and 8 MHz. This provides coverage of about 3.5 to 8.5 MHz without the need for a band switch. This tuning range includes 80 and 40 meters and a number of popular shortwave bands. It's possible to operate the mixer at higher frequencies, but more-complicated oscillator circuits are needed to achieve the required frequency stability. A preselector consisting of Q1 and related components precedes the converter. The preselector allows the use of simple wire antennas. T1 and C1 form a tuned circuit providing receiver front-end selectivity that helps minimize images. R2 at the gate of Q1 reduces the T1/C1 tuned-circuit Q and sufficiently broadens the tuning so that a vernier drive is not required with C1.

The incoming-signal level can be attenuated by R1, a 1-kΩ potentiometer. Providing attenuation control is important with the SA602 mixer. Overloading the mixer creates many unwanted mixing products that produce considerable audio hash. The loop antenna used in the original SLR makes it very difficult to overload the mixer. That's one of the reasons the apparent sensitivity and selectivity of the SLR receiver design are so good.

Broadband transformer T2 converts the single-ended low-impedance output of Q1 into a fairly well balanced 3-kΩ input impedance required by the mixer. As I found with the SLR receiver, the SA602 works

[1]Notes appear on page 4-22.

A front-panel view of the OCR II.

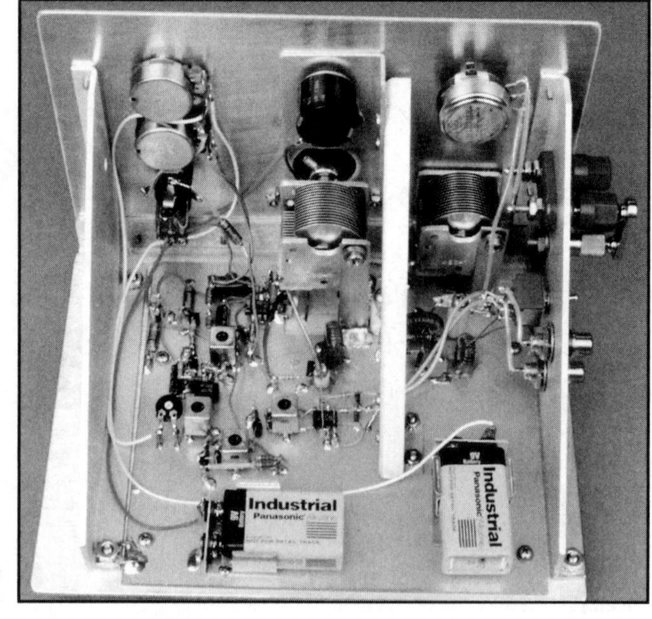

To a builder's eye, the inside of the OCR II is as attractive as the outside.

considerably better when used with balanced inputs and outputs. I found this to be true even though I expended considerable effort trying to provide proper single-ended terminations. Note that the preselector is essentially an impedance-matching buffer and provides no gain, thus it has little chance of oscillating.

As with the original SLR, the preselector circuit can be removed from the signal path and a tuned loop antenna used in its place. When a loop antenna is used, connections to the mixer are made via C4 and C5. In general, I find that there is little difference in receiver performance with the loop antenna or the combination of the preselector and a modest wire antenna. However, by properly positioning the loop antenna, you can null local noise sources and strong broadcast stations—the wire antenna alone cannot accomplish this feat.

U1's output is terminated in the primary of T3, a 455-kHz IF transformer. (I use these IF transformers wherever possible because they're inexpensive and allow a good range of impedance-matching flexibility.) T3's secondary is terminated by R7. This presents an approximate 3-kΩ termination impedance to U1. Q2 and T4 form a tuned 455-kHz amplifier. Q3 is used as an impedance-matching stage between T4 and T5. This is necessary because the secondary of T5 is terminated in a relatively low (and variable) impedance of regeneration controls R13 and R14. If not for the buffering action of Q3, this would impact the IF amplifier and could result in unwanted oscillations.

The 455-kHz energy is coupled to linear optocoupler U2 via the secondary of T5. An Agilent (formerly Hewlett-Packard) HCPL4562 linear optocoupler is the heart of the OCR. Although its operation is fully described in the original OCR article, a brief explanation of how it works is worth mentioning here. The 455-kHz RF energy is coupled to the cathode of U2's LED via T5. This energy modulates the current flowing through the LED. Photons from the LED provide the base current for the optocoupler transistor. The transistor in U2 is configured as a 455-kHz Colpitts oscillator using L2 and associated components. The current flowing through the LED controls the circuit oscillation creating an ideal regenerative oscillator. The magic in this design is that by virtue of the LED, both the RF energy and the regeneration control are *totally isolated* from the sensitive areas of the oscillator, such as the tank circuit. This technique delivers a very well behaved regenerative detector with none of the infamous regenerative detector problems. An infinite impedance detector (Q4) recovers the audio, as opposed to a transformer and RF-choke scheme often employed with regenerative circuits.

The detected audio is band-pass filtered by C22, C25, C28 and R20 and R22. These components along with Q5 form the audio preamplifier. A ubiquitous LM386 (U4) is used as the headphone or loudspeaker amplifier.

Regulated voltage is supplied by U3, an LM78L05 three-terminal regulator. The regulated voltage is used at U1, U2 and tuning diode D1.

Construction Details

One of the more difficult tasks in designing a project such as the OCR II is component selection. When building a single unit, one-off parts, such as those found at flea markets or in junk boxes, are okay to use. But when developing a design to be copied, every effort must be made to use readily available parts. This, in turn, often forces design decisions that may appear arbitrary. An example of this process is the trade-off required when deciding how to implement the tuning in the OCR II. Some of the various options included a band switch, pluggable coils or an external VFO. Each choice has its own set of complications including part availability and cost. For this project, I decided to use common 365-pF air-dielectric variable capacitors to eliminate more complex band-switching circuits that require good-quality switches. Because such switches cost about the same as the capacitor, I used the latter. A vernier dial is used with the **BANDSET** capacitor. Besides making the tuning easier, it provides a calibration scale.

For the **TUNING** control, however, I decided in favor of a low-cost tuning diode and 10-turn potentiometer. Both of these decisions are based upon the availability of reasonably priced components from at least one reliable source.[4] In general, I have taken a minimum-component design approach, consistent with the desired receiver performance. No components can be eliminated and still retain good circuit performance. The bulk of the parts are available from standard suppliers. The HCPL-4562 (U1) is stocked by Newark Electronics.

The frequency-dependent portion of the U1 oscillator design can be scaled for other frequencies of interest in the lower HF region. However, for operation above about 10 MHz, consider using an external, well-shielded VFO for improved stability.

Instead of the PC board, you may use any of the standard construction techniques such as point-to-point wiring on a copper-

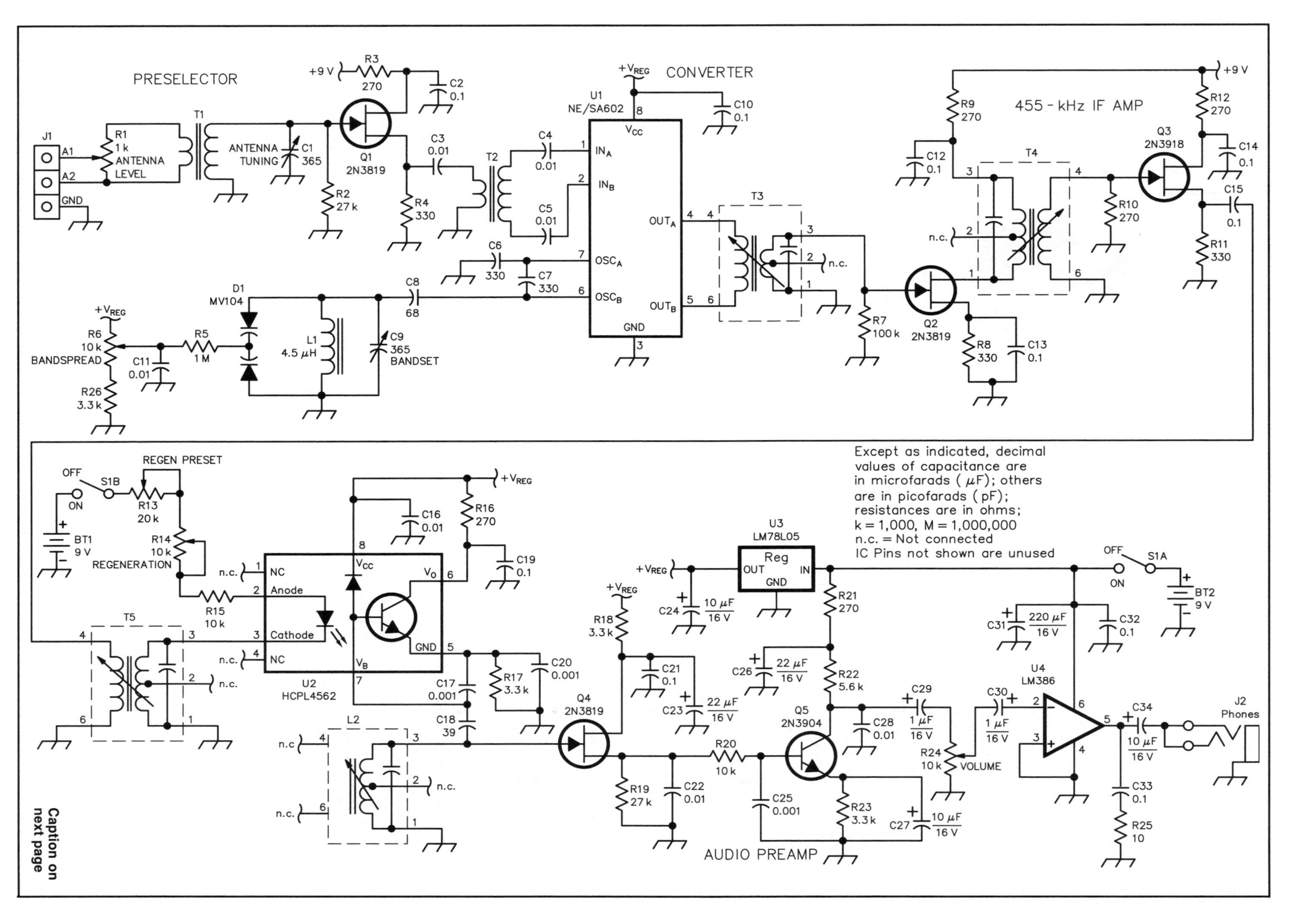

Caption on next page

Figure 1—Schematic of the OCR II circuit. Unless otherwise specified, resistors are $^1/_4$-W, 5%-tolerance carbon-composition or metal-film units. RS part numbers in parentheses are RadioShack. Other suppliers include Digi-Key Corp, 701 Brooks Ave S, Thief River Falls, MN 56701-0677; tel 800-344-4539, 218-681-6674, fax 218-681-3380; http://www.digikey.com and Newark Electronics, 4801 N Ravenswood Ave, Chicago, IL 06040-4496; tel 800-463-9275, 312-784-5100, fax 312-907-5217; http://www.newark.com. Also, see Notes 3 and 4. Equivalent parts can be substituted; n.c. indicates no connection.

BT1—9 V
BT2—9 to 12 V; see text.
C1, C9—365 pF air-dielectric variable
C6, C7—330 pF, 5% NP0; see text.
C8—68 pF, 5% NP0; see text.
C24, C27, C34—10 µF, 16 V electrolytic
C23, C26—22 µF, 16 V electrolytic
C29, C30—1 µF, 16 V electrolytic
C31—220 µF, 16 V electrolytic
D1—MV104 varactor
J1—Spring-action terminal strip (RS 274-315)
J2—Three-conductor phone jack (RS 274-249)
L1—Approx. 4.5 µH, 28 turns #24 enameled wire on a T-68-2 core
L2—0.64 mH variable coil Toko RMC-2A6597HM (Digi-Key TK1302)
Q1-Q4—2N3819, MPF102 (RS 276-2035)
Q5—2N3904, 2N2222
R1—1-kΩ, linear-taper pot (RS 271-1715)
R6—10-kΩ, linear-taper, 10-turn panel-mount pot; see text.
R13—20-kΩ, linear-taper, PC-board mount pot; see text.
R14—5-kΩ, linear-taper, panel-mount pot; see text.
R24—10-kΩ audio-taper pot (RS 271-1721)
S1—DPDT (RS 275-614)
T1—Pri: 2 turns #26 enameled wire; sec: 35 turns #26 enameled wire on T-68-2 core
T2—Pri: 10 turns #26 enameled wire; sec: 25 turns #26 enameled wire on T-50-43 core
T3, T4, T5—0.64 mH variable coil, Toko RMC-502182NO (Digi-Key TK1305)
U1—NE/SA602 double-balanced mixer and oscillator
U2—HCPL4562 optoisolator (Newark #HPCL-4562)
U3—LM78L05 5 V, 100 mA positive voltage regulator
U4—LM386-4 audio amplifier
Misc: PC board (see Note 3); two-inch diameter vernier dial, $^1/_4$-inch shaft (Philmore #S50); 9-V battery snap connectors (RS 270-234); 9-V battery holders (RS 270-326); enclosure; hardware

The OCR II can be built to use either wire antennas or tuned loop antennas. You can add the ability to use either antenna by adding a DPDT switch to select the preselector circuit (used with the random-length wire) or the loop antenna.

clad perfboard, or "ugly" ("dead-bug") construction on a bare copper PC board. The only critical area is the SA602 oscillator circuit. NP0 (C0G) capacitors are used here to enhance frequency stability. Use short, direct leads in this area. Make the circuit as mechanically robust as possible to help ensure stability. The hardware used for antenna connector J1 and switch S1 may be whatever you prefer. You can fashion an enclosure from PC board, aluminum or use a manufactured enclosure. A fully closed case is not required for good operation. Several prototype OCR receivers built as open-frame units perform very well over reasonably constant temperature ranges.

One of the more useful and interesting features of the OCR design is the **REGENERATION** control. This not only controls the amount of detector oscillation, but also controls the detector Q, which sets its bandwidth. With careful adjustment, bandwidths of a few tens of Hertz are achievable just before the detector starts to oscillate. To take advantage of this control, the pot used for regeneration control must have a fairly good resolution. Although a multiturn pot could be used here, I took a more cost-effective approach. A board-mounted 20-kΩ pot (R13) is used to preset R14, a 5-kΩ front-panel-mounted pot, for regeneration control. R13 is adjusted so that regeneration starts with R14 at about 75% of its maximum range. Used this way, R14 gives very good control over the regeneration. This scheme works well because the oscillator frequency is fixed, and the regeneration point is quite constant. Because the total current used by the LED in U2 is only about 400 µA, the change in battery voltage related to battery aging is very slow and therefore, only infrequent R13 readjustment is required.

Checkout and Operation

When you finish circuit assembly, carefully check your work for wiring errors and cold or missing solder joints. Verify that the components have been installed correctly before applying power. Note that BT1, used for the regeneration circuit, should be a 9-V battery. This reduces noise in the detector. BT2 may be a 9-V battery for headphone operation or a 12-V battery for loudspeaker operation.

Once all has been checked, plug your headphones into J2 (phones with an impedance of 16 Ω or greater give the best results) and apply power to the receiver. The headphones used should be of good quality. With the **VOLUME** control (R24) set at about midrange, advance the **REGENERATION** control (R14) to about 75% of its range. Adjust the **REGEN PRESET** potentiometer R13 until you hear a gentle but distinct increase in background noise. This indicates that U2 is oscillating and all is well with the OCR detector circuit.

Set the detector frequency by listening for the 455-kHz signal from U2 in a general-coverage receiver. Adjust L2 to set the operating frequency of U2. Alternatively, a frequency counter can be used to set the frequency. (Many inexpensive digital multimeters now have frequency counters usable to 10-MHz.) To measure the frequency of U2, connect the frequency counter to pin 5 of U2. Although the signal level is lower at pin 5 than at L2, measuring at this point avoids incorrect readings caused by the probe loading the tank circuit. (Similarly, the converter-oscillator frequency can be measured at pin 7 of U1; this is discussed later.) As the detector frequency is adjusted, the regeneration controls may have to be retouched to keep U2 oscillating. It is not critical that U2's oscillator frequency be set to *exactly* 455 kHz because there are no narrow filters used in the receiver. It is only necessary that all the IF transformers be within adjustment range of each other.

Verify that the mixer oscillator is operating over the correct range by listening for its signal in a general-coverage receiver or using a frequency counter. The converter-oscillator frequency can be adjusted by adding a turn to, or removing a turn from L1. Remember to subtract the IF from the mixer-oscillator frequency. For example, the required mixer-oscillator frequency for receiving a 3.5-MHz signal is 3.045 MHz. Once this is done, connect a 15- or 20-foot-long wire test antenna to terminal **A1** on connector J1. Connect J1 terminal **A2** to the **GND** terminal. If an earth ground is available, connect it to the **GND** terminal also. With the detector oscillating, use the **BANDSET** control to find a signal. Peak the signal with the **ANTENNA TUNING** capacitor, C1. Next adjust the tuning slugs in T3, T4 and T5 for maximum signal strength. There is little interaction between these adjustments. The tuning of T5 is very broad, and an obvious peak is hard to discern. I generally place the tuning slug at the midpoint of its adjustment range. Finally, verify that the **ANTENNA LEVEL** control works and that the **BANDSPREAD** tuning is functional. That's it for tune up!

Using the OCR II

If this is the first regenerative receiver to which you've been exposed, tuning the OCR II will take some practice. The most sensitive operating regions of the detector for AM-signal reception is the area just *before* oscillation and for CW, just *at* oscillation. For SSB reception, the best operating point is found with just a bit more regeneration than that required for CW reception. After using the **REGENERATION** control for a short time, you'll get the feel of the receiver. The interaction between the **REGENERATION** control setting and the gain and selectivity of the detector will become quickly apparent. You may find yourself digging out CW and SSB signals from beneath the AM stations in the 40-meter band—signals you could never hear on other simple receivers! Those who have tried other regenerative receivers will notice that there is virtually no interaction between the received signal strength and the regeneration setting required. And, since the detector is at a fixed frequency, the regeneration level can be maintained over the entire tuning range of the receiver. This is a radio that is great fun to use because you have virtually total control of the receiver performance.

On 80 meters, the **BANDSPREAD** is fairly limited, covering only about 20 kHz or so. I use the **BANDSET** to tune the band and the **BANDSPREAD** as a "fine tuning" control. About 25% of the total tuning range is used to cover 3.5 to 4 MHz, so using it as a "main tuning" control works well with the vernier dial. At 40 meters, the **BANDSPREAD** covers the entire band. When tuning the 40-meter band, insure that the preselector is tuned to 7 MHz. It will also peak up at the image frequency around 6 MHz. This adds even more QRM to the band!

When the conditions are good, use the antenna **LEVEL CONTROL** to reduce the signal level. I have found that if the input signal from the antenna can not be reduced to the level that no signal can be heard on the receiver, the antenna is too big and can overload the converter section when the **LEVEL CONTROL** is set at its minimum.

To receive AM stations, I use the following procedure: Set the regeneration as for CW reception and "zero beat" the AM station. Next, reduce the regeneration just to the point where the oscillation stops. Keeping the regeneration level as high as possible allows the maximum detector sensitivity and provides the tightest audio passband. Depending on the strength of the station and the QRM present, the regeneration level can be reduced. This improves the fidelity of the signal because of the increased detector bandwidth. This technique is possible on the OCR II for two reasons. First, there is virtually no interaction between the received frequency and the regeneration control. Additionally, there is no frequency "pulling" by strong stations. Therefore, a weak station next to a strong station can be easily received.

The measured CW receiver sensitivity is less than 1 μV (by my ear) when driven by a laboratory-grade 50-Ω signal generator. The AM sensitivity is a little more difficult to measure since it depends upon the amount of regeneration being used, but it's about 2 or 3 μV.

Antennas

As mentioned earlier, the OCR II can be built to use a wire antenna or a tuned loop. For versatility, you can add a switch to choose the preselector circuit for the wire antenna or the loop antenna. I did this on a prototype with very good results.

The preselector has two antenna terminals (**A1** and **A2**) and a chassis ground terminal (**GND**). This allows maximum flexibility when using simple wire antennas. For the simplest random-length wire antennas, connect the antenna to terminal **A1**. Connect terminal **A2** to the **GND** terminal. If an earth ground is available, always connect it to the **GND** terminal as well. An antenna length of 20 or 25 feet will give good results. I've found that when an earth ground is available, a simple wire antenna just a few feet long works very well.

If you use a balanced antenna, connect one antenna leg to terminal **A1**, the other to terminal **A2**. Again, if an earth ground is available, connect it to the **GND** terminal. Don't be afraid to experiment with the antenna connections to find the best combination for your antenna. Remember: Overloading the OCR II mixer degrades overall receiver performance. Use the **ANTENNA LEVEL** control to reduce overloading when using large antennas or when very strong shortwave stations are encountered. The **ANTENNA LEVEL** control in conjunction with the **REGENERATION** control make a powerful combination to improve shortwave listening.

Using a tuned loop provides the receiver with front-end selectivity. (Loop-antenna designs are presented in the SLR article; see Note 1.) Generally, the loop should be designed for the lowest operating frequency. For 3.5 MHz, a square loop about 18 inches on a side is a good minimum size. A shielded or unshielded design can be used. My rule of thumb for calculating the inductance of a small wire loop is to estimate the inductance at 26 nH per inch. Thus, the small loop of 18 inches per side will have an inductance of about 1.87 μH. To tune this loop to 3.5 MHz, a capacitance of about 1100 pF is required. At 8.5 MHz, you'll need 187 pF. A combination of fixed-value and variable capacitors can be used to tune a loop over this frequency range. (Here's a good application for that triple-section, 365-pF-per-section variable capacitor you've been saving because it's "too good to throw out!") Of course, the loop can be made a bit larger or a multiturn loop can be used to reduce the capacitance required to tune the loop antenna.

A shielded loop antenna is shown with the OCR II in the title photo. An earlier version of this loop appeared on the cover of *QST* for October 1997. That loop was constructed of 22-gauge wire as described in the SLR article. However, that antenna and the receiver were borrowed so often that replacing the wire loops became a weekly task! I rebuilt the loop using 3/32-inch-diameter tubing sold at model and hardware stores. Copper tubing is used for the active portion of the loop, with lower-cost brass tubing used for the shield loops. This material is sold in 12-inch lengths, solders easily and is quite rigid. The loop made with loops of tubing has proven to be very durable.

This particular loop is small, only nine inches on a side; therefore it is used on frequencies above 6.5 MHz. The loop inductance is approximately 0.95 μH and a capacitance of about 550 pF is required to resonate it at 7 MHz. Tuning the loop is accomplished with a 365-pF air-dielectric variable capacitor in parallel with a fixed-value capacitance of 220 pF. The fixed-value capacitor can be switched in and out, providing two tuning ranges. The lower range covers 7 to 12 MHz, while the upper range covers 8 to 30 MHz. The loop is connected to the receiver using short lengths of low-cost audio cable and standard phono connectors.

Summary

The OCR II receiver is a simple, all-mode multiband receiver. It retains the best features of its predecessors, the SLR and OCR receivers. With sensitivity equal to that of the SLR and the good selectivity provided by the OCR regenerative detector, the OCR II offers performance greater than the sum of its parts. I have enjoyed designing this radio, and have had great fun operating it. I thank those builders of the SLR and OCR receivers who have sent me mail and inspired the design. I hope others are inspired to try their hand at "homebrewing" this and other projects.

Notes

[1]Daniel Wissell, N1BYT, "The 40M SLR—a Shielded-Loop Receiver," *QST*, Oct 1997, pp 33-38.

[2]Daniel Wissell, N1BYT, "The OCR Receiver," *QST*, Jun 1998, pp 35-38.

[3]Jade Products, Inc,, PO Box 368, East Hampstead, NH 03826-0368; tel 800-523-3776, fax 603-329-4499; **jadepro@jadeprod.com**; **http://www.jadeprod.com/**. Jade also offers components used in this project.

[4]Additional parts sources include: Dan's Small Parts and Kits, Box 3634, Missoula, MT 59806-3634; tel and fax: 406-258-2782; **http://www fix.net/dans.html**, for variable capacitors and multiturn potentiometers; The Xtal Set Society, PO Box 3026, St. Louis, MO 63130; tel 800-927-1771; **xtalset@midnightscience.com**; **http://www.midnightscience.com/crystal.html**) is one source of 365-pF air-dielectric variable capacitors.

Dan Wissell, N1BYT, was first licensed as WN2WGE and upgraded to Extra in 1984 as N1BYT. He's been with Compaq Computer Corporation (formerly Digital Equipment Corporation) for 20 years. He is currently a Principal Member of the technical staff, designing RF and analog systems. You can contact Dan at 7 Notre Dame Rd, Acton, MA 01720; **n1byt@arrl.net**.

Chapter 5

Accessories

By Phil Salas, AD5X

From *QST*, January 2003

A Compact 100-W Z-Match Antenna Tuner

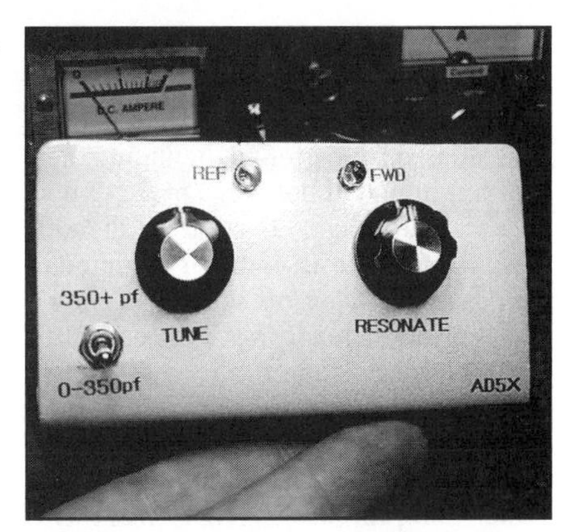

The nice thing about the Z-Match tuner is that it will match just about anything on the HF bands and it uses only two controls. Here's a 100 W version using toroid inductors.

After reading about Z-Match antenna tuners for quite a while, I eventually capitulated and bought an Emtech ZM-2 QRP Z-Match tuner kit.[1] After building it to go along with my Yaesu FT-817, I became a real believer in the Z-Match design. The folks who use these tuners speak very highly of them, but it had always appeared to me that acquiring the necessary air-wound inductors and variable capacitors for a higher power (100 W) version was more trouble than it was worth. In addition, air-wound inductors implied larger enclosures and I was interested in a tuner that was compact enough for me to easily take along with my portable HF setup that used an ICOM IC-706.

I later discovered an excellent article on Z-Match tuners by Charles Lofgren, W6JJZ.[2] In that article, the author suggested using a toroidal core inductor. This idea effectively solved the inductor size problem for me. I then found that 440 pF (per section) variable capacitors were available from Fair Radio Sales.[3] Similar capacitors are available from other sources, although it might take a bit of hunting. All of my excuses for not building a 100 W version of the Z-Match had vanished!

Construction

The final circuit shown in Figure 1 is based on W6JJZ's article, with the output transformer changed to a single 8-turn output link. Also, 440 pF per section variable capacitors were used for C1 and C2. So far, I haven't found anything I can't match and that's from 80 through10 meters!

I built the tuner in a 5¼×3×5 inch (HWD) aluminum box. Toroid L1 is supported by its own leads, as shown in Figure 2. Some hot glue is used between this inductor and the frame of C2, one of the variable capacitors. I also put a little hot

[1]Notes appear on page 5-3.

385 - 440 pF, per Section
C1B
S1
C1A
J1
RF In
SWR
Circuit
22T
L1
11T
5T
8T
J2
RF Out
Banana Jack
S2
Close for Coax
C2A
C2B
385 - 440 pF,
per Section
L1: T157 - 6 Toroid core wound with
#20 enameled wire.
Pri: 22t tapped at 5 and 11t.
Sec: 8t (centered around primary 5t tap).

Figure 1—Schematic of the 100-W Z-Match tuner. AA = Amidon Associates (www.amidon-inductive.com); RS = RadioShack (www.radioshack.com); FRS = Fair Radio Sales (www.fairradio.com).

C1, C2—385-440 pF 2-sec variable capacitor (FRS APS-440 or equivalent).
C3—2-20 pF variable capacitor (RS 900-5850).
C4—100 pF capacitor (RS 272-123).
C5-8—0.01 µF capacitor (RS 272-131).
C9—4.7 µF capacitor (RS 272-1012 or 272-1024).
D1—Red LED, high-intensity (RS 276-307).
D2—Green LED, high-intensity (RS 276-304).
D3, D4—1N4148 (RS 276-1122).
L1—T157-6 toroid (AA).
L2—FT37-43 toroid (AA).
R1—150 Ω, ¼ W.
R2—3.3 kΩ, ¼ W.
R3—2 kΩ, ¼ W.
R4—4.7 kΩ, ¼ W.
R5—100 kΩ, ¼ W.
R6—2.2 kΩ, ¼ W.
R7—1.5 kΩ, ¼ W.
S1, S2—SPDT mini-toggle switch (RS 275-634).
U1—Display driver IC, LM3914 (RS 900-6840).
U2—10-LED bar graph display (RS 276-081).

Misc

J1, J2—SO-239 connectors (RS 287-201).
Enclosure (RS 270-253).
Perforated board (RS 276-1394).
#20 solid enameled wire.
#26 solid enameled wire.

Figure 2—L1 is mounted with hot glue to the bracket of C2, but also supported by its leads.

glue between the inductor and the side of the enclosure.

Because both variable capacitors must be insulated from ground, including their shafts, I mounted both on a piece of perforated board that was cut to fit the aluminum case. The assembly is mounted in the case with stand-off screws, as shown in Figure 3. I made my own capacitor shaft couplings from a 1/8-NPT brass nipple, available in the plumbing section of most hardware stores. These nipples have a ¼-inch inside diameter. Cut a 1-inch long nipple in half to make two couplers. Drill and tap holes for two #6 set screws in each coupling. The completed shaft coupling is shown in Figure 3. For the insulated shafts, I used ¼ inch diameter nylon rods, which are also available from most hardware stores.

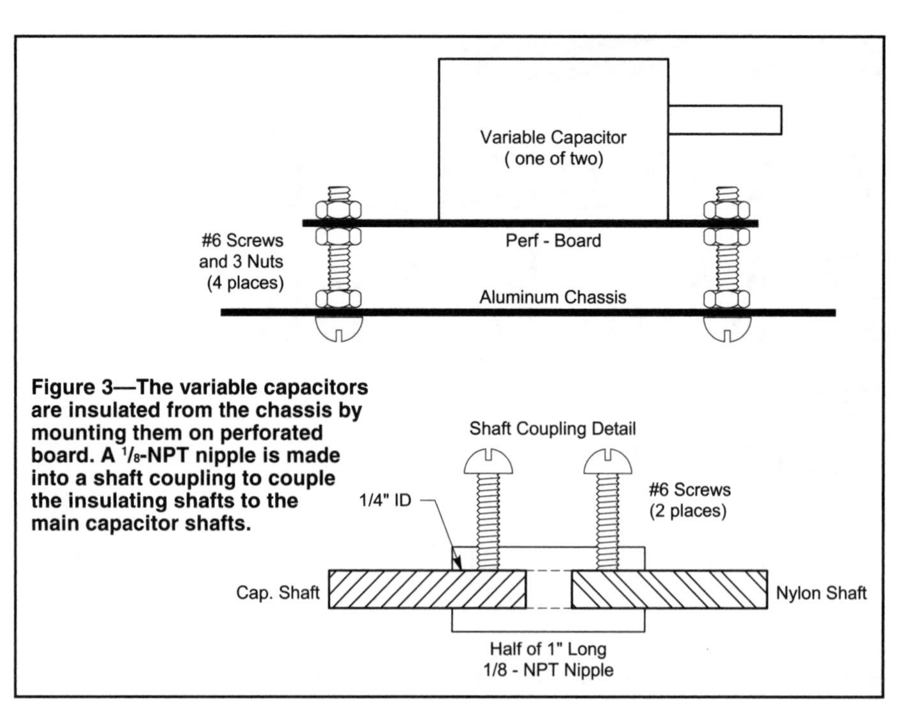

Figure 3—The variable capacitors are insulated from the chassis by mounting them on perforated board. A 1/8-NPT nipple is made into a shaft coupling to couple the insulating shafts to the main capacitor shafts.

Operation

Tuning the Z-Match tuner is very easy. First, adjust the resonating capacitor C2 for maximum receiver noise. Then apply some RF power and adjust C1 and C2 for minimum SWR. If you need more capacitance for matching, use S1 to switch in extra sections for C1. Balanced feed lines, which are terminated in banana plugs, plug into the center pin of the output SO-239 and the adjacent banana jack. To feed coax, ground one end of the output link with switch S2.

Optical HF SWR Meter

You can use an external SWR meter with the Z-Match tuner, but I built a convenient optical (LED) SWR meter into the same case. It works well with the newer high intensity LEDs that are currently available. The schematic is shown in Figure 4A. I built the circuit on a small piece of perforated board and mounted it into the Z-Match tuner enclosure. This can be seen in Figure 5. I also added a bit of hot glue between the perforated board assembly and the back of the chassis.

This broadband circuit works well at the 100 W level through at least the 10 meter amateur band. With short leads, it should work well through 6 meters. The transformer is an FT37-43 ferrite core wound with 10 bifilar turns of #26 enameled wire. The primary is just the single wire passing through the center of the toroid. To calibrate the SWR bridge, connect the output to a resistive 50 Ω load. Apply RF power on any HF band and adjust the 20 pF variable capacitor until the REFL LED goes out.

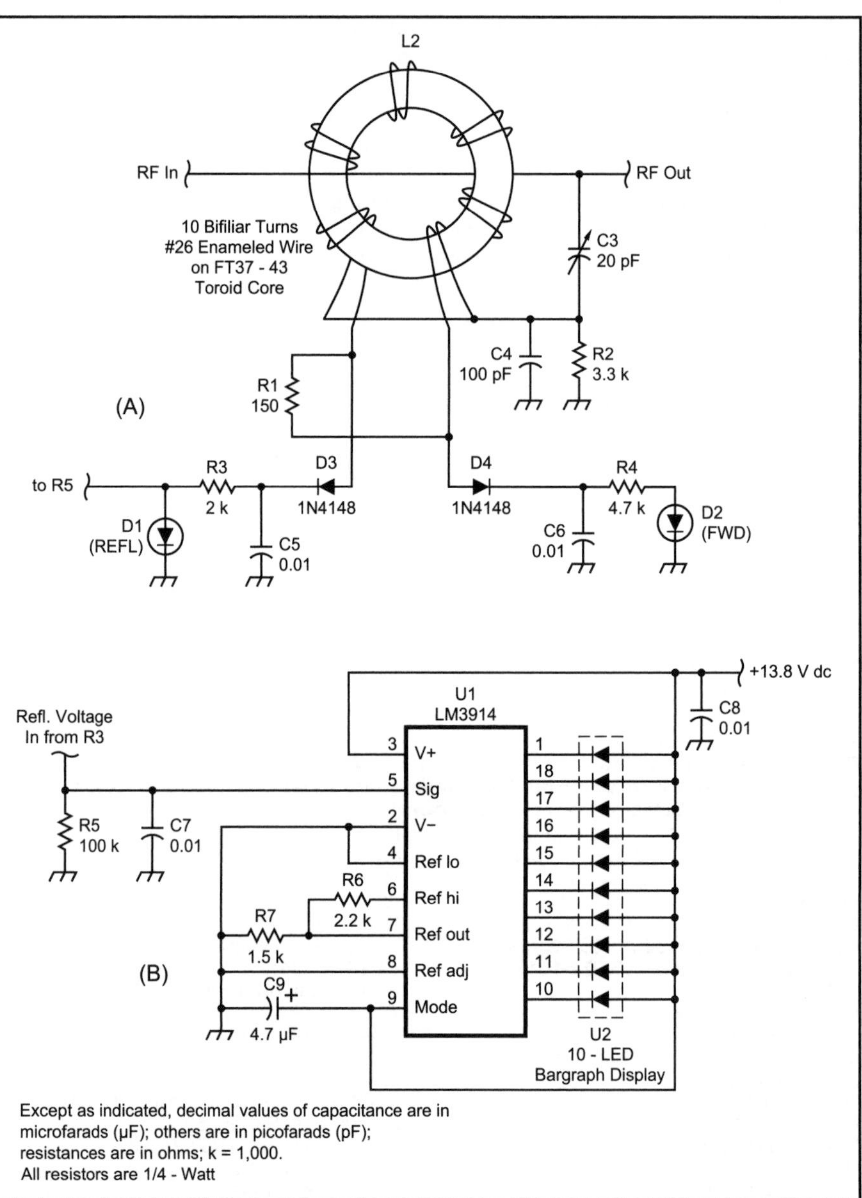

Figure 4—Optical SWR meter schematic with the bar graph display modification (see text). The basic LED version is shown in A and the bar graph addition in B.

Figure 5—The SWR Meter circuit is soldered directly to the RF Input connector and attached to the back panel with hot glue.

Figure 6—The bar graph display of reflected power makes tuning easier and more intuitive.

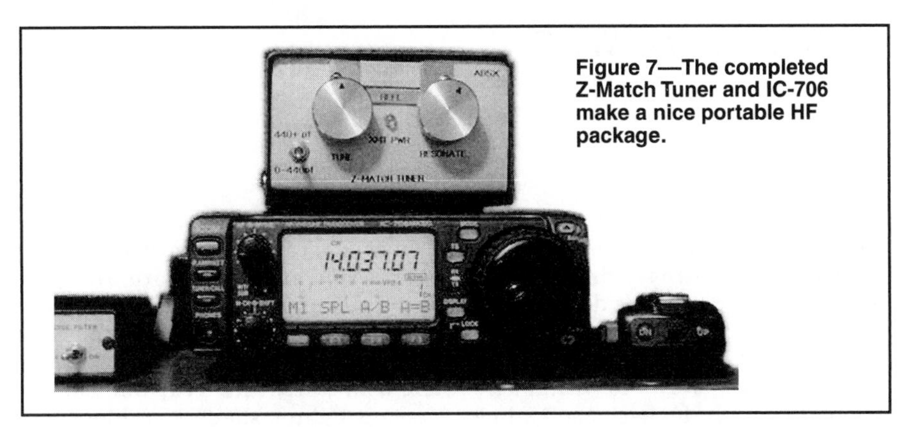

Figure 7—The completed Z-Match Tuner and IC-706 make a nice portable HF package.

Adjust the Z-Match tuner for minimum brightness of the REFL LED. When this occurs, the SWR will be something less than 1.5:1. The brightness of the FWD LED is an indication of transmitter forward power output. If the green LED is too bright, increase the value of the current limit resistor in the FWD circuit. If desired, you can eliminate this LED completely and use only the REFL LED.

If you can supply dc power to your Z-Match Tuner and SWR meter, you may want to add a bar graph display for the SWR reflected power. This is shown, schematically, in Figure 4B and physically, in Figure 6. It uses an LM3914 LED display driver to drive a bar graph display and takes its input from the reflected voltage output of the SWR sampler. If you do use the bar graph display, you can remove LED 1 and connect pin 5 of the LM3914 input to the output of R3. The nice thing about the bar graph display is that it seems easier to null the reflected power, as this display brings the operator closer to the intuitive "feel" of a classic analog meter pointer, but it does require an external dc voltage.

Conclusion

The completed Z-Match tuner on top of the IC-706 is shown in Figure 7. It is very easy to adjust. The biggest obstacles to construction of a compact 100 W version, the necessity of using large air-wound inductors and finding cost-effective multi-section air-variable capacitors, were overcome. The result is an inexpensive, wide-band, easily adjustable tuner for portable or base station operation.

Notes

[1] Emtech, 1127 Poindexter Ave W, Bremerton, WA 98312, tel 360-405-6805; **www.emtech.steadynet.com**.

[2] C. Lofgren, W6JJZ, "An Improved Single-Coil Z-Match," *The ARRL Antenna Compendium, Vol 5*, p 194.

[3] Fair Radio Sales, 2395 St Johns Rd, Lima, OH 45802, tel 419-227-6573; **www.fairradio.com**. Similar capacitors in this range are available from other sources.

Phil Salas, AD5X, has been a ham for 38 years. He is an ARRL Life Member, and is currently the Director of Hardware Engineering at Celion Networks in Richardson, Texas. You can reach the author at **ad5x@arrl.net**.

RF

By Zack Lau, W1VT

A Miniature HF 50:200 Ω Balun

This 50:200-Ω balun was designed for serious QRPers who want a lightweight HF balun using commonly available parts. It should handle 3 W on 20 meters, more than enough for "milliwatt" types who never run more than 1 W. While there are designs with less loss, rarely do they use popular parts found in most junk boxes. This balun uses the FT-37-43; perhaps the most popular ferrite toroid used in amateur construction projects.

The design is shown in Fig 1. First, a 4:1 unbalanced to unbalanced transformer steps up the impedance to 200 Ω. Next, a 1:1 transformer provides the desired unbalanced-to-balanced function. For simplicity, both transformers use 10 turns of #28 AWG enameled wire, bifilar wound. Bifilar transformer winding is quite simple if you know how. There is also a clever way to wind the transformers—if you want to eliminate a solder joint—that I'll cover later.

First, wind each winding side by side with no crossovers (see Fig 2). Next, strip the insulation from the two "center" conductors closest to each other and verify with an ohmmeter that they are different windings. Finally, solder them together.

The balun is even easier—the input wires go in together and they come out together. Most of the time it doesn't matter if you swap the output wires—so it isn't necessary to keep track of the wires. If you are feeding a phased array, however, phasing does matter. Swapping the wires may result in an unexpected 180° phase shift.

The clever way to wind the transformers is to use one 12-inch length and two six-inch lengths of wire, instead of four six-inch lengths. This way, after you wind the first bifilar winding, you have enough wire to wind the second core, with the longer piece of wire. I'd recommend winding the two six-inch lengths of #28 AWG enameled wire first, on separate cores. Then wind the 12-inch length on both cores. This is shown in Fig 2—the 12-inch wire is black. The six-inch wires are red, and should have a lighter shade in a black and white picture. Alternately, you can slip some thin Teflon tubing over a wire before it is spliced to the other core. The tubing can then be slid over the splice. With a bit of skill or luck,

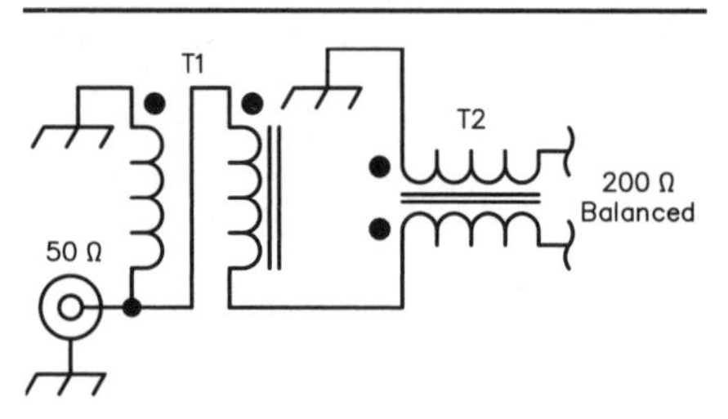

Fig 1—Schematic of a low power 50:200-Ω balun.
T1, T2—10 turns #28 AWG enameled wire bifilar wound on a FT-37-43 toroid.

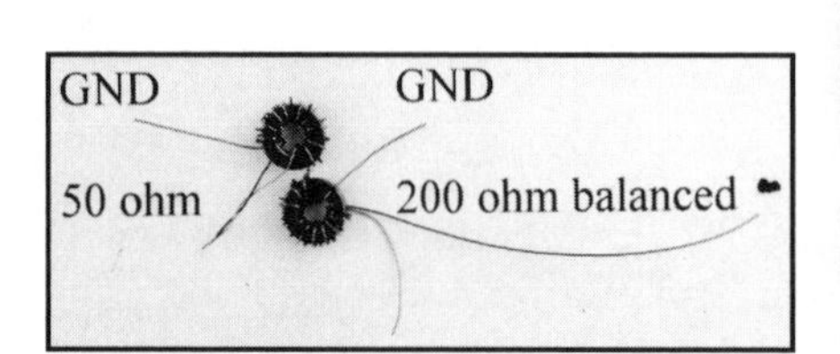

Fig 2—Details of the FT-37-43 toroid core windings.

Fig 3—#10 Hardware installed in ½-inch CPVC pipe.

Fig 4—A machined CPVC pipe cap showing the carved hole (lower).

225 Main St
Newington, CT 06111-1494
zlau@arrl.org

the splice will be just the right thickness to hold the tubing snugly in place. I often use 18-gauge lightweight spaghetti from Small Parts.[1]

I've found that 1/2-inch CPVC tubing is just big enough to house this balun. The tough fit wasn't the ferrite cores, but the screw terminals. Number 8-32×1/2 screws just barely fit inside the pipe caps—I needed to angle one of the mounting holes so that the second screw would fit. With a mounting hole properly modified, even #10-32×1/2 screws are useable, as shown in Fig 3. Fortunately, CPVC is soft and easy to carve with a sharp knife. This is shown in Fig 4. Notice the lower hole: Plastic is carved away so it doesn't obscure the hole. I put on a pair of safety goggles and used a hobby knife to do the carving.

Choosing the location of the holes is a tradeoff between having enough clearance for the screw head and having a reasonably flat surface for the lockwasher. I made my holes about 0.2 inches from the interior flat surface of the cap. I put the cap on a section of 1/2-inch pipe and drilled them together. The pipe can be easily held in a vise, unlike the awkward shape of the cap. Unlike copper caps, the walls of the cap vary in thickness significantly. The holes in the pipe provide reference marks for milling notches for the screw heads. To get the pipe to fit around the screw heads, I milled 0.33-inch wide notches.

Mounting the UG-1094A/U connector was also a challenge. I decided the pipe cap was too small to properly tighten the 3/8×32 mounting nut. Instead, I first used a 3/8-inch bullet-shaped drill bit to cut a relief for the connector—just a little more than 50 mils should work fine. Otherwise, the connector won't screw in flush against the pipe cap. Next, I drilled a Q-sized hole (0.3320 inch) and tapped a 3/8×32 thread for the BNC connector. This extra-fine tap may be difficult to find—you may need to get it from a machine shop supply. I applied some Loctite thread-locker to the connector and firmly threaded the BNC connector into the pipe cap. When installing threaded BNC sockets, it helps to plug in an old BNC plug for protection. It also provides a knurled surface designed for gripping firmly with fingers.

I used solder lugs wrapped around the screw heads to attach the balun wires. As shown in Fig 5, bending them in a **U** maximized clearance for the short section of pipe cap. You might imagine installing them facing downward and then bending them so the tabs face upwards. I found it necessary to bend them before installation, however, it is too difficult to bend them in place. I soldered the balun to the solder lugs first. Then I slid the tubing over the balun, and firmly attached the pipe cap.

The final tricky part is soldering the wires to the BNC connector. I used a 60-W temperature controlled iron to solder the wire directly to the body of the connector, after first tinning the connector and wire. This is why I used the longer connector—the shorter UG-1094/U would be more challenging to solder. Finally, I soldered the other wire to the center conductor. I really doesn't matter which is the ground side and which is hot, so you may find it convenient to swap wires when making these final two connections, depending on the wire lengths.

This may seem like a lot of work to machine the three CPVC parts shown in Fig 6—it is. It is much easier to make a larger balun, perhaps using 3/4 or 1-inch plastic tubing and end caps. The

[1]Notes appear on page 5-7.

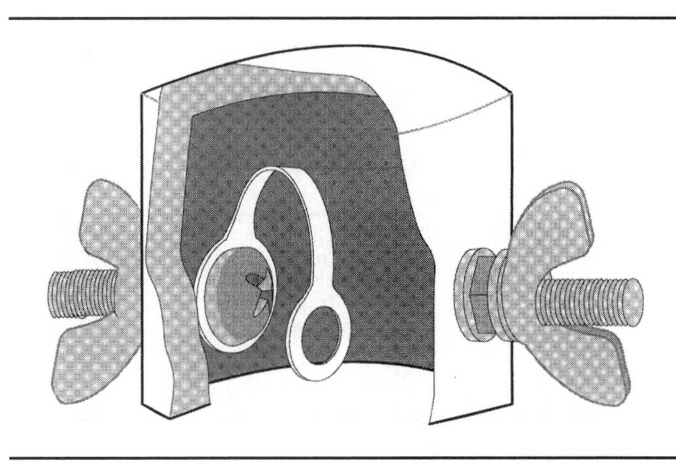

Fig 5—The arrangement of the screw terminals and solder lugs.

Fig 6 —CPVC pipe parts prepared to build the balun.

Table 1—Measured balun performance

Frequency MHz	*Transformer loss (dB)*	*1:4 Balun loss (dB)*	*Unbalanced Fig 3A (dB)*	*Unbalanced Fig 3B (dB)*	*Low Z loss (dB)*	*High Z loss (dB)*	*Balun RL (dB)*
1.8	0.29	0.66	1.06	1.06	0.18	1.31	22
3.5	0.21	0.43	0.68	0.67	0.17	0.75	27
7	0.16	0.32	0.47	0.47	0.24	0.53	33
10	0.15	0.29	0.42	0.42	0.38	0.51	34
14	0.15	0.27	0.38	0.39	0.67	0.54	31
28	0.15	0.29	0.38	0.39	1.34	1.31	24
50	0.18	0.39	0.41	0.43	2.32	2.76	18
60	0.19	0.39	0.46	0.48	2.38	3.38	17

result wouldn't be as small or light, however.

The performance is quite good when properly terminated, considering the low cost and lightweight of this design. The balun weighs just 0.070 ounces without the CPVC parts. The loss from 3.5 to 60 MHz is under 0.43 dB, for over 90% efficiency. The return loss is better than 24 dB from 3.5 to 28 MHz, degrading to 19 dB at 50 MHz, when terminated directly in a 1/4-W 200-Ω carbon-composition resistor. The results shown in Table 1 are for the packaged balun in Fig 7. Not surprisingly, the extra lead length for the screw terminals degraded 6-meter performance slightly. Thus, this should work well on all HF bands when terminated in a 200-Ω load. It may even be useful on 6 meters.

The insertion loss was measured using a pair of baluns back to back, as shown in Fig 8. The loss was also measured with points A and B alternately grounded. Ideally, the loss wouldn't change. The RF signal generator was a Marconi 2041. The power meter was a HP 437B with an 8482A sensor. The return loss was measured with a Marconi signal generator, a Mini-Circuits ZFDC-20-5 directional coupler and a HP 8563E spectrum analyzer, as shown in Fig 9. The measurement data is shown in Table 1. The numbers shown represent half the total loss measured, since most people want to know what the loss is for a single balun or transformer. The accuracy of the HP power meter and sensor is ±0.02 dB, enough to justify the precision shown in the table.

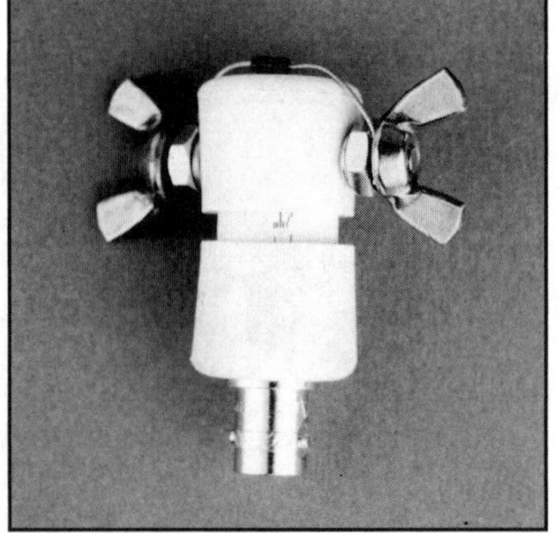

Fig 7 —The completed balun with a 200-Ω 1/4-W carbon-composition resistor used for return-loss measurements. The balun is in a typical mounting position, with the load resistor across its top.

The last two columns are an attempt to characterize balun performance when the unbalanced input impedance is either 12.5 or 200 Ω. The impedance step-down was measured by simply swapping the input and output connections. The impedance step-up was a little more difficult—I measured the balun along with yet another 4:1 bifilar transformer—10 turns of #28 AWG on an FT-37-43 core. I subtracted the loss of the extra transformer for the table entries. The balun works quite well from 7 to 14 MHz, showing no additional loss. However, high-frequency performance was significantly degraded at both high and low impedances. The impedance step-down actually enhanced low-frequency performance. This is the result of the transformers operating more efficiently—while the wires used in the balun are short enough to avoid causing an impedance mismatch.

The loss can be used to estimate power handling. As described on page 2-21 of *QRP Power*,[2] the power loss required to raise the temperature of a toroid core is:

$$P_{loss}(mW) = (\text{surface area in cm}^2)(\Delta T)^{1.2} \quad \text{(Eq 1)}$$

An FT-37 core has a surface area of 5.7 cm^2, which means that 0.27 W are required to raise its temperature 25°C. If the core has a loss of 0.4 dB, 9% is lost as heat. Thus, the power handling capability is 0.27 Watts/0.09 or 3 W.

While amateur operation is intermittent, rather than 100% duty cycle, it may not be safe to significantly increase the rating if the cores are sealed in a plastic tube that prevents heat dissipation. On the other hand, a little wind will significantly increase the power handling—just ask anyone who has tried to solder outdoors on a windy day.

Further research is needed to determine the power handling capability on 6 meters. While the insertion loss is nearly 3 dB, much of it is the result of impedance mismatching—2.76 dB. It is likely that a tuner will allow the balun to handle more than 0.5 W, but exactly how much is unclear.

You need to be very careful when using baluns with phased arrays. A properly designed balun has outputs that are virtually indistinguishable from each other, except for relative phase between the input and output. This can be disastrous—an unexpected 180° phase shift can turn a high-gain array into a low-gain nightmare. Why doesn't it work? An incorrectly phased balun or antenna is an all too common answer. Thus, it is important to identify and mark balun output ports intended

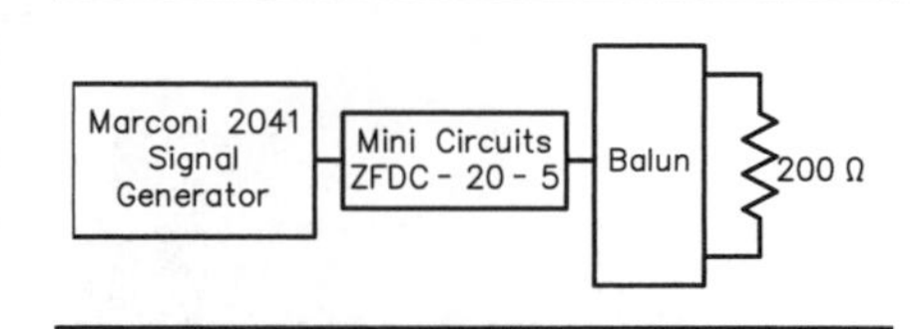

Fig 9—Schematic of the setup used to measure return loss of the balun when terminated in a 200-Ω carbon-composition resistor.

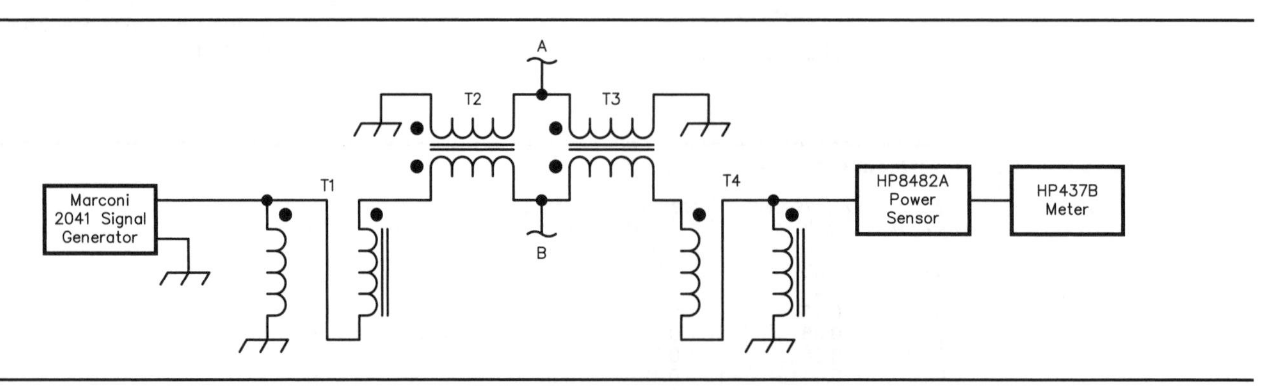

Fig 8—Schematic of the power-loss test setup using a signal generator and power meter.

for use in phased arrays, so they can be used properly. Antennas also have phase—it makes a significant difference if you flip yagis over.

Theoretically, it may be possible to construct identical baluns if you use colored windings and pay careful attention to detail. Realistically, most people need to measure constructed baluns and properly identify the output terminals. Fortunately, this is quite easy to measure, if you have the appropriate phasing harness. After all, if an in-phase harness works properly, any output terminal will be either in-phase or 180° out of phase relative to any other terminal. Thus, you can just connect a resistor of the appropriate wattage across any two terminals and see whether it heats up. If it does, the terminals are out of phase. If the resistor is stone cold, the terminals are in phase. I'd use the same resistance used to terminate the baluns properly. Don't touch the resistors while RF power is applied, always turn off the transmitter first.

There are two caveats to this simple test procedure. First, the resistor needs to be of low inductance. If you use a wirewound resistor, it may act as a RF choke, remaining stone cold no matter how it is hooked up. Thus, I'd make sure that the resistor heats up properly when connected to terminals that are suppose to be out of phase—don't just look at the in-phase connections.

The second is the influence of SWR-foldback circuitry. Theoretically, it may be possible for an aggressive foldback circuitry to reduce power enough to make the heating effect too small for conclusive results. Fortunately, such transmitters usually have an SWR or power indicator that will indicate the SWR change. In this case, a rise in SWR or drop in power should correspond to the out-of-phase connection. The in-phase connection should not change the power or SWR.

Notes

[1]Small Parts Inc, 13980 NW 58th Ct, PO Box 4650, Miami Lakes, FL 33014-0650; tel 800-220-4242 (Orders), 305-557-7955 (Customer service), fax 800-423-9009; e-mail **smlparts@smallparts.com**; Web site **http://www.smallparts.com/**.

[2]J. Kleinman, N1BKE, and Z. Lau, W1VT, Eds. *QRP Power,* (Newington, Connecticut: ARRL 1996; ISBN: 0-87259-561-7) Order No. 5617, $12. ARRL publications are available from your local ARRL dealer or directly from the ARRL. Check out the full ARRL publications line at **http://www.arrl.org/shop/**.

By Mike Aiello, N2HTT

From *QST*, April 1999

The QRP Buddy

Here's everything you need to do battle with Murphy in the field—all in a lightweight, compact package!

Recently, while planning a business trip to the Midwest, I decided to bring a QRP station with me. Although time is usually tight during these trips, this stay was to extend over several days—the odds favoring at least one evening's opportunity for some hotel-room QRP! Luggage space was at a premium because I was carrying quite a bit of work-related gear and documentation. So, I pared my QRP station to a minimum: tiny rig, tiny Transmatch, tiny key, tiny power supply, headphones borrowed from a personal stereo and a spool of 28-gauge enameled wire to use for antenna fabrication. The entire package was impressively... tiny. It all nestled comfortably in a corner of my checked luggage.

As it turned out, the first day of the trip presented the only opportunity to play radio. Having the evening off because of an unexpected change of schedule, I set up my station in the room. Everything went smoothly, including the improvisation of a 20-meter wire dipole, half in the room and half out the window. (Well, it wasn't *exactly* a dipole, but the antenna tuner works really well.) I adjusted my headphones, flipped the power switch and was welcomed by the rush of...absolutely nothing! As I stared at the station in disbelief, I realized that I had *no way* to tell me what the problem was or where it lay—I hadn't brought a multimeter with me. I did what I could to troubleshoot the failure, including moving lamps around the room to make sure the wall outlet I was using was live. In the end, I was pretty sure it was a "no power" problem, but whether the fault was in the power supply, the cabling or some failure in the radio, I couldn't tell.

The next day, I borrowed a DMM from the lab I was visiting and determined that the power supply was fine. It was the fuse in the power cable that turned out to be the culprit. Although the fuse *looked* okay, it exhibited a very high resistance, but wasn't completely open. I guess it was damaged, perhaps by vibration or mechanical shock. (I wonder what they did to that suitcase?) The paring-down of my QRP kit had done away with the plastic bag containing spare fuses, and my schedule prevented me from shopping for some. I never did get on the air that trip.

Reflecting on the experience afterward, I decided it would be a good idea to have a compact, lightweight troubleshooting aid that was in scale with my station—something that could give simple go/no-go indications of some critical parameters. The instrument should provide for checking fuse continuity and have a visual indication that the dc supply voltage was okay. I also wanted visual confirmation of RF output when keying the rig. Without being overly complex, a tool like that would give me enough information to isolate the cause of a no-go situation. With these goals in mind, I put together an electronic multipurpose tool that I dubbed my QRP Buddy.

Voltage Checker

My QRP rig is happy with a supply voltage between 10 and 16 V dc, and I wanted an indicator that showed when the input voltage was between these limits. A minimal display, with one LED illuminating at the lower voltage limit and a second lighting up to indicate too high a voltage, would do the trick. A circuit to accomplish this ought to be straightforward.

My first thought was to use two comparators. A comparator is a differential amplifier: a circuit which outputs a voltage when the voltage you are testing exceeds a reference voltage in magnitude. I discounted this idea though, because I thought I'd need a really high supply voltage to cover the voltage range of interest. I envisioned the voltage checker requiring a pair of 9 V batteries in series—definitely not what I had in mind. As I discovered later, comparators turned out to be the way to go, but at first I spent considerable time playing with Zener diodes.

Zener Experimentation

A Zener diode does not conduct when reverse biased until the reverse voltage reaches a specific value. At that point, the diode becomes a low-resistance current path. Why not use a Zener diode as a voltage-controlled on/off switch? I tried a circuit consisting of a 10 V Zener diode, an LED and a series current-limiting resistor.

It worked—sort of. Unfortunately, the current characteristics of a Zener diode are such that the diode doesn't just suddenly start conducting when the breakdown voltage is reached. Instead, the LED in the circuit glowed dimly at very low voltages, got brighter as the voltage increased, then suddenly got very bright as the Zener voltage was achieved. This was not what I had in mind.

I then tried adding a transistor to act as a switch. I hoped that by using the Zener to bias an NPN transistor, with the LED grounded by the conducting transistor, the LED turn-on would be crisper. This idea worked a little better than the first circuit, but the transistor turned on too soon at the 16 V upper limit. Again, the trickle of current as the Zener approached breakdown was enough to forward bias the transistor and light the LED at about 12 V.

Resigning myself to using comparators, I sought the help of a friend and colleague, who has the advantage of being a *real electrical engineer*.[1] He provided me with the key idea to build the comparator circuit without requiring an 18 V supply. See Figure 1.

My friend suggested I use a very low-voltage Zener diode (D1) as a voltage reference, and compare it not to the input voltage directly, but to the input voltage scaled down by a voltage divider. Using a low-voltage Zener ensures that the diode is fully in the conducting state when the input voltage is in the range of interest, guaranteeing a stable reference voltage.

The voltage dividers consist of two sets of resistors (R1/R17 and R6/R18) connected between the input-voltage terminal and ground. The ratio of the resistance values is the same as the ratio between limit voltage I'm testing for and the Zener reference voltage. The Zener diodes I had on hand were 3.6 V units, so my resistance network needed a ratio of: 10 V / 3.6 V = 2.78 for the 10 V lower limit, and 16 V / 3.6 V = 4.45 for the 16 V upper limit.

A little playing around with standard resistor values got me pretty close to the required ratios. I used a variable resistor in each divider (R17 and R18) to allow for final tweaking of the limit voltages that illuminate the LEDs. The absolute magnitude of the resistances used in the network should be high—in the neighborhood of 1 MΩ—because you don't want a lot of current flowing in the divider network. Using high resistance values allows you to employ low-power resistors, as the power dissipated in the network is negligible.

LM339s are readily available and contain four independent comparators. I use only two of them, one for each voltage limit; the other two comparators are unconnected. Although any pair of comparators will do, I used comparators A and D of U1 because their input pins are on opposite sides of the chip.[2]

Because the LM339 is an *open collector* configuration, the pin floats high until the comparator circuit pulls it to ground, and allows current to flow through the LED. If I had connected the Zener reference voltage to the comparator's V– input, and the test voltage to the V+ input, the LED would light when the voltage was below the limit, and extinguish when the voltage went above. So rather counterintuitively, the Zener reference connects to V+ and the test input voltage to V–. Note that the circuit uses the test input voltage to power U1 and the LEDs. No batteries are required.

Zener diodes are excellent noise generators. Disconnect the voltage checker from the power supply before operating your rig, or you may be hearing your QRP Buddy and little else.

[1]Notes appear on page 5-10.

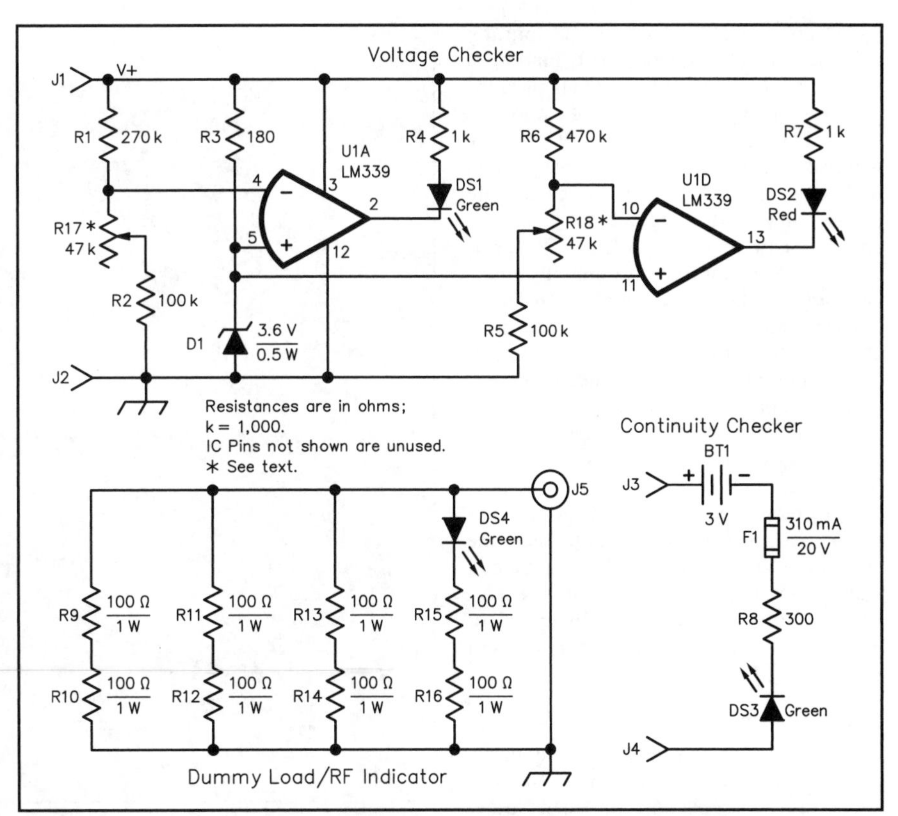

Figure 1—Schematic of the QRP Buddy. Unless otherwise specified, resistors are 1/4 W, 5% tolerance carbon-composition or film units. RS part numbers in parentheses are RadioShack. Equivalent parts can be substituted. All resistors used in this project (except for R9-R16) were obtained from a package of 100 assorted 1/4 W resistors (RS 271-306).

D1—1N5334B 3.6 V, 5 W Zener diode used here, but 0.5 W unit suitable (ECG5006A, 1N747A, 1N5227B, etc), available from sources such as Digi-Key Corp, 701 Brooks Ave S, Thief River Falls, MN 56701-0677; tel 800-344-4539, 218-681-6674, fax 218-681-3380; **http://www.digikey.com**; Mouser Electronics, 958 N Main St, Mansfield, TX 76063; tel 800-346-6873, 817-483-4422, fax 817-483-0931; **sales@mouser.com**; **http://www.mouser.com**, and others.
DS1, DS3, DS4—Green LED (RS 276-022)
DS2—Red LED (RS 276-041)
F1—310 mA, 20 V, fast-acting, 5×20 mm glass fuse (RS 270-1046)
R1—270 kΩ
R2, R5—100 kΩ
R3—180 Ω
R4, R7—1 kΩ
R6—470 kΩ
R8—300 Ω
R9-R16, incl—100 Ω, 1 W metallic-oxide film resistor (RS 271-152)
R17, R18—47 kΩ pot (RS 271-283)
U1—LM339 quad comparator (RS 276-1712)
Misc: PC-board-mount fuse clips (RS 270-744); lithium battery holder (RS 270-430); lithium battery CR2032 (RS 23-162); PC board (RS 276-150); banana jacks (RS 274-725); female BNC, chassis mount (RS 278-105); dual IC perfboard (RS 276-159); project box (RS 270-211); LED snap holders (RS 276-079).

RF Indicator

Thinking back to stories my Elmer had told me about using a light bulb as a visual indicator that RF energy is present, I decided to build an illuminated dummy load. A dummy load presents a 50 Ω, purely resistive load to a transmitter. Low-power (1 W) metallic-oxide resistors are inexpensive, readily available and present a purely resistive load at HF. By combining several higher resistance values in parallel, you can make a load capable of handling several watts that still presents a 50 Ω load. To dissipate *n* watts, you need *n* resistors in parallel, each with a resistance of *r*, where

$$r = n \times 50 \quad \text{(Eq 1)}$$

I wanted the load to dissipate 4 W continuously, so it could consist of four 200 Ω, 1 W resistors in parallel. I used four parallel pairs of two series-connected 100 Ω resistors.

To make the dummy load light up, I inserted an LED (DS4) into one leg of it, so the 200 Ω resistance of that leg does double duty: It contributes to the total 50 Ω resistance and acts as a current-limiting resistor for the LED. The resulting circuit is a safe dummy load that lights up when RF is pumped in. It's 100% solid state, and a lot more compact and rugged than a light bulb. (True, the SWR is slightly off from 1:1 because of the presence of the LED, but not enough to matter.)

Continuity Checker

This part was easy. A voltage source

(BT1), an LED (DS3) and a current-limiting resistor (R8) are all you need. Finding a voltage source that met my goals of compactness and light weight turned out to be a little tricky, though. Of course, a 9 V battery would do the job, but would add weight and bulk to the project. I thought I might be able to use a 6 V photographic battery, as they are tiny and readily available. Unfortunately, they do not fit any commonly available holder. They have a slightly larger diameter than common AA or N batteries. Then I noticed that RadioShack[3] sells 3 V lithium button batteries, (intended for computer-memory backup) and a little plastic holder for them as well. These batteries have a very long shelf life, and are well-suited to intermittent use in a continuity checker.

To prevent potentially hazardous battery damage if I inadvertently connect the continuity checker to a voltage source, I added a fuse (F1) to my continuity checker. The LED draws but a few milliamperes of current, so I used the lowest-current fuse I could find (0.31 A).

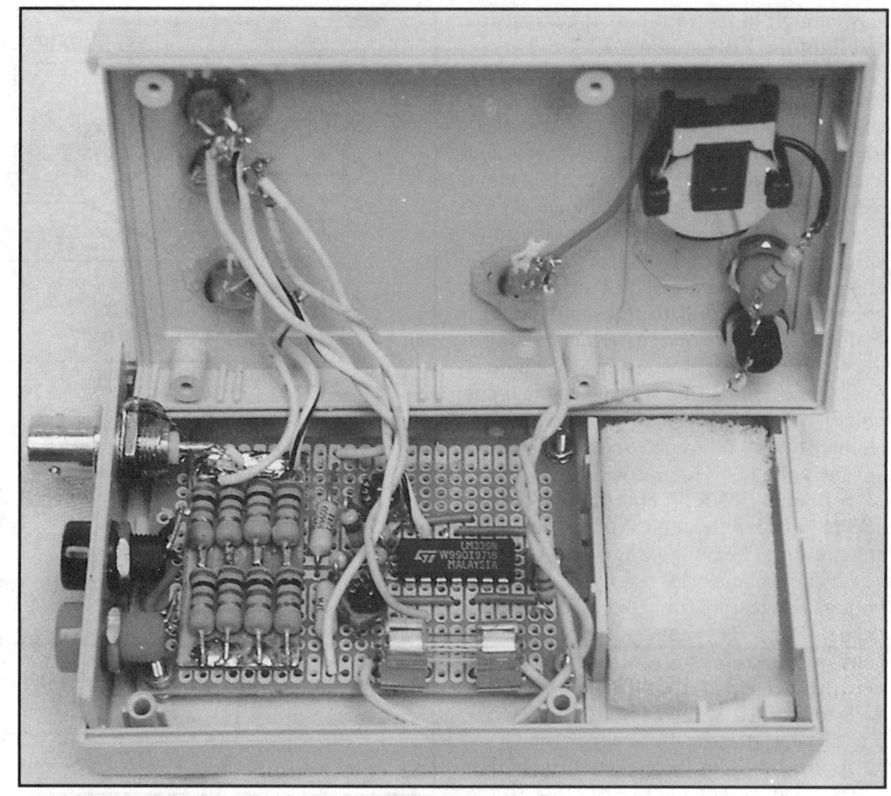

An inside view of the QRP Buddy. Secured to the top cover are BT1 (at the far right), J3 and J4, along with LEDs DS1-DS4. J1, J2 and J5 are mounted on the left-hand panel. Immediately behind the panel is the dummy load/PC board module. The two pots are just visible beneath the wires connecting the perfboard to the LEDs. F1 is in the foreground.

Construction

The QRP Buddy has two distinct parts: the dummy load and everything else. The "everything else" portion does not have any critical layout requirements, and any construction approach will do. I built the voltage comparator and the continuity checker on a piece of perfboard.

The dummy load (R9-R16 and DS4) is another matter. Resistance network dummy loads are usually constructed on copper-clad board, in a way that minimizes lead inductance and capacitance. This usually means keeping lead lengths to an absolute minimum and keeping the resistors as close to the ground plane (copper foil) as possible. I built my dummy load on a piece of double-sided PC board approximately 1 inch square, then glued it as a module to the perfboard for installation in a box. First, trim the resistor leads to length, then glue each resistor in place on the PC board before soldering it. Krazy Glue (a cyanoacrylate) works well for this purpose because it sets almost instantly. (Caution: These glues bond instantly to skin. Wear eye protection when using such adhesives and keep a bottle of the debonder near.) Solder one end of each pair of 100 Ω resistors to the copper foil. Connect three of the free ends together and attach the RF input lead from J5 to them. Solder the LED leads between the RF input jack and the free end of the last resistor pair. Solder the ground lead of the dummy load directly to the copper foil. After the dummy load module was finished and tested, I used cyanoacrylate glue to attach it to the main perf board.

I chose the smallest practical RadioShack box for the QRP Buddy. This box has a hatch on the back to allow access to the batteries, which is a plus. The voltage comparator and continuity checker use banana jacks as inputs (J1/J2 and J3/J4, respectively), sharing one pair of test leads between them. I use a BNC connector at J5; it occupies less bulkhead space than an SO-239 connector. I carry PL-259-to-BNC adapters anyway, because my rig sports a BNC antenna jack.

Calibration

The voltage comparators must be calibrated before use. If you have access to a variable-voltage dc power supply, fine. Set the supply for the lower limit voltage and adjust R17 until the low-limit LED (DS1) just lights. If it is not possible to get the LED to extinguish (or turn on) as you adjust the pot, you may need to switch the pot to the other side of the resistance network, or change the value of one of the resistors to alter the voltage ratio. After calibrating the low-limit LED, repeat the process for the high-limit comparator (R18 and DS2), with the supply set for the maximum safe voltage your rig can tolerate.

In the absence of a variable-voltage bench supply, you can get the voltages you need by using a collection of batteries and a voltmeter. Because the comparator circuit draws little current, any combination of common household batteries (AA, C, D, 9 V) can be connected in series to make up the voltages you need.

Summary

Although not quite the same as a complete electronics test bench, the QRP Buddy delivers essential information when troubleshooting problems in the field. Its compactness makes it easy to take along without overburdening a miniature QRP kit. Since I built mine, I haven't had any problems with my kit in the field that I couldn't resolve. Of course, I've learned my lesson: I don't travel anywhere without a bag of spare fuses!

Notes

[1] I am just a software guy who likes to play with radios. Thanks to my good friend Rich Galante for his invaluable guidance on the comparator circuit.

[2] You can use an LM393 8-pin dual comparator in place of the LM339.

[3] Valuing convenience over cost-effectiveness, all the parts I used to build this project are from RadioShack, except for the 3.6 V Zener diodes. These are common parts, and may be mail-ordered from suppliers such as Digi-Key, Mouser and others (see the caption of Figure 1).

Mike Aiello, N2HTT, has held an Advanced license since 1988, and enjoys working all modes on HF. He's been a software developer since 1969, working originally in finance and insurance applications, more recently in medical systems. Mike's areas of expertise are in embedded systems, robotics and analyzer/information system interfaces. Presently working on an MS in computer science at Polytechnic University of NY, Mike graduated from Cooper Union in 1972, with a BE in chemical engineering, followed by an MS in Management Technology from Polytechnic University in 1994. You can contact Mike at 7 Old Albany Post Rd, Croton, NY 10520; **n2htt@bestweb.net.**

By Larry Coyle, K1QW

From *QST*, August 2003

A "Clamped-Bandwidth" Gyrator Audio Filter

What's a gyrator? This little audio processor can improve the performance of a less-than-ideal receiver.

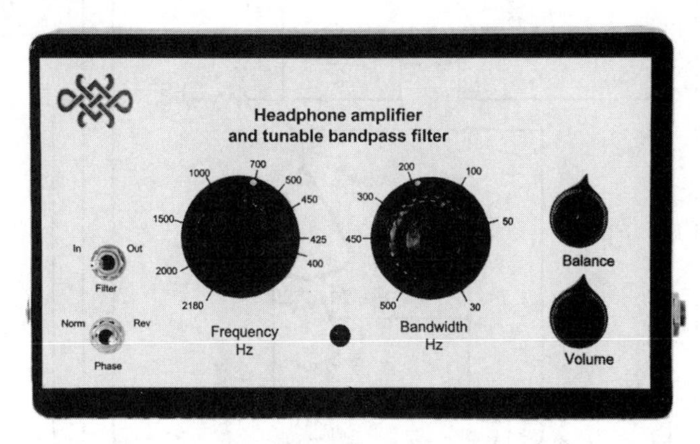

While waiting for enough cash to accumulate in my bank account to upgrade my ham equipment, I had to make do with a cheap, bottom-of-the-line receiver setup. Any of you who have been in that situation know the problems—noise, poor selectivity, etc. A few months ago, I decided to buckle down and put together an audio processor box. I wanted something that would plug in between the audio output jack from my receiver and my headphones and give me a little boost in performance.

In particular, I decided that I needed the following:

- a tunable band-pass filter
- a means of changing the balance between the left and right sides of the headphones (so I could compensate for a slight hearing loss in one ear)
- some additional gain

After a little more thought I came up with a few more nice-to-have features:

- a switch to reverse the phase between both sides of the headphones
- a constant or "clamped" bandwidth circuit that, once set, remains fixed, regardless of center frequency

This last feature would make the unit a lot more pleasant to use than the typical constant-Q band-pass filter, where the bandwidth gets wider as the center frequency is increased.

After perusing a few handbooks and asking around a bit, I decided that the heart of my little audio processor should be a variable band-pass filter based on a circuit called a *gyrator*. The gyrator is a useful circuit that doesn't seem to be as familiar as it should be, so I'll explain a little bit about it before jumping into the other details of the processor.

The Gyrator

A gyrator is a unique circuit that uses a couple of operational amplifiers, some resistors and a capacitor to mimic the properties of an inductor. A basic gyrator circuit is shown in Figure 1. The nice thing about a gyrator, aside from being small and inexpensive, is that the effective inductance can be quite large and can be varied over a wide range by simply changing the setting of a variable resistor. So, a gyrator inductor in parallel with a capacitor gives a fine imitation of a variable tuned circuit at audio frequencies, without the need for a huge iron-core inductor. More information on this useful circuit is available elsewhere.[1]

A gyrator with the circuit values shown in Figure 1 covers the inductance range of 600 mH to 27 H simply by rotating the 500 kΩ potentiometer (R1) over its full range. With a 0.01 μF capacitor in parallel, the circuit's resonant frequency can be varied from 300 Hz to 2000 Hz.

The Complete Circuit

Figure 2 shows the complete schematic for the audio processor. The gyrator inductor consists of op amps U1C and D, ca-

[1]P. Horowitz and W. Hill, *The Art of Electronics*, 2nd ed., Cambridge University Press, 1989, p 281.

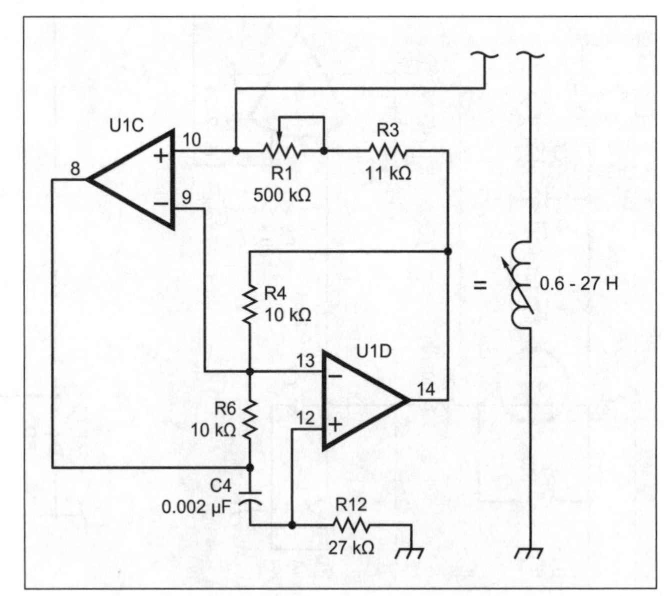

Figure 1—The basic gyrator circuit. This circuit looks like an inductor but minus the heavy iron. As the 500 kΩ pot is varied over its range, it mimics a 0.6 H-27 H variable inductor.

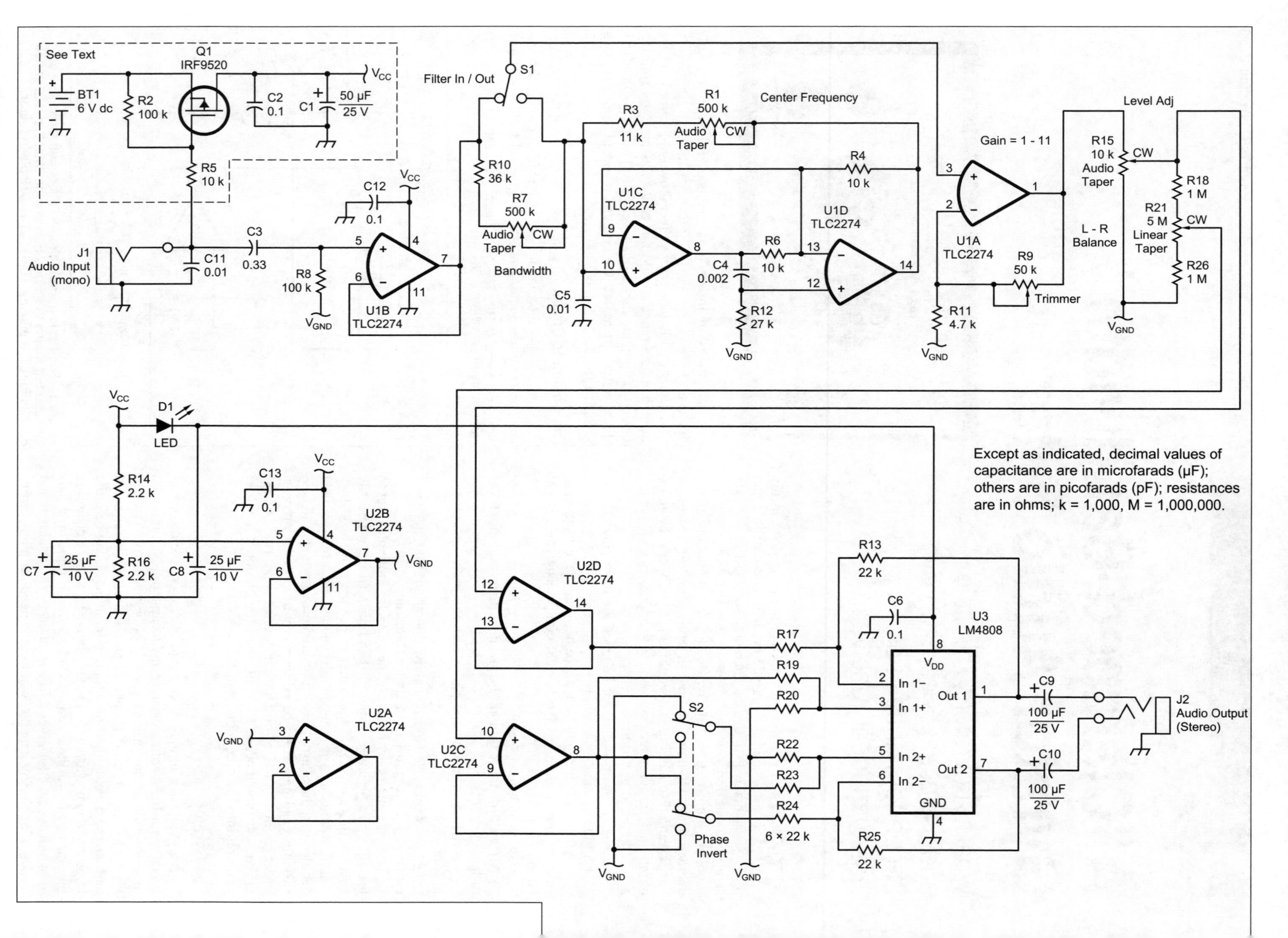
See Text
Q1 IRF9520
BT1 6 V dc
R2 100 k
C2 0.1
C1 50 µF 25 V
V_{CC}
R5 10 k
J1 Audio Input (mono)
C11 0.01
C3 0.33
R8 100 k
C12 0.1
U1B TLC2274
V_{GND}
Filter In / Out
S1
R10 36 k
R7 500 k
Audio Taper
CW
Bandwidth
C5 0.01
R3 11 k
R1 500 k
Audio Taper
Center Frequency
U1C TLC2274
C4 0.002
R6 10 k
R12 27 k
R4 10 k
U1D TLC2274
Gain = 1 - 11
U1A TLC2274
R11 4.7 k
R9 50 k
Trimmer
Level Adj
R15 10 k Audio Taper
L - R Balance
R18 1 M
R21 5 M Linear Taper
R26 1 M
D1 LED
R14 2.2 k
R16 2.2 k
C7 25 µF 10 V
C8 25 µF 10 V
C13 0.1
U2B TLC2274
U2A TLC2274
U2D TLC2274
U2C TLC2274
Except as indicated, decimal values of capacitance are in microfarads (µF); others are in picofarads (pF); resistances are in ohms; k = 1,000, M = 1,000,000.
R13 22 k
C6 0.1
U3 LM4808
V_{DD}
In 1−
In 1+
In 2+
In 2−
Out 1
Out 2
GND
R17
R19
R20
R22
R23
R24
6 × 22 k
R25 22 k
S2
Phase Invert
C9 100 µF 25 V
C10 100 µF 25 V
J2 Audio Output (Stereo)

Figure 2—Schematic diagram and parts list for the tunable-bandpass headphone amplifier. Parts are available from a number of sources, including Digi-Key Corp, 701 Brooks Ave South, Thief River Falls, MN 56701; tel 800-344-4539; **www.digikey.com**; Mouser Electronics, 1000 N Main St, Mansfield, TX 76063; tel 800-346-6873; **www.mouser.com** and Ocean State Electronics, PO Box 1458, 6 Industrial Dr, Westerly, RI 02891; tel 800-866-6626; **www.oselectronics.com**. Common resistors, capacitors and toggle switches are available at RadioShack; **www.radioshack.com**.

BT1—Battery, 4-AA in holder (6 V dc).
C1—50 μF, 25 V electrolytic capacitor.
C2, C6, C12, C13—0.1 μF, 50 V ceramic capacitor.
C3—0.33 μF, 50 V ceramic capacitor.
C4—0.002 μF, 50 V ceramic capacitor.
C5, C11—0.01 μF, 50 V ceramic capacitor.
C7, C8—25 μF, 10 V electrolytic capacitor.
C9, C10—100 μF, 25 V electrolytic capacitor.
D1—LED, red, panel-mount.
J1—¼ inch, mono phone jack.
J2—¼ inch, stereo phone jack.
Q1—IRF9520 MOSFET.
R1, R7—500 kΩ potentiometer, audio taper.
R2—100 kΩ, ¼ W, 5%.
R3—11 kΩ, ¼ W, 5%.
R4, R5, R6—10 kΩ, ¼ W, 5%.
R8—100 kΩ, ¼ W, 5%.
R9—50 kΩ, 10% trim potentiometer.
R10—36 kΩ, ¼ W, 5%.
R11—4.7 kΩ, ¼ W, 5%.
R12—27 kΩ, ¼ W, 5%.
R13, R17, R19, R20, R22, R23, R24, R25—22 kΩ, ¼ W, 5%.
R14, R16—2.2 kΩ, ¼ W, 5%.
R15—10 kΩ potentiometer, audio taper.
R18, R26—1 MΩ, ¼ W, 5%.
R21—5 MΩ potentiometer, linear taper.
S1—SPDT miniature toggle switch, panel mount.
S2—DPDT miniature toggle switch, panel mount.
S3—SPST miniature toggle switch, panel mount (optional power switch, see text).
U1, U2—IC, TLC2274CN quad op-amp, TI (Digi-Key 296-7129-5-ND).
U3—IC, LM4808M dual headphone amp (Digi-Key LM-4808M-ND).

pacitor C4 and resistors R1, R3, R4, R6 and R12. The 0.01 μF capacitor C6 resonates with the gyrator to form the equivalent of an LC tuned circuit. As R1 is varied over its range, the center frequency varies from 300 Hz to about 2060 Hz. R7 and R10 are in series with the tuned circuit and control the bandwidth, which can vary from 30 Hz to 420 Hz over the range of R7.

Note that with this arrangement, the circuit Q changes along with the resonant frequency, so that once the bandwidth is set by R7, it remains at that setting regardless of where the center frequency is set—one of my design goals.

Op amp section U1A acts as a buffer to prevent the level control, R15, from loading the filter and changing its electrical characteristics. I also included a gain-setting trimmer resistor, R9, in this stage, in case it became necessary to boost the receiver audio. R9 could be just a fixed resistor, if you don't need to fool much with the maximum gain.

Audio Output Stage

Following this buffer amplifier are the front panel level and balance controls. The outputs from these pots are buffered by op amp voltage followers and fed into the LM4808 dual audio amplifier, U3. Switch S2 reverses the phasing of one of the headphone channels and illustrates a very interesting phenomenon. In one position of S2, you will hear the sound in your headphones as two separate sources, one at each ear. With the phase reversed, the sound appears to come from somewhere inside your head.

Audio Input and Power Control

Audio is ac coupled from the input jack to a unity-gain buffer stage, U1B, to isolate the filter bandwidth control from the outside world. Across the input jack is a 0.01 μF capacitor to prevent local high-powered AM broadcast stations from getting into this circuit.

The whole circuit draws less than 11 mA and runs very nicely off four AA alkaline cells. To avoid wasting battery energy (and to save wiring in a power switch), I used a P-channel MOSFET, Q1, to switch power to the circuit. When the input cable is connected to the receiver audio output jack, R5 is pulled toward dc ground, turning on Q1 and applying power to the circuit. (If you decide to use this feature, you must be sure there's a dc path to ground through your receiver's audio output jack. Mine didn't have one, so I had to install a 10 kΩ resistor inside the receiver, across the jack. Alternatively, you could just wire in a power switch in place of Q1 and its associated components.) [An LM78L05 voltage regulator with two 1N916 diodes in series with the regulator's ground (center) pin (cathodes toward ground) will give 6.3 V dc at up to 100 mA from a 12-14 V dc station source. Place a 0.33 μF capacitor at the regulator input and a 0.1 μF capacitor at its output. Put the power switch before the regulator circuit.—*Ed.*]

Finally the LED, D1, serves a double purpose. It drops the supply voltage to IC U3 so as not to exceed the 5.5 V maximum supply voltage rating of the LM4808, and it also shows that power to the circuit is on. Because the current drawn by U3 varies with audio level, the LED blinks in time with the CW signals being received. I haven't yet found any practical use for this effect, but it does look cool. [It could be a good indicator of a CW signal centered within the preset passband, while tuning.—*Ed.*]

Construction

You can use any wiring technique you like for this project, as there is really nothing critical about parts layout or lead dressing. The 4 AA cells that power the unit mount inside the box, or an external 6 V dc or 12 V dc supply with regulator can be used, as described earlier.

Almost all parts are of the through-hole variety, one exception being the LM4808 output amplifier, which is available only in a surface mount package. A well-stocked junk box is a good start for most of the routine components, and ICs and pots are all readily available through the major parts suppliers.

Conclusion

This little box won't turn a third-rate receiver into a top of the line model, but it will help you dig out those CW stations from deeper in the pileups. It's easy to use, inexpensive to build, and when you crank the bandwidth down, it's a real joy to hear how the QRM just melts away!

Larry Coyle, K1QW, was first licensed in the late 1950s, while in high school. He operated CW and phone back then, using homebrew gear on 40 and 10 meters. Inactive for some time while pursuing an MSEE degree and a career, and raising a family, he reentered the amateur ranks after earning an Amateur Extra class license. Larry is looking forward to operating the HF CW bands again, whenever he can take time away from his EE consulting business. You can reach him at 100 Rolling Ln, Needham, MA 02492; **lmcoyle@attbi.com**. QST

By Dave Benson, NN1G

From *QST*, Decemer 1998

FREQ-Mite—A Programmable Morse Code Frequency Readout

This simple, one-evening project delivers a rig's frequency with dots and dashes!

When I incorporated a Morse-based frequency readout in a recent design,[1] I received a number of inquiries about its use in other applications. The PIC microcontroller family lends itself beautifully to a wide variety of Amateur Radio projects, and the temptation to modify that readout design for general-purpose use proved irresistible!

General Description

For experimenters and even occasional "home-brewers," a simple frequency counter often comes in mighty handy. At a mere 0.4×1.75×1.25 inches (HWD), the FREQ-Mite is small enough to get lost in your shirt pocket! For many of the QRP crowd, portable and backpacking rigs attract a lot of interest, and the compact form factor of this counter is aimed at ease of incorporating these gadgets into field-ready gear.

Morse code is unbeatable for hardware simplicity. To get this compact, this counter's output is delivered as an 800 Hz Morse code audio signal. The audio signal can be injected into the AF portion of a receiver, transceiver or, for stand-alone applications, connected to a piezo annunciator or speaker.

In the title photo you'll see a dual row of connector pins. By properly interconnecting them, these pins define an operating offset for the counter in the event it's embedded in, say, a superhet transceiver. In a superhet configuration, a local oscillator (LO) and mixer convert the incoming signal to an intermediate frequency (IF). In the transmitting mode, the same LO signal is mixed with an IF signal to form the transmitter's output frequency.

A wide variety of IF choices exists in kit-based and home-brewed gear. As a result, a general-purpose counter such as this needs a means of providing a frequency offset for a number of the possible IF values. In earlier designs, this process was handled by using analog trimmer potentiometers,[2] preprogrammed offset look-up (for the most popular frequencies used in kits) and even by keyer-paddle input. In this counter, the rows of jumpers (10 in all) serve as a compact way to enter any offset value (0-999) for superhet transceiver use. Because you already know which amateur band you're operating in (I hope!), there's little to be gained by announcing the megahertz digit(s). In the transceiver mode, then, the counter simply outputs the frequency in hundreds, tens and units of kilohertz. For example, with an operating frequency of 7.112 MHz, you'd hear **112** in Morse when requesting a readout.

These inputs also serve to select the general-purpose mode: Jumpering the leftmost seven pins on the connector strip corresponds to an invalid offset and the readout is in full four or five-digit fashion.

The counter is programmed to count any input frequency up to 32.767 MHz. Sensitivity is good. Figure 1 shows the minimum input signal required as a function of frequency. Accuracy is quite acceptable as a general-purpose tool—it's within 2 kHz of measurements made with other counters over this frequency range. At the other extreme, I've tested the circuit with inputs of up to about 10 V P-P input without incident. Somewhat higher drive levels (within reason) shouldn't harm this circuit. The maximum usable input-signal level begins to fall off around 25 MHz as parasitic effects become more prominent, but reducing the RF-input signal to a more modest level takes care of this situation.

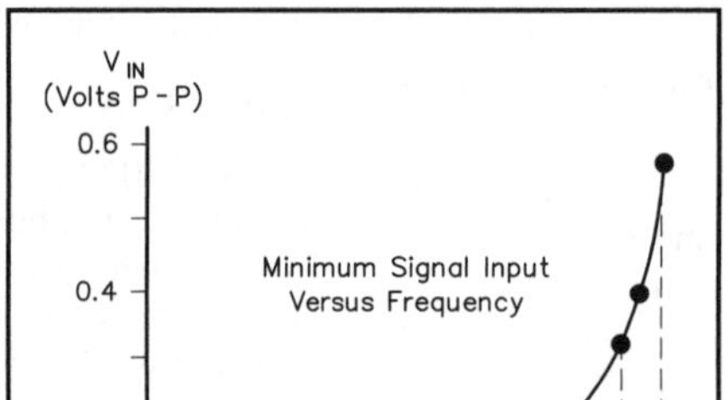

Figure 1—Input sensitivity versus frequency.

Circuit Description

As a glance at the schematic (Figure 2) reveals, there's not a lot of complex hardware in this design! The Microchip 16C621 PIC[3,4] (U1) is a single-IC, 8-bit microcontroller. U1 requires 5 V dc, which is ensured by the combination of R5 and Zener diode D1. Q1 is an emitter follower. It presents a high-impedance load to the given RF source. This is important if the counter is connected directly to a VFO as you don't want the counter loading (pulling) the oscillator frequency. Q2 is configured as a high-gain amplifier and converts

[1]Notes appear on page 5-16.

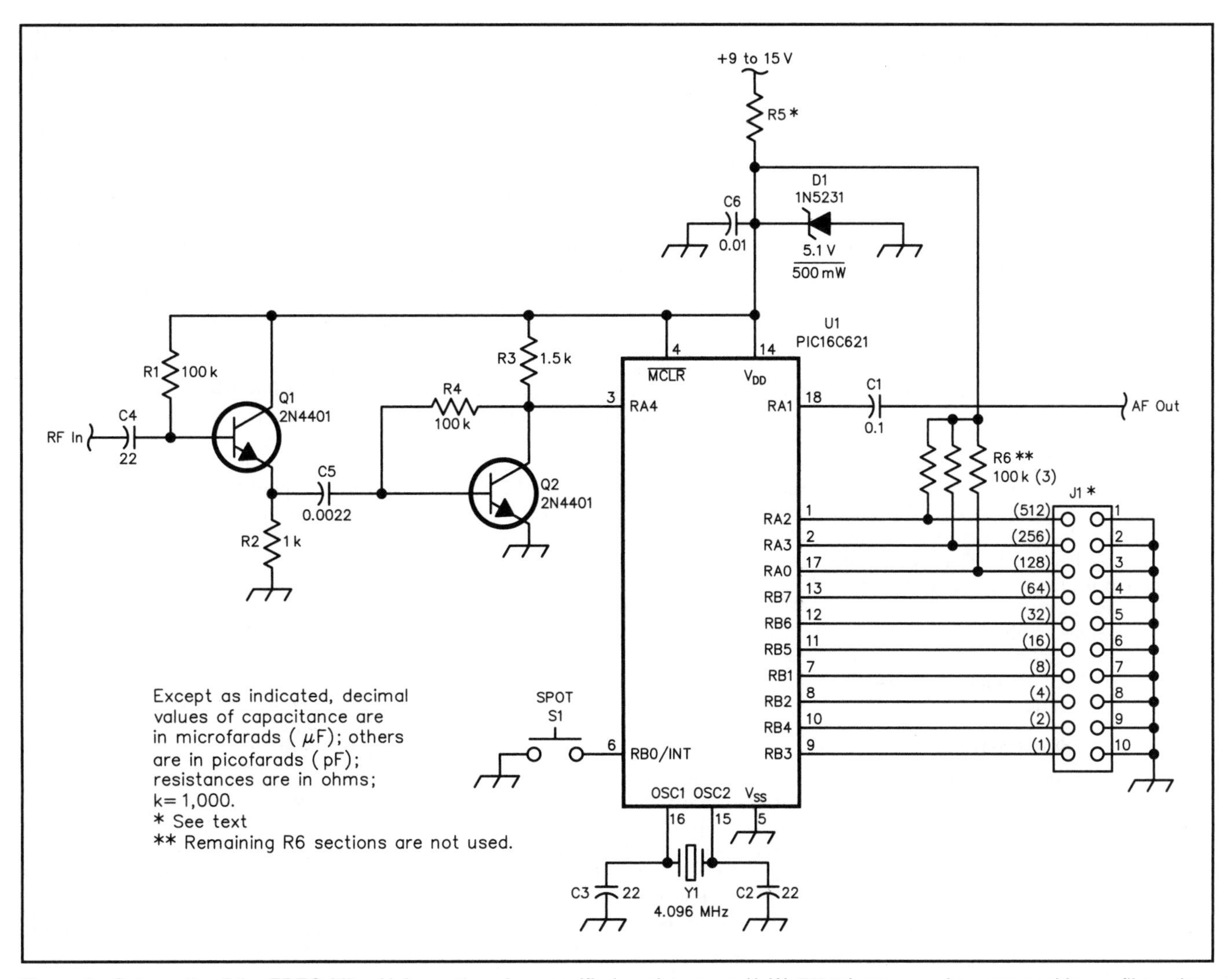

Figure 2—Schematic of the FREQ-Mite. Unless otherwise specified, resistors are 1/4 W, 5% tolerance carbon-composition or film units. Equivalent parts can be substituted.

C1—0.1 μF monolithic cap
C2-C4—22 pF NP0, 5%-tolerance ceramic disc
C5—0.0022 μF disc ceramic
C6—0.01 μF disc ceramic
D1—5.1 V, 500 mW Zener diode (1N5231)
Q1, Q2—2N4401 NPN
R1, R4—100 kΩ
R2—1 kΩ
R3—1.5 kΩ
R5—330 Ω or 470 Ω (see text)
R6—100 kΩ resistor network, three sections used (6-lead/5-resistor SIP such as Digi-Key 770-61-R100K [CTS] and Q5104 [Panasonic]); Digi-Key Corp, 701 Brooks Ave S, Thief River Falls, MN 56701-0677; tel 800-344-4539, 218-681-6674, fax 218-681-3380; **http://www.digikey.com**.
U1—*Preprogrammed* PIC 16C621 (see Note 3)
Y1—4.096 MHz, HC-18/U holder (20 pF)
Misc: 18-pin IC socket, hardware.

the RF signal to a 5 V P-P wave compatible with the PIC's I/O requirements. Audio output is taken from pin 18 of the PIC and consists of an 800 Hz square-wave tone sequence. This output is activated only on power-up and on closing S1(**SPOT**) for a frequency reading. Otherwise, the output is turned off to minimize "thump" when this project is tied to an audio circuit.

When power is applied, U1's code performs a number of initialization chores. First, it examines each of the J1 inputs (jumper positions) to form a value used later to calculate IF offsets. (How these jumpers are set is covered later under **Construction, Setup and Use**.) U1 then asks for input: readout speed (slow or fast) and for normal or inverted readout, if for transceiver use. Following these steps, U1 goes to sleep.

When you press the **SPOT** switch, U1 wakes up and counts the number of LO cycles for a period of one millisecond, in effect, the LO frequency in kilohertz. (Actually, a divide-by-128 prescaler function is used and the main timebase counts for 128 milliseconds.) If the offset jumpers have established that the chip is in a transceiver (three-digit) mode, it adjusts the count result for the programmed IF. If the IF offset is jumper-programmed to an invalid offset (greater than 1008 [decimal]), the counter output switches to a full four- to five-digit readout and no offset is applied. A math routine within U1 converts the count result from its native hex format to binary-coded decimal (BCD). These data are the values of the 10 MHz, 1 MHz, 100 kHz, 10 kHz and 1 kHz digits. These digits are then converted to Morse strings using a look-up table. The strings are fed to a routine that toggles U1's I/O port RA1 (pin 18) at an 800 Hz rate; the audio is then sent to the associated equipment's AF amplifier. Output audio is programmed at 13 WPM (slow) and 26 WPM (fast); speed selection is discussed later.

Construction, Setup and Use

PC boards, *programmed* PICs and kits of parts are available (see Note 3). When building your FREQ-Mite, note that the value of R5 should be 330 Ω for use with a 9 V power supply and 470 Ω with a 12 to

15 V supply. Make sure that the power supply is disconnected before you install U1. *This part is static-sensitive. Ground yourself while handling and installing this device.* One easy way to do this is to use a conductive wrist strap.

In receivers and transceivers, the point at which the FREQ-Mite's output is injected into the radio's audio chain is typically just prior to the radio's audio-output stage. A 100 kΩ to 10 MΩ resistor connected in series with C1 in the output line to the host audio stage reduces the signal to acceptable levels. You'll likely have to experiment with the resistance value as the required audio level is rig-dependent.

Programming

The FREQ-Mite can be programmed for use with most transceivers. Transceivers generally use a superhet configuration in which the LO is shifted by the IF to yield the operating frequency. The FREQ-Mite can be set to any three-digit offset (0 to 999) by means of jumper block J1. The row of header locations is binary-weighted. The right-most position (position 10) has a code of 1, the one adjacent to that is 2, then 4, etc. The left-most position has a code of 512.

Transceiver (Three-Digit) Use

The offset code should be set equal to the three-digit value of the IF in kilohertz. *You may need to vary this value slightly (±1 kHz) to compensate for offsets within the rig.*

Example: If the IF is 8.192 MHz, 8.192 = 192. The binary-weighted code for 192 = 128+64, the jumper coding (from left to right on the board) is 0011000000. You simply add jumpers at the locations denoted by 1 and leave the 0 locations open.

Programming Example for Inverted Operation

The White Mountain-75 SSB rig uses a 9830 kHz IF. The offset coding is set to 830, and the pushbutton should be pressed after the **I?** prompt. For instance, for an LO of 6005 kHz, the FREQ-Mite calculates (9830 – 6005) and outputs 825 to indicate an operating frequency of 3.825 MHz.

Naturally, there are exceptions: In the case of a rig with a 455 kHz IF and an LO *above* the operating frequency, the programming must be reversed. For example, if the IF is 455 kHz and the LO is 4055 kHz, the operating frequency is 3600 kHz. The correct offset value is (1000 – 455) or 545.

Sanity Check

After you've checked your FREQ-Mite for correct assembly, install jumpers at J1 positions 3 and 4 (the 128 and 64 binary-weighted positions). Connect the audio output of the FREQ-Mite as detailed earlier. With no RF source connected to the FREQ-Mite, apply power to the circuit. The FREQ-Mite will output **S?** in Morse. Wait until U1 sends an **AR** character to let you know the initialization is complete. Briefly press S1 and the Morse readout you hear should equal **192**. If you want to check out the various binary codes, you'll need to cycle power to the board each time you change jumpers; the offset is read only during power-up.

General-Purpose Use

When you're home-brewing an oscillator or checking a transmitter's output, it's useful to know the operating frequency complete with the megahertz digit(s). To enter this mode, add shorting jumpers to the left-most seven shorting locations of J1 (positions 1 through 7) and apply power. This results in an offset code of 1008 (decimal), and the chip recognizes this as an invalid offset, suppresses the offset calculations, and outputs four or five digits. Here are some examples: For an input frequency of 455 kHz, U1 sends **0455**; for an input frequency of 7.110 MHz, U1 sends **7110**; for an input frequency of 21.106 MHz, U1 sends **21106**.

Initialization

Readout Speed

On power-up, FREQ-Mite sends an **S?**. If you press S1 within about two seconds, U1 switches to a higher speed (26 WPM) readout. It acknowledges this entry with an **R**. If you do nothing, U1 maintains its default (13 WPM) readout speed.

Normal/Inverted Readout

If—and only if—you've programmed a legitimate offset for transceiver (three-digit) mode, FREQ-Mite inquires if you want an inverted IF by sending **I?**. If you press S1 within about two seconds, the chip switches to calculating an inverted frequency. (This is necessary if the IF is *above* the LO because the operating frequency is the *difference* between these two frequencies rather than the sum.) In this case, U1 acknowledges this entry with an **R**. If you don't press the pushbutton, the chip maintains its default (summing) operation. In this case, the chip sends an **AR** prosign after several seconds to let you know it's completed initialization.

Acknowledgments

Wayne Burdick, N6KR, deserves much credit for the concept of annunciating operating frequency in Morse code. Thanks also to Tony Fishpool, G4WIF, and Jim Hossack, W7LS, for their constructive comments and suggestions during development.

Notes

[1]Dave Benson, NN1G, "A Single-Board QRP SSB Transceiver for 20 or 75 Meters," *QST*, Apr 1997, pp 29-33.

[2]Neil Hecht, "A PIC-Based Digital Frequency Display," *QST*, May 1997, pp 36-38.

[3]This project requires a *programmed* 16C621 PIC. Programmed PICs and assembly instructions are available for $12 postpaid; a complete kit of parts, including a double-sided PC board, all on-board parts, mounting hardware and instructions is available for $20 postpaid from Small Wonder Labs, 80 E Robbins Ave, Newington, CT 06111; **bensondj@aol.com**. Only personal checks or money orders are accepted. Source code is not available.

[4]See John A. Hansen, W2FS, "Using PIC Microcontrollers in Amateur Radio Projects," *QST*, Oct 1998, pp 34-40.

Dave Benson, NN1G, is well-known to the QRP community. He's been publishing articles for specialty journals since the early '90s, and his work has appeared in QST *and* The ARRL Handbook. *Dave owns Small Wonder Labs, a supplier of QRP-oriented Amateur Radio kits. When not dreaming up new projects, his other interests include gardening, camping and playing guitar. Dave also reports—with considerable relief—completing an addition to his home.*

You can contact Dave at 80 E Robbins Ave, Newington, CT 06111; **bensondj@aol.com**.

By Wes Hayward, W7ZOI, and Bob Larkin, W7PUA

From *QST*, June 2001

Simple RF-Power Measurement

PHOTOS BY JOE BOTTIGLIERI, AA1GW

Making power measurements from *nanowatts* to 100 watts is easy with these simple homebrewed instruments!

Measuring RF power is central to almost everything that we do as radio amateurs and experimenters. Those applications range from simply measuring the power output of our transmitters to our workbench experimentations that call for measuring the LO power applied to the mixers within our receivers. Even our receiver S meters are power indicators.

The power-measuring system described here is based on a recently introduced IC from Analog Devices: the AD8307. The core of this system is a battery operated instrument that allows us to directly measure signals of over 20 mW (+13 dBm) to less than 0.1 nW (–70 dBm). A tap circuit supplements the power meter, extending the upper limit by 40 dB, allowing measurement of up to 100 W (+50 dBm).

The Power Meter

The cornerstone of the power-meter circuit shown in Figure 1 is an Analog Devices AD8307AN logarithmic amplifier IC, U1. Although you might consider the

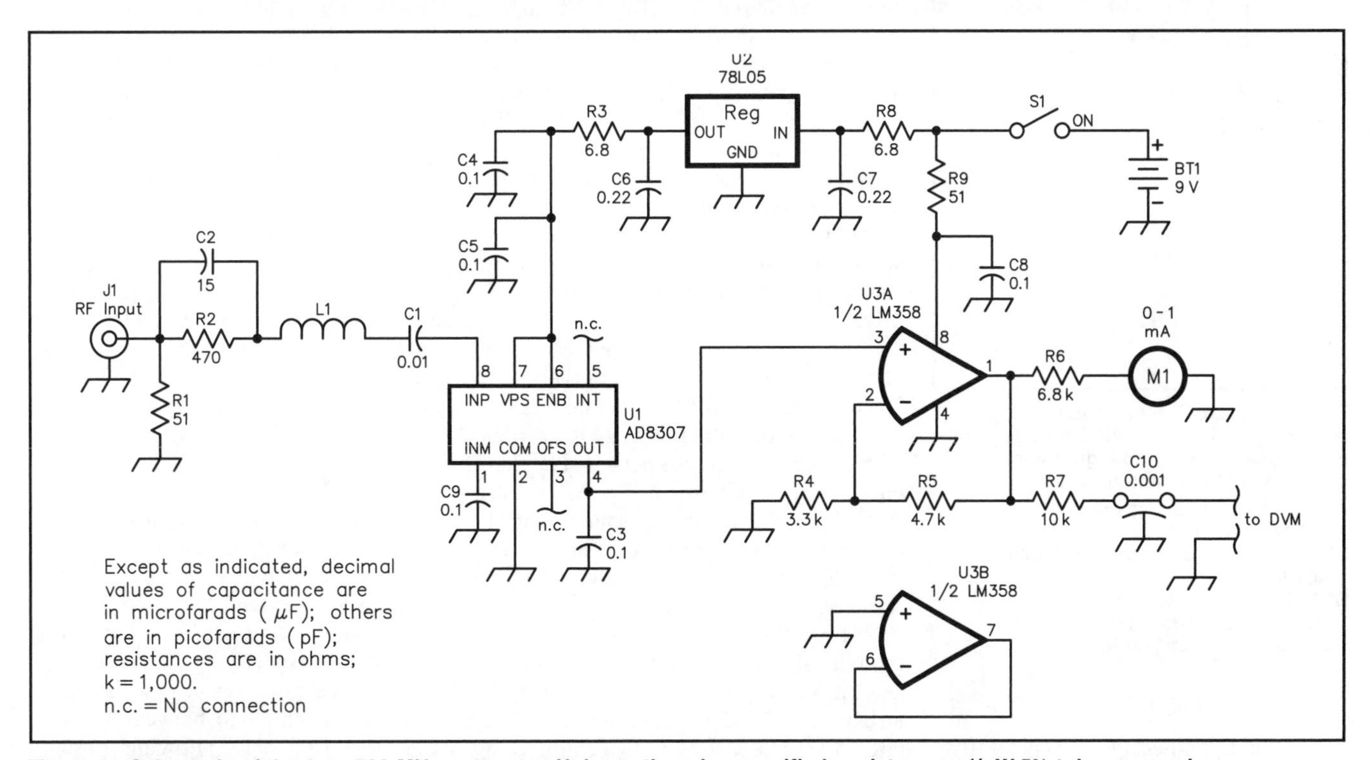

Figure 1—Schematic of the 1- to 500-MHz wattmeter. Unless otherwise specified, resistors are ¼-W 5%-tolerance carbon-composition or metal-film units. Equivalent parts can be substituted; n.c. indicates no connection. Most parts are available from Kanga US; see Note 2.

J1—N or BNC connector
L1—1 turn of a C1 lead, 3/16-inch ID; see text.
M1—0-15 V dc (RadioShack 22-410); see text.
S1—SPST toggle
U1—AD8307; see Note 1.
U2—78L05
U3—LM358
Misc: See Note 2; copper-clad board, enclosure (Hammond 1590BB, RadioShack 270-238), hardware.

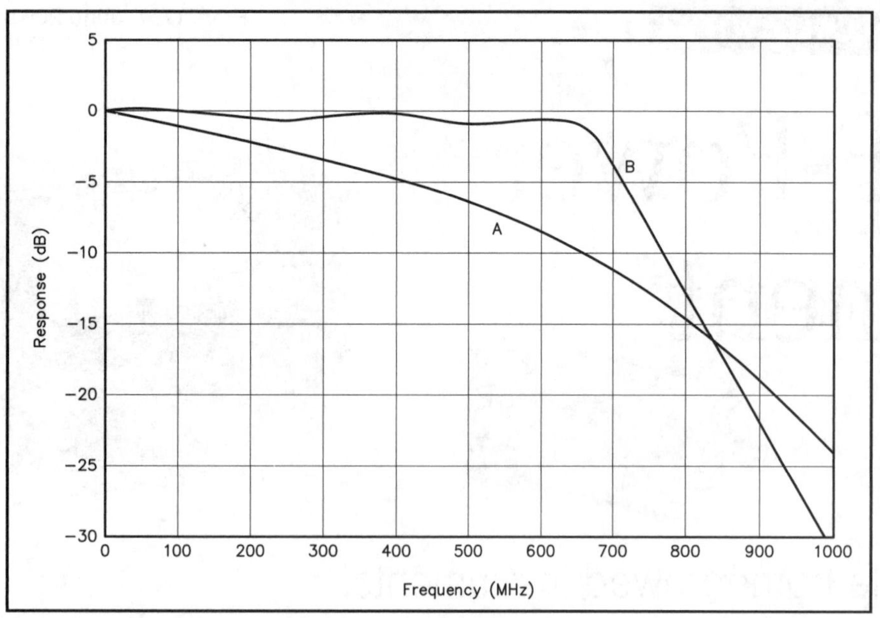

Figure 2—Response curves for the power meter before (A) and after (B) addition of R2, C2 and L1.

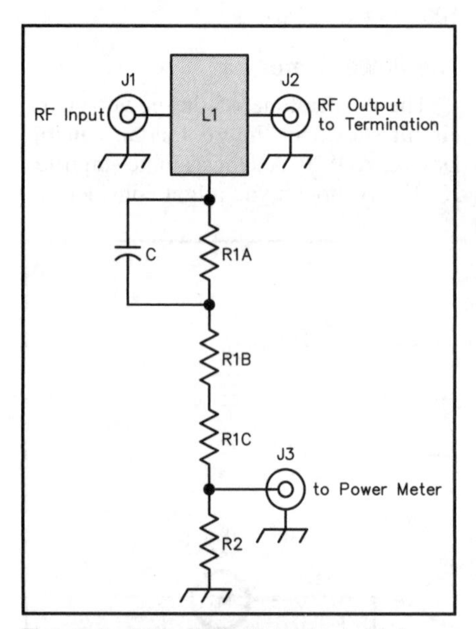

Figure 3—A tap that attenuates a high-power signal for use with the power meter. See the text and Figure 4 for discussion of the capacitor, C.

J1, J2—N connectors; see text.
J3—BNC connector
L1—1×1½-inch piece of sheet brass; see text.
R1A-R1C—Three series-connected, 820-Ω, ½-W carbon film
R2—51-Ω, ¼-W carbon film

IC as slightly expensive at about $10 in single quantities, its cost is justified by the wide dynamic range and outstanding accuracy offered. You can order the part directly from the Analog Devices Web site, which also offers a device data sheet.[1,2]

[1]Notes appear on page 5-22.

U1's power supply can range from 2.7 to 5.5 V. A 5-V regulator, U2, provides stable power for U1. U3, an op amp serving as a meter driver completes the circuit. U1's dc output (pin 4) changes by 25 mV for each decibel change in input signal. The dc output is filtered by a 0.1 µF capacitor and applied to the noninverting input of U3, which is set for a voltage gain of 2.4. The resulting signal with a 60 mV/dB slope is then applied to a 1-mA meter movement through a 6.8-kΩ multiplier resistor. When the circuit is driven at the 10-mW level, U3's output is about 6 V. U3's gain-setting resistors are chosen to protect the meter against possible damage from excessive drive.

U1 has a low-frequency input resistance of 1.1 kΩ. This combines with the resistances of R1 and R2 to generate a 50-Ω input for the overall circuit. R2 in parallel with C2 form a high-pass network that flattens the response through 200 MHz. L1, a small loop of wire made from a lead of C1, modifies the low-pass filtering related to the IC input capacitance, extending the response to over 500 MHz.

M1 is a RadioShack dc voltmeter. Although sold as a *voltmeter*, it actually is a 0-1 mA meter movement supplied with an external 15-kΩ multiplier resistor. The 0 to 15 V scale is used with a calibration curve that is taped to the back of the instrument to provide output readings in dBm. The dBm units can be converted to milliwatts by using a simple formula, although dBm readout is generally more useful and convenient.[3]

An auxiliary output from C10, a feedthrough capacitor, is provided for use with an external digital voltmeter or an oscilloscope for swept measurements.[4] We use the DVM when resolution is important. The analog movement can be read to about 1 dB, which is useful when adjusting or tuning a circuit. Enterprising builders might program a PIC processor to drive a digital display with a direct reading in dBm.

The first power meter we built did not include R2, C2 and L1. That instrument was accurate in the HF spectrum and useful beyond. Adding the compensation components produced an almost flat response extending beyond 500 MHz with an error of only 0.5 dB.The compensation network reduces the sensitivity by about 3 dB at HF, but boosts it at UHF. If your only interest is in the HF and low-VHF spectrum through 50 MHz, you can simplify the input circuit by omitting R2, C2 and L1. The responses before and after compensation are shown in Figure 2.

The power meter is constructed "dead-bug" fashion without need of a PC board. It is breadboarded on a strip of copper clad PC-board material held in place by the BNC input connector. R1 is soldered between the center pin and ground with short leads. U1 is placed about ¾ inch from the input in dead-bug fashion (leads up) with pins 1 and 8 oriented toward J1. The IC is held to the ground foil by grounded pin 2 and bypass capacitors C3, C4 and C5. R2 and C2 are connected to the J1 center pin with short leads. L1 is formed by bending the lead of C1 in a full loop. Use a 3/16-inch-diameter drill bit as a winding form. None of the remaining circuitry is critical. It is important to mount the power meter components in a shielded box. We used a Hammond 1590BB enclosure for one meter and a RadioShack box for the other, with good shielding afforded by both. *Don't use a plastic enclosure for an instrument of this sensitivity.*

Higher Power

Transmitter powers are rarely as low as the maximum that can be measured with this power meter. Several circuits can be used to extend the range including the familiar attenuator. Perhaps the simplest is a resistive tap, shown in Figure 3. This circuit consists of a flat piece of metal, L1, soldered between coaxial connectors J1 and J2, allowing a transmitter to drive a 50-Ω termination. A resistor, R1, taps the path to route a sample of the signal to J3, which connects to the power meter. R2 shunts J3, guaranteeing a 50-Ω output impedance. Selecting the values that comprise R1 establishes the attenuation level.

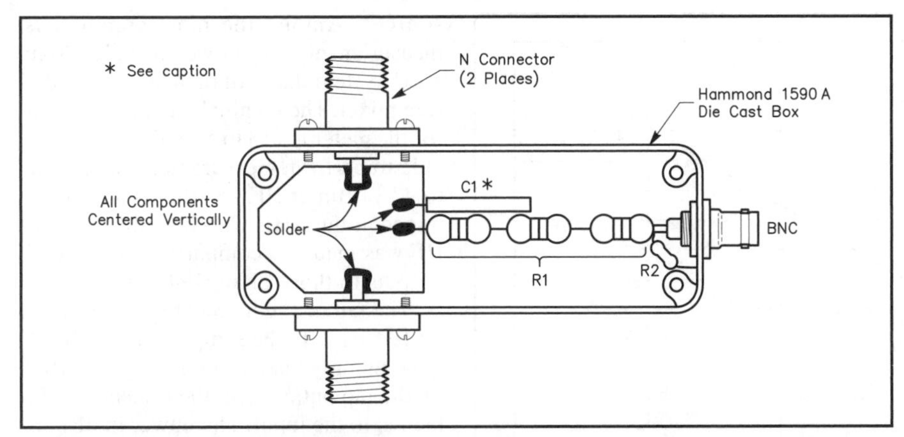

Figure 4—Drawing of the 40-dB power tap assembly shown schematically in Figure 3. The center conductors of the two N connectors (RF INPUT and RF OUTPUT) are connected by a 1×1½-inch piece of sheet brass with its corners removed to clear the pillars in the Hammond 1590A die-cast aluminum enclosure. C1 is made from a piece of #22 AWG insulated hook-up wire; it extends 0.6 inch beyond the edge of the tinned metal piece and almost rests against the two resistor bodies.

The tap extends the nominal +10 dBm power meter maximum level to +50 dBm, or 100 WPower dissipation becomes an issue at this level, so R1 is built from three series-connected half-watt, carbon-film resistors.

The tap is built with the J1 to J2 connection configured as a 50-Ω transmission line as shown in Figure 4 and the accompanying photographs. Adjustments were performed with an HP-8714B network analyzer. The analyzer was used to adjust the value of C for an attenuated path to J3 that is flat within 0.1 dB up to 500 MHzThe tap can then be used with a spectrum analyzer or laboratory grade power meter.

It is not realistic to achieve 0.1-dB accuracy through UHF without a network analyzer for adjustment. However, if the tap is duplicated using the mechanical arrangement of Figure 4, you can expect the tap to be flat within about 1 dB through 500 MHz. The low-frequency attenuation is determined merely by the resistors, so can be guaranteed with DVM measurements. If the primary interest is in measurements below 150 MHz, you can replace the N connectors with BNC connectors. The tap is housed in a Hammond 1590A box.

Calibration

We read and use our meters in one of two ways. The DVM output is recorded and used with an equation that provides power in dBm. Alternatively, we read the panel meter and look at a chart taped to the back of the instrument. In both cases, we need a known source of RF power to calibrate the tool.

The easiest way to calibrate this instrument is with a calibrated signal generator. Set the generator for a well-defined output and apply it to the power meter. We did our calibration work at 10 MHz and levels of –20 and –30 dBm. The two levels provide a 10-dB difference that establishes the slope in decibels per DVM millivolt. One of the two values then provides the needed constant for an equation. The signal generator can be stepped through the amplitude range in 5- or 10-dB steps to generate data for the meter plot. Figure 5 shows a plot of DVM output Vs power meter reading. The meter plot is similar.

If you don't have access to a quality signal generator, you can calibrate the power meter using a low-power transmitter. A power level of 1 to 2 W at 7 MHz is fine. Attach the transmitter through the tap to a dummy load where the output voltage can be read directly using a diode detector and DVM as shown in Figure 6. If the power output is 1 W, the peak RF voltage will be 10 V. The detector output will then be about 9.5 V and the signal to the power meter is –10 dBm. Adding a 10-dB pad, as shown in Figure 6, at the meter input drops the power to –20 dBm for the second calibration point.

Applications

There are dozens of applications for this little power meter, a few of which are shown in the accompanying figures. Some applications are obvious and practical while others are more elaborate and instructive. Most of these measurements are *substitutional*, where the power meter is substituted for a load in a circuit. In contrast, most measurements with an oscilloscope are *in situ*, performed *in place* within a working circuit.

Figure 7A shows power measurement for early stages of a transmitter, very low power transmitters, or signals from other

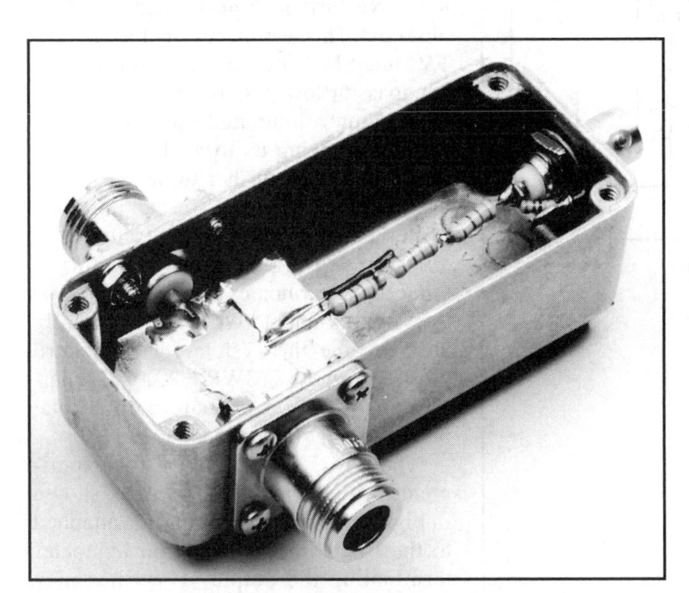

An inside view of the simple attenuator shown in Figure 3.

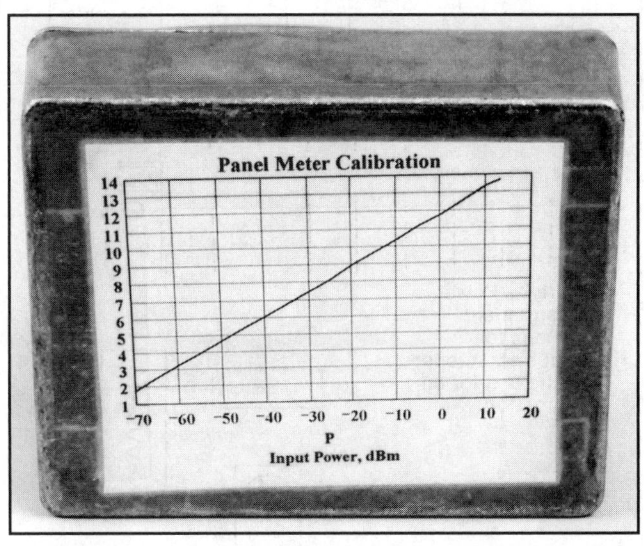

The calibration curve taped to the back of the instrument provides output readings in dBm relative to the meter's 0 to 15 V scale.

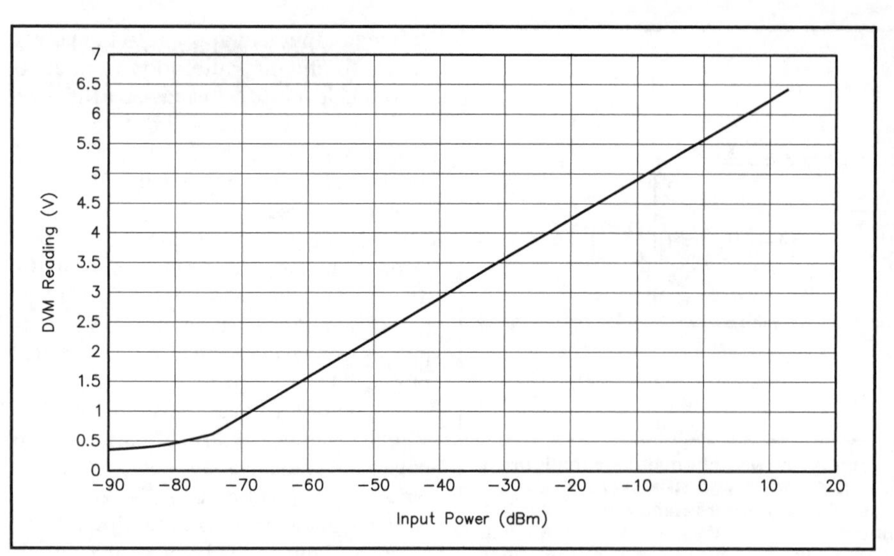

Figure 5—Plot of the DVM output versus generator power.

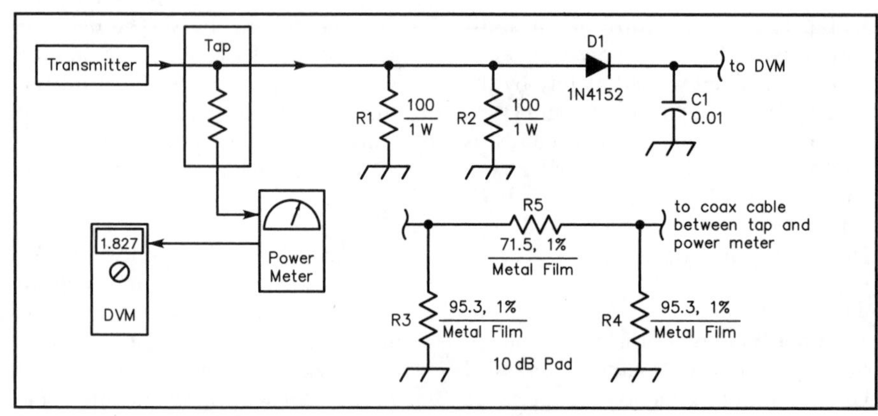

Figure 6—If a signal generator is not available, calibration can be done using a low-power transmitter. Resistor values for a 10-dB pad are shown.
C1—0.01 µF disc ceramic
D1—1N4152
R1, R2—100 Ω, 1 W
R3, R4—95.3 Ω, 1% metal film
R5—71.5 Ω, 1% metal film

Figure 7—Power measurement schemes for various situations; see the text for an explanation.

sources. Among the most common is measurement of the power available from a LO system that will then drive a diode-ring mixer. The nominal maximum power for the meter is +13 to +16 dBm. We were able to perform measurements nearly up to +18 dBm at HF, but this is not maintained at the VHF. Careful calibration at HF was made by comparing our meters' outputs to those of an HP435A.

The tap of Figure 3 extends transmitter testing with the setup of Figure 7B. A good dummy load (termination) is placed on the tap output with the transmitter attached to the input. The power in dBm is now that read on the meter in dBm plus the tap attenuation in decibels.

We sometimes wish to measure power during an operating session. This can be done with the setup of Figure 7C. A typical application might be a QRP station where the operator experiments with significantly reduced, but variable power.

The power meter is useful for a variety of applications with bridge circuits. Figure 8 shows the meter as the detector for a return loss bridge (RLB) driven by a signal generator.[5] In this example, we use the system to adjust an antenna tuner. Because of the excellent sensitivity of the power meter, the generator need not have high output power. For example, we often make these measurements with a homebrew generator delivering +3 to +10 dBm.

The use of such low power can complicate the measurements, as we discovered when we tried the experiment of Figure 8. The exercise began without either of the filters shown. When the generator was turned on, the power meter indicated –4 dBm from the RLB. We tuned the matching circuit, but could only achieve –25 dBm, indicating 21-dB return loss.[6] No further improvement could be observed. This was the result of local VHF TV and FM broadcast interference. A bandpass or low-pass filter in the power-meter input eliminated the residual response, allowing us to achieve a 45-dB return loss with further tuning. But this was also a limit where no further improvement seemed possible. Adding a low-pass filter to the signal generator output reduced the harmonics there, allowing further improvement. We were eventually able to tune the system for an absurd 60-dB return loss (SWR = 1.002), generally impossible to measure with normal bridges using diode detectors.

The power meter is ideal for experiments with various RF filters as shown in Figure 9. A signal generator is attached to the filter input with the power meter terminating the output. The filter may then be tuned or swept. Temporarily re-

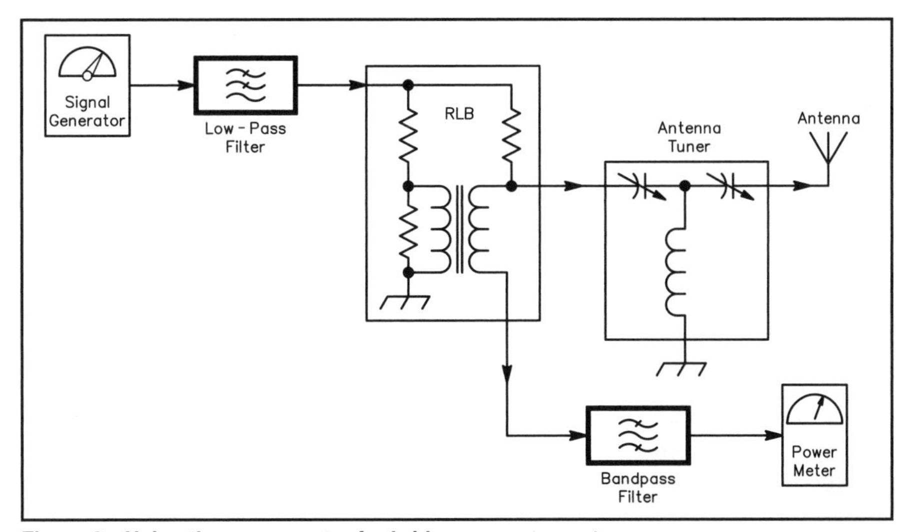

Figure 8—Using the power meter for bridge measurements.

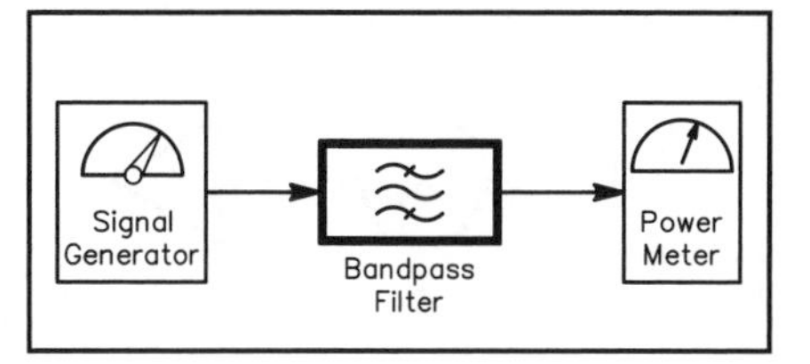

Figure 9—Filter measurements with the power meter.

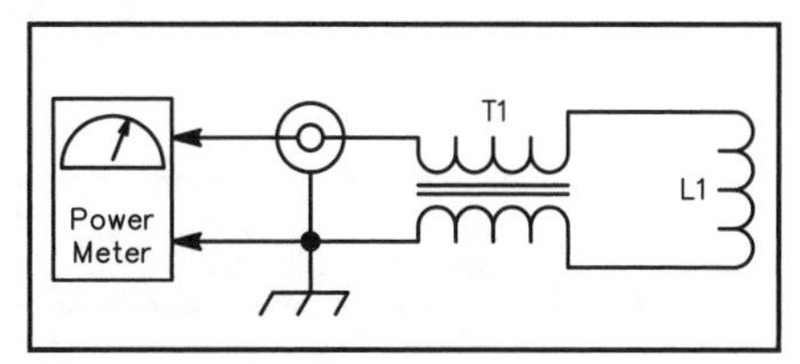

Figure 11—An RF sniffer probe that allows observation of relative RF levels. This probe allows you to see self-oscillation in amplifiers, or proportional responses in a receiver or transmitter.

L1—Two turns of insulated wire, 1/4-inch ID.

T1—Several ferrite beads on a length of coaxial cable; see text.

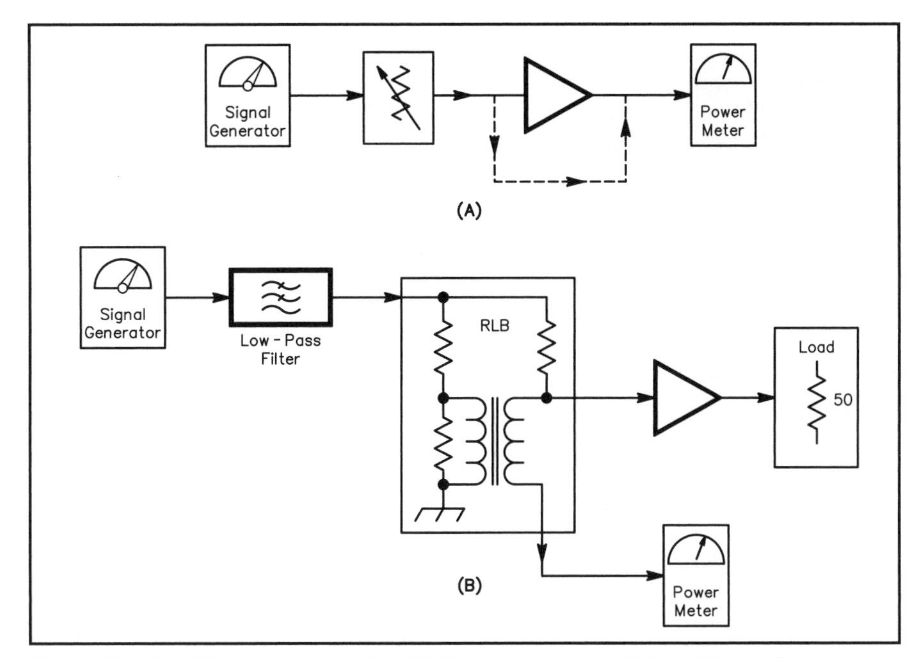

Figure 10—Amplifier measurements with the power meter. At A, making gain measurements. At B, a method to determine the input-impedance match. Reversing the amplifier terminals allows investigating reverse gain and output match.

placing the filter with a coaxial through connection allows you to evaluate filter insertion loss. Both the power meter and the signal generator are 50-Ω instruments, so the filter will need matching networks if it is not a 50-Ω design.

As with the previous example, measurement anomalies can be observed when investigating filters. For example, after using the power meter to adjust a 7-MHz bandpass filter we were able to easily measure the second-harmonic content of the signal generator when tuned to 3.5 MHz.

Figure 10A shows the power meter used with a signal generator to study amplifier circuits. A step attenuator is shown with the generator, allowing the power level to be reduced while preserving a 50-Ω environment. Generally, a drive level of –30 dBm is low enough with typical circuits. The system is initially set up with a through connection, indicated by the dotted line. Then the amplifier is inserted and the output power is measured. The difference between the two responses, each in dBm, is the gain in decibels. An interesting and easily performed related measurement is that of amplifier reverse gain. Merely swap the amplifier terminals, attaching the generator to the *output* and the power meter to the *input*. The measured *gain* will now be a negative decibel number.

Amplifier investigation continues with the setup of Figure 10B where we use an RLB to measure the input impedance match. Although a simple bridge will not provide the actual input impedance, it will tell you how close the circuit is to a perfect match. Adjustments can be done to achieve a match. Again, reversing the amplifier allows examining the output. We included a low-pass filter in the generator output, a precaution that may also be useful with the gain-determination setup. The measurements of Figure 10 provide the information normally provided by a scalar network analyzer.

The power meter can serve as the detector for a number of simple instruments. Figure 11 shows a simple RF sniffer probe, handy for examining circuit operation. The probe consists of a small inductor attached to the end of a piece of coaxial cable (RG-58, RG-174, or similar). A few ferrite beads of about any type are placed over the outside of the cable near the coil. The probe can be placed close to an operating circuit to look for RF. The smaller the link diameter, the greater the spatial resolution can be. This is a scheme that actually lets you see self-oscillation in an amplifier, much more useful than a speculation that a circuit "might be oscillating."

The power meter can be used with other probes. One might be a simple antenna that would allow field-strength determinations. Another is a resonance-indicating probe that would provide traditional dip meter-like measurements, but with improved ac-of experimental frequency synthesizers.

Concluding Thoughts

The traditional view of a power meter is as an instrument that examines transmitter output. But it can be much more than that. The AD8307 allows you to build a power meter that turns a common Amateur Radio station into the beginnings of a RF measurement lab.

Our thanks to Barrie Gilbert of Analog Devices Northwest Labs for providing the AD3807 IC samples.

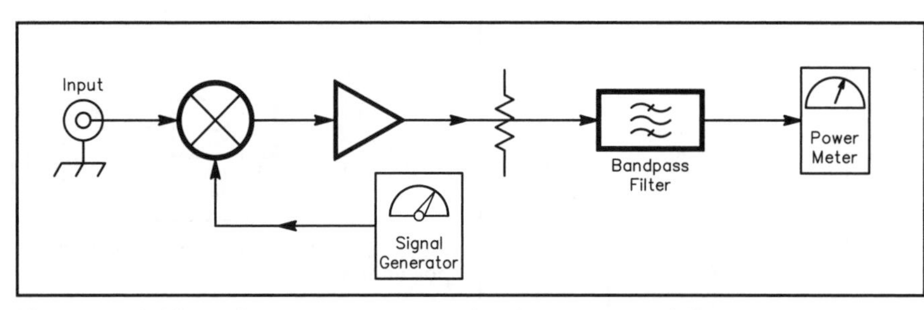

Figure 12—Adding a few components to a signal generator and the power meter creates a measurement receiver. The kinds of measurements possible depend on the filter used. See the text for some possibilities.

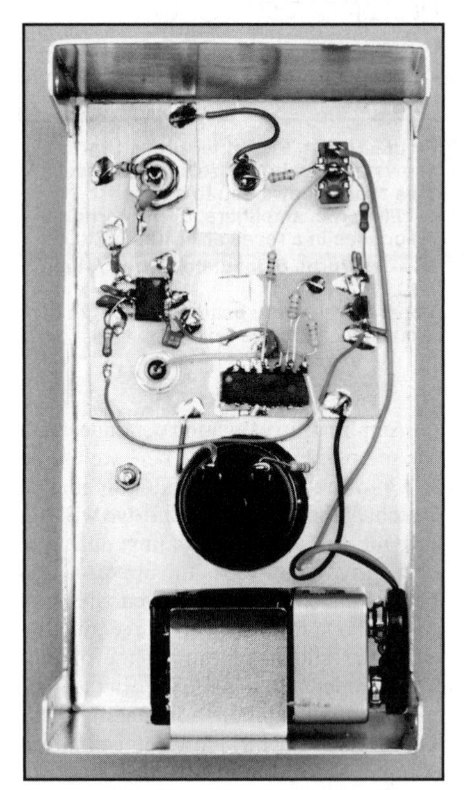

An inside view of the wattmeter showing its "dead-bug" construction and simplicity.

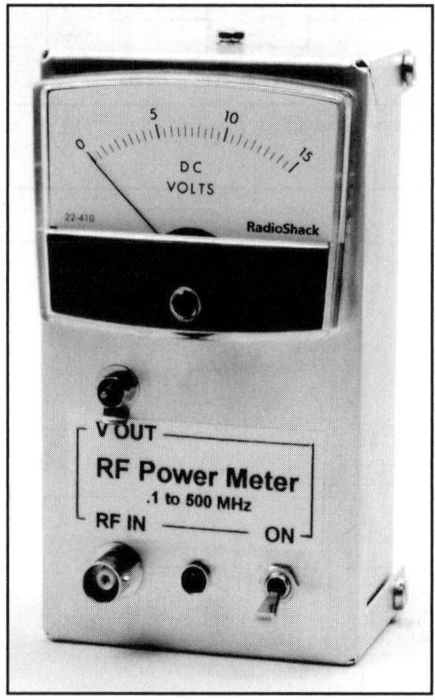

This version of the power meter is built in an inexpensive RadioShack utility box.

curacy and sensitivity.[7]

A recent *QST* project developed by Rick Littlefield, K1BQT, uses an AD8307 as a relative RF indicator.[8] That instrument, with the probe described in the sidebar by Ed Hare, W1RFI, is aimed at examining conducted electromagnetic interference (EMI). Our power meter should function well with that probe. There is great potential for small portable instruments for the study of both conducted and radiated EMI.

Figure 12 shows an example of some simple instruments that can be built using the power meter as a foundation. Here, the signal generator becomes the LO for a mixer such as the popular diode ring. This drives an optional amplifier and attenuator, followed by a bandpass filter. The power meter measures the filter output. The result is a custom measurement receiver.

We have built two variations of this project. The first uses a three-resonator LC bandpass filter tuned to 110 MHz, while the signal generator tunes from 50 to 250 MHz.[9] A Mini-Circuits MAV-11 is used for the amplifier. The resulting receiver can then be used to measure signals over the entire spectrum up to 360 MHz with sufficient resolution to examine transmitter spurious responses.

The second measurement receiver uses a homebrew 5-MHz crystal filter with a 250-Hz bandwidth. The signal generator is a homebrew unit with extreme tuning resolution, or bandspread. This instrument was used to measure SSB-transmitter carrier and sideband suppression and IMD, and for examining spurious output

Notes

[1]**www.analog.com**. The data sheet includes an extensive discussion of the theory of operation of the logarithmic detector and applications beyond the scope of this article.

[2]Kanga US offers a collection of most of the parts for this project, excluding the meter, copper-clad board and enclosure. For specifics, contact KANGA US, Bill Kelsey, N8ET, 3521 Spring Lake Dr, Findlay, OH 45840; tel 419-423-4604; **kanga@bright.net**; **www.bright.net/~kanga/**.

[3]$P_{mW} = 10^{dBm/10}$

[4]Feedthrough capacitors are available from Down East Microwave Inc, 954 Rt 519, Frenchtown, NJ 08825; tel 908-996-3584, fax 908-996-3702; **www.downeastmicrowave.com/**.

[5]See Wes Hayward, W7ZOI, and Doug DeMaw, W1FB, "Solid-State Design for the Radio Amateur," p 154, ARRL, 1977. Directional couplers are also useful in this application, such as that used in the classic W7EL power meter, Roy Lewallen, W7EL, "A Simple and Accurate QRP Directional Wattmeter," *QST*, Feb 1990, pp 19-23 and 36.

[6]A return loss of 21 dB corresponds to a SWR of 1.196, already a great match for most practical antenna situations.

[7]See Wes Hayward, W7ZOI, "Beyond the Dipper," *QST*, May, 1986, pp 14-20. Also, the signal-generating portion of that instrument is useful as a simple, general-purpose RF source.

[8]Rick Littlefield, K1BQT, "A Wide-Range RF-Survey Meter," *QST*, Aug 2000, pp 42-44; see also Feedback, Oct 2000, p 53.

[9]Wes Hayward, W7ZOI, "Extending the Double-Tuned Circuit to Three Resonators," *QEX*, Mar/Apr 1998, pp 41-46. The band-pass filter was then used in the instrument described in Wes Hayward, W7ZOI, and Terry White, K7TAU, "A Spectrum Analyzer for the Radio Amateur," *QST*, Aug 1998, pp 35-43; —Part 2, Sep 1998, pp 37-40.

Over the years, Wes Hayward, W7ZOI, has provided readers of QST, The ARRL Handbook and other ARRL publications with a wealth of projects and technological knowhow. His most recent article, The Micromountaineer Revisited (which he wrote with K7TAU), appeared in July 2000 QST. You can contact Wes at 7700 SW Danielle Ave, Beaverton, OR 97008; **w7zoi@easystreet.com**.

Bob Larkin, W7PUA, is a consulting engineer for communication companies. His last article, "An 8-Watt, 2-Meter 'Brickette'" appeared in June 2000 QST. You can contact Bob at 2982 NW Acacia Pl, Corvallis, OR 97330; **boblark@proaxis.com**.

By Art Rideout, WA6IPD

From *QST*, June 1996

A Simple LED SWR/Power Meter

Most analog SWR/power meters display *average* power readings. Their mechanical meter movements simply cannot react fast enough to show power fluctuations that take place in fractions of a second. When you speak into the microphone of an SSB transceiver, for example, the output power bounces up and down in sync with your voice. Maximum output occurs at split-second amplitude peaks. You'll never see these with an ordinary SWR/power meter. If you want to get a handle on the true output performance of your transceiver, you need a *peak-reading* meter.

"Oh, great—something else I have to buy!"

Not so! Why not add peak-reading capability to your existing SWR/power meter? The peak-reading LED display I'm about to describe works best when it is used to completely replace your analog indicators. (When the LED display is connected to the meter circuit, its accuracy may suffer if the analog indicators are used at the same time.) However, you can opt for a versatile combination if your analog meter features a **FWD/REF** switch. You can set the analog meter to read average reflected power while the LED meter displays peak forward power, or vice versa. They won't interfere with each other in this configuration.

I use my LED meter with a Drake L-4B linear amplifier, which has a meter that can be switched to indicate grid current, plate voltage or SWR/power. I usually monitor grid current with the analog meter and use the peak-reading LED display for SWR/power.

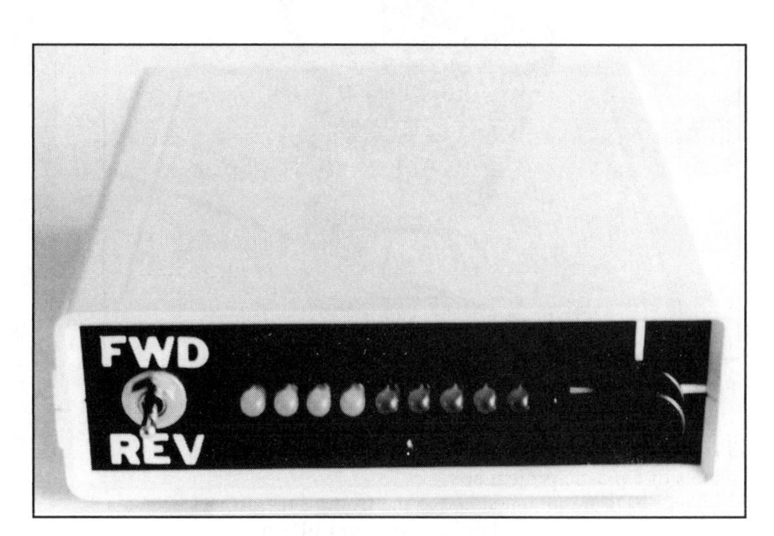

The meter works equally well with 100 W or 1500 W. The lowest power level for satisfactory operation is 15 W. The cost of parts can vary greatly, so I suggest you price shop. This project should cost under $20 to build.

Construction

As you can see in Figure 1, this is a very simple project with many parts readily available from Radio Shack. At its heart is U1, an LM3914 dot/bar display driver. I mounted U1, C1 and R3 on a Radio Shack prototyping board (Figure 2), but a printed-circuit board is

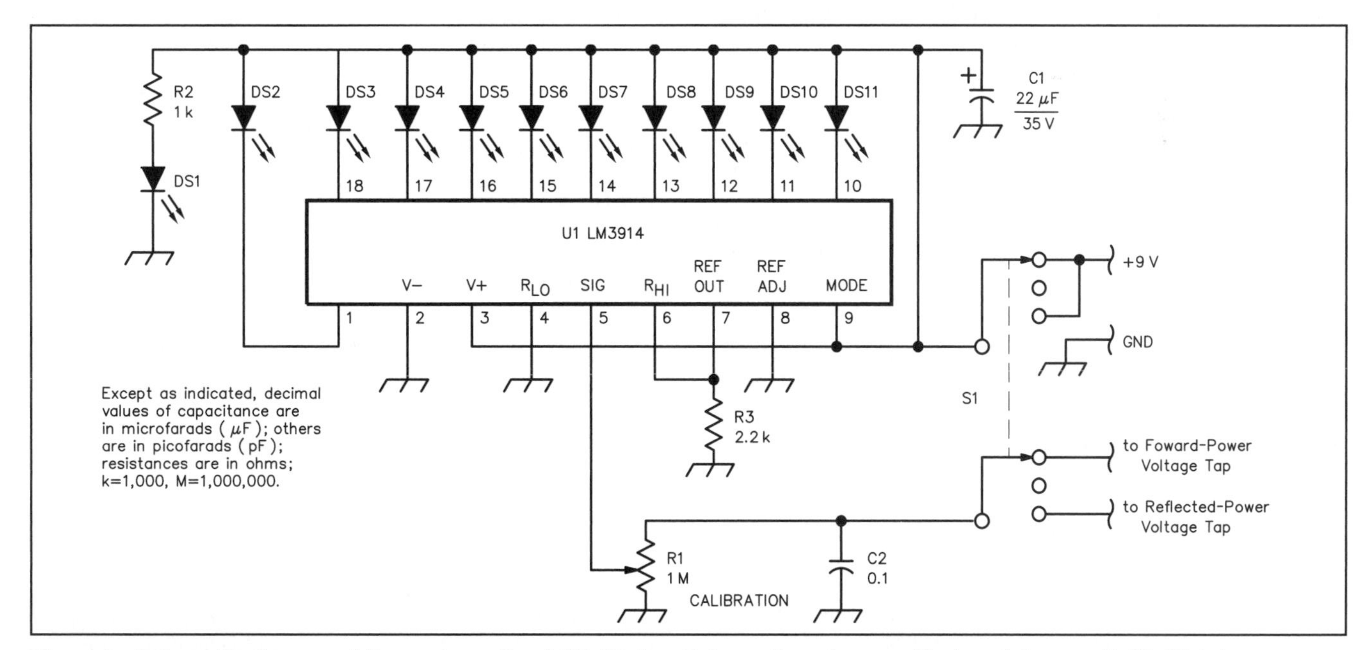

Figure 1—Schematic diagram of the peak-reading LED display. Unless otherwise specified, resistors are 1/4-W, 5% tolerance carbon composition or film.

C1—22 μF electrolytic capacitor, 35 V (Radio Shack 272-1026)
C2—0.1 μF ceramic disc capacitor (Radio Shack 272-135)
DS1, DS11—Red LEDs (Radio Shack 276-041)
DS2, DS3, DS4, DS5—Green LEDs (Radio Shack 276-022)
DS6, DS7, DS8, DS9, DS10—Yellow LEDs (Radio Shack 276-021)
R1—1 MΩ potentiometer (Radio Shack 271-211)
R2—1 kΩ resistor (Radio Shack 271-1118
R3—2.2 kΩ resistor (Radio Shack 271-1121)
S1—DPDT toggle switch (Radio Shack 275-1545)
U1—LM-3914 display driver (Hosfelt Electronics, 2700 Sunset Blvd, Steubenville, OH 43952; tel 800-524-6464)

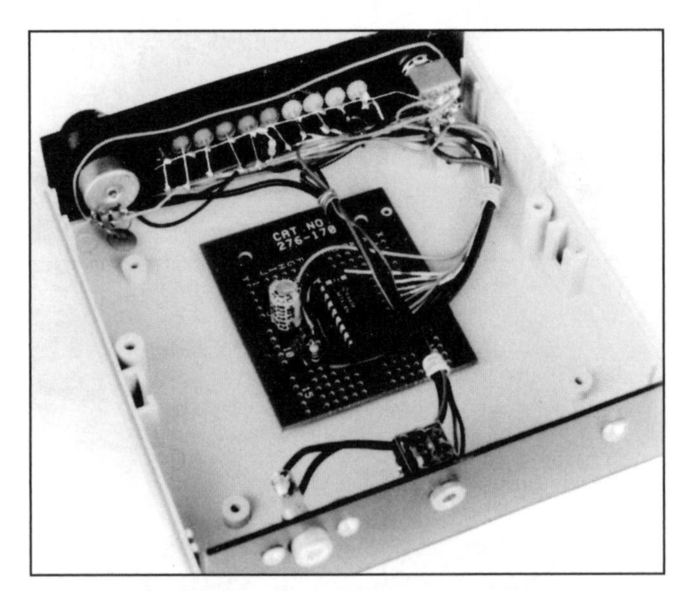

Figure 2—Internal view of the author's LED display.

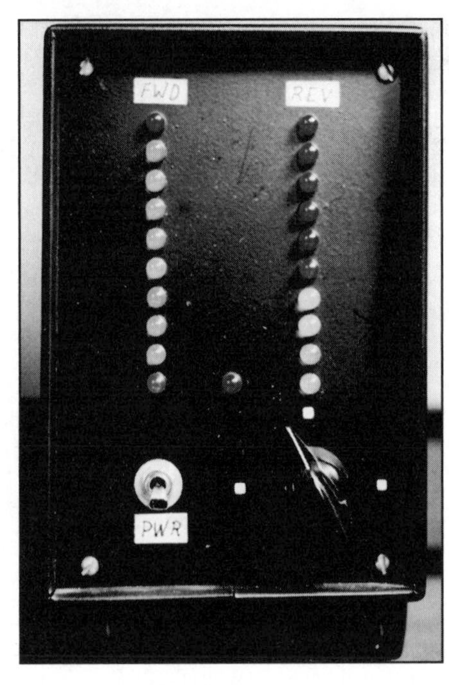

Figure 3—Why settle for one display when you can monitor forward and reflected power simultaneously? The next step for the curious experimenter is a *dual* LED display such as this one.

also available.[1] All the other components are installed on or near the panels of a plastic project box.

Ten LEDs are mounted across the front of the project box in a particular left-to-right order: The first four LEDs are green, the next five are yellow and the last one is red. When installed correctly, the LEDs will light from left to right. If the meter is used to measure reflected power, any reading past the fourth green LED indicates an SWR greater than 3:1.

I suggest that you use a straight edge and carefully mark the LED holes prior to drilling. The alternative is to create a paper template and tape it onto the front of the case as a drilling guide. Your goal is to drill a series of 10 holes in a neat horizontal row. You'll need an additional hole below the row for the **POWER** LED (DS1). Its limiting resistor, R2, connects nearby. Push each LED through its appropriate hole from the inside. A little dab of Superglue or epoxy along the rear edge will hold it in place.

You should also install the **CALIBRATE** potentiometer (R1) on the front panel, along with the **FWD/REV** switch, S1. S1 is a center-off DPDT (double-pole, double-throw) switch that is used for dc power switching and for selecting the forward or reflected source voltages from the SWR meter. Note that capacitor C2 is wired across the terminals of R1.

I placed my circuit board in the center of the box, but this will vary, depending on the type of enclosure you use. You can secure the board with epoxy, or with screws and standoffs. Placement of the wiring between the PC board and the LEDs, switch, potentiometer and jacks isn't critical.

Speaking of jacks, I used a 1/8-inch stereo phone jack for the connection to the SWR meter. A common phono plug is used for the 9-V power connector. Both jacks are installed on the rear panel.

Connecting Your SWR Meter

Remove the cover of your analog meter and look inside for two diodes. These diodes change the RF "sampled" from your coax into dc voltages for the meter(s). When you find the diodes, examine their connecting circuitry carefully. You'll discover that only one wire from each diode connects to small RF bypass capacitors. The points where these wires connect to the capacitors are where you must tap the forward and reflected source voltages.

I strongly recommend that you use a shielded cable to connect the LED display to your SWR meter. I used a two-conductor-plus-shield audio cable with an 1/8-inch stereo plug on one end. Solder the conductors of the bare end of that cable to the tap points, and connect the shield braid to the meter case or circuit ground.

Calibration and Operation

Calibrating the LED meter is simple. With both the analog meters and the LED display connected in tandem, switch the LED display off (the center switch position). Place your transceiver in **CW, TUNE,** or whatever mode is necessary to achieve maximum *continuous* output. For our hypothetical example, we'll say that maximum output is 100 W. Then adjust the rig's RF power control until you measure 100 W on your analog power meter. Now, patch the LED meter into the circuit by switching S1 to the **FWD** mode. Adjust the **CALIBRATE** control (R1) until the red LED on the far right just lights. The peak reading on the LED meter now represents 100 W output. Switch S1 to the **REV** position and you'll display reflected power.

If your analog meter reads 100 W and your LED display barely flickers, try throwing S1 to the opposite position. You may have tapped the voltage sources "backward," mistaking the forward power voltage source for reflected, and vice versa. There is no need to rewire your switch unless you've already attached your **FWD/REV** labels to the box.

SWR readings are a snap. Adjust your transceiver output for a full-scale reading in the FWD position and then switch to REV to read reflected power. Four green LEDs indicate an SWR of 3:1, three LEDs indicate 2:1, and so on.

This indicator does not automatically calibrate itself. So, *don't forget to recalibrate every time you change your transceiver output settings.*

Double Your Pleasure

The next obvious step is to build a unit that monitors both forward and reflected power *simultaneously*. We can eliminate S1 and bring both source voltages into a dual-potentiometer version of R1. Of course, you need two sets of display drivers, LEDs and so forth. This time, however, another switch may have to be added to remove the analog meter completely from the circuit when the LED meter is in use.

My unit has LEDs running vertically (see Figure 3). Those on the left indicate forward power and are all green except for the top one, which is red. The LEDs on the right read reflected power. The first two are green, the second two are yellow and all the rest are red. The **POWER** LED is mounted in the center.

2235 Gum Tree Ln
Fallbrook, CA 92028
Photos by the author

[1] A PC board for this project is available from FAR Circuits, 18N640 Field Ct, Dundee, IL 60118-9269; tel 847-836-9148 (voice and fax). Price: $3.75 plus $1.50 shipping for up to four boards. Visa and MasterCard accepted. This PC board includes mounting points for the LEDs to simplify assembly.

By Wayne McFee, NB6M

The NB6M QRP Paddles

When you're operating QRP in the field, the last thing you need is a cumbersome key. These Lilliputian paddles will go anywhere!

The best part about these tiny CW paddles is that you can make them yourself in just a couple of hours, start to finish. They are made from readily available materials, and cost almost nothing to build.

All you need are some scraps of double-sided PC board material, a piece of single-sided PC board material for the base (measuring $1^{7}/_{16} \times 2$ inches), two optional phono jacks, two short pieces of hookup wire, four 4-40 brass nuts, and two 4-40 × $^{1}/_{2}$-inch long steel screws. One more 4-40 steel screw, one inch long, is used to position the adjustment screw supports for soldering, and is then put back in the junk box.

The base could be made out of double-sided PC board, if you want. The phono jacks are optional, because you could very well just solder the wires of the connecting three wire cable directly to the paddle set, which is what I did when I made my first one.

Let's Get Started

The tools you will need for this project are: a hack saw (just the blade will do), a file for rounding and smoothing edges, a small hand (or electric) drill and appropriate bits, and a low-wattage soldering iron and solder. For enlarging the tension adjustment hole in the paddle, you can use a tapered reamer (available from RadioShack) or a small rat-tail file.

Using the Figure 1 and the accompanying photographs as guides, first cut out and shape the five parts made from PC board material. It will be far easier to drill the holes for phono jacks in the rear frame, the initial hole for tension adjustment in the paddle, and the holes for adjustment screws *before* cutting the relatively small pieces from the board. First, outline their shapes in the material, drill the appropriate holes, and then cut the pieces from the board. Now use the file to round corners and smooth the edges of the pieces.

I soldered one 4-40 brass nut to one side of each of the adjustment screw supports, in order to provide the threads for the screws to fit into. You could simply drill and tap (with a 4-40 tap available at any hardware store) the PC board material and use just the lock nuts. However, in the interest of strength and durability, I recommend soldering a nut to each support.

Note that the copper foil is cut in two places on each side of the

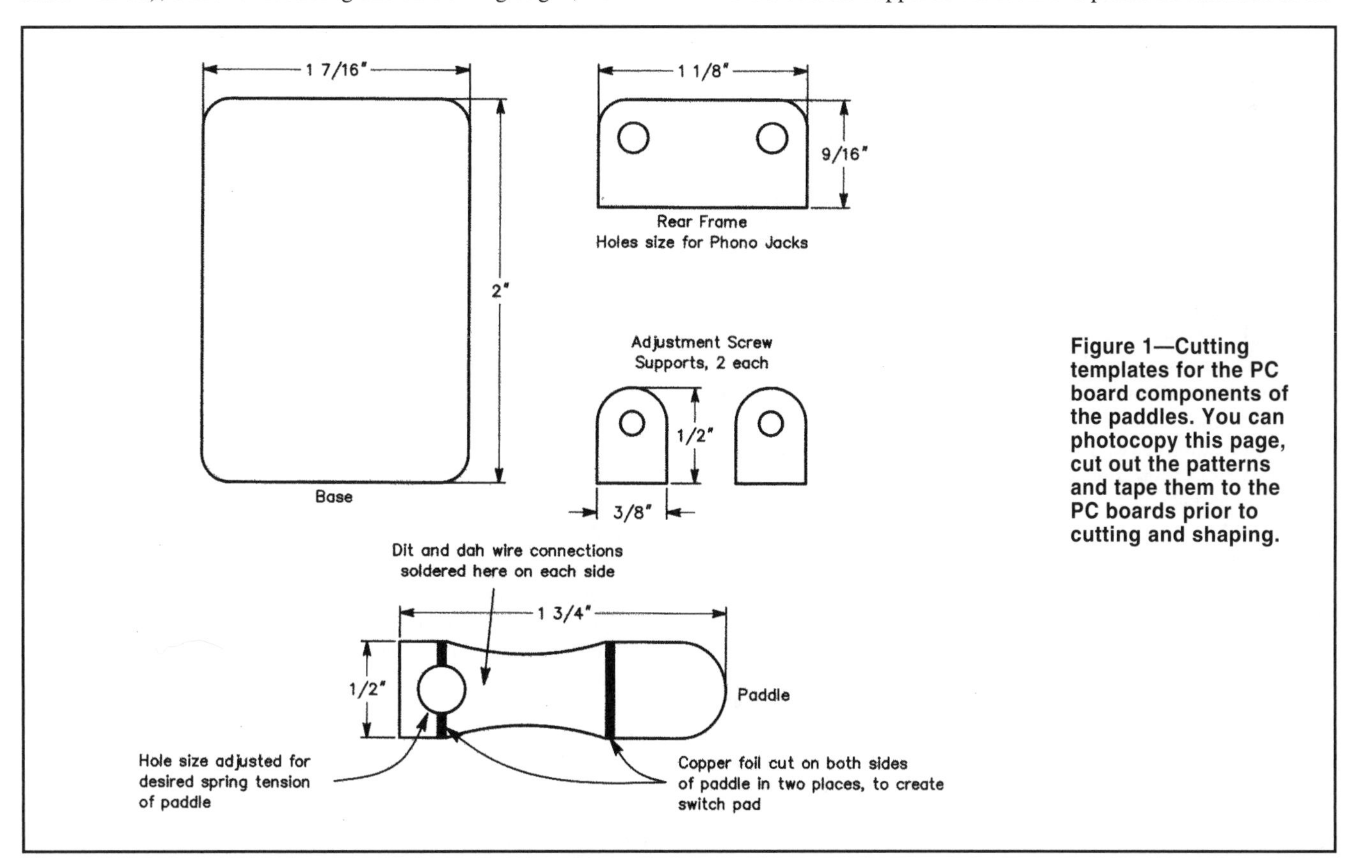

Figure 1—Cutting templates for the PC board components of the paddles. You can photocopy this page, cut out the patterns and tape them to the PC boards prior to cutting and shaping.

paddle. This can be done with either the edge of a small file or with a hacksaw. Cut just enough to be sure you have separated the copper foil nicely. What you are doing is creating the switch contact pads for the *dit* and *dah* sides of the paddle. It is necessary to cut the foil in two places on each side of the paddle, as shown in the drawings, so that static electricity and other stray electrical currents from your skin won't cause erratic keying.

The tension adjustment hole in the paddle is enlarged (thereby removing material from the paddle itself) to provide whatever spring tension you desire. This is done before the phono jacks and connecting wires to the switch pads are installed, and *after* the paddle set itself is soldered together and the adjustment screws are installed and adjusted for whatever switch gap feels good to you. So, initially, cut about a 1/8-inch hole.

Assemble the Paddles

First, set the rear frame in place, centered and about a quarter inch in from one end of the base, and tack solder one lower edge of the rear frame to the base. Check visually for proper placement and make sure that the rear frame is perpendicular to the base. Heat the solder tack, move the rear frame as necessary with a finger, and let the solder tack cool. Then tack the other side before running a bead of solder all along each lower side of the rear frame.

When you are running a bead of solder between two surfaces which are at 90° angles, the trick is to prop the unit up so that the two surfaces form a **V**, with the apex at the bottom and the two sides about 45° from the vertical. This way, the melted solder will run along both sides of the joint, and form a strong, nice-looking connection.

Next, solder one 4-40 nut to one side of each adjustment screw support. The trick to doing this is to screw the nut onto a screw, put the end of the screw through the hole in the support, rest the support in a horizontal position, heat the nut and its adjoining copper foil, and wick the solder underneath the nut. Once the nut has been soldered in place, and the unit has cooled, simply unscrew the screw from the nut, and do the same operation on the other support.

Position the two adjustment screw supports on the paddle base and solder them into place. The trick to positioning them is to screw the one-inch long 4-40 brass screw through both supports, leaving about 3/8 inch between the two soldered-on nuts, which should be on the inside of each support, facing each other. Use the paddle as a guide to how far away from the rear frame to position the supports. The front edge of the supports should be about even with the foil cuts separating the finger-contact portions of the paddle from the switch pads. The far end of the paddle, with the tension hole in it, will butt up against the rear frame, and *after* the adjustment screw supports are soldered in place and the one-inch screw removed, it will be installed permanently.

The adjustment screw supports should be soldered along both lower sides of each support. Again, tack solder one side, then the other, and then run a bead of solder along the lower edge of each. This will ensure that they don't move during the soldering process.

Next, remove the one inch screw from the adjustment screw supports, install the two lock nuts, one on each screw, running the lock nuts right up to the screw heads, and screw the adjustment screws into their respective supports. Leave enough space between the two to fit the paddle between them.

Place the paddle in position between the two adjustment screws and butted up against the rear frame. The lower edge of the paddle should be at least 1/16-inch above the surface of the base.

Finger tighten the two adjustment screws against the paddle, which will hold the paddle pretty well in position while you then tack solder each side of the paddle against the rear frame. Run a bead of solder along the edge of each side of the paddle where it butts against the rear frame.

Now loosen the two adjustment screws slightly and adjust them for whatever switch gap feels good to you. Tighten the lock nuts to maintain that gap.

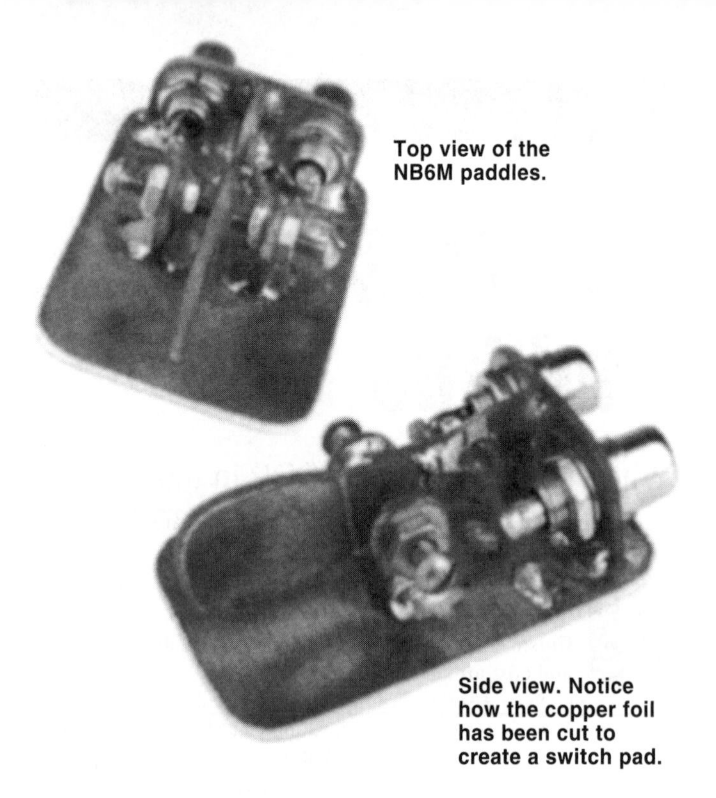

Top view of the NB6M paddles.

Side view. Notice how the copper foil has been cut to create a switch pad.

Before installing the two phono jacks and the two short pieces of hookup wire that connect from the center of each to their respective switch pads, open up the tension adjustment hole as desired. Remove material a little at a time until you have whatever paddle spring tension feels good to you. Remember that you can always remove more material. It is hard to put it back.

When you have the paddle spring tension set to your liking, install the two phono jacks and solder the two short pieces of hookup wire to the center connections of each and to their respective switch pads. Remember to leave a small amount of slack in each wire so that the paddle can move easily.

Hooking It Up

You will need a three wire cable (or two shielded audio cables with a stereo miniature phone plug installed on one end and two phono plugs on the other) to run from the paddle set to your keyer or rig. For my own installation I used a pair of shielded audio cables. I cut the phono plugs from one end of each and installed a stereo miniature phone plug, soldering the shield of each audio cable to the ground connection of the plug—one center wire to the ring connection and one to the tip connection of the plug.

Before you install the stereo plug, cut several short (3/16 inch) lengths of the outer insulation from some scrap RG-58 or 59 (or RG-8X or whatever you have) and slip these over the two audio cables, spacing them about every three inches. These make great cable ties and they look good with black audio cables.

That's all there is to it! You now have a portable, small, lightweight and, best of all, *cheap* set of paddles.

Operation

I operate my paddles by holding the set in the palm of one hand and working them with the thumb and forefinger of the other. Alternatively, you could hold the paddle set down on a tabletop with your middle finger placed over the junction of paddle and rear frame, and the tip of your forefinger placed next to the adjustment screw support on the near side of the paddle base. This leaves plenty of room to operate the paddle with the thumb and forefinger of the other hand.

Enjoy!

909 Marina Village, #163
Alameda, CA 94501; nb6m@mcr.win-net.org

By Dale Botkin, NØXAS

From *QST*, December 2003

The PicoKeyer—An Ultra Low Power CW Memory Keyer

NØXAS redefines "low power" for us with a keyer that draws 4 nA (that's 0.004 microamperes!) at idle.

For most of the time I have been active on the HF bands, I have used various electronic keyers with a Vibroplex paddle. While I have spent time operating with a straight key, I have never really enjoyed it or gotten good at it. I have used built-in keyers in several brands of gear, including a few from well-known low-power kit manufacturers. All have worked well...some better than others.

I recently had the opportunity to upgrade my home HF station to a Kenwood TS-930S/AT. While I enjoy the rig, it lacks an internal CW keyer—so it was back to the straight key. I felt sorry, frankly, for some of those operators on 40 meters who were subjected to my fist during those first few weeks. I wasn't getting better, and I decided it was time to go back to the paddle and the keyer. Being a hacker at heart, however, I wasn't going to go out and buy what I could build in an evening! I decided I would build a very simple, barebones keyer to do nothing more than generate dots and dashes with proper timing and spacing. Like many other projects, this was a good idea until it got out of hand.

I knew I had all the parts I would need. One of my hobbies involves building projects with embedded controllers, mostly using Microchip's PIC processors.[1] For me, the perfect project is one that involves an absolute minimum of parts. First of all, I like to be able to build without ordering anything, and my "junk box" is pretty well stocked, but it doesn't have everything. Ideally, I like to use a small microcontroller with as close to zero external components as can be achieved. I am also a big fan of designing for as little current consumption as possible, mainly because I don't like changing (or buying) batteries and because power supplies are always a pain to build. I guess I've gotten lazy in my old age, but it does make for some interesting design challenges!

The Design

I wanted to use nothing more than one chip for the keyer, with a single pushbutton for any features that got added along the way. I also wanted to play with one of the latest 8-pin designs from Microchip, the 12F629. These little processors cram a lot of features into a very small package! 1024 words of program memory, 64 bytes of RAM, 128 bytes of nonvolatile EEPROM, interrupts, counter/timers, an internal oscillator, an analog comparator, and more. The 12F675 shares the 12F629's features and adds a 10 bit, four channel analog to digital converter. These chips are a major leap ahead compared to the older 8 pin 12C5XX-type devices and their power requirements show it as well. The new pair will operate down to 2.0 V, and their SLEEP mode lets them power down and run on not much more than a cool breeze.

The hardware design was the simple part, requiring nothing more than connections from the key to the PIC and a single transistor to key the rig. I added the SPST pushbutton switch and decided to use one pin for an LED during debugging, leaving one pin unused. I assembled the project on a solderless breadboard and wired up the paddle. Now came the tough part—the software—the heart of the project.

One of the first steps was to determine the timing to be used when sending Morse code. An hour or so of reading various pages found with a simple Internet search and I had the "magic numbers" needed to determine the dot and dash timing for any given code speed. A couple of hours of programming later and the chip was generating properly spaced code from 5 WPM to as fast as I could send, with the upper limit set at 60 WPM. The next step was to add a setup mode for changing the code speed. While I was at it, I added functions for setting the weight and selecting Mode A or Mode B timing.

When I was reasonably satisfied with the performance, it was time to move from the solderless breadboard to a "real" setup and try it out on the air. I used half of a small RadioShack perfboard and wired it up as a prototype assembly. On-air results were good, but there were a few key things still missing.

The project took on a life of its own

[1]Notes appear on page 5-29.

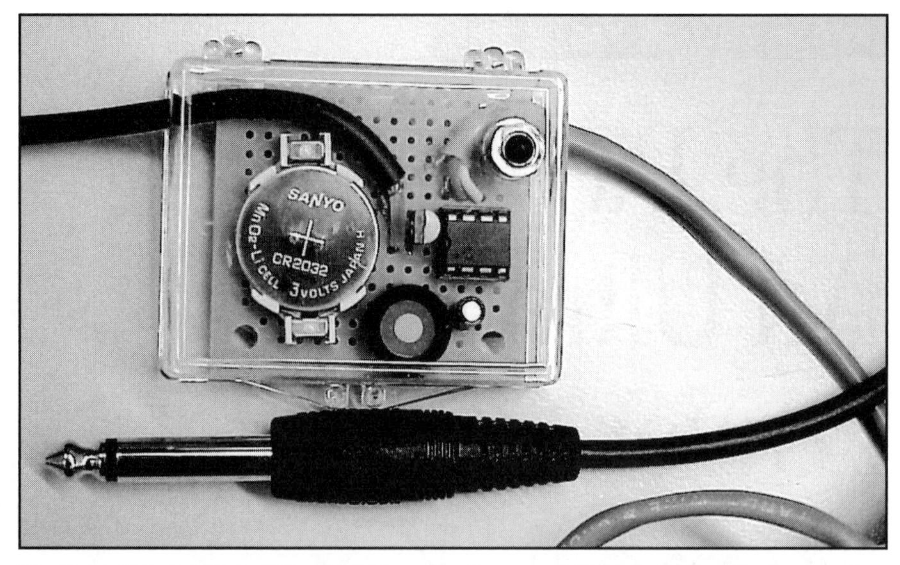

Figure 1—The completed PicoKeyer in a plastic case. A ¼ inch phone plug is shown for comparison.

over the next couple of weeks. The LED was replaced with program-generated audio sidetone that could be turned on and off, and I managed to find a tiny speaker with low enough current requirements to drive directly from one of the PIC pins. I added message memory, beacon mode with variable repeat delay, then variable pitch for the sidetone, then paddle reversing for left-handed ops or miswired paddles. A few bugs were uncovered and fixed along the way, as I logged more operating hours with the keyer.

Redefining "Low Power"

Along the way, I got curious about the current consumption of the chip. I had the PIC in its low power sleep mode any time it was not actively doing something like keying the rig or generating a sidetone. Although it has now been updated, at the time, the data sheet for the 12F629 was pretty sketchy about sleep current. I knew it would be low, but I needed to determine whether the CR2032 PC motherboard clock lithium cell I was using would last long enough to be practical as I salvaged the battery from a defective old PC motherboard I was scrapping. While it was still reading a healthy 3.1 V, I was planning to switch to a couple of AAA alkaline cells if needed. After connecting my Fluke 77A meter in series with the battery I was surprised to see it reading 0.00 mA while the chip was idle. Thinking the meter was simply not accurate at low currents, I tried my Micronta bench meter on the 3 µA range. I was quite surprised to see it read 0.004 µA or 4 nA while sleeping! I abandoned the thought of alkaline cells—that little lithium battery would last quite a while! I've been using that same salvaged cell for a few months now, and it still reads a tad over 3 V when the keyer is idle.

Construction

I built my keyer on a small piece of RadioShack perfboard. The few connections required are pretty simple and straightforward. Power to the PicoKeyer chip, U1, is supplied on pin 1, while pin 8 is ground. The photo on the previous page shows the completed keyer as it appears before packaging. Figure 1 is a view of the finished product in a mini plastic case. Its size relative to a standard ¼ inch phone plug reveals how small this keyer really is—and that's complete with power supply and speaker! The complete schematic and parts list is shown in Figure 2.

Pin 2 is the sidetone output. If you are building a standalone keyer you will probably want a small piezo or low current magnetic speaker connected to the chip through a capacitor. I used a small magnetic speaker with a drive requirement of only 15 mA. Other piezo or magnetic speakers can be used—but be careful not to exceed the 20 mA current capacity of the PIC output pin. If you are building the keyer into a rig, you can simply inject the sidetone signal into the radio's audio stages. A volume control might be a good idea. If you need more audio than the PIC can provide directly, a small audio amp like an LM386 could be used for more punch. This would of course require a more robust power source. The signal produced is a square wave; some filtering can be used to approximate a sine wave, if desired.

Q1 is a 2N7000 MOSFET, and is used to key the radio. You can use a 2N2222A or a similar NPN transistor with a current limiting resistor on the base or emitter. This works, but it requires more current. Since the MOSFET is a voltage, rather than current, operated device, it is ideal for super low power designs such

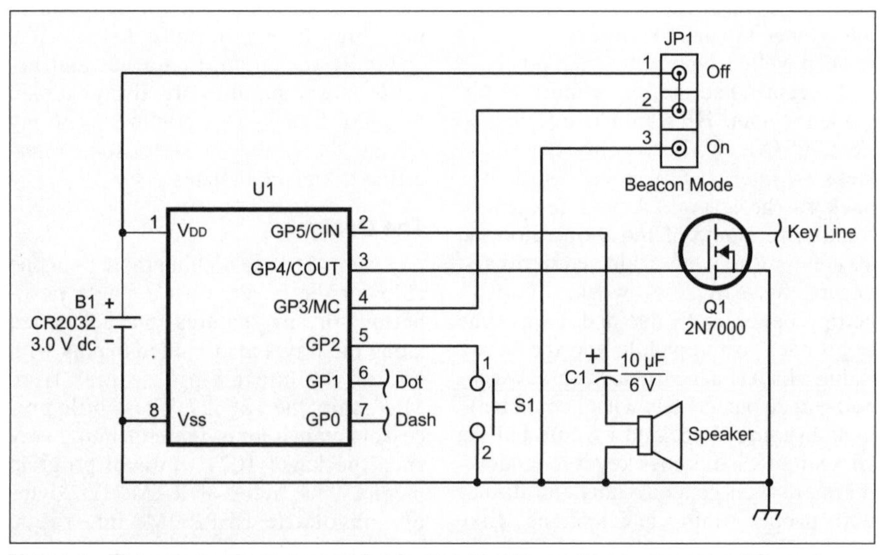

Figure 2—The schematic and parts list for the PicoKeyer. The battery is a PC clock motherboard cell, but any 3 V dc cell can be used.

B1—CR2032 lithium battery with holder or any 3-5 V dc cell (AA or AAA cells okay).
C1—10 µF, 6 V electrolytic capacitor.
JP1—Beacon mode switch or jumper (see text).
Q1—2N7000 MOSFET, Digi-Key 2N7000FS-ND (Digi-Key Corp, 701 Brooks Ave South, Thief River Falls, MN 56701; tel 800-344-4539; www.digikey.com).
S1—SPST pushbutton, momentary.
SPK—Soberton GT-111PS, Digi-Key 433-1023-ND (see text).
U1—Microchip PIC12F629 with PicoKeyer code. Available blank from Digi-Key (part no. PIC12F629-I/P-ND) or preprogrammed from the author.[2]

as this one. Q1's gate is connected to pin 3 of the keyer chip, with its source connected to ground. The drain lead is used to provide positive keying for the radio.

Pin 4 is used to select beacon mode. If you do not intend to use the beacon feature, you may simply connect pin 4 to the positive supply on pin 1. If you wish to be able to use beacon mode, install an SPDT switch or a jumper to connect pin 4 to either the positive supply or to ground.

The pushbutton switch is connected to pin 5 of the chip. Since the 12F629 has internal pull-up resistors, no external resistor is needed to keep the input from floating.

Pins 6 and 7 are the paddle inputs. When grounded (or driven low by other circuits), the keyer will produce a continuous stream of dits or dahs with proper spacing. If both inputs are grounded, alternating dits and dahs are sent.

Once all the connections are made, apply power to the keyer. You should hear the keyer send 73 in Morse code at 13 WPM via the sidetone only. This indicates that the chip is powered up and healthy. If you do not hear 73, recheck your wiring and reapply power. If everything is okay, you're all set to start using the keyer! The default settings are 13 WPM, normal weighting, message memory empty, Mode B timing and a beacon delay of five seconds.

If you have experience programming PIC processors, the HEX code is available at the ARRL Web site: **www.arrl.org/files/qst-binaries/picokeyer.zip**. A programmed version of the processor with the latest revisions is available from the author.[2]

Operation

Operating the PicoKeyer is simple and straightforward. To use it as a normal iambic keyer, simply apply power and connect the paddles. Dual or single lever paddles can be used. To send the contents of the message memory, press and quickly release the pushbutton once. While the message is being sent, any paddle or pushbutton input will immediately end the transmission.

To enter setup mode, press and hold the pushbutton. The keyer will step through the setup menu choices, with about a one second delay in between. When you hear the character for the item you want to check or change, release the button immediately. You may then verify or change that item. A short press of the button will exit setup mode, or you may press and hold the button to keep cycling through the choices. When you exit the menu the keyer will end with the Morse prosign SK.

Menu Selections

The menu selections are as follows:

U (tUne/straight key mode): In this mode, either paddle input will act as a straight key. This is useful for sending a steady carrier for tuning.

S (Speed): Use the paddles to raise or lower the speed. After each paddle hit the keyer will send the new setting.

T (Tone): Turns the sidetone output ON (Y), OFF (N), or sets RIG (R) mode. In either the ON or OFF settings, the rig is not keyed when in setup mode. RIG mode is useful if you wish the transmitter output to be active even in setup mode.

M (Message): Hitting either paddle will play the current message, followed by the Morse prosign AR. To record a new message, hit either paddle again. The keyer sends K and waits for your input. Enter the message, making sure to exaggerate word spacing. When you're finished sending the message, press and release the pushbutton, and the keyer will send R to confirm. You may then replay the message by hitting either paddle. It may take a couple of tries to get the hang of the timing, but it's not too picky unless you send the characters too slowly.

W (Weight): Adjusts the weight. Adjustment range is from 0 (50% "light") to 9 (50% "heavy"), with 5 being the normal weight setting.

C (Curtis mode): Selects Mode A or Mode B timing. This determines the behavior of the keyer when the paddles are released after a "squeeze." In Mode A, the keyer will complete the element (dit or dah) currently being sent. In Mode B, the keyer will "remember" seeing the other paddle, and will send one last element. For example, say you are sending the letter C. In Mode A, you would not release the dit paddle until the start of the last dit. In Mode B you would need to release both paddles as soon as the second dah has started. The keyer will "remember" to send the last dit.

P (Paddle): Selects the dit paddle. Simply hit the paddle you wish to use for dits. Note that in the menu mode, the "normal" dit and dah paddle inputs are used regardless of whether the paddles have been reversed or not.

B (Beacon delay): Sets the delay between message transmissions while in beacon mode, from 0 to 99 seconds.

A (Audio tone): Sets the audio sidetone frequency. After each paddle hit, a slightly long dash is sent.

When you are finished altering the keyer settings, press and release the pushbutton. The keyer will save all settings to its internal nonvolatile EEPROM memory and send SK. At that point you are back in iambic keyer mode, ready to go!

I have been using this keyer for a few months now and several other hams have built them as well. One of the things I like is the ability to change features very quickly; it took about an hour of revising the program code to make a version pin compatible with the popular RockMite transceiver.[3] Another couple of hours and I had a version that used a potentiometer to set the speed instead of the menu; it was even less time to produce one to replace another 8-pin keyer chip that used a different pinout. While I still enjoy soldering, I also like this facet of our wonderfully diverse hobby. Besides, I can write code while in an airline seat—an environment in which a soldering iron is somewhat impractical!

Notes

[1]**www.microchip.com**.

[2]A pre-programmed version of the 12F629 PIC processor with the author's latest HEX code is available from the author for $8.95 (**www.hamgadgets.com**).

[3]D. Benson, K1SWL, "The RockMite—A Simple Transceiver for 40 or 20 Meters," *QST*, Apr 2003, pp 35-38.

Photos by the author.

Dale Botkin, NØXAS, was first licensed in 1981 as KA5MSS, although that call and HL9CA both lapsed without ever having been used. Dale was relicensed as a Technician in 1994, with subsequent upgrades to General and Amateur Extra. He enjoys casual operation using low power CW and PSK31 on the HF bands. He also operates VHF and UHF mobile and makes an occasional FM satellite contact. As is obvious, Dale is an avid radio experimenter, especially with embedded microcontrollers. You can contact him at 16624 Elm St, Omaha, NE 68130 or at **n0xas@arrl.net**.

QST

Phil Salas, AD5X

From *QST*, March 2005

A Compact Battery Pack for the SG-2020

While designed for the SG-2020, this portable pack can be used with most low power transceivers—think reasonable cost!

I really enjoy operating the HF bands with a portable transceiver. My favorite portable rig is probably the SGC SG-2020. I like this radio because it gives me the ability to switch to 20 W of RF output when necessary. Twenty watts is a nice compromise between QRP (low power) and QRO (high power). (Well, relative QRO—it's an S-unit up from 5 W, and about an S-unit down from 100 W.)

For portable operation, being able to run off a battery pack can be a benefit in many situations. Reasonable operating time, however, at the 20 W level, requires a pretty substantial battery pack. And, as many SG-2020 owners know, the SG-2020 is often unhappy by the time battery voltage drops to below 12 V. The '2020 is specified to operate from 10-18 V dc, but this should, in reality, be more like 12-18 V dc, at least for older SG-2020 examples.

The Battery Pack

Recently, I've been looking at nickel metal hydride (NiMH) radio-control (R/C) type battery packs since they tend to be pretty inexpensive, especially on Web auction sites. The popular voltages are 7.2 and 9.6 V dc. The 7.2 V dc packs are very interesting because two of these in series will give you 14.4 V dc. And reasonably priced packs have up to 3000 mAh of capacity.

I purchased a pair of 7.2 V dc, 3000 mAh packs, along with a fast/smart charger for $56.95.[1] The fast charger will charge 6, 7.2 and 9.6 V dc NiMH batteries, output 2 A of charge current, and then automatically switch to 50 mA of trickle charge when it detects a full battery charge. The charger is relatively small, as can be seen in Figure 1. For these 3000 mAh packs, the charge time is less than 2 hours for a fully depleted pack. Incidentally, I added additional electrical tape around the batteries, as there was some exposed terminal metal on the batteries that I received.

Again, the nominal series voltage is 14.4 V dc, with a maximum (fully charged) voltage of just over 16 V dc, and a discharged voltage level of 12 V dc. This is perfect for the SG-2020! While you're at it, buy a pair of male R/C connectors and a female R/C connector. The male R/C connectors are needed if you want to interface everything inside your battery box. The female R/C connector is used to make an R/C to 5.5 mm charging adapter cable. You can buy these R/C connectors at RadioShack (the two 7.4 V dc R/C repair kits called out in the parts list provide everything necessary).

The next problem is finding a box to house these batteries. My solution was a Serpac model 171 plastic enclosure. This 4.88×6.88×1.51 inch enclosure fits the R/C battery packs perfectly. These cases are available from Mouser Electronics for $8 each.[2] Actually, the case shown in the figures is their model 271-B, which has a 9 V battery door (I found it locally at Fry's Electronics). The Model 171, however, is less expensive and provides a less cramped battery fit.

[1]Notes appear on page 52.

Building the Battery Pack

Figure 2 shows the schematic of the battery pack and Table 1 contains a complete materials list. The two battery packs have to

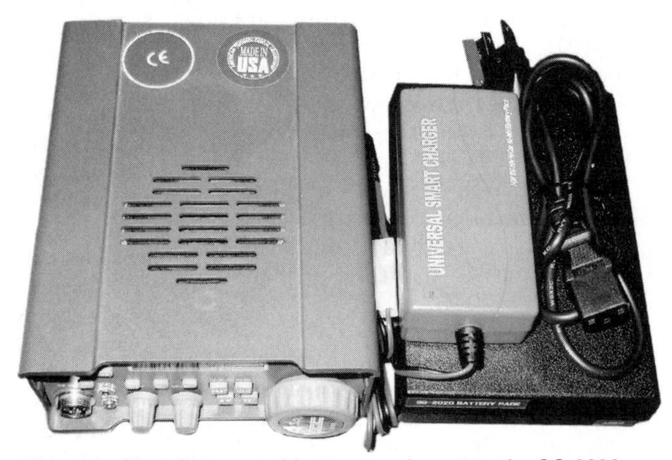

Figure 1—The charger and battery pack next to the SG-2020 transceiver.

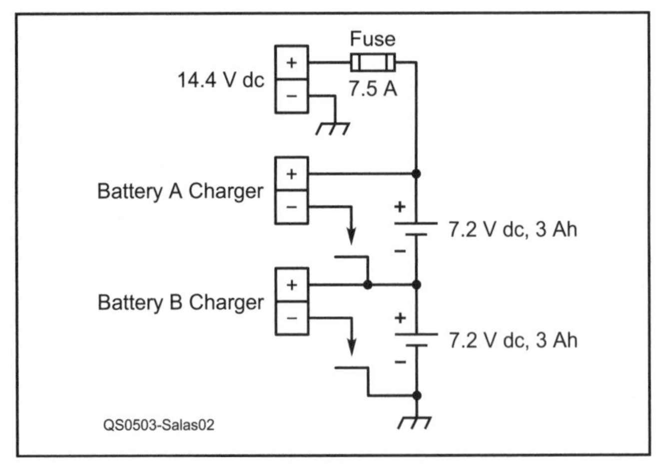

Figure 2—The schematic of the NiMH battery pack.

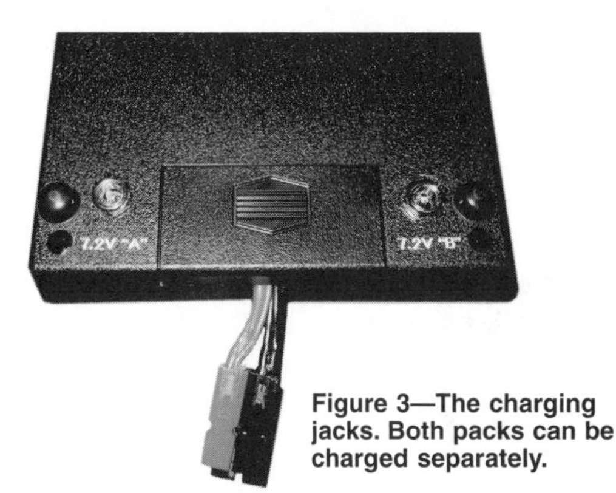

Figure 3—The charging jacks. Both packs can be charged separately.

Table 1
Materials List, Battery Pack

Batteryspace—**www.batteryspace.com**
Mouser—**www.mouser.com**
RadioShack—**www.radioshack.com**

Quantity	*Item*	*Source*	*Price*
1	2 NiMH batteries/charger	Batteryspace	$56.95
1	Serpac 171-B box	Mouser 635-171-B	$8
2	5.5×2.1 mm jacks	RadioShack 274-1582	$1.99
1	5.5×2.1 mm plug	RadioShack 274-1569	2/$1.99
2	7.2 V R/C repair kit	RadioShack 23-444	$1.99
1	Mini-blade fuse holder	RadioShack 270-1237	$1.99
1	7.5 A mini-blade fuse	RadioShack 270-1092	3/$1.59

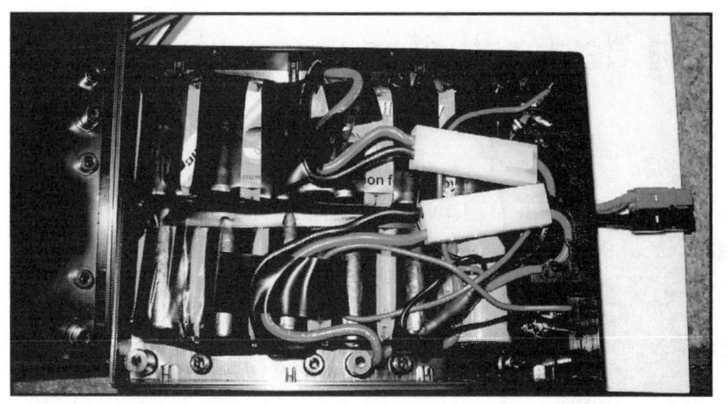

Figure 4—An internal view of the pack. The NiMH batteries fit snugly in the enclosure.

Table 2
SG-2020 Battery Current (A)

Transmit Power (W)	*160-20 M*	*17 M*	*15 M*	*12 M*	*10 M*
20	3.8 A	4.2 A	4.8 A	4.5 A*	4.6 A*
10	3.0 A	3.4 A	3.7 A	3.8 A	3.5 A
5	2.5 A	2.7 A	2.9 A	3.0 A	2.7 A

*On 12/10 meters, the maximum output was 15 W.
Receive current: 540 mA with ADSP2 on or off.

Figure 5—The battery pack and the transceiver. The pack furnishes sufficient voltage for over 1½ hours of operation at the 20 W level. Note the output voltage of the battery pack; the SG-2020 can meter the dc input voltage on its display.

be wired in series, but you also have to be able to charge them individually. For charging purposes, I used 2.1×5.5 mm jacks from RadioShack connected across each battery. Use the RS274-1582 jacks with internal switches, as these isolate the collar of the jack from the negative side of the battery until a charging plug is plugged in. This eliminates any possibility of accidentally shorting things if you put something metallic across these jacks. The jacks are placed as shown in Figure 3. Their positions are not critical, and you can determine where to put them once you position the batteries inside the box. The batteries fit snugly in the box as shown in Figure 4. You can also see in this drawing that I interfaced the wires such that the batteries can be easily removed. A small piece of foam over the batteries helps to keep them snug and immobile when the cover is in place.

I also put rubber feet on the bottom of the case, and labeled the case and charging connectors using "white on clear" Casio labeling tape.

Finally, don't forget to fuse the unit. These batteries can source up to 30 A for a short period of time! You can see the final unit, on top of the transceiver, in Figure 5. Incidentally, the total weight of this battery pack and charger is less than 2 pounds.

Operation

So how long can you go between charges? Over a period of two days, I made 5 contacts for a total QSO-only time of a little over 95 minutes, all running a full 20 W of CW on 40 meters. There was also some additional "in-between QSO time" when I was looking for contacts, but I didn't count this in the total operating time. My "end-of-charge" time was defined to be when the battery voltage fell to 12 V dc. I monitored this voltage on the SG-2020 (pressing CMD-SPEED brings up the voltage display). I did find that when the metered voltage drops to around 13 V, it's really time to consider recharging the batteries, as you are getting very close to a depleted pack.

I also measured the current drain of my SG-2020 so as to give you an idea of how you might do at different power levels. Obviously, you'll get longer operating time at lower power levels. For my ADSP2-equipped SG-2020, the measured key-down current drain data is as shown in Table 2.

Conclusion

The SGC SG-2020 is a great little portable rig that gives you the ability to operate up to 20 W when necessary. Since "portable" often means "battery powered," this article describes an effective, and relatively inexpensive, battery pack for this radio. And, since this battery pack can be recharged quickly, you may even start using it as the primary power supply for your SG-2020, just as I have! Finally, this battery pack may be of interest to you for other QRP portable transceivers. But you need to add at least two series power diodes in order to drop the voltage below 15 V dc when fully charged (15 V dc is the typical maximum voltage for many radios—the SG-2020 is an exception at 18 V dc maximum).

Notes

[1]**www.batteryspace.com**.
[2]**www.mouser.com**.

Photos by the author.

An ARRL Life Member, Phil Salas, AD5X, has been a ham for over 40 years. He received a BSEE degree from Virginia Tech and an MSEE from Southern Methodist University. Currently, he is the VP of Engineering at Celion Networks. Phil shares his station with his wife Debbie, N5UPT, and daughter Stephanie, AC5NF. He can be reached at **ad5x@arrl.net**.

Phil Salas, AD5X

From *QST*, June 2005

Input Voltage Conditioner—and More—for the FT-817

Make a versatile rig even more so with a device that incorporates several useful mods.

The Yaesu FT-817 is a compact all-mode 1.8-450 MHz (except 222 MHz) radio that is ideal to take along on trips, vacations and backpacking. I've used my FT-817 for over three years and have really enjoyed it.

My only concern is the lack of fusing and reverse battery protection. Further, there have been reports that some switching power supplies put out momentary voltage spikes up to 18 V when first turned on. Since the finals aren't disconnected from the dc input even when the radio is turned off, a voltage spike exceeding the 15 V maximum spec of the FT-817 can be trouble!

My goal was to condition the input power supply voltage to ensure that it could not harm the radio in any way. Also, I wanted to do something about the 4 mm × 1.7 mm dc power connector, which I've never cared for.

Interesting Observations Lead to a Plan

A notable fact about the FT-817 is that it draws the same amount of current across most of the useful voltage range for a given output power. That is, at 5 W output power, the FT-817 typically draws 1.9 A, whether the supply voltage is 9.6 V or 13.8 V! So all you're doing at the higher voltages is dissipating more power inside the radio (8 W difference between 9.6 V and 13.8 V).

Once I realized this, I decided to limit external voltage to the FT-817 to 9.6 V. Additionally, I wanted to add dc input low- and high-frequency decoupling along with my goals of reverse polarity protection, 15 V overvoltage protection and an in-line fuse. And finally, I wanted to change the power supply interface to an Anderson Powerpole connector.

Electrical Design

Figure 1 is the schematic of the power conditioning circuit. The heart of this circuit is the STM LD1085V low drop-out adjustable voltage regulator. This regulator is packaged in a TO-220 case, and is rated at 3 A of output current. At the full 3 A current limit, this device only needs a 1.2 V drop input/output differential voltage, so you can easily run the radio from an external 12 V power supply. The voltage drop is closer to 1 V at the 2 A normally drawn by the FT-817 at 5 W output power. You can adjust the regulator output for other voltages if you wish, by adjusting the value of the two resistors, or you can just replace the two resistors with a 1 kΩ potentiometer with the wiper connected to the ADJ pin of the regulator.

For both overvoltage and reverse voltage protection, a 15 V/600 W voltage transient suppressor diode is used. Aside from taking care of any voltage transients above 15 V, this transient suppressor diode looks like a forward biased diode under reverse-voltage conditions, and will blow the fuse. The fuse is a miniature leaded fuse soldered in place. I also included a switch to bypass the regulator when I'm using external voltages less than 10.5 V (like with 9.6 V NiMH R/C batteries), or if higher voltage (+13.8 V) is needed for charg-

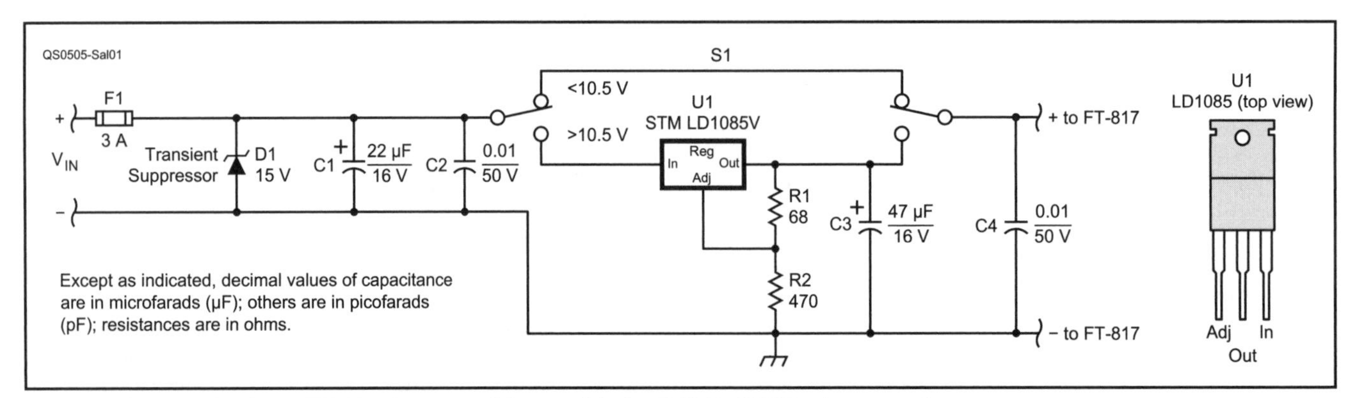

Figure 1—Schematic of the FT-817 voltage conditioner, with the STM LD1085V voltage regulator.

ing internal NiMH batteries from the internal FT-817 charger.

Mechanical Design

The interesting part of this project was the mechanical design. As I discussed earlier, I wanted to get away from the 4 mm × 1.7 mm dc plug. Also, I needed to package everything neatly. I found a nice little plastic box that mounts nicely up against the back of the FT-817.

Drill two holes, as shown in Figure 2, on the bottom of the plastic box. The larger 5/16 inch diameter hole fits over the ground post on the FT-817. The smaller 3/16 inch diameter hole should be a tight fit to the threaded shaft of the 4 mm × 1.7 mm dc plug. You may have to enlarge this hole slightly, but keep the fit snug. Discard the plastic thread-on collar, and thread this plug partially into the 3/16 inch diameter hole, keeping it as straight as possible. Now insert this plug/plastic box assembly so that the plug mates with the dc jack on the FT-817. Adjust the length of the plug (by screwing it in or out of the plastic box) such that the plug fully mates with the jack when the bottom of the box is flush with the back of the FT-817. Once you've determined the correct length, epoxy both sides of the 4 mm × 1.7 mm plug to the plastic box. See Figure 3.

The last two items to be mounted on the box are the slide switch and the Powerpole connector. Since the box is soft plastic, the rectangular cutouts are easily made with a hobby knife, assuming you have a little patience. Figure 4 shows the cover

Parts List

Qty	*Description*	*Source*
1	STM voltage regulator	Mouser 511-LD1085V
1	3 A fuse	Mouser 576-0251003.M
1	15 V/600 W Voltage Protection Device	Mouser 511-P6KE15A
1	47 µF/16 V electrolytic capacitor	Mouser 140-HTRL16V47-TB
1	22 µF/16 V electrolytic capacitor	Mouser 140-HTRL16V22-TB
2	0.01 µF/50 V capacitor	Mouser 80-C315C153K5R
1	4×1.7 mm dc plug	Mouser 171-3219
1	470 Ω/1/4 W resistor	Mouser 71-CCF07-J-470
1	68 Ω/1/4 W resistor	Mouser 71-CCF07-J-68
1	1.97"×1.38"×0.67" box	Mouser 546-1551GBK
1	DPDT 3 A slide switch	Mouser 629-GF1263011
1	TO-220 mica mounting kit	Mouser 534-4724
1	Heatsink grease	RadioShack 276-1372
1 pr	Anderson Powerpole connectors	West Mountain Radio

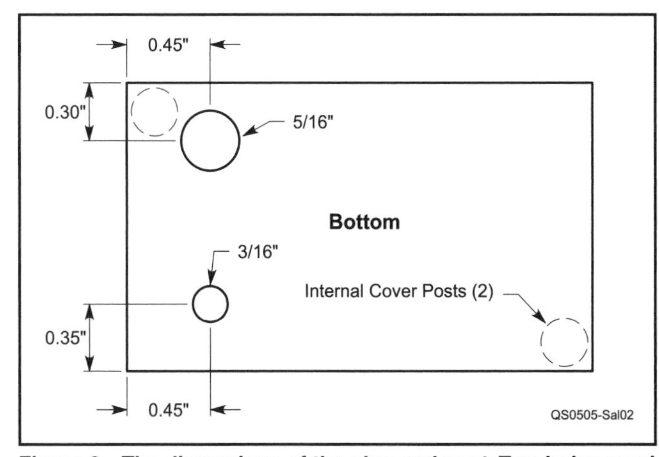

Figure 2—The dimensions of the plug and post. Two holes need to be drilled in the bottom of the box.

Figure 3—The photo shows the 4×1.7 mm dc plug affixed to the box with epoxy.

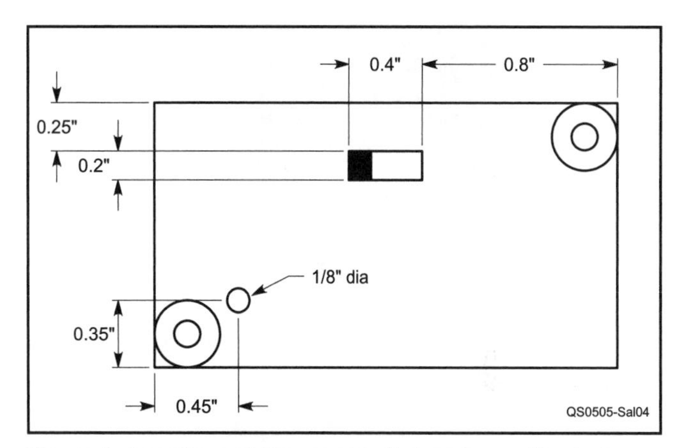

Figure 4—Locations of the cover switch and mounting hole.

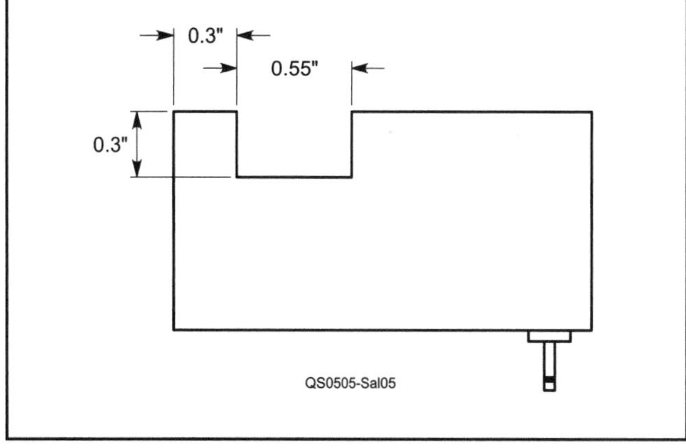

Figure 5—A side view showing the Powerpole cutout.

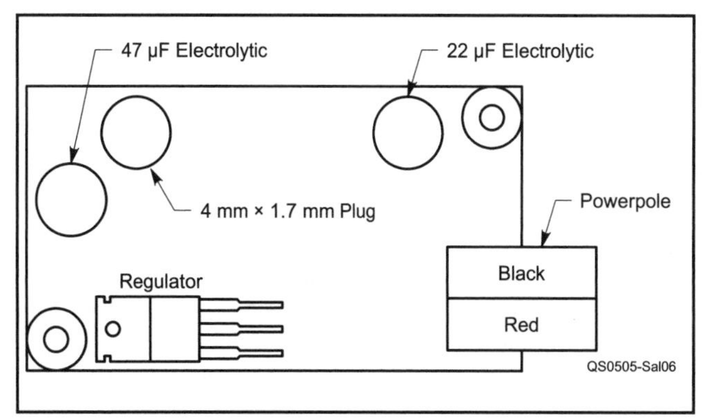

Figure 6—A top view, showing placement of the major internal components.

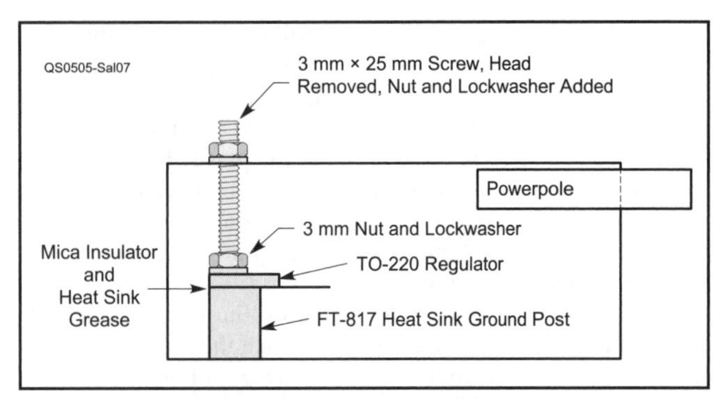

Figure 7—A side view, showing the regulator mounting.

Figure 9—This photo shows the completed voltage conditioner mounted to the rear of the author's FT-817 transceiver.

Figure 8—This photo shows the internal wiring and parts placement.

mounting locations for the switch and the screw hole necessary for holding the assembly to the FT-817. Figure 5 shows the dimensions for mounting the Powerpole connector on the box.

Internal Wiring and Assembly

To wire everything up, first temporarily mount the TO-220 voltage regulator with a screw, nut and washers to simulate the height of the heat sink post it will eventually be mounted to. Then wire everything point-to-point. One of the 0.01 µF capacitors was soldered directly across the 4 mm × 1.7 mm plug terminals. I found that this completely eliminated the need for the external ferrite required by earlier FT-817 radios when externally powered and using the attached whip antenna on 440 MHz. Refer to Figure 6 for the component locations of the major parts, primarily the 47 µF and 22 µF electrolytic capacitors.

When you are finished, remove all the temporary hardware from the regulator and then mount the regulator as shown in Figures 7 and 8. The regulator tab must be electrically insulated from the ground post, so use a TO-220 mica insulator kit and heatsink grease as shown in Figure 7. Verify that the tab is insulated from ground with an ohmmeter, just to be sure. Note that I used a 3 mm × 25 mm screw with the head cut off to mount the regulator. This creates a 3 mm × 25 mm stud for mounting the regulator, plus it extends through the cover and holds the entire assembly to the FT-817 chassis. And there is enough length left to add another nut and lockwasher should you wish to still use this for a ground connection.

Will It Work with Similar Radios?

The input voltage conditioner can be used for any low-power rig that doesn't exceed the current rating of the regulator circuit. The IC-703 is another good example—it puts out 10 W at 12 V, or 5 W at 9 V—so there's no need to put a full 13.8 V on it.—*AD5X*

Notice that with the regulator mounted directly to the FT-817 heatsink, power normally dissipated internal to the FT-817 at higher voltages is now dissipated directly into the heatsink before it even gets into the radio! Figure 9 shows the complete assembly mounted in place on the back of the FT-817. All labeling is done using Casio "white-on-clear" labeling tape (Casio XR-9Axs).

Operation

Operation is very simple. Just connect your Powerpole dc input to the Powerpole on the voltage conditioner. If the input voltage is below 10.5 V, flip the switch to the "< 10.5 V" position to bypass the 9.6 V regulator. Also use this switch position when you are using the internal FT-817 charger to charge an internal battery pack (you need +13.8 V input to have sufficient current to charge internal batteries). For input voltages greater than 10.5 V, put the switch in the "> 10.5 V" position. In both cases, the overvoltage, reverse voltage and fusing is on-line and protecting your FT-817.

Remember that the FT-817 automatically switches to 2.5 W output power when the input voltage is below about 11 V. However, you can easily set the output power to 5 W through the FT-817 menu (Function row 9). Note that even when an external power supply is used, when the regulator is in-line you will draw current from the internal battery until its voltage drops below the 9.6 V regulator output. So always make sure the internal battery is charged just before operating battery-portable.

Summary

I've described a bolt-on input voltage conditioner assembly for the FT-817. This assembly is compact, yet provides overvoltage and reverse voltage protection, fusing and improved power dissipation at voltages over 10.5 V. Additionally, the dc interface is now the standard Anderson Powerpole connector. Build one of these units for your FT-817 and sleep a little easier!

Photos by the author.

Phil Salas, AD5X, has been licensed over 40 years and primarily enjoys HF CW operation. He is an ARRL Life Member, and holds BSEE and MSEE degrees from Virginia Tech and Southern Methodist University, respectively. Phil retired recently as the Vice President of Engineering for Celion Networks in Richardson, Texas. Phil may be contacted at **ad5x@arrl.net**.

By Phil Salas, AD5X

From *QST*, November 2003

An FT-817 Compact Fast Charger

Don't wait 20 hours to recharge those FT-817 cells. Build your own versatile fast charger and have 16 more hours to operate!

I really enjoy using my Yaesu FT-817 low power tranceiver in portable applications. With an internal NiMH battery pack, it can give me several hours of operation even at the 5 W level under normal operating conditions. Unfortunately, that few hours of operation results in 20 hours of charging time when using the internal FT-817 charger. The One Plug Power (OPP) option by W4RT Electronics helps get around this.[1] The OPP provides an 1800 mAh NiMH battery pack and a new battery cover for the FT-817 that includes a 2.1 mm × 5.5 mm charging jack. You can now fast-charge the batteries from an external fast charger without having to remove the batteries from the radio, but…what do you do about the fast charger?

The Idea

I have a small, portable kit that includes my FT-817 and some accessories (MFJ-4103 power supply, multiband dipole, antenna tuner, paddle and some RG-174 coax). I also wanted a fast charger to carry along with my portable kit to give me the option of fast-charging the OPP. Unfortunately, most of the fast chargers available are physically large. Probably the most popular is the Maha C777/C777+.[2] But this fast charger is almost the size of the FT-817 and all of its accessories put together! It occurred to me that there's really no reason that the charger must be self-contained—it does not have to run directly from 117 V ac if you already have access to a 13.8 V dc power source.

[1]Notes appear on page 5-36.

So, I decided to build a very small fast charger that would run from any 13.8 V dc source, like the MFJ-4103[3] miniature switching power supply or your car cigarette lighter socket.

The Design

The compact charger is built around the Maxim MAX712/713 Fast-Charge Controller IC.[4] Either the MAX712 or MAX713 can be used with NiMH batteries. These are very versatile ICs that permit fast charging of many different types of battery packs (NiCd, NiMH) with many different voltages. In this case, however, I was just interested in the 8 cell, 9.6 V NiMH battery pack used in the FT-817, which results in a very simple circuit.

Figure 1 shows the final design. Again, it is designed specifically for an 8 cell, 9.6 V NiMH battery pack. The fast-charge current is about 600 mA, and it will charge a depleted pack in 3 to 4 hours. It is also designed for a time-out of 4.4 hours. This controller automatically senses when the battery is charged and switches to trickle. The green LED is just a "power" indicator. The red LED is on during fast charge, and turns off when the charger is in "trickle" mode.

All the parts, except for the MAX712/713, are available from All Electronics.[5] All resistors are 1/4 W. The MAX712 can be purchased directly from Maxim.[6] You can also download data from the Maxim site for other charging conditions or numbers of batteries.

I built the entire unit into a tiny plastic enclosure with a metal cover, with most of the parts being mounted on the All

Figure 1—The completed fast charger unit ready to be plugged in. Note the power input and charge indicator LEDs. The charge LED remains lit until the charger reverts to "trickle" mode.

Figure 2—The interior view of the charger showing the wired perf-board. Transistor Q1 must be heat sinked. It is bolted to the top plate.

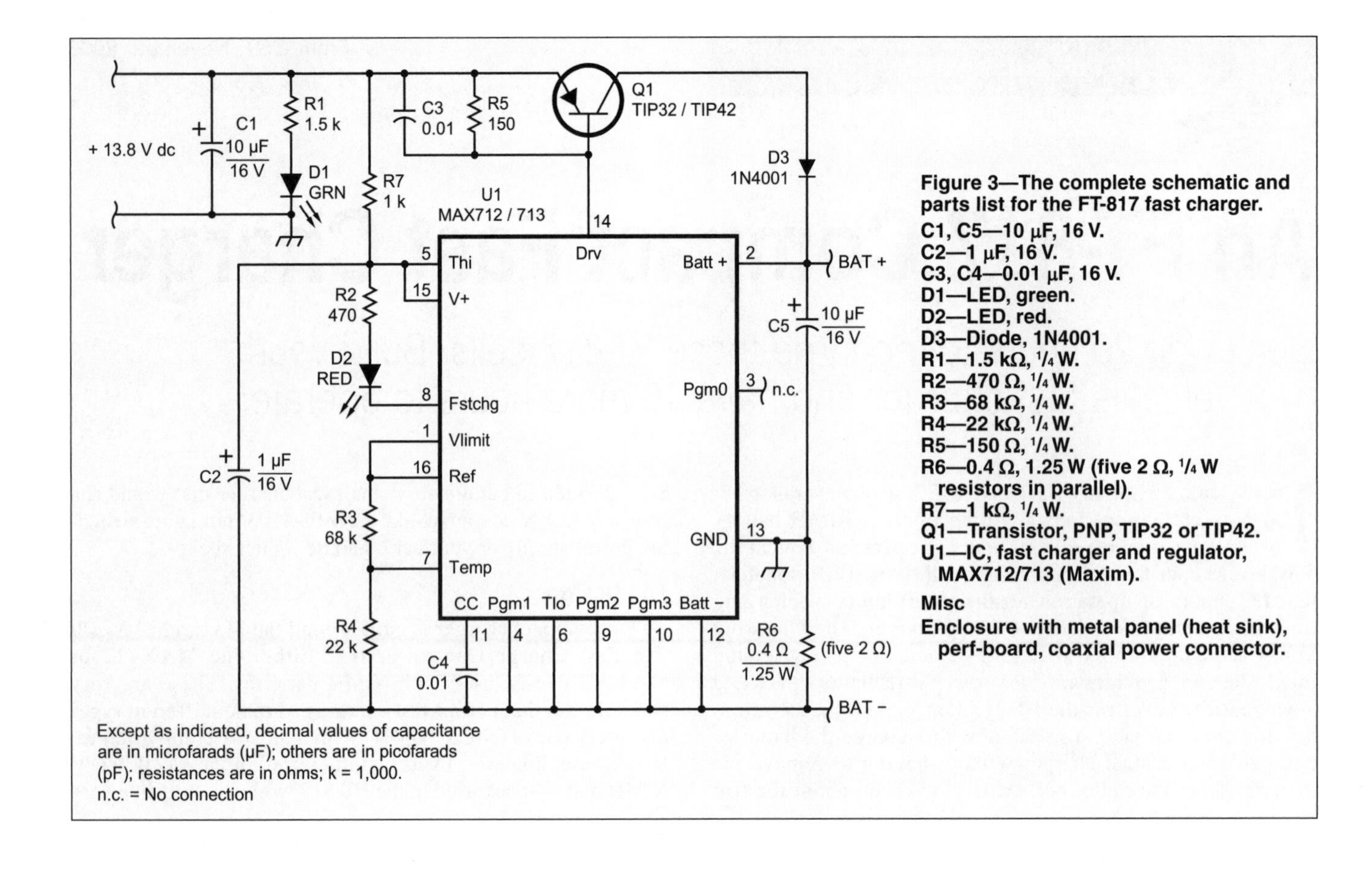

Figure 3—The complete schematic and parts list for the FT-817 fast charger.
C1, C5—10 μF, 16 V.
C2—1 μF, 16 V.
C3, C4—0.01 μF, 16 V.
D1—LED, green.
D2—LED, red.
D3—Diode, 1N4001.
R1—1.5 kΩ, 1/4 W.
R2—470 Ω, 1/4 W.
R3—68 kΩ, 1/4 W.
R4—22 kΩ, 1/4 W.
R5—150 Ω, 1/4 W.
R6—0.4 Ω, 1.25 W (five 2 Ω, 1/4 W resistors in parallel).
R7—1 kΩ, 1/4 W.
Q1—Transistor, PNP, TIP32 or TIP42.
U1—IC, fast charger and regulator, MAX712/713 (Maxim).
Misc
Enclosure with metal panel (heat sink), perf-board, coaxial power connector.

Electronics PC-1 perf-board that matches their TB-1 enclosure.[7] The TIP32 transistor must be mounted to the metal cover so as to be able to dissipate its heat during high-current charging. I mounted a 2.1 mm dc socket[8] on the plastic box for the 13.8 V input, and I used a pendant cable with a 2.1 mm dc plug[9] for the output that plugs into the W4RT OPP 2.1 mm jack. As you can see from Figure 2, the perf-board is mounted with the components facing down into the box. Figure 3 is the complete schematic and parts list. Obviously, you can use this charger with other vendor battery packs depending on how you may wish to adapt and access them.

Finally

With just a few hours of assembly time and less than $20 you can build a compact fast charger for your FT-817 internal battery pack. Keep that Maha 777 at home for all your other battery backs—use this compact fast charger to keep your FT-817 travel pack of accessories compact and lightweight.

Notes

[1]W4RT Electronics, 3077-K Leeman Ferry Rd, Huntsville, AL 35801; **info@w4rt.com**; **www.w4rt.com**.
[2]**www.mahaenergy.com**.
[3]**www.mfjenterprises.com**.
[4]**www.maxim-ic.com**.
[5]**www.allelectronics.com**.
[6]See Note 4.
[7]All Electronics TB-1 enclosure (see Note 5).
[8]All Electronics DCJ-1 (see Note 5) or RadioShack 274-1563 (**www.radioshack.com**).
[9]All Electronics dcSID (see Note 5).

Phil Salas, AD5X, has been licensed for 38 years and enjoys all facets of HF operation. He is currently the Vice President of Engineering for Celion Networks in Richardson, Texas. Phil may be contacted at **ad5x@arrl.net**. QST

By Sam Ulbing, N4UAU

From *QST*, July 1997

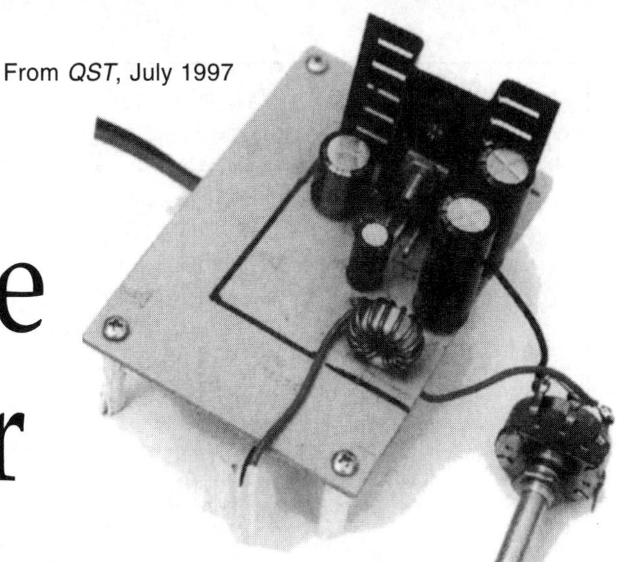

My All-Purpose Voltage Booster

This simple state-of-the-art switching regulator is a learning tool—and a handy gadget to have around!

Did you ever wish you could run your QRP rig using four NiCds instead of a heavy 12-V battery? Boating hams: Did you ever want to recharge your laptop's or hand-held's batteries from your boat's system, without having to run the engine to get a full 13.5 V? Do you need 24 V dc for that nifty relay you bought at the hamfest? Do you want to use a 6-V battery to get 12 V? With the voltage booster described here, you can do all of this.[1]

My booster gives you a dc voltage *greater* than the input dc voltage. The booster is quite flexible, efficient and easy to build. Dc-dc converters are usually built for a specific output voltage, and often for a limited output-current range. My voltage booster is different. It has an *adjustable* output voltage (up to 65 V) and works with loads drawing from less than 20 mA to over 3 A.

Why Use A Switching Regulator?

Switching regulators are becoming dominant in the electronics industry and their sales are forecast to increase markedly[2] because they offer several benefits over linear regulators. First, they are much more efficient. If you use a *linear* regulator (such as a 7805) to get 0.5 A at 5 V from your car's 12-V battery, you will find the 7805 gets *very* hot because it is dissipating 3.5 W. The efficiency of this regulator is less than 42%—the other 58% is lost as heat. *Switching* regulators, on the other hand, are usually 80% to 90% efficient. Further, switching regulators can do things no linear regulator can—such as *increase* the 5 V output of your 4-cell NiCd battery pack to 12 V!

Switching regulators are not perfect. They are noisier than linear regulators, but with proper design, the noise effects can be minimized.

[1]Notes appear on page 2-4.

Designing My Switching Regulator

Several years ago, I wanted to build a switcher so I could learn about them. But after I bought the IC, I read: "Unfortunately switching regulators are also one of the most difficult linear circuits to design. Mysterious modes, sudden, seemingly inexplicable failures, peculiar regulation characteristics and just plain explosions are common occurrences."[3] I was not exactly encouraged and gave up the project.

Then last fall, I read a comment that switching regulators were becoming smaller, more efficient and easier to build. The part that really got my attention was this: "National Semiconductor's Simple Switcher Modules for example...remove a lot of pain and aggravation for novice power-supply designers."[4] I decided to try again and see if I could design and build one. It turned out to be a piece of cake!

First I went to National Semiconductor's Web home page[5] and downloaded the data sheet for a module that appeared to do what I wanted—the LM2587. I also downloaded National's *free* design software, and within just a few days, I had designed and built a regulator that is exactly what I wanted: a general purpose step-up converter. (See the sidebar "Designing a Switching Regulator: That Was Then—This Is Now.")

The Circuit

Figure 1 shows my voltage booster circuit and the title photo shows the prototype. The IC operates like all "boost mode" switching regulators. Pin 4 is connected to an internal switch—a high-speed MOSFET that switches on and off at about 100 kHz. When the switch is closed, supply current builds up through it and the inductor. When the switch opens, the voltage across the inductor rises above the input voltage (remember that inductors generate a back EMF to try to keep the current flowing when the connection is broken) and current is forced to flow through diode D1 to the capacitor C_{OUT}. Energy is transferred from the inductor to the capacitor, but at a *higher* voltage than the input voltage.

The output voltage is controlled by the amount of time the switch is closed: $V_{OUT} = V_{IN} \times [T \div (T - T_{ON})]$. T is the time of a cycle (0.00001 second for a 100-kHz switch) and T_{ON} is the time the switch is closed. T_{ON} is determined by an internal error amplifier and a feedback loop formed by R1 and R2. The output voltage of the circuit is regulated at $V_{OUT} = 1.23 \times (R1+R2) \div R2$, as long as the input voltage is less than the output voltage. By using the combination of a 10-kΩ potentiometer and a 5.6-kΩ resistor for R1, I was able to vary the voltage from 6 to 14.5 V.

C_{OUT} filters the switching pulses as well as storing energy from the inductor. C_{IN}, the input capacitor, reduces the current surge demand on the input source. C_F and R_F prevent the regulator from becoming unstable. C_S and L_S comprise an optional output filter that reduces the RFI generated by the circuit.

You might wonder—as I first did—how this design can be efficient when the switch shorts current directly to ground. The efficiency depends (in part) on the switch's resistance. If it were zero (a perfect switch), there would be no power lost across the switch ($P=I^2R$), so all the power must be stored in the inductor ($I^2L/2$). The regulator's internal MOSFET is designed to have a low resistance when it is *closed*. The critical part of the operation is when the switch changes from fully closed to fully open. As the switch opens, its resistance changes from very low to very high. When the switch is fully open, the current through it is zero, so there is no power loss. But *during transition*, there is a power loss due to the decreasing current flow and in-

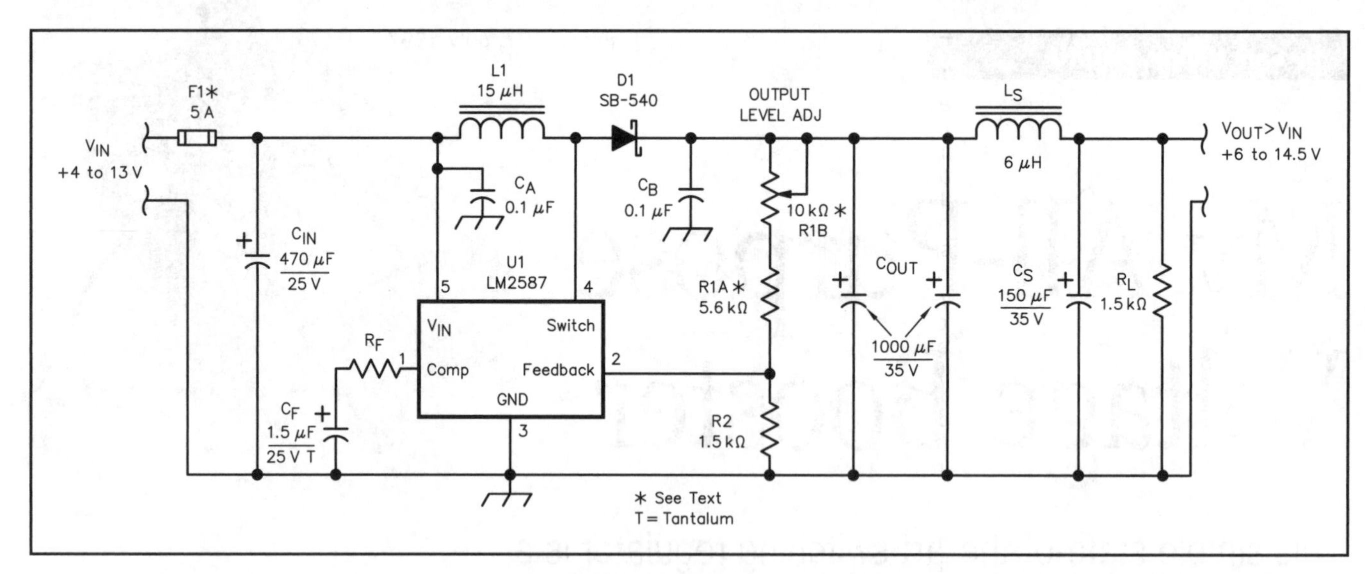

Figure 1—Schematic of the voltage-booster circuit. Unless otherwise specified, resistors are ¼ W, 5% tolerance carbon-composition or film units. Use low-ESR capacitors such as those recommended in the parts list. Equivalent parts can be substituted. Most parts are available from Digi-Key Corp, 701 Brooks Ave S, Thief River Falls, MN 56701-0677 tel 800-344-4539, 218-681-6674; fax 218-681-3380; **http://www.digikey.com**. Parts kits are available from the author (see Note 1). DK part numbers in parentheses are Digi-Key.

C_A, C_B—0.1 µF, 35 V monolithic
C_F—1.5 µF, 25 V tantalum (DK P2044); Nichicon TAP155K025
C_{IN}—470 µF, 25 V electrolytic; Panasonic ECA-1EFQ471 (DK P5704); Nichicon UPL1E471MPH
C_{OUT}—Composed of two 1200 or 1000 µF, 35 V electrolytics; Panasonic ECA-1VFQ122 (DK P5746); Nichicon UPL1V102MHH
C_S—150 µF, 35 V electrolytic; Panasonic ECA-1VQ151 (DK P5732); Nichicon UPL1V151MPH
D1—5-A Schottky; SB-540 (DK SB540CT)
F1—5-A, fast-acting fuse
L1—15 µH; 17 turns #18 wire on a Micrometals T-68-52A core (Micrometals Inc, 5615 E La Palma, Anaheim, CA 92807-2019, tel 800-356-5977, 714-970-9400; fax 714-970-0400; e-mail **ironpowder@aol.com**. Micrometals has a $25 minimum order. The cores identified here are also available from Amidon Associates, PO Box 25867, Santa Ana, CA 92799, tel 714-850-4660, fax 714-850-1163; e-mail **amidon@earthlink.net**; **http://websites.earthlink.net~sold/amidon/** —*Ed.*
L_S—6 µH; 14 turns #18 enameled wire on a Micrometals T-50-18B core
R1B—10-kΩ trimmer potentiometer
R_F—2 kΩ
U1—LM2587, National Semiconductor Simple Switcher 5-A flyback regulator IC; (available from Digi-Key by special order)

creasing resistance. To keep the transition loss to a minimum, it is necessary to use a very fast switch. The rapidly changing current through the switch generates RFI; this is why switching regulators are noisier than linear regulators.

Although this is a very flexible regulator, there are limits to its use. Maximum output for this circuit (using a different value for R1) is about 17 V, although output voltages of up to 65 V are possible with only small changes to the circuit component values. The IC requires a 4-V minimum input. The switching current through the IC must be limited to 5 A, but you will not always get 5 A out. When boosting 5 V to 12 V, you can draw a maximum of 1.5 A before you exceed the 5 A switch-current limit. When boosting 12 V to 13.5 V, you can draw 3.5 A before you exceed the switching limit. Internal current and thermal limiting circuits protect the IC.

It is also important to remember that even though you can get more *voltage* out than you put in, you cannot get more *power* out than you put in. In fact, you get about 10% to 20% less. This is important to realize, especially if you plan to use NiCds as the power source. An output of 1.5 A at 13.5 V is 20 W, which needs 25 W input at 80% efficiency, or 5 A at 5 V. On the other hand, if you want 3 A output at 13.5 V (40 W) from your 12-V lead-acid battery, you need only 4.2 A input, which that battery can easily provide.

Building the Booster

To obtain efficient operation and avoid the hazards described earlier, it is necessary to carefully select the circuit components. Use good components, not junk box residents. Capacitor ESR (effective series resistance), ripple current rating and voltage rating are important. Inductors must be selected for proper saturation current, inductance value, etc. The diode must be a fast recovery type (Schottky). National Semiconductor literature advises:" . . . keep the length of the leads and traces as short as possible. Use single-point grounding or groundplane construction for best results. Separate the signal grounds from the power grounds. Keep the programming resistors as close to the IC as possible." It is a good idea

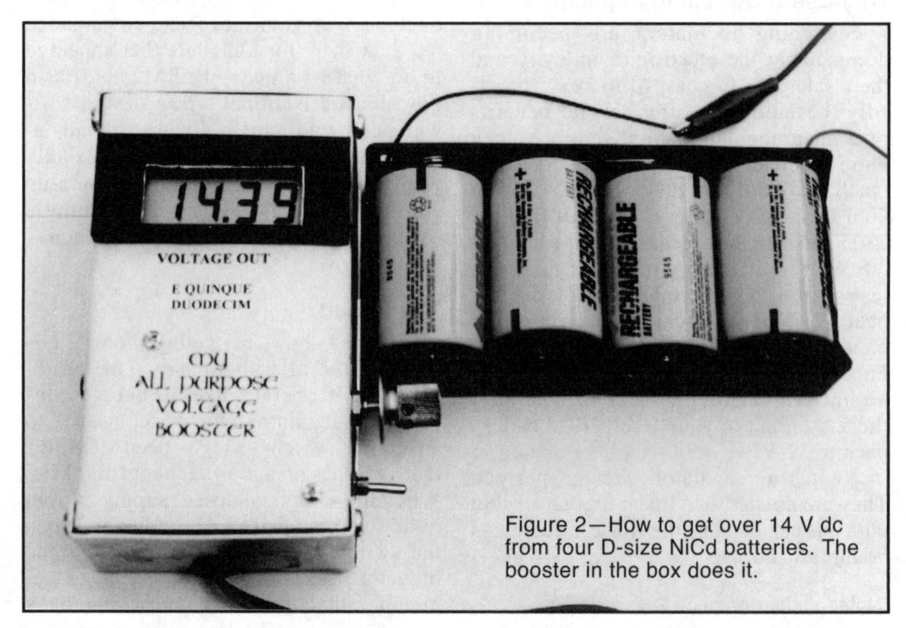

Figure 2—How to get over 14 V dc from four D-size NiCd batteries. The booster in the box does it.

Designing A Switching Regulator: That Was Then—This Is Now

It may appear that switching regulator design has become easy, but the *only* reason it is easy is because National's software does all the calculations! Data sheets I have seen from most other switcher-IC manufacturers still require *you* to do the calculations. To give you a feel for the process, let me describe my experiences. You may want to use National's software yourself, but it is not necessary for this project.

To get my design, I started by entering the input and output voltages, the required current draw and percentage of allowable ripple. The program gave me a circuit design. Because the program is so easy to run, I started playing with it, trying other parameters. I soon found many parameter sets delivered similar designs, ie, the same inductance value, but slightly different capacitor values. I discovered that I could force the software to use component values other than it offered, so I changed the values that were different between the designs and made them equal. The program identified the acceptable operating limits for each component change. By using different component-value combinations, it soon became apparent that with a 15-µH inductor, a fixed set of capacitors and resistors, and a potentiometer at R1 (of Figure 1), I could build a regulator that works over a wide range of currents and voltages. My All Purpose Voltage Booster was born!

Compare this process with what was required *without* the software:

That Was Then

The 76-page Application Note (AN) that came with the IC was full of equations and graphs.* I had to learn about pulse-width modulators, error amplifiers, frequency compensation, output diode losses, inductor and transformer design.

Inductor: Trade-offs are size, maximum output power, transient response, input filtering and, in some cases, loop stability. The AN gave eight pages of equations to "help" me select the proper inductor!

Output Capacitor: Criteria are "low ESR . . . a reasonable procedure is to let the reactance of the output capacitor contribute no more than 1/3 of the total peak-to-peak output voltage ripple."

Frequency Compensation: Three pages of instructions including comments such as: "Inject a signal and check the 'phase margin'; several large-signal dynamic tests should also be done; check the startup overshoot."

This Is Now

The data sheet for the LM2587 is 25 pages long but, instead of having a lot of equations, there are some suggested circuits with *specific* parts recommended for building them. Better yet, the software gave me specific answers:

Inductor: Use a T-68-52A toroid core with 17 turns of #17 wire.†

Output Capacitor: two Panasonic ECA-1VFQ122 in parallel.

Frequency Compensation: C_F = 1.5 µF tantalum; R_F = 2 kΩ, 1/4 W.

As you can see in the photo, my booster is not particularly small. That's because I wanted it to handle a lot of power. Had I wanted a smaller regulator, I could have designed for a lower maximum current and the software would have specified smaller capacitance and inductance values.

The software did not help me with *circuit layout*, but the data sheet noted that layout is "very important" because "CMOS devices ...can cause incredible noise.... typically made up of crosstalk, power supply spiking, transient noise and ground bounce." In the process of building my switcher, I tried six different PC board layouts. They all worked well, but the ones with the least EMI were those that minimized the "switching loops" and reduced the length and area of the connections at the ICs switching node (pin 4). As all hams know, loop antennas are efficient radiators and, in the case of my booster, there are two major loop antennas. The *output* loop consists of the diode, output capacitor and the IC switch. The *input* loop is the inductor, input capacitor and IC switch. For boost-mode regulators, the output loop is the most critical. By reducing this loop size and connecting the diode and inductor as close as possible to pin 4, I was able to make significant reductions in RFI.

A post filter consisting of C_S and L_S made a significant RFI reduction when I put it on a breadboard, but when I moved it to the PC board, the noise returned. I found this was because I was using the PC board groundplane instead of a direct (point to point) return to pin 3 of the IC. I corrected this with a jumper wire, and the noise went away (I guess layout is important!). To reduce leakage from the toroid, I rewound it with #18 wire and kept the windings tight to the core and spaced evenly. This also reduced RFI. While #18 wire has 25% greater resistance than #17, I felt it was not significant (the resistance increased from 8 milli-ohms to 10 milliohms) except at the highest load currents, and it was much easier to bend the wire and get it close to the core. Some literature suggests using several small-diameter wires in parallel for the windings as another way to get better conformation to the toroid, but I have not tried this.

Designing Other Switchers

Using the software, you can modify my regulator to get 24 V at 300 mA from 12 V by simply changing the values of L1 and some resistors. You can get up to 65 V out, but be sure to use capacitors and a diode with higher voltage ratings. National Semiconductor makes several switcher ICs optimized for different functions. The IC used for my voltage booster can also be used in a *flyback* configuration to provide more than one output voltage. For instance, you can build one that gives positive and negative voltages. Other switchers, such as *buck regulators*, let you *reduce* voltage efficiently. If you want to try one of these, all the data is available at National Semiconductor's home page. The software specifies commercial transformers for flyback designs. Such transformers are sometimes difficult to locate, although the serious experimenter can usually obtain a sample from the manufacturer.† Since all the pertinent design data is specified by the software, it is also possible to wind your own toroid transformer if you know how to select the proper toroid-core material. I don't, but perhaps a technically gifted *QST* reader can help me (and others) by explaining—in *simple* terms—how to select the proper core given the required inductance, peak current, power output, etc.—*Sam Ulbing, N4UAU*

*Linear Technologies, *Applications Note 19*.

†Odd wire sizes are available from motor repair shops.

to add a fuse (5 A) to the input circuit. Although the IC has many safety features, a dead short at the output will cause dc to flow directly from the power source through the inductor and diode to the output. For heavier loads, use a heat sink on the IC. I originally put the regulator in a metal case for shielding (see Figure 2). Later, I moved it to a plastic box so I could add an inexpensive DVM to display the output voltage.

Using the Booster

I have used my booster in a number of situations and it works as promised. When I used NiCds to power my NorCal 40A QRP rig, I had this voltage booster three inches away from the rig and heard no noise. The station I was working said my signal sounded good and my antenna tuner indicated I was putting out the full power. By changing the booster voltage from 12 to 13.5 V, I could see the output power increase. (Remember, I was powering my rig from four D-cell NiCds!) The booster has also worked well powering my Switched Capacitor Audio Filter[6] and Uncle Albert's Keyer[7] with no noise.

My tests show some noise is apparent on the higher bands at higher loads, but generally this noise is not very objectionable as long as I use good RFI grounding techniques. At lighter loads, the noise is hardly noticeable. While drawing 2.5 A for an extended period, the booster stayed cool with a small heat sink on it (see the title photo). One nice feature of this IC is that it has a "soft start" feature, which limits the initial current inrush that is common with older regulators.

Finding the Parts

Most of the parts for this project are not hard to find. Digi-Key and other large suppliers have most of them. My greatest difficulty was locating the proper inductor. Inductors and transformers for circuits such as these are often available only directly from their manufacturer and sold only in quantities of hundreds or thousands. That's one reason I selected the LM2587 for my booster: The software indicated that I could wind the inductor on a toroidal core. I had learned how to do this while building QRP rigs and knew where to get toroids.

Summary

It may take you a little while to get used to thinking in terms of *boosting* rather than *dropping* voltage, but once you do, you will realize this switching regulator has many uses. At home with an inexpensive 5-V computer power supply at its input, you'll have a power supply variable up to 17 V (or more, with minor changes). Think about using the booster as a lamp dimmer for a 12-V lamp, or as a motor-speed controller. In the field, you can use NiCds or other low-voltage batteries as a power source. If you want to design a different regulator, all the data you need is available at the referenced locations. Keep in mind that switching regulator technology is improving very rapidly, so you may find faster, smaller, more-efficient modules available. I am interested in hearing the thoughts, ideas and experiences of anyone who does explore this area.

Notes

[1]A PC board, all board-mounted parts and detailed instructions for this project are available from the author for $21, plus $2 for shipping in the US and Canada, $4 elsewhere. Foreign orders please include an international money order or a check in US currency payable at a US bank. Charge cards are *not* accepted. Foreign orders are shipped by air, small packet. Florida residents please add sales tax. A template package is *not* available.

[2]*Electronic Business News*, Oct 28, 1996, p 24.

[3]Linear Technologies *App Note 25*, Sep 1987; **http://www.linear-tech.com**

[4]*Electronic Engineering Times*, Sep 30, 1996, p 28.

[5]To obtain the free software and get more information on the LM2587 and other Simple Switcher modules, visit the following National Semiconductor Web page: **http://www.national.com/sw/SimpleSwitcher/0,1043,0,00.html**; you can download the individual software versions from **http://www.national.com/sw/switch/sms421.exe** and **http://www.national.com/sw/switch/sms33.exe**. Also see **http://www.national.com** and **http://www.nsc.com**, or contact them via e-mail **support@tevm2.nsc.com**. —*Ed.*

[6]Sam Ulbing, N4UAU, "An Active Audio CW Filter You Can Build," *QST*, Oct 1992, pp 27-29.

[7]Sam Ulbing, N4UAU, "Uncle Albert's Keyer," *QST*, Jan 1994, pp 42-44.

Sam Ulbing, N4UAU, has contributed a number of project articles to QST, QEX *and* 73 Amateur Radio Today Magazine. *Most of these articles have been low-power, 12-V-based projects. This is because Sam is one of the growing number of sailors who like to take their ham gear along when they sail.*

Sam became a ham after spending a winter on his sailboat in the Bahamas and meeting other boaters who are hams. The advantages of having a ham radio on board were immediately apparent: Hams are able to get vital information such as weather reports daily on nets like the Waterway Net. In addition, Sam found that boaters who have their ham stations on board tend to become much closer friends because they can keep track of each others' location and are thus able to meet often. Sam reluctantly uses SSB. His favorite mode is CW, which—with its low power consumption—is ideal for use on a boat. When he's not on his boat, you can contact Sam at 5200 NW 43rd St, Suite 102-177, Gainesville, FL 32606; e-mail* **n4uau@afn.org**.

Photos by the author.

* *The Waterway Radio and Cruising Club meets daily on 7268 kHz at 0745 Eastern.*

QST

By Daniel R. Kemppainen, N8XJK

From *QST*, November 2004

A 12 V dc Boost Regulator for Battery Operation

A dc-dc boost switching converter is the answer to low voltage battery problems for mobile, portable or emergency-power operation.

Battery low charge state conditions, combined with voltage drops in wiring, can cause reduction in output power, transmit signal distortion or even total shutdown in many radios. One solution to this problem is to build a switch-mode power supply (SMPS) to maintain the dc input voltage. An SMPS can offer boosted power levels and allow longer operating times from a given battery. This article describes how to build and test one from both new and recycled parts for about $50.

Overview

This SMPS is a simple boost supply, designed to make up the difference between battery voltage and a preferred output voltage level at the cost of some additional current draw from the battery. It was designed for an output current of about 25 A. When turned off, the battery voltage (less one diode voltage drop) is present at the output terminals of the supply. No power transfer relays or switches are required. The supply can be set up to operate on demand or continuously, depending on user requirements. A switch or relay contact is used to switch the power supply control power off when not in use. This reduces power consumption during periods of inactivity or when voltage is sufficient to power the radio.

Two "on demand" inputs are provided to enable the voltage boost function. One of the inputs is a simple remote enable input, and requires only a battery voltage signal. This can be used in conjunction with a control signal from a radio to key the supply or it can be enabled by a toggle switch for manual operation. The other input is an RF detector. The RF detector can be used to monitor the RF output of the attached radio and allow the voltage boost to take place when the radio is transmitting. The RF detector attaches directly to the antenna lead of most radios using a coax T fitting or a coupling transformer. This design has been tested with radios transmitting from several watts to 100 W. Operation at higher power levels may require some circuit modifications. The completed supply is shown in Figure 1.

Circuit Description

The SMPS uses a push-pull design topology. Its schematic appears in Figure 2. The positive battery terminal is connected to the center tap of the primary of the switching transformer T1. The secondary of T1 is also a center tapped winding, with its center tap also attached to the battery voltage. The voltages seen on the secondary legs of T1 are the battery voltage plus the voltage of the transformer windings. This configuration allows the transformer to supply only the difference between the output and battery voltages. In addition, the power requirements of the transformer and switching transistors are reduced. This also allows battery voltage to be present at the output of the supply when it is switched off.

Figure 1 (above)— The completed boost regulating switching supply. It is shown with an aluminum sheet that acts both as a mounting base and a heat sink.

MOSFET transistors Q5 and Q6[1] alternately switch the legs of the primary winding of T1 to ground, creating an ac flux waveform in the transformer. The secondary legs of transformer T1 are rectified by the dual Schottky diode D7. Inductor L1 and eight 3300 µF capacitors form a low pass filter to smooth the rectified waveform.

A switch-mode power supply controller, U1, handles the voltage regulation. The controller used in this supply is an LM3524D.[2] The LM3524D uses pulse width modulation to control the time that switching transistors Q5 and Q6 are turned on. By varying the pulse width, the ac voltage of the transformer is varied and the output voltage is maintained.

A simple battery voltage monitor circuit is used to monitor low battery conditions. The low voltage protection circuit shuts down the LM3524D in the event that battery voltage falls below a minimum level. The protection voltage is jumper selectable to 9, 10 or 11 V. The circuit uses an LM339[3] quad comparator in conjunction with a +5 V dc reference voltage provided by the LM3524D controller IC. When the protection circuit is tripped, the supply boost function is disabled and battery voltage is present at the output of the supply. A reset of the battery protection circuit is accomplished by cycling the power switch.

Collecting the Parts

The inductor and transformer are custom parts that will need to be made for

[1]Notes appear on page 5-45.

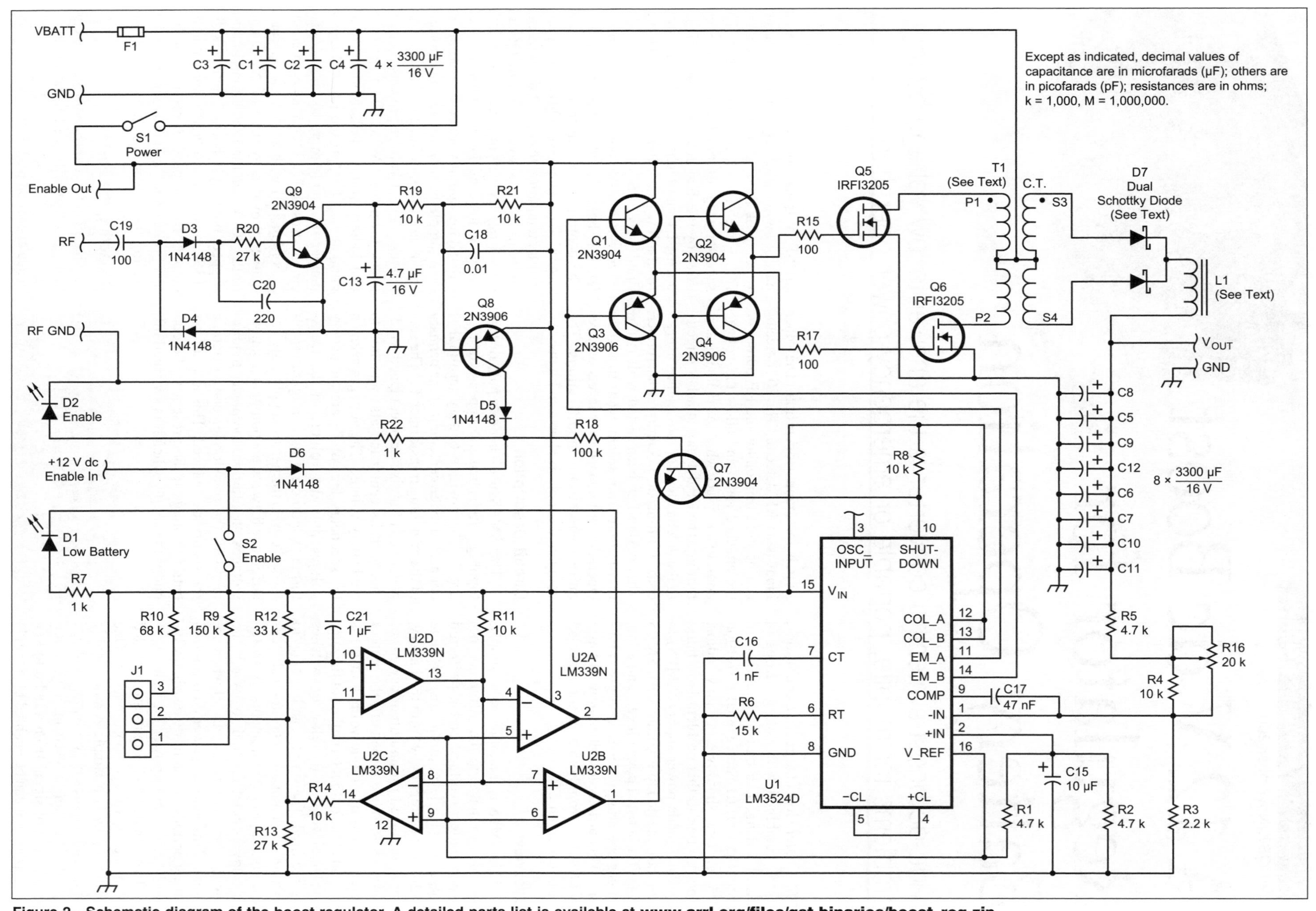

Figure 2—Schematic diagram of the boost regulator. A detailed parts list is available at **www.arrl.org/files/qst-binaries/boost_reg.zip**.

the supply. A good source of these components is an old PC power supply. For this SMPS, an AT-style computer power supply was chosen. The inductor and transformer will need to be disassembled and rewound before building this supply. Details can be found on the ARRLWeb.[4] The dual Schottky diode (D7) can also be salvaged from the same surplus power supply. Capacitors C1 through C12 are specific, and will need to be ordered new. The rest of the parts are common and can be replaced, provided that close matches can be found.

Detailed directions for disassembly of the transformer core, transformer winding calculations, directions for winding the transformer and inductor, the PC-board layout, a complete parts list, and construction preparation and assembly instructions can also be found on the ARRL Web site (see Note 4).

A good reference for ferrite core applications and design is the *Ferrite Core Design Manual*, available from Magnetics, PO Box 11422, Pittsburgh, PA 15238; **www.mag-inc.com/ferrites/fc601.asp**.

The Filter Inductor, L1

The only inductor used in the circuit is L1. The inductor is used in conjunction with eight 3300 μF high frequency electrolytic capacitors to make the output low pass filter section of the power supply. The approximate value of L1 is 9 μH and it must be capable of handling currents of up to 25 A. Nine turns of wire on the salvaged toroid core will produce the 9 μH inductance. Ten paralleled 24 gauge copper wires work well for the winding. Each strand needs to be about 18 inches long. Plan on a total length of about 15 feet of wire for the winding. Strip, twist and tin the leads before inserting the inductor into the circuit board.

Supply Boost Enable Circuit

The supply enable circuit is designed to allow the supply to boost voltage when required. Two inputs allow the circuit to function. The first is a simple 12 V input and the second is an RF detection input. The ENABLE input feeds current into D6, supplying current to the base of Q7 through R18 and to the ENABLE LED through R22. The ENABLE function can also be manually forced by closing S2, the ENABLE switch. Feeding current into the base of Q7 allows comparator U2B to pull the shutdown pin low, thus enabling the LM3524D.

The RF INPUT enables the supply when RF is present. The RF INPUT feeds current into the base of Q9 through D3, causing C13 to be discharged and half of the supply voltage to appear at the base

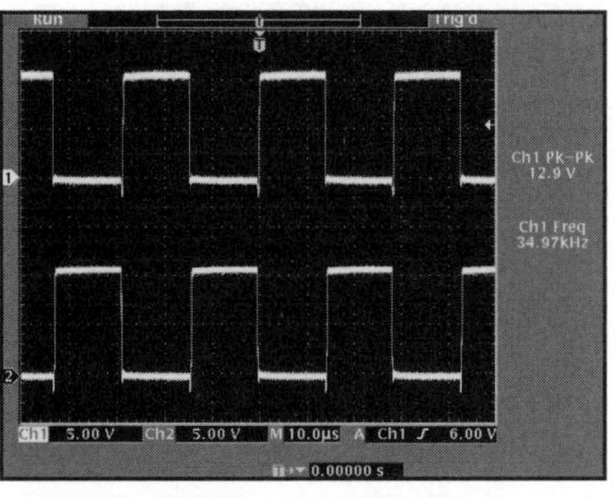

Figure 3—MOSFET gate driver waveforms shown with no MOSFET devices attached. Note the 180° phase shift between phases.

of Q8. Transistor Q8 then feeds current into D5, enabling the supply just as the ENABLE input would. The charge time of the 4.7 μF capacitor, C13, sets up a small delay that keeps the supply enabled after the RF signal is removed.

To test the ENABLE circuit, apply power to the supply and close the power switch. Supply a 12 V signal to the ENABLE input or close the ENABLE switch. The ENABLE LED should come on. Remove the 12 V feeding the ENABLE input, and the ENABLE LED should go off.

To test the RF INPUT, apply an RF signal to the RF INPUT. It is important to note that a 2 V peak-to-peak signal will be the minimum needed at the RF INPUT to enable the supply. The supply ENABLE LED should light up in the presence of RF, and go off about a quarter of a second after removing the RF signal. The turn-off time may vary slightly, depending on the strength of the RF source.

Battery Low Voltage Protection Test

The battery low voltage protection circuit was designed to protect the battery from being discharged too far. The circuit works by comparing a sample of the battery voltage to a reference voltage. The LM339 comparator is used to compare the battery voltage to the reference provided by the LM3524D. The controller circuit is shut down when a low voltage condition has been detected. R12 and R13 form the voltage divider used to sample battery voltage. R9 and R10 and the jumper J1 act in parallel with R12 to vary the voltage divider ratio. U2D compares the sample battery voltage to a reference voltage. When the voltage falls below the reference, U2D pulls its output pin to ground. U2C then seals the circuit by pulling the voltage going to U2D down even further. U2A acts as a simple switch turning on the low voltage LED. U2B allows Q7's emitter to go high, allowing R8 to pull up the shutdown pin on the LM3924 and thereby disabling the supply.

To test the low voltage protection circuit, simply apply power and a ground to the supply and close the power switch. With the power switch closed, remove the 12 V input from the supply. After a few seconds, the LOW BATTERY voltage LED should light up and stay on. Before depleting the voltage stored in the input capacitors C1 through C4, reconnect power to the supply. The LOW BATTERY voltage LED should still be on. Turn the power switch off for a few seconds, and then back on again. This should reset the battery protection circuit and the LOW BATTERY LED should be off.

Next, test the threshold voltage for the low battery protection circuit. Start by connecting a voltmeter or oscilloscope across the battery input to the supply. Set the battery protection voltage to 11 V by removing the voltage select jumper J1. Connect power to the supply, and cycle the POWER switch OFF and then ON. Remove the input voltage to the supply, and watch for the LOW BATTERY LED to come on. Take note of the voltage at which the LED turns on. This voltage should be very close to 11 V. Reconnect power and cycle the power switch. Repeat this test for the 10 V (jumper toward fuse) and the 9 V (jumper away from fuse) threshold settings.

Switching Regulator IC

The switching regulator is the heart of the circuit. The regulator uses pulse width modulation to vary how long the switching transistors Q5 and Q6 stay on for each switching cycle. By adjusting the pulse width of the switching transistors, the output voltage can be kept at a constant level. The switching regulator monitors the output through a voltage divider as a 2.5 V dc signal. Resistors R3, R4, R5 and R16 form the voltage divider that provides the feedback voltage. R16 is variable, allowing for adjustment of the output voltage. The reference signal comes from the LM3524D's internal 5 V reference and is divided down by the voltage divider formed by R1 and R2.

The next test will verify that the switching frequency is correct, the feedback network is working, and the gate driver transistors are operating correctly. After D7, Q5 and Q6 are removed from the supply, tie the supply positive input to the supply positive output. Turn variable resistor R16 fully clockwise. This will help to ensure that the output voltage setting is above the input voltage. Attach a voltage source to the supply, and close the power switch. Check that the input voltage to the supply is around 12-13 V dc.

Using an oscilloscope or a frequency

counter, check the frequency at the nodes between R15 and Q5 and R17 and Q6. This should be about 35 kHz, with a 50% duty cycle for each transistor. If the frequency is not within a few kilohertz of 35 kHz, the timing capacitor or resistor (C16 or R6) will need to be adjusted. The waveform should have fast rising and falling edges. Figure 3 shows a sample of the gate driver waveforms without Q5 and Q6 attached.

With the oscilloscope or frequency counter still attached to the nodes between R15 and Q5 or R17 and Q6, turn the voltage adjust resistor all the way to the left. This will set the minimum output voltage to around 9 V dc, well below the input voltage. The switching drive signals should be a steady 0 V signal, indicating that the output voltage is above the current voltage setting. When the test is complete, center the output voltage potentiometer. This should set the supply to around 13.3 V dc.

Putting It All Together

After operation of the switching regulator IC, RF detect and enable inputs and the battery voltage protection circuit have all been verified, mount the switching transistors (Q5 and Q6) and rectifier diode (D7). It is important that a heat sink is attached to the switching transistors and rectifier diode. The transistors come in an electrically isolated package and need only heat sink grease between them and the sink. The diode is not an electrically isolated package and it requires electrical isolation between its package and the heat sink. The insulating pad originally used to isolate the diode from the heat sink in the PC should also be used.

The first test with the switching transistors and rectifier diodes should be done with a low current fuse (5 A) or a pair of 100 W, 12 V light bulbs in series with the battery. If light bulbs are used, set the battery protection circuit to 9 V. Hook a voltmeter and a small 12 V automotive tail lamp or similar small load to the output of the supply. Hook the supply to the 12 V battery. With the power switch off, the 12 V load should be energized, and the meter should read the battery voltage minus one diode voltage drop (0.5-0.6 V).

Close the power switch. No LEDs should be on. If the battery protection LED is on, check the input voltage to the supply and cycle the power switch. Enable the supply with the RF DETECT input or ENABLE input and watch the output voltage. The voltage should jump up slightly when the ENABLE LED comes on. If no change in voltage is detected, turn R16 clockwise until an increase is detected. As long as the battery protection LED is off, the ENABLE LED is on, and the output voltage setting is higher than the input voltage, the supply should be running. Set the output voltage to a desired level by adjusting R16. The supply is now running, and will regulate the minimum output voltage.

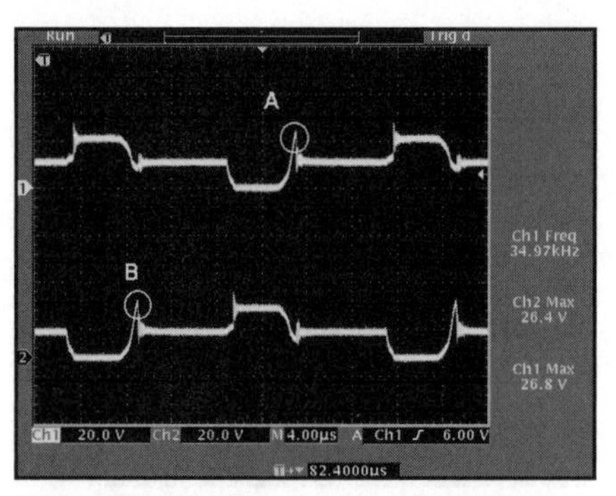

Figure 4—The drain to source voltages of Q5 and Q6 while operating the supply at a load of 300 W. The input voltage was between 10 and 11 V dc. The output voltage was set to 14 V dc. Note points A and B showing the transients generated during MOSFET turn-off.

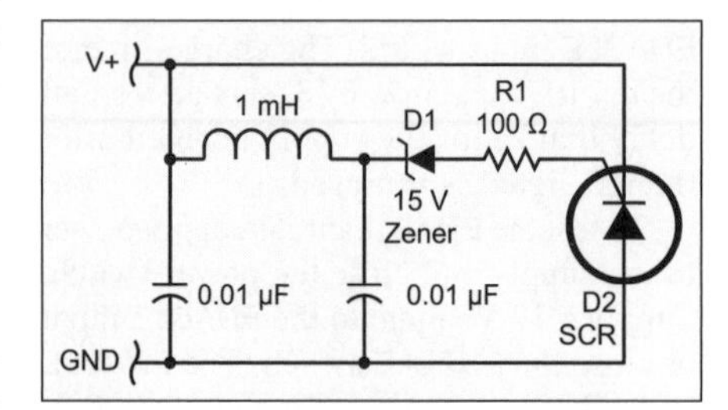

Figure 5—A sample crowbar circuit. Values shown are recommended starting points. Determine the exact values experimentally, depending on the SCR and Zener diode used and the clamp voltage desired.

A final check of the supply should now be made. The transient voltages on the switching transistors now need to be checked. This check is important and should not be skipped. This test will be run several times, with an increase in the load on the supply each time. The minimum peak voltage that will normally be seen is double the supply voltage. This is because the switching transformer acts as an autotransformer when each transistor is ON. This voltage should not be of any concern, but the voltage that is of concern is the transient voltage generated when the switching transistors turn OFF. This occurs when the voltage across the transistor starts to rise from the 0 V level (see Figure 4). The changing current in the transformer, combined with any leakage inductance, cause these transients to be generated. The peak voltages need to be below the rated 55 V of the switching transistors.

Attach an oscilloscope or peak detecting voltmeter from the drain lead (center lead) of Q5 and Q6 to ground, and attach a 1 A load to the supply. Enable the supply and check the peak voltage across each of the transistors. If peak voltages are close to the transistor breakdown voltage (55 V) under light loading, stop the testing immediately. The switching transformer will need to be rewound with tighter spacing between all of the windings.

Repeat the previous test with a large load on the supply. Remove the current limiting from the battery, and place a 30 A fuse in the fuse holder. Attach a load of around 100 W to the supply. A 100 W automotive light bulb works well for this. You may have trouble with the battery protection circuit when trying to enable the supply with the 100 W light bulb attached. If so, leave the voltage on the ENABLE input to the supply, then cycle the power switch. The supply has a built-in soft-start circuit that will bring the output voltage up slowly when the ENABLE input is powered before closing the power switch. It may be necessary to use this feature when testing the supply with large light bulb loads. With the 100 W load attached, check the transient voltages to make sure they are below the 55 V rating of the transistors. If the transients are low, repeat the test with 200-300 W of load. If any of the tests reveal transients close to the 55 V maximum, the transformer will need to be rewound with tighter spacing between the windings.

Final Notes

The power supply can be mounted in an enclosure, provided that enough cooling is available for the switching transistors. The level of power drawn and the state of charge of the battery will determine the heat sink requirements. Basically, if the heat sink or cases of the transistors are too hot to touch, a bigger heat sink or more air flow is required.

A set of optional high frequency "snubbers" can be added to the switching transformer and the switching transistors. The snubbers are basically a series RC network used to reduce the high frequency ringing that can occur during switching transitions. Place the snubbers from each leg of the transformer to the center tap of the transformer, and from the drain lead (center) of the switching transistors to ground. A 220 pF ceramic capacitor in series with a 220 Ω resistor is a good starting point for each snubber. It is best to determine the exact values experimentally. Find values that reduce the ringing without dissipating excessive heat in the resistors.

It is highly recommended that a crowbar circuit be included on the output of the supply. If a malfunction were to occur in the supply feedback circuit, the output voltage could rise enough to cause damage to any attached equipment. The crowbar circuit watches the output voltage of the supply and shorts the output—blowing fuse F1—if the voltage gets too high. The simple circuit shown in Figure 5 uses a Zener diode in series with the gate of an SCR. A small current limiting resistor and an RF filter are included. The current limiting resistor prevents damage to the gate of the SCR, and the filter prevents RF on the 12 V line from tripping the crowbar circuit. Values for the capacitors, inductors, resistors and the Zener diode will depend on the particular SCR used and the crowbar voltage desired. Although the values shown in the circuit are a good starting point, it is best to determine the exact values experimentally. When testing the crowbar circuit, use a large 12 V light bulb in series with the battery or voltage source. This can save the cost of several fuses.

It may be desirable to add additional filtering and shielding to the supply. No testing of the electromagnetic interference (EMI) generated by this supply has been done. The generated EMI should not be very great, but it may be strong enough to be received by an attached radio. This, of course, will depend on your antenna and mode of operation. For this reason, it is recommended that you use the RF enable circuit, whenever possible. Many radios have problems with low input voltage only when transmitting and not when receiving. The RF detect circuit has been designed to supply full power to the radio when transmitting, and it allows pass-through battery voltage to be supplied while receiving. This reduces the EMI generated by the supply and received by the radio.

Figure 6 shows oscilloscope voltage traces of the supply in operation. The supply was powering a radio that was being used to transmit SSB voice. In this case, the RF input was used to enable the supply and boost the rig voltage only when transmitting. Trace 1 shows the output voltage of the supply. The corresponding RF envelope is shown in trace 2. The terminal voltage of the battery was approximately 11 V dc, while the boosted output voltage was set at 13.8 V dc. The supply regulation action can be clearly seen as the higher levels of voltage shown in trace 1 after RF excitation turned the supply on. In this case, the supply was being tested from a weak battery, similar to that of typical operating conditions.

This simple supply can be built by anyone—no background in switch-mode power conversion is required. To simplify the parts list, the sometimes hard-to-find magnetic components are recycled from a

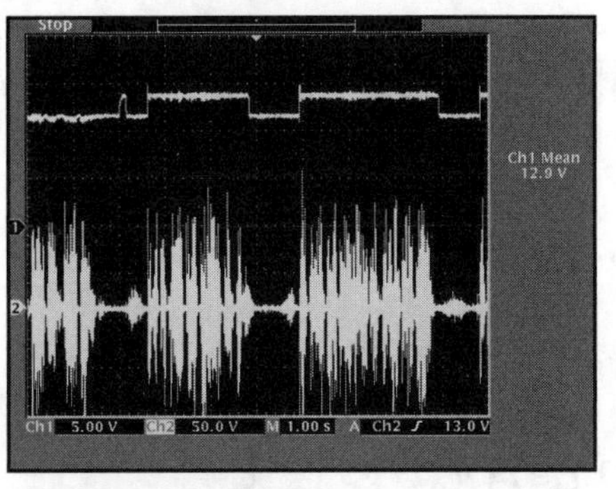

Figure 6—The supply in operation. Trace 1 shows supply output voltage. Trace 2 shows the RF voltage at the input of the RF detector. The time base is set to 1 s/cm. The zero reference baseline for trace 1 is at arrow 1. The battery voltage is about 11 V dc (a weak battery) and the boost voltage is set to 13.8 V dc. The first two seconds the supply is OFF. It comes on with RF excitation, as shown by the lower trace. It can also be enabled with 12 V dc to the ENABLE input or by operation of the ENABLE switch.

common power supply. The rest of the components are available from a single distributor, to make ordering easy. Even the most difficult task (winding the switching transformer) should take no more than an hour. Anyone who has had some experience with soldering and coil winding can build and test this power supply.[5] A little extra money, a few spare parts and some free time are all that are required.

A special thank you to John Kemppainen, N8BFL, and Jim Carstens, W8LTL, for their article help, testing and photography.

Notes

[1]IRFI3205 power MOSFET; International Rectifier (**www.irf.com/product-info/datasheets/data/IRFI3205.pdf**).

[2]LM3524D regulating pulse width modulator; National Semiconductor (**www.national.com/pf/LM/LM3524D.html**).

[3]LM339 low voltage quad comparator; National Semiconductor (**www.national.com/pf/LM/LM339.html**).

[4]**www.arrl.org/files/qst-binaries/boost_reg.zip**.

[5]The author may be able to supply a professionally built PC board. If interested in procuring a board, write or e-mail the author at the address shown at the end of the article. If enough requests are received, details as to cost and availability will be supplied.

With a lifelong love for science and electronics, Daniel Kemppainen, N8XJK, has been a ham for more than 10 years. As is obvious from this article, his area of interest includes switching power conversion, but he's also interested in high power audio amplifier design. Daniel is a design engineer dealing with analog and digital electronics, data acquisition systems and programming for the Windows *operating system. He has both Associate's and Bachelor's degrees in Electrical Engineering Technology. You can reach him at 25403 E Acorn St, Calumet, MI 49913 or at* **drk@pasty.com**. QST

From *QST*, December 2005

The SuperPacker HF Amplifier

Turn your QRP rig into a 100 W powerhouse with this project.

Jonathan Gottlieb, WA3WDK, and Andy Mitz, WA3LTJ

Low power (QRP) operation is a lot of fun, but let's face it: there are times when a few more watts are welcome. Perhaps you want the extra power to help you crack a DX pileup, or maybe the waning solar flux is the real motivator. Either way, this homebrew project will add some welcome kick to your QRP rig when you want to have it. You might even let your QRP rig do double duty as your full power base station. To ensure the success of this homebrew project, it has been designed as a collection of small, manageable projects carefully packaged into one box. So, get out your drill bits and soldering iron—it is time to build your own 100 W HF amplifier.

In developing this project, our goal was to make an HF amplifier that was easy to build and (nearly) foolproof. The basic idea was simple. Use existing, proven designs as much as possible. We looked for circuits that have been published, tested, and ones with readily available parts. The result is an amplifier called the SuperPacker. It uses a popular Motorola amplifier board design, a low pass filter (LPF) and a programmable integrated circuit (PIC) SWR meter from past *QST* articles, three relays, some spare parts, and ideas shamelessly pilfered from other great projects.

The result is the all-band HF amplifier shown above. The SuperPacker delivers up to 100 W output with just a few watts of drive. Its low pass filter ensures a spectrally clean signal and its built-in SWR meter provides a direct digital display of the antenna SWR while automatically protecting the amplifier from even a serious mismatch.

Design

A good HF amplifier needs a stable gain stage, proper low-pass filtering to attenuate harmonics, TR switching and some way to protect against unexpected loads such as a disconnected or broken antenna. The heart of the SuperPacker is a robust bipolar amplifier design developed by Motorola in 1976 and still widely used today. A high quality circuit board, the original Motorola Semiconductor Products Applications Note, and all the parts needed to build the amplifier stage of the SuperPacker are available from Communications Concepts (CCI).[1]

The LPF is a solid design developed by Jim Valdes, WA1GPO, as part of the FARA HF Project featured in the June 2003 issue of *QST*.[2] All the construction details for the LPF, including the original *QST* article and an updated parts list, are available on the Falmouth Amateur Radio Association's Web site, **www.falara.org/tektalk/tektalkfs.html**. The PC board is available from FAR Circuits.[3] Just remember to let them know you want the LPF board to build a SuperPacker.

A good amplifier also needs good transmit to receive switching. The SuperPacker TR switching is accomplished using two new circuit boards designed specifically for this project—the *relay board* and the *control board*. These two boards are also available from FAR Circuits. They are not complicated and are described in detail in this article.

Now for the icing on the cake: The SuperPacker includes a digital SWR meter based on the Bert Kelley, AA4FB, PIC SWR meter project that was featured in the December 1999 issue of *QST*.[4] This SWR meter not only provides a direct readout of the SWR, but the PIC controller has been updated and reprogrammed to provide automatic power cutback whenever the SWR gets too high. The circuit board for the PIC SWR meter, including a pre-programmed PIC processor and the digital display, are all available from FAR Circuits. In short, the entire SuperPacker project is at its heart a modular design based on a collection of simple projects.

[1]Notes appear on page 5-52.

Description

The SuperPacker is housed in a 3.06×8.25×6.13 inch aluminum enclosure sold by Ten-Tec. This enclosure is not only sturdy and attractive, but it is inexpensive and makes for a professional looking finished project. The only two controls are the BAND switch and the HI/LO POWER switch. RF INPUT, RF OUTPUT, 12 V POWER and PTT connectors are all mounted on the rear of the enclosure. The front panel also has a two digit SWR display and red power level indicator LED. As described in the original PIC SWR article, the display shows SWR values from 1:1 to 10:1. Above 10:1 the display indicates HI. If the output power drops below about 8 W the display indicates LP. At a SWR above 4:1 the modified SWR program activates a circuit to reduce the amplifier drive by 10 dB, giving about 8 to 10 W of output. The red high power LED indicator turns off under the high-SWR condition. The arrangement is ideally suited for using a manual antenna tuner.

The amplifier requires a suitable power supply able to deliver 12 to 13.5 V at about 20 A. Internally, the 12 V power is split between the power amplifier board and the remaining circuitry. Each branch is separately fused and protected from accidental polarity reversal. The power transistors idle at about 250 mA. The PIC SWR meter can draw another 50 mA or more, so the amplifier requires around 300 mA when idling. A large heat sink on the top of the case keeps

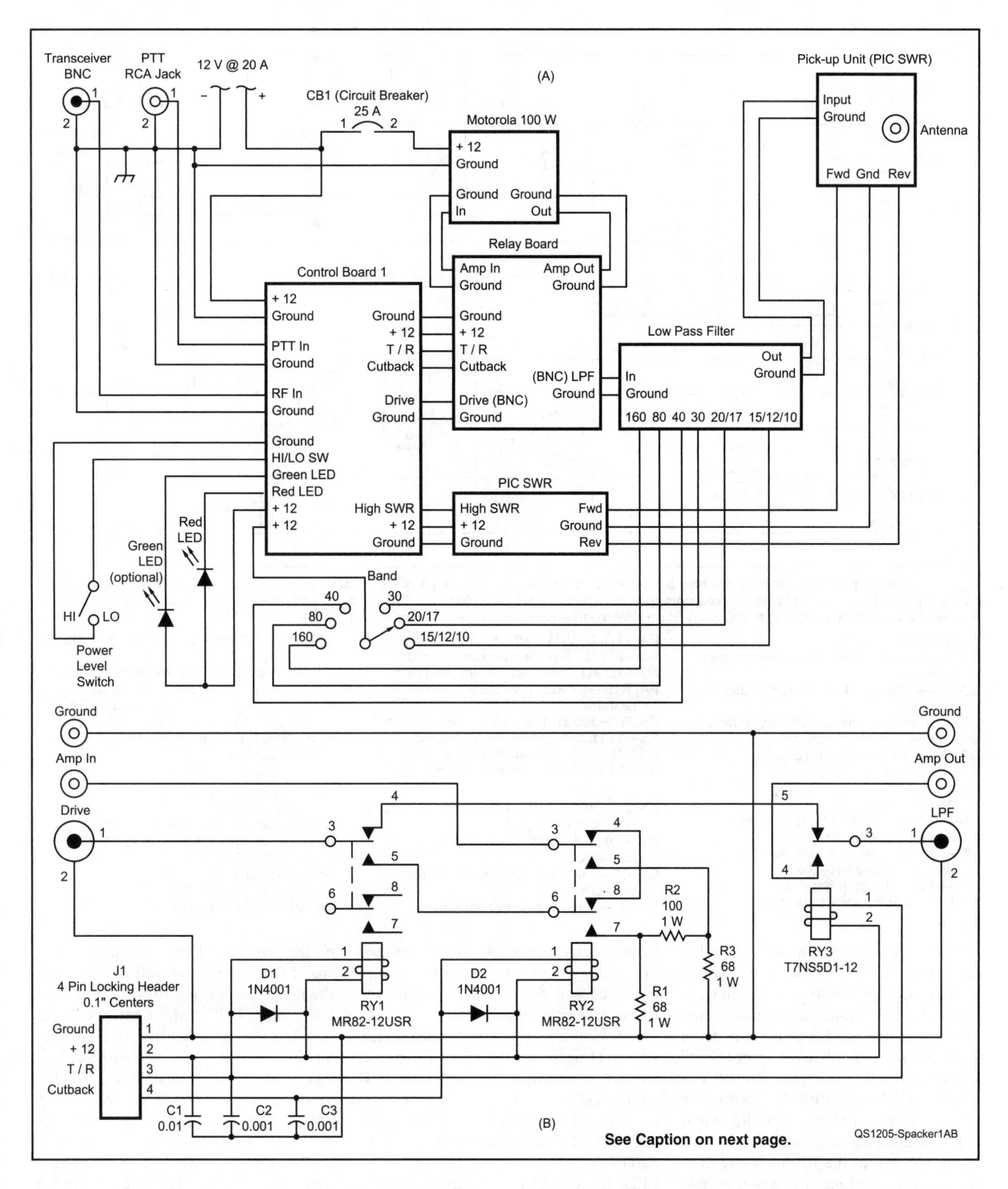

the RF power transistors cool during typical use. With 5 W input, the output can be as high as 100 W. The amplifier is not intended for 100% duty cycle operation, but would tolerate the extra heat with an added fan.

RF coming into the SuperPacker goes directly to the control board. The control board has a very sensitive VOX circuit. The VOX circuit activates TR switching relays (on the relay board) at the start of transmission. You can, however, use the push-to-talk (PTT) input instead. Just ground the PTT input to transmit. The PTT line draws about 20 mA of current. PTT control of the amplifier speeds up TR switching and is ideally suited for contest operation. A jumper (JP1) on the control board can be used with PTT operation to eliminate the VOX delay.

In addition to driving the VOX circuitry, the input RF passes through an attenuator formed by R5 through R7 on the control board. Almost any QRP rig can be connected to the input as long as attenuator (R5 to R7) is set to provide a maximum of 2.5 W output from the at-

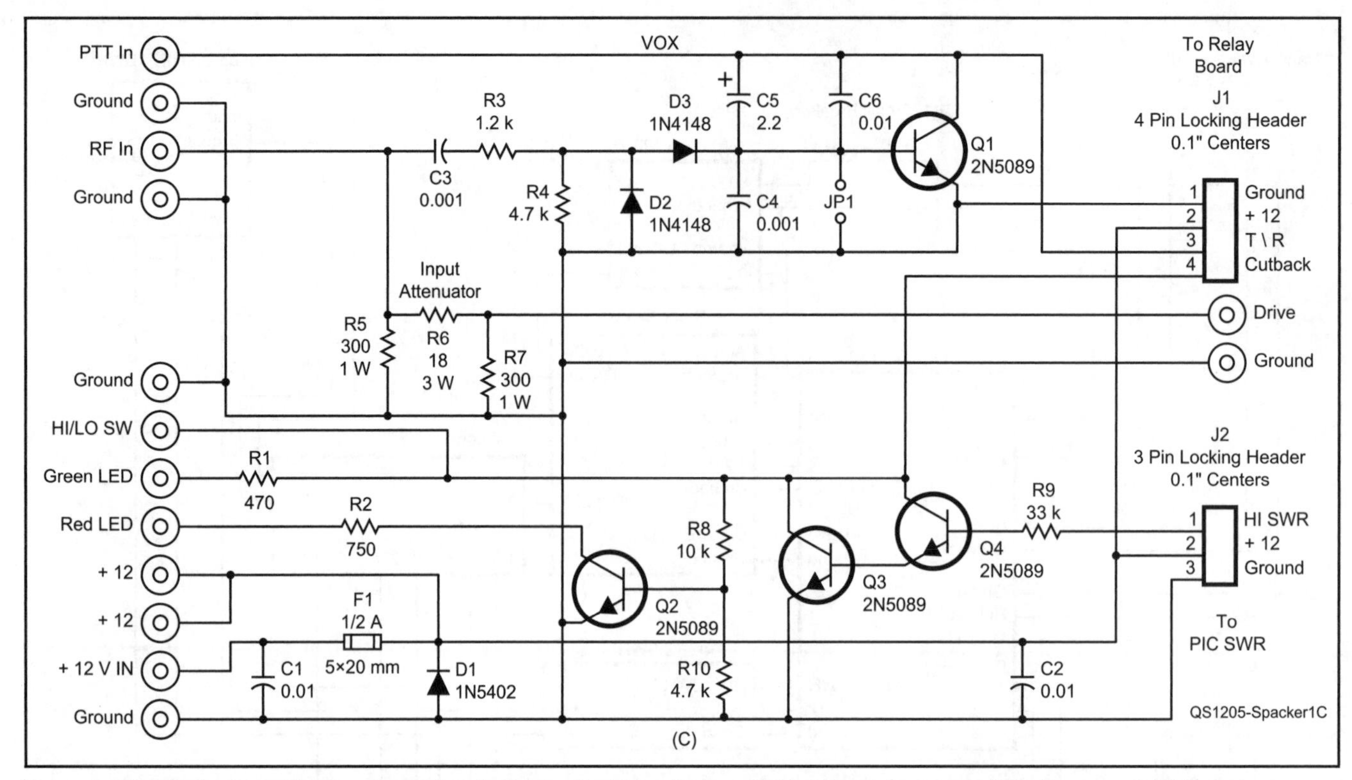

Figure 1—Detailed schematic diagram and parts list for the SuperPacker. At 1A is the interconnection diagram, at 1B the relay board and at 1C the control board. Most components are stocked by major distributors such as Digi-Key (www.digikey.com), Mouser (www.mouser.com) and Ocean State Electronics (www.oselectronics.com). Resistors are 5%.

Control board parts:
C1, C2, C6—0.01 µF, 100 V ceramic disc capacitor.
C3, C4—0.001 µF, 100 V ceramic disc capacitor.
C5—2.2 µF, 50 V electrolytic capacitor.
D1—1N5402 power rectifier.
D2, D3—1N4148 silicon rectifier.
F1—1/2 A, 5×20 mm fuse.
Holder for F1—5×20 mm fuse holder (Mouser 534-3527).
J1—4 pin, 0.1 inch locking header (Mouser 538-22-23-2041).
J2—3 pin, 0.1 inch locking header (Mouser 538-22-23-2031).
JP1—2 pin, 0.1 inch header.
Q1-Q4—2N5089 NPN transistor.
R1—470 Ω, 1/4 W carbon film resistor.
R2—750 Ω, 1/4 W carbon film resistor.
R3—1.2 kΩ, 1/4 W carbon film resistor.
R4, R10—4.7 kΩ, 1/4 W carbon film resistor.
R5, R7—300 Ω, 1 W carbon film resistor.
R6—18 Ω, 3 W carbon film resistor.
R8—10 kΩ, 1/4 W carbon film resistor.
R9—33 kΩ, 1/4 W carbon film resistor.

Relay board parts:
BNC1, BNC2—BNC connector, PC mount (Mouser 161-9317).
C1—0.01 µF, 100 V ceramic disc capacitor.
C2, C3—0.001 µF, 100 V ceramic disc capacitor.
D1, D2—1N4001 power rectifier.
J1—4 pin, 0.1 inch locking header (Mouser 538-22-23-2041).
R1, R3—68 Ω, 1/4 W carbon film resistor.
R2—100 Ω, 1 W carbon film resistor.
RY1, RY2—DPDT 12 V miniature relay (Mouser 551-MR82-12USR).
RY3—SPDT 12 V relay (Mouser 655-T7NS5D1-12).

Other parts:
30 A Anderson PowerPole connectors, 2 red, 2 black.
Bracket set for PowerPole, 1462G1.
25 A circuit breaker ETA 1610-92-25A (Digi-Key 302-1245).
Socket for circuit breaker (Mouser 441-R347B).
Enclosure (Ten-Tec TG-38).

tenuator. The values shown on the schematic provide 3 dB of attenuation, which is suitable for a 5 W transceiver like the Yaesu FT-817. Both the attenuator and VOX circuitry are taken directly from K5OOR's "HF Packer-Amplifier" project, which is a refinement of *The 2001 ARRL Handbook* design by WA2EBY. See K5OOR's Web site (**www.hfprojects.com**) for details.

In addition to the RF input circuitry, the control board has a power fuse, and the dc amplifier for power cutback. When the PIC SWR meter detects a high SWR condition, it drives the base of Q4 through R9. Q3 and Q4 form a Darlington pair amplifier for activating relay RY2 on the relay board. RY2 adds R8 through R10, another π-section attenuator, in series with the power amplifier board's input. R8 through R10 introduce 10 dB of attenuation, corresponding to a 10:1 reduction in drive power. These resistor values are fixed and do not depend upon the transceiver being used.

Looking at the control board again, when Q3 activates the relay for power cutback it also lights up the green LED circuit. Q2 acts as an inverter and lights the red LED when Q3 is off (no power cutback). The green and red LEDs can be part of a single common-anode red-green LED, separate LEDs, or you can use just one color (as we did). R1 and R2 determine the brightness of the green and red LEDs, respectively. The schematic and parts list are shown in Figure 1.

Construction and Testing—The Control Board

The control board is simple to build and easy to test. Wire all the connections on the left side of the schematic, and a coax cable to the DRIVE output. This cable should be made using RG-174 miniature coax with a BNC male connector on the end. For now, connect the BNC connector through a low power wattmeter to a 50 Ω dummy load. The dummy load only needs to handle about 5 W. Two 100 Ω, 3 W resistors in parallel (of the same type as R6) can be used.

Check to see if the full power of your transceiver remains below 2.5 W at the wattmeter. Attach a 1 kΩ resistor between your 12 V power supply and the PTT IN connection. Connect a high-impedance voltmeter between ground and the PTT IN connection. A very small amount of RF from the transceiver should make the voltage at PTT IN drop from 12 V to nearly zero. This test verifies the operation of the VOX circuitry. Remove the 1 kΩ re-

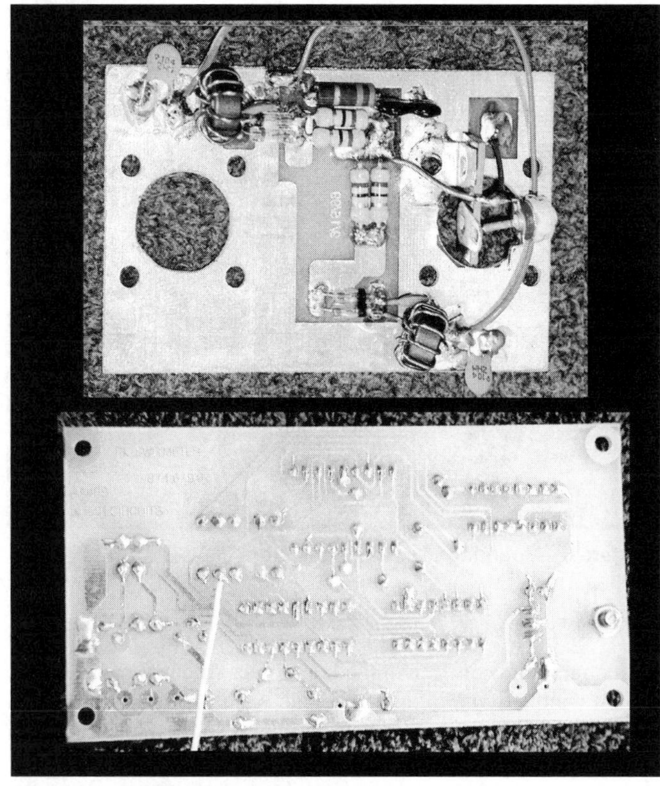

Figure 2—Modified PIC SWR boards.

Figure 3—Amplifier board with major portions assembled.

Figure 4—C6 on amplifier board.

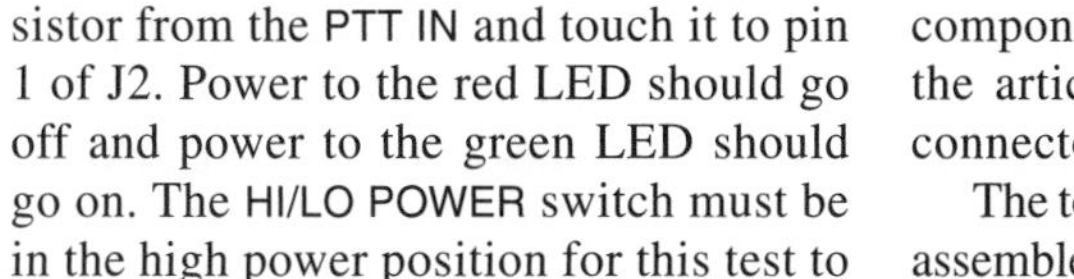

sistor from the PTT IN and touch it to pin 1 of J2. Power to the red LED should go off and power to the green LED should go on. The HI/LO POWER switch must be in the high power position for this test to work.

PIC SWR Boards

The PIC SWR requires two printed circuit boards. The main PIC SWR board has all the control and display circuitry. The pick-up unit has the RF sensing circuit. Build the main PIC SWR board exactly as described in the original article. Then, add a wire to the board by soldering it to the side of R6 that connects to the PIC (lower board in Figure 2). This wire will be the HIGH SWR signal. The HIGH SWR signal requires using a revised version of the *PIC SWR* program, so be sure to order the PIC chip programmed for the SuperPacker.[5] The original PIC chip (16C71) is out of production, but the replacement chip (16C711) is widely available. The new program supports the 16C711 and is fully backward compatible with the original PIC SWR circuit. The source code and Hex files for the PIC program can be downloaded from the ARRL Web site.[6]

Building the PIC SWR pick-up board for the SuperPacker is a bit different from the original article. Do not use a pick-up board purchased before May 2004; the silkscreen on the board does not match the copper traces. Mount most of the components on this board as directed in the article, but do not mount the input connector.

The top board in Figure 2 shows a partly assembled pick-up board. Look at the terminal strip (upper right). The terminal strip replaces the INPUT connector (J1). It is used to support C10 and the input end of the coaxial cable (not yet on the board). The mounting scheme differs from the original article.

Low-Pass Filter Board

The SuperPacker uses the LPF board from the FARA project. It works well, but the coils can get warm when filtering a 100 W amplifier. We redesigned the circuit values with larger toroid cores. Table 1 shows the new values. The cores are available through Amidon Associates.[7] Testing at the ARRL Laboratory verified that the new coil set provides excellent harmonic suppression.

Amplifier Board

The most critical board is the amplifier board. Most of its parts, including a high-quality circuit board, can be purchased from CCI. The board is supplied with the original Motorola Applications Note and some additional helpful instructions. Build the board with care. There are a few surface mount capacitors. If this is your first try at surface mount, see the *QST* articles by Sam Ulbing, N4UAU, for detailed instructions.[8] Figure 3 shows the assembled board without transistors. Figure 4 will help with the placement of C6, made from two surface mount capacitors and a regular silver mica capacitor.

The amplifier board was designed for use with IRF454 transistors from Motorola, and the transistors must be a matched pair. CCI will sell you a matched pair of transistors. The other option is to use the less expensive Toshiba 2SC2290 transistors, available from RF Parts Company as matched pairs.[9] These have about 2 or 3 dB less gain and slightly thicker leads. The lower gain requires a change to the input attenuator. In fact, the transistors we tested worked with a Yaesu FT-817 running 5 W without any attenuation (R5 and R7 removed, R6 shorted). The thicker leads require a small modification to the printed circuit board.

Because the amplifier board mounts on the inside of the case, and the heat sink is on the outside side of the case, the transistors must be mounted lower than usual. To accomplish this deeper position the transistor leads are bent upward (Figures 5 and 6); see instructions below. You must enlarge the holes in the amplifier board to let the 2SC2290 power transistors pass through freely. We did not need to enlarge the holes for the IRF454 transistors. Transistor mounting is the most critical part of assembly, and mounting will go smoothly only if the heat sink is drilled

Table 1
New Capacitor and Coil Values for the FARA LPF

Band	*Turns (L1/L2)*	*C1/C3 (pF)*	*C2 (pF)*	*Core*	*AWG*	*Wire length (inches)*
160(A)	26	1500	2700	T80-2	24	24
80(B)	20	910	1600	T80-2	24	18
40(C)	15	470	820	T80-2	24	14
30(D)	14	360	600	T80-6	22	14
20-17(E)	9	180	360	T80-6	22	10
15-10(F)	8	91	180	T80-10	22	8

Figure 6—Mounting of power transistors.

Figure 5—Detail of bending of the power transistor connections.

and tapped with care. The power cable and both coaxial cables must be soldered to the amplifier board before it is mounted. RG-174 type small coaxial cable is used for the input circuit. RG-58 type cable is used for the output (see Figure 7). The cables can be cut to length after the board is mounted.

Wrapping Up

Finishing the enclosure is relatively simple and will not require any special tools to result in a clean, professional look. The panel labels were made following Bill Sepulveda's, K5LN, instructions in his December 2002 *QST* article.[10]

Assembling the Top of the Chassis

Final assembly begins with mounting the heat sink and relay board on the top cover, and then mounting the amplifier board onto the heat sink. There should be holes (cut-outs) in the top cover that line up with the power transistors and with diode D1 on the CCI amplifier board. These holes expose the heat sink so the transistors and diode can be screwed directly to it. Wire both coaxial cables and the power cables to the amplifier board prior to mounting the board to the top cover. Number 8 nuts are perfect spacers for mounting.

Inspect the space under that board to make sure nothing from the board touches ground except for diode D1. D1 is held in place against the heat sink with a 4-40 screw. Instead of having the screw head push down on the circuit board, we enlarged the screw hole in the board and used a screw with a small head. The screw pushes directly on D1 and holds it firmly to the heat sink. You must put silicone thermal grease (heat sink compound) on this diode for good thermal conductivity with the heat sink. Now mount the board as shown in Figure 7.

The next steps are the most critical for the project. Look at the two power transistors. The lead (gold tab) that is cut off at an angle on each transistor is the collector. It must face in the direction of the large output transformer (see Figure 6). Very carefully bend all four leads toward the top of the transistor as shown in Figure 5. Make this bend as sharp as possible without stressing the leads or the case of the transistor. Too much stress will break the sealing compound or crack the case. Put the transistor into place through the cutout of the printed circuit board. Do not use any heat sink compound yet.

Slowly screw the transistor into place. The oval metal bottom of the transistor must press *firmly* against the heat sink. If the tapped holes in the heat sink are not lined up properly, it is possible to bend a lead too sharply and damage the transistor. If the transistors line up properly, bend the leads flat onto the top of the circuit board where they will eventually be soldered. Do not solder them yet!

Now, remove the transistors. Use your soldering iron and a touch of solder to coat the solder pads of the circuit board where the leads of the transistors will go. Coat the bottom of each transistor with a thin layer of heat sink compound. Mount the transistors again making sure the collector is still pointed in the proper direction. Tighten the mounting screws firmly.

Follow this procedure for soldering each transistor lead: Press the transistor lead to the circuit board with your hot soldering iron. Apply solder to the iron and then the lead to get solder flowing on the top of the lead. Continue to heat the lead for a few moments until solder flows on the circuit board. Let go of the solder and grab a flat blade screwdriver. Press the lead down with the screwdriver and remove the hot iron. Hold the screwdriver in place until the solder cools. You can breathe now. For safety, make sure the bias potentiometer is set fully clockwise before going on to the next assembly step.

Assembling the Bottom of the Chassis

Mount all the rear-panel parts except the 12 V power connector (see Figures 8 and 9). The rear-panel parts include the PIC SWR pick-up board. Hold the main PIC SWR circuit board in place on the front panel and measure and cut all the wires for it. Three wires go to the pick-up board and three wires go to a connector that will plug into the control board. Wind the three pick-up board wires twice through a T37-43 toroid before connecting them. Before mounting the PIC SWR main board you might want to assemble the test circuit (Figure 4 of the original article) and use it to adjust R1 and R2. Otherwise, you will have to unmount the PIC SWR main board later to adjust R1 and R2.

Once you have mounted the PIC SWR board, mount the HI/LO POWER switch and mount one or both LEDs. Don't mount either LED too close to the top of the case. You will need clearance for a connector on the relay board during final assembly.

Assemble the rear panel 12 V power connector next. Cut one of the in-line fuse socket wires to 4 inches and strip off 1/2 inch of insulation. Strip 1/2 inch from some red

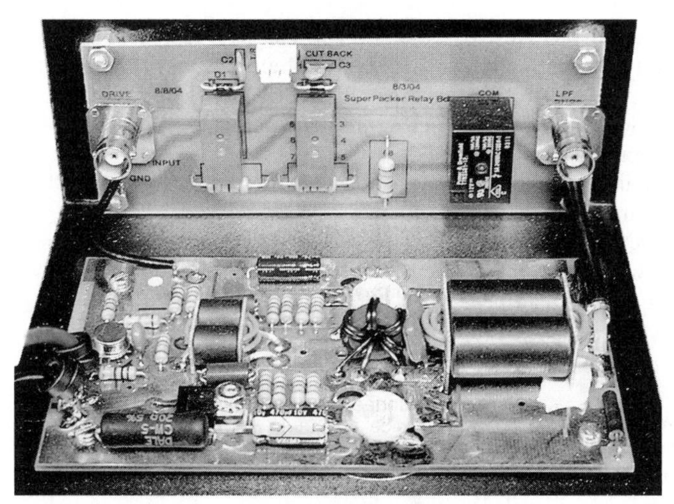

Figure 7—Inside of the top of the amplifier box.

Figure 8—Inside of the bottom of the amplifier box.

20 gauge stranded wire and twist the red wire and the larger fuse socket wire together. Tin, trim, and solder them to the PowerPole connector pin. Solder some 20 gauge and 16 gauge black hookup wire to a PowerPole pin in the same manner. Link together a black and red PowerPole housing and snap the completed pins into the housings. Next, measure 4 inches from the free end of the in-line fuse socket. Strip and solder this end to a pin; snap it into a red housing. Trim the free end of the 16 gauge black wire so it can be assembled to a black housing linked with this red one. The fuse holder will be in the positive lead. Mount the PowerPole pair with double wires onto the rear chassis. The free black wire is connected to the chassis via a ground lug. Keep this lead very short.

Install the band switch and LPF circuit board next. Note the positions of the input and output of the board. The output of the LPF circuit board connects directly to the input terminal strip of the PIC SWR pick-up board. Use a short length of RG-58 coax cable.

The input cable is also made of RG-58, but has a BNC connector on the far end. The BNC connector must reach the LPF input on the relay board mounted to the chassis lid. Excess cable is stuffed inside the box when the lid is closed, so do not make this cable any longer than necessary. A length of 10 inches is about right. Once all the connections are wired, firmly bolt the LPF to the bottom of the chassis using 1/4 inch metal standoffs.

The control board is wired into the chassis after the LPF is installed. Orient the board near its final position. Trim and strip the wires that will solder to the board, then solder the wires into place before mounting the board with screws. A three pin cable comes from the PIC SWR board and a four wire cable goes to a connector that will plug into the relay board. A coaxial cable made with RG-174 also goes to the relay board.

Preassembly for Testing

The amplifier and relay boards on the top half of the box (Figure 7) are connected to the boards on the bottom half of the box (Figure 8) through four cables—a small coax cable and four wire cable to the control board, a larger coax cable from the LPF board and the power cable. In constructing the SuperPacker, these four cables should be kept as short as possible. Make the small coax cable and four wire control cable just long enough to lay the top half of the box along the right side of the bottom half. Cut the other two cables just long enough to assemble the two halves of the case together. You might want to make extension cables for testing. The power extension cable is made from heavy gauge wire and four Power Pole connectors. The RF extension cables are best made using a length of RG-58 cable with a male BNC connector on each end. Use a double female BNC connector (barrel connector) to convert the double male cable into an extension cable.

Adjusting and Testing the Unit

Careful adjustment and testing can save you a lot of grief in the long run, so plan to spend some time making sure everything is just right. In fact, it will probably pay off to spend some time checking all the connections and looking to see that transistors, diodes and resistors are all oriented properly in the printed circuit boards. Use an ohmmeter to confirm that the polarity of the power connections are correct from the power supply all the way to the circuit boards.

The power amplifier board has one potentiometer that adjusts the idling current of the amplifier transistors. We recommend an idle current of 240 to 300 mA.

Remove the fuse from the control board so power to the SuperPacker energizes only the amplifier circuit board. Place a digital ammeter or multimeter in line with your power supply. It must be able to accurately read 350 mA (0.35 A). Follow the directions that come with the amplifier circuit board.

Check out the TR switch operation next. Remove the 25 A circuit breaker and put the control board fuse back in place. Power up the SuperPacker and connect your transceiver to the input (BNC) connector. Connect the output of the SuperPacker to a dummy load. The dummy load only needs to tolerate 5 W for this test. Transmit a signal and listen for the antenna relays to switch. Convince yourself that the relays switch on with the start of speech and off again a moment after you stop talking. If you are unsure about the relay sound, try grounding the PTT input—that should also switch the relays.

Now we will verify the LPF switching and wiring. Leave the transceiver in receive mode and connect an antenna to the SuperPacker. Listen on 40 meters. Tune in a station, if possible. Switch the band switch on the SuperPacker to 40 meters, then to 80 meters. The signals received on 40 meters should be attenuated when the SuperPacker is set to 80 meters or 160 meters. Try each band, and verify that when the SuperPacker band switch matches your transceiver band, there is little or no noticeable loss of signal strength. When the SuperPacker is set to any band below the transceiver's frequency, there should be an appreciable loss of signal. These tests help verify the LPF board is working properly.

If you want to test the LPF board more completely, disconnect the LPF board from the Relay board. Connect a 100 W rig to the LPF board; if your rig does not have a

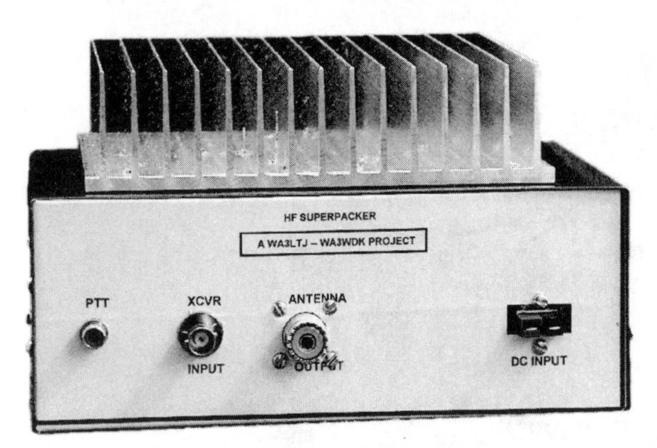

Figure 9—Amplifier rear panel.

built-in SWR meter, place one between the rig and the LPF board. Connect the output connector of the SuperPacker to a 50 Ω dummy load. Using between 20 and 50 W, check the SWR reading on each band. A high SWR on any band indicates a problem with that section of the LPF. The most common fault is not energizing the relays for a given band.

The PIC SWR may be calibrated using the same test arrangement that was used for the LPF. Refer to the original PIC SWR meter article for the calibration steps.

The PIC SWR meter is a digital circuit that generates its own high frequency signals. Sometimes these spurious signals can be heard in the transceiver connected to the SuperPacker. Check for a "birdie" at 7.250 MHz. If you can detect the signal, try reducing it by moving the wires that come off the main board of the PIC SWR meter. If the noise is strong, make sure you wrapped the leads around a ferrite core (see above), and then try adding a ground wire from the PIC SWR board to the LPF board or control board. We were able to reduce the birdie to near the background noise level.

Final Assembly and Operation

The top section connects to the bottom section with two coaxial cables, the power cable, and one four wire control cable. Make sure none of these wires gets trapped or crimped as you gently lower the top of the box into place. Make sure the LED doesn't interfere with the coaxial connector just above it. Install the four case screws. That's it—you're ready for operation!

Connect the amplifier input to your QRP rig, the output to a suitable dummy load and the power to a 12 V supply capable of at least 20 A. Select your favorite band both on the QRP rig and on the SuperPacker band switch. Switch the HI/LO POWER switch to HI. The SWR display will indicate LP (power too low for a reading) until you transmit. Try transmitting a carrier. The amplifier should switch from receive to transmit automatically. If all is well, the digital display will indicate an SWR reading close to 1.0 (assuming a good 50 Ω dummy load). If you have a power meter in line with the dummy load, it should read between 80 and 110 W when the drive is maximum.

Switch the amplifier to LO and transmit again. The red LED should turn off (the green LED will turn on, if you have one installed) and the output power will drop to 10 or 20 W. Check every band on high power to make sure the amplifier is working properly. You can safely test the automatic power cutback circuitry using an antenna tuner and dummy load as described earlier. When the SWR is above 4:1, the red LED will turn off and power will drop until six seconds after the SWR returns to a safe level. If everything is working properly you are ready to go on the air. The amplifier is almost indestructible. (There is one danger, however. If you drive the amplifier with much over 10 W, you will burn out the power transistors.)

If you use the SuperPacker for contesting or other rapid operations, the carrier operated TR switch can be circumvented with the PTT input. The PTT input operates by shorting it to ground for transmit. Install jumper J1 on the control board to eliminate the VOX delay.

The SuperPacker has a HI/LO POWER switch, but it has no provision to run the transceiver straight through to the antenna (i.e., "barefoot"). Disconnecting the power supply disconnects the antenna, because the LPF is in the signal path during receive and the LPF relays are all de-energized without power. You could wire the LPF between the amplifier board output and the relay board. With the power off, barefoot operation is limited only by the input attenuator. If the input attenuator is placed between the amplifier board input and the relay board, this limitation is also lifted. If you are not using much input attenuation, you might simply connect jumper J1 to a switch. The switch will inhibit carrier-operated TR switching and let you run barefoot. Power must be on to the SuperPacker and the band switch must be set correctly.

At the time of this writing the FCC still bans the manufacture and sale of HF amplifiers for low power radios. Whether or not the ban is lifted, it makes sense to build your own. Now that you have an easy, robust and well-tested project to work from, there is every reason to sharpen those drill bits, warm up the soldering iron and get started!

Notes

[1]Circuit board, power transistors, most other parts, and helpful construction hints for the Motorola AN762 are available from Communication Concepts Inc, 508 Millstone Dr, Beavercreek, OH 45434, tel 937-426-8600, **cci.dayton@pobox.com**.

[2]J. Valdes, WA1GPO, "The FARA HF Project," *QST*, Jun 2003, pp 35-39.

[3]The following SuperPacker parts are available for $48 from FAR Circuits, 18N640 Field Ct, Dundee, IL 60118-9269, tel 847-836-9148 (voice and fax), **www.farcircuits.net**—the FARA low-pass filter board, the control and relay PC boards, the two PIC SWR boards with programmed PIC, crystal, 7445, 7805 and LED display.

[4]B. Kelley, AA4FB, "A PIC SWR Meter," *QST*, Dec 1999, pp 40-43.

[5]See Note 3.

[6]Drill templates, panel drawings and other construction information is available in SuperPacker.ZIP, and the updated *PIC SWR* software code (.ASM and .HEX) is available in PICSWRS.ZIP at **www.arrl.org/files/qst-binaries**.

[7]Amidon Associates Inc, 240 Briggs Ave, Costa Mesa, CA 92626, tel 800-898-1883, **www.amidon-inductive.com**.

[8]S. Ulbing, N4UAU, "Surface Mount Technology—You Can Work With It!" Parts 1 to 4, *QST*, Apr 1999, pp 33-39; May 1999, pp 48-50; Jun 1999, pp 34-36; Jul 1999, pp 38-41.

[9]Matched pairs of 2SC2290 transistors are available through RF Parts Company, 435 S Pacific St, San Marcos, CA, 92078, tel 760-744-0700, **www.rfparts.com**. If you use these transistors select R5-R7 on the control board for 4 to 5 W of drive to the amplifier board. That is, use about 2 to 3 dB less attenuation on the input.

[10]B. Sepulveda, K5LN, "Panel Layout with Microsoft PowerPoint," *QST*, Dec 2002, p 61.

Jonathan Gottlieb, WA3WDK, has been a licensed ham since 1973. After taking a 25-year break, he became active in ham radio again in 2001. He operates DX phone at home and loves to take a QRP rig camping or traveling. Over the past few years he has been bitten by the homebrew bug and the SuperPacker was an outgrowth of this newfound love. Jonathan is a practicing lawyer, which he says qualifies him as a test subject to see if the SuperPacker project will be easy to build. You can reach Jonathan at 9317 W Parkhill Dr, Bethesda, MD 20814; **jwg@xebec.cc**.

Andy Mitz, WA3LTJ, has been a licensed ham since 1968. He enjoys operating the HF bands and is active in public service, but is more often found in his basement shop working on a new design or restoring an old radio. Andy's technical skills span from 1920s vacuum tube receivers to current-day embedded microcontrollers. He maintains the FM Only radio collector's Web site, **www.somerset.net/arm/fm_only.html**. *In addition to a Master's degree in electrical engineering, Andy has a PhD in brain research and works as a scientist at the National Institutes of Health. You can contact Andy at 4207 Ambler Dr, Kensington, MD 20895;* **arm@gnode.org**.

Chapter 6

Antennas

By Dave Benson, K1SWL

From *QST*, March 2002

Taming the Trap Dipole

A self-supported dipole for 10/15/17 meters can be a fine thing—if it's designed right.

After our recent move from a city location to several acres of wooded bliss, it was only natural that a young man's fancy would turn to thoughts of . . . antennas! I've experimented with any number of antenna configurations over the years, but multiband operation always seemed to involve tuners used to press non-resonant wires into service. With the "clean slate" afforded me with the new location, I decided I wanted to pursue the "hook up the coax and forget it" approach. I'm also reluctant to spend my limited discretionary funds on commercial antennas when the homebrew approach works well.

One approach to a multiband dipole design is the so-called "fan dipole" wherein a separate electrical half-wavelength of wire is added in parallel at the feedpoint for each band of interest. This can become mechanically cumbersome after the first several bands and interaction between bands becomes noticeable, at least with close wire spacings. I elected instead to pursue the trap approach. This article describes the development of a self-supported 10/15/17 meter trap dipole.

This project moved from the back burner to the "gotta try it" category when I found that the local home-improvement emporium carried 8-foot lengths of $^3/_8$-inch aluminum C-channel stock. This material has one important advantage: all surfaces are flat, which eases a number of construction details. The joints between the element sections need to be an insulating material and of sufficient strength to carry the weight of the outboard sections. The ideal material for this application turned out to be $^3/_8$-inch square black Delrin (plastic) stock, which has good tensile strength properties.[1] This material is also available in sizes up to 4 inches square (at daunting prices) for applications where higher strength is required.

Figure 1 shows the dimensions of the trap antenna. The innermost dipole section (10 meters) is decoupled from the rest of the antenna by a pair of traps tuned to 28.1 MHz. The next pair of sections is decoupled from the outer wires by a pair of traps adjusted to 21.1 MHz. Although the dimensions shown are for the 10/15/17-meter bands, there's nothing to prevent you from developing other combinations.

[1]Notes appear on page 6-3.

The traps themselves are quite simple—a parallel-resonant tuned circuit adjusted to the center of each amateur band of interest. I constructed each of these from iron-powder toroidal cores and a pair of silver mica capacitors. Each trap uses two 1 kV-rated capacitors in series and T94 cores, the largest that would fit in the "low-profile" trap enclosures I chose. I used Serpac C-series enclosures available from mail-order distributors, and a number of choices are also available through RadioShack. Figure 2 shows the construction details—a pair of machine screws exits through the rear wall of the trap enclosure and passes through holes drilled through the insulator stock and the aluminum C-channel.

The traditional tool for adjusting traps has been a grid-dip meter, and this has been supplanted more recently by antenna

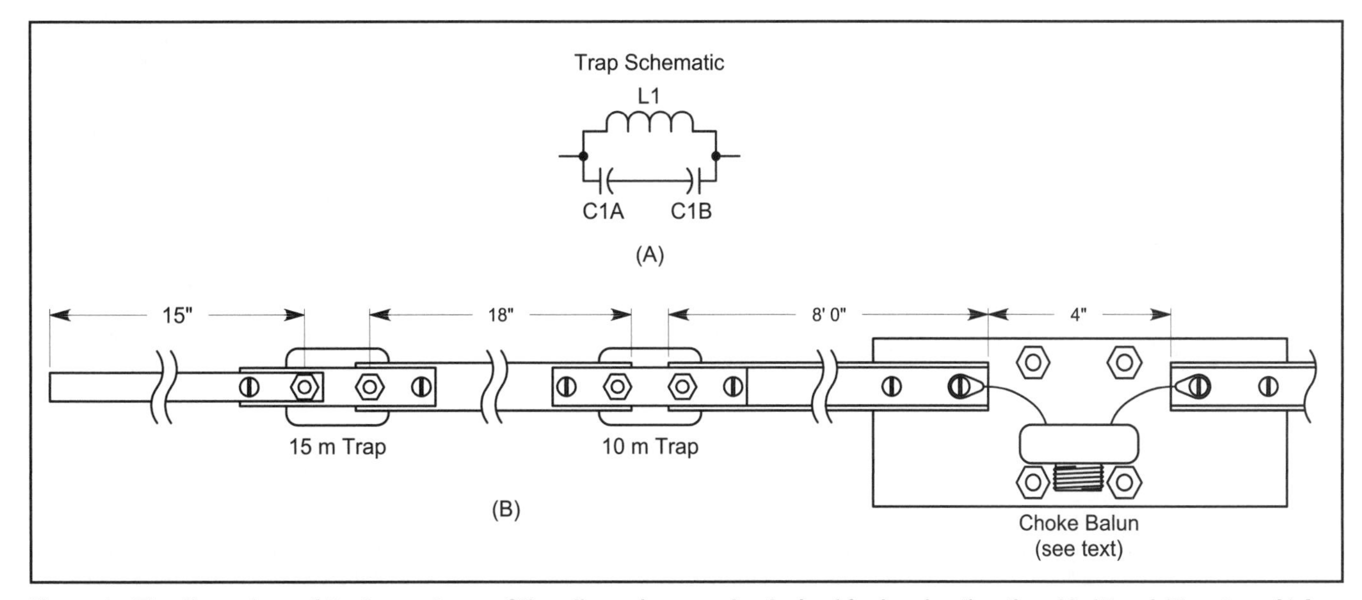

Figure 1—The dimensions of the trap antenna. Other dimensions can be devised for bands other than 10, 15 and 17 meters. At A, the schematic of the trap. At B, dimensions for one side of the dipole antenna.

C1A, C1B—100 pF, 1 kV silver mica capacitor.

L1—*10 meters*: 9 turns on a T94-10 toroidal core; *15 meters*: 11 turns on a T94-6 toroidal core. The coils must be tuned to resonance.

The Noise Bridge

Diode D1 is a source of broadband noise. This noise is amplified to useful levels by the two-stage circuit comprising Q1, Q2 and associated components. Although there's no attempt made to frequency-compensate this noise source, there's plenty of signal for our purposes—its output level ranges from S9+20 dB at 1.8 MHz to S7 at 30 MHz. In practice, when the impedances connected to points B and U are equal, this "bridge" circuit is in a balanced condition and output to the receiver is at a null. The only "tricky bit" in this circuit consists of the trifilar winding T1. [The circuit board project offering uses color-coded wire for this toroid, so hookup is pretty much foolproof.]

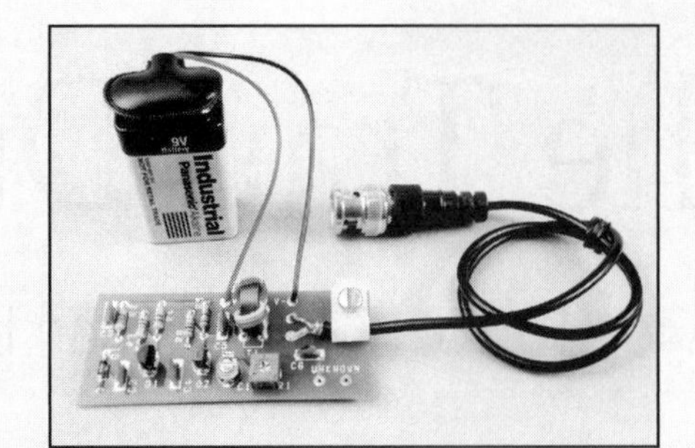

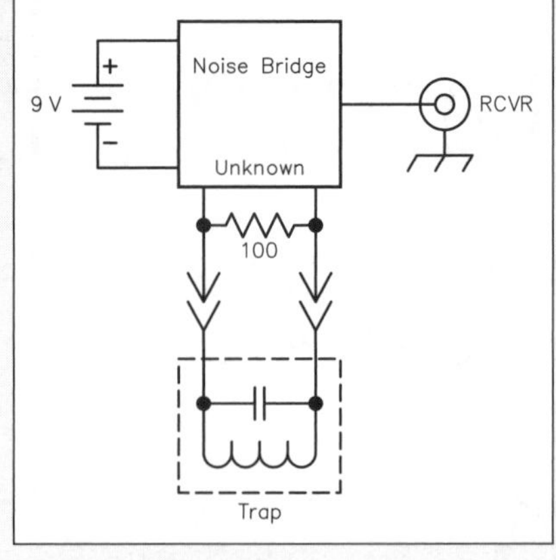

Figure B—How to hook up the noise bridge.

So Now What?

Let's put this to practical use: Connect a 100-Ω ¼ W resistor across the "unknown" terminals and connect to your receiver with a length of coax. Apply dc power (8-15 V) to the noise bridge circuit and you should hear a loud rushing noise in the receiver. Adjust control R1 for minimum S-meter indication and then C1. Once these are both adjusted carefully, the noise level in the receiver should drop to its internal noise level alone. The noise bridge is now adjusted for a null—the impedance presented by the 100-ohm resistance and stray capacitance is now balanced by the bridge's R1 and C1 settings.

Putting it all Together

If you add the trap—a parallel L-C circuit—at its resonance frequency across that 100-ohm resistor, there'd be no disturbance to the null since its impedance at the intended operating frequency is theoretically infinite. Away from the resonance frequency, the noise level will rise as the receiver is tuned off to either side. Finding the trap's resonant frequency amounts to tuning your receiver until you've located the noise null. This null will be fairly broad; however, it should be easy to locate using 1-MHz and then 100 kHz tuning steps.

Once you've found the null, bunch the toroid turns together to lower the trap resonance frequency or spread the turns apart to raise the resonance frequency. There's a fair amount of adjustment possible without resorting to changing the toroid turns count—the 21 MHz traps, for instance, could be tuned in this manner to cover a range of 19-22 MHz.

Caution

My initial attempts at repeatable resonance measurements were inconsistent—the "casual" approach using clip leads yielded well over a MHz of variation in resonance frequency at 25 MHz! It's critical to make the leads from the "unknown" terminals on the bridge to the traps as rigid as is practical. I used 2-inch lengths of no. 20 magnet wire to the 100-Ω parallel load and installed solder lugs outboard of that resistor. This allowed the traps to be added and removed with a minimum of change in stray capacitance, which affects the resonance measurement significantly. Once these precautions were taken, the measurements became reassuringly repeatable. *Note: Once these trap hookup connections are ready to go and prior to adding the traps, be sure to readjust C1 for a noise null—this effectively tunes out the test setup stray capacitance.*

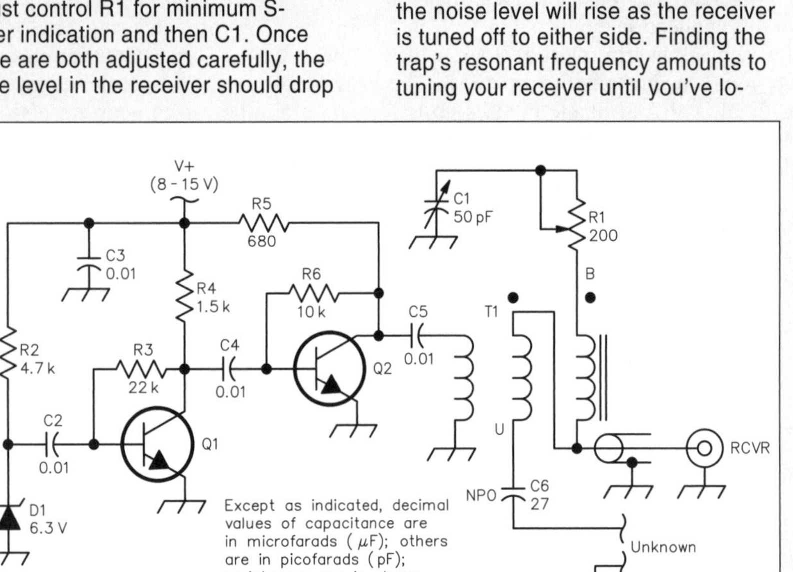

Figure A—The schematic diagram of the noise bridge, based on a design that appears in *The ARRL Antenna Book.* All resistors are 5%, ¼-W carbon composition.
D1—6.3-V, 0.5-W Zener diode, 1N753A or equiv.
Q1, Q2—High-speed NPN switch, PN2222A, 2N4401 or equiv.
T1—4 turns trifilar-wound on FT37-43 toroid; observe phasing.

analyzers. If you don't have access to either of these tools, though, despair not! If you have an HF transceiver with general coverage capability, you've already got most of what you need.

The remaining piece of equipment required is a noise bridge. Despite the arcane-sounding name, this is a simple circuit that is easily duplicated. The sidebar shows the schematic diagram for this circuit, and this is taken largely intact from *The ARRL Antenna Book.*[2] A printed circuit-board kit was developed as a club project and is available to interested builders.[3]

Antenna Adjustment

This antenna was developed by starting with the innermost (10-meter) section and working outward one band at a time. With a 4-inch spacing between the ends of the 8-foot channel sections, the 10-meter antenna simply worked on the first try. Resonance for this dipole was at 28.1 MHz and SWR characteristics were fairly broad due to the element thickness.

Upon adjustment of a pair of 10-meter traps, these were added to the element

Figure 2—Construction details of the trap. See text.

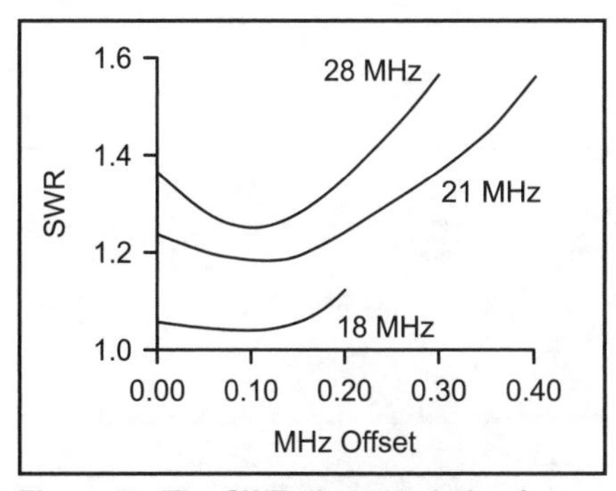

Figure 3—The SWR characteristics for the trap dipole. Since the author operates primarily CW and data modes, the lengths are optimized for the lower end of each band.

ends and outboard sections for 15 meters were added. Rather than use the C-channel material on the initial adjustments, I found it much more convenient to install outboard sections of ¼ inch aluminum rod stock. This material proved to be quite easy to trim to length with a pair of bolt-cutters! Tune-up was done at an initial height of 20 feet. Element lengths are adjusted using an SWR bridge and transmitter to determine the frequency at which SWR is minimum and adjusting accordingly. *A gentle suggestion: It's much easier to start "long" and subtract material rather the reverse!*

You'll find that the outboard lengths for each additional lower-frequency band do not meet the familiar formula for computing dipole lengths. The traps themselves present a very high impedance at their design frequency but below this frequency are inductive. This has the effect of shortening the resonant length of the antenna. [It has a modest effect in lowering feedpoint resistance as well, but is not significant within the context of this application.] With the trap components I chose, each outboard section length was shortened by 30-35% over the expected values for a dipole. For the adventurous, this length may be estimated by calculating the effective impedance of the trap at the lower band and applying it to any of several tools. This information is found in graphical form in *The ARRL Antenna Book*[4] or by use of *EZNEC*.[5]

With the length of the 15-meter section under control, I replaced the temporary rod sections with C-channel and added a pair of 15-meter traps. With the addition and adjustment of the outer 17-meter sections, this completed the design for my applications, so I elected to leave the outer antenna ends in the form of ¼-inch rod stock to reduce weight and lower the antenna's visible "profile."

The center insulator/mounting block is constructed from a ³/₈ × 3 × 12-inch block of Delrin plastic. This provides sufficient rigidity for this antenna, although if the concept is extended to lower bands you'd probably want thicker plastic material. A small plastic box at the feedpoint contains a choke balun. I constructed this using a short length of RG-174 coax looped three times through a group of six FT37-43 ferrite toroids. There's nothing magical about this approach—any of a number of other methods can be used to achieve the same goal.

Construction

All fastening hardware for the trap dipole should be of stainless steel, and toothed lock washers are needed to maintain integrity of the tightened joints. Once the traps are adjusted to the desired resonance frequencies, the trap enclosures are sealed shut with an edge-bead of model airplane cement and resonance was re-checked. This final check ensures that adding the enclosure covers has not disturbed the trap frequencies—a possibility given the tight quarters afforded by the enclosures I chose.

Results

The SWR characteristics for this antenna are shown in Figure 3. I operate primarily CW and data modes, so my interest is in the lower end of each band; the lengths in this article reflect that preference. Whatever frequency you choose, you know you've done a careful job tuning the traps if the addition of these traps and outboard sections has no effect on resonance frequency of the inner antenna portion. Their presence, though, will narrow the effective SWR bandwidth as you move away from resonance—the trap-antenna bandwidths are lower than for that of a "plain-vanilla" dipole.

Trap Losses

Although I normally operate at 5 W output or less, that's not everyone's "cup of tea." I've tested this antenna at 100 W without incident. *EZNEC* analysis using the published "Q" values for the toroid trap material shows antenna gain at 28 MHz to be 0.8 dB down from the expected free-space values, and 0.9 dB down at 21 MHz. At 18.1 MHz, the loss is approximately 0.25 dB. These values would be somewhat improved with the use of higher-Q inductors. This design has traded "compact" and "low-profile" for modest gain penalties—proof indeed of the old adage about "no free lunch."

A point of interest—I calculate the peak voltage across the traps at that power level to be over 1 kV. This is no place for junkbox capacitors of questionable pedigree! A high-quality NP0 capacitor type is a "must"—the types typically available from your local electronics emporium may be quite lossy at high frequencies, and this will translate into considerable component heating and disappointing performance. The 500-V silver mica capacitors available from the large distributors are sufficient for lower-power (QRP) operation.[6]

I installed this antenna at the 35-foot level above my roof and have been very pleased with its performance. After years of "low-profile" QRP operation, my success rate snagging contacts on the first call has improved markedly. To a large extent, the old maxim of "Put it up high and in the clear" applies here! As a final "food-for-thought" consideration, the trap-construction scheme I've described would lend itself nicely to multiband vertical and ground-plane antennas.

Acknowledgments

Special thanks to Seabury Lyon, AA1MY, for his assistance with the noise bridge project.

Notes

[1]Delrin plastic may be purchased in small quantities from McMaster-Carr, **www.mcmaster.com**; see "raw materials."

[2]*The ARRL Antenna Book*, 19th Ed., p 27-24.

[3]A noise bridge kit consisting of double-sided/silkscreened printed-circuit board, on-board parts and RG-174/U cable with BNC connector and instructions is available from the New England QRP Club for $17 ($20 overseas) postpaid. Checks or money orders payable to S. Lyon, AA1MY, 99 Sparrowhawk Mtn Rd, Bethel, ME 04217.

[4]Dean Straw, N6BV, Ed., *The ARRL Antenna Book*, 19th Ed., p 6-28.

[5]*EZNEC* is available from Roy Lewallen, W7EL, **www.eznec.com**.

[6]Toroids are available from Amidon Associates (tel 714-850-4660) or Palomar Engineers, **www.palomar-engineeers.com**. 1-kV silver mica capacitors are available from RF Parts Co (**www.rfparts.com**; tel 800-737-2787).

Dave Benson, K1SWL, is a frequent contributor to QST. *He can be reached at* **dave@smallwonderlabs.com**.

By Jim Valdes, WA1GPO

From *QST*, December 2004

The FARApole

A portable HF antenna for 6 through 20 meters that's ideal for toting.

The Falmouth Amateur Radio Association (FARA) is one of the largest and most active Amateur Radio groups on Cape Cod, Massachusetts. The group has a number of amateurs who enjoy the construction phase of our hobby—they are affectionately known as "hackers." Several of FARA's projects have been published and are available on the Web.[1] FARA's latest project is a low cost portable HF antenna, the FARApole—it's ideal for low power, multiband transceivers, like the Yaesu FT-817.

There are a number of commercially available HF portable antennas. Most, however, are relatively costly and they are frequency limited, factors that might discourage "casual" operating. One of our goals was to have an antenna as versatile as the FT-817 transceiver itself, and the pictures illustrate the flexibility of that design.

The antenna is compact; it will easily fit in a suitcase along with a UHF-style magnet mount for weekend getaways. Although primarily intended for low power operation, power levels up to the 100 W level can be tolerated when it is operated as a dipole or in a mobile configuration, provided the radio is located a satisfactory distance from the transmitter. This is necessary for safety and to minimize RF feedback at the

[1]Notes appear on page 6-6.

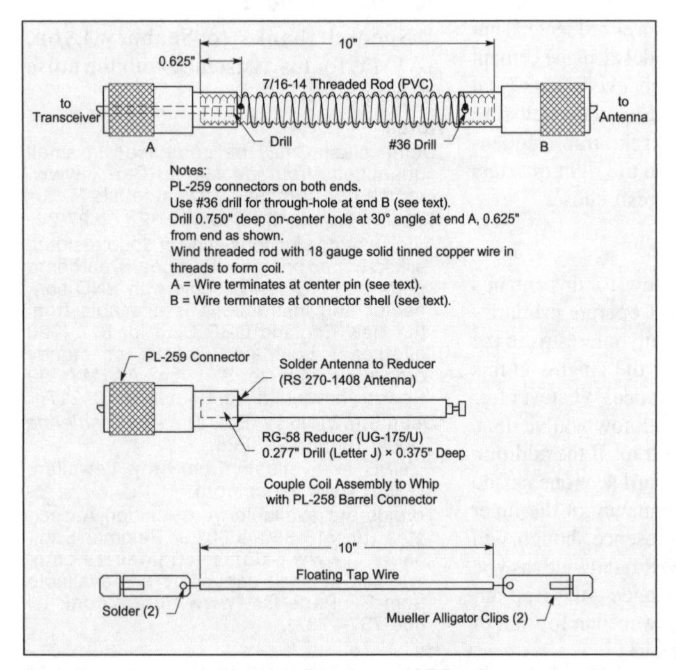

Figure 1—Construction details of the FARApole antenna. Note that the coil connects to the PL-259 center pin on one end, but to the connector shell at the other end. Construction of a dipole requires two "B" terminations for one element (see text).

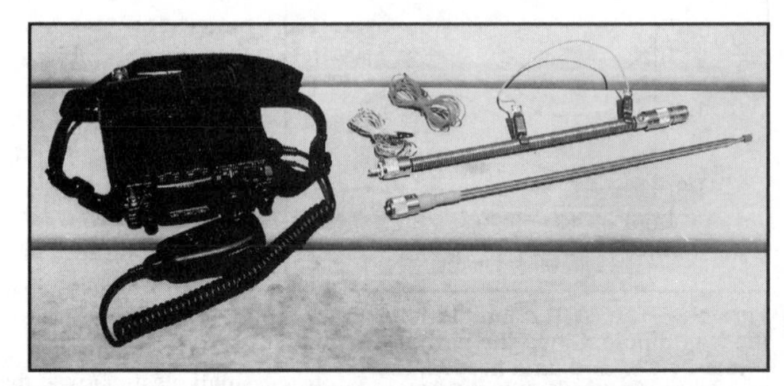

Figure 2—The completed FARApole antenna, dismantled, but ready for installation. The wire bundles are counterpoise radials that will be attached to the transceiver's ground terminal. Note the right-angle UHF connector at the rear of the transceiver. This will ensure that the vertical antenna is, indeed, vertical.

Figure 3—The antenna assembled and mounted on the transceiver. The counterpoise radials can be seen running to the left and right of the transceiver.

100 W power level. The operator is cautioned to observe recommended safe RF exposure limits. [This is always prudent when operating close to an antenna at moderate power levels. An excellent ARRL reference text, *RF Exposure and You*, contains effective safe guidelines for operation at various power levels with respect to frequency and distance.[2]—*Ed.*]

This is a portable antenna design utilizing readily available components and it is easy to construct—only simple hand tools are needed for fabrication. The antenna is base loaded with a telescoping whip and a "wandering" lead to tap a loading coil for various bands. The overall length is approximately 7 feet. With all parts on hand, it can be constructed in less than 15 minutes.

Figure 4—A dipole version of the antenna can be constructed (see text). The portable mast is made of PVC pipe, joined by a PVC coupling. The elements are mated using a UHF T connector, supported by a PVC T section that is cut in half and joined with electrical tape. One of the elements requires two "B" terminations, as outlined in the text and shown in Figure 1.

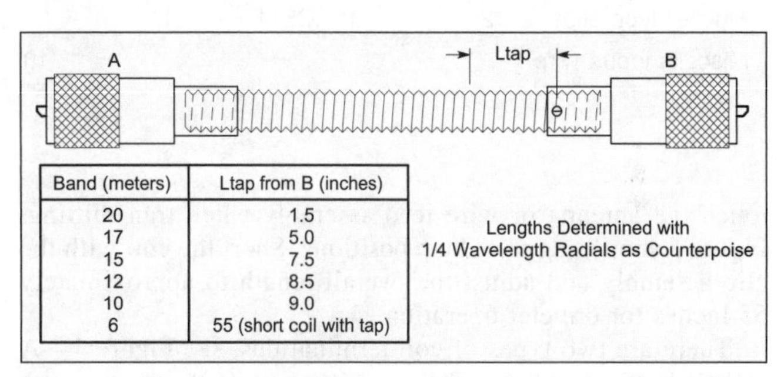

Band (meters)	Ltap from B (inches)
20	1.5
17	5.0
15	7.5
12	8.5
10	9.0
6	55 (short coil with tap)

Figure 5—The approximate tap positions for the loading coil. For 6 meter operation, the coil is shorted with the tap wire and the overall antenna length (including the tap wire) is adjusted to 55 inches.

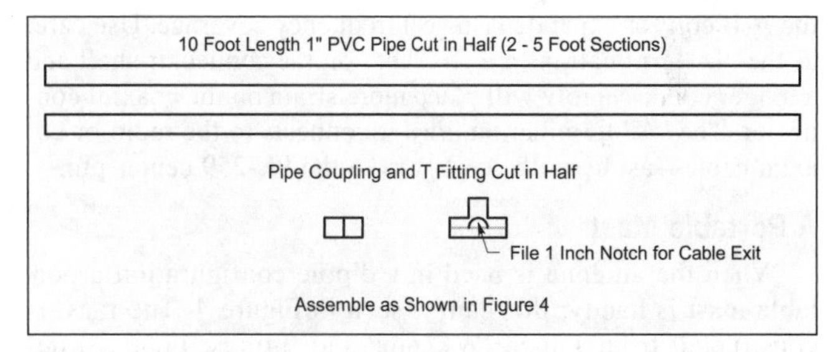

Figure 6—Details of the portable mast fabrication. The mast consists of two 5 foot PVC sections, joined with a coupling. A PVC T, cut in half, is used to support (with the help of electrical tape) the dipole version of the antenna. A notch filed in the T serves as an exit hole for the cable.

Assembly Hints

The construction of the antenna is detailed in Figure 1. Table 1 is the parts list and the completed antenna (dismantled) is shown in Figure 2, next to an FT-817 transceiver.

The loading/matching coil is wound with 18 gauge tinned solid copper wire. Stretch and straighten the wire by pulling it over the round end of a rake handle or other similar tool to remove any kinks. Temporarily install the PL-259 connectors and drill the holes in the 7/16 inch threaded PVC rod prior to starting the final assembly. Remove the connector near the 30° hole—bend one end of the wire into a "J" shape and work it through the 30° hole and out the end of the rod. Straighten the wire and reinstall the PL-259 connector.

Install the 4-40 hardware in the through-hole on the other PL-259 connector—do not tighten at this time. Carefully wind the wire onto the threaded rod using the threads as a coil form. Wind the coil as tightly as possible; this will take about 5 minutes.

At the far end of the windings, loop the wire over the 4-40 screw, pull it tight and secure the hardware. Refer to Figure 1 during the assembly. Trim the wire and solder the center pin of the first PL-259. You're almost done!

Place the RG-58 reducer in a vise and carefully enlarge the hole to accommodate the base of the telescoping element.[3] Drill very slowly as the brass fitting tends to seize and grab the drill bit. Use a 0.277 inch (letter size J) drill bit, as shown (Figure 1). Access to a small lathe is desirable, but is not essential. *Caution*—keep your fingers away!

Use a little emery paper to buff the lower end of the whip prior to assembly. A little rosen flux (don't use acid flux) will improve the solderability of the whip. Again, clamp the modified reducer in a vise, insert the whip (extend the upper sections so as not to overheat), and sweat-solder the whip with a butane torch. Install the reducer into the remaining PL-259 and secure the threads with some Loctite 242 thread-locking compound. A little shrink tubing over the assembly will enhance its appearance.

Lastly, solder the Mueller clips to the ends of the 10 inch piece of tap wire. This serves as a wandering lead to tap the coil for selecting the necessary loading in order to bring the antenna into resonance on various bands. That's it—you're done!

Operation

All loaded antennas represent some compromise—in this case it's the ease of construction. A center loaded antenna with an air-core inductor would likely be more efficient, but much more difficult (and expensive) to build. In effect, this antenna functions as a base loaded 1/4 wavelength vertical. As such, it *requires* some form of counterpoise (in this case, radials) to complete the "other half" of the dipole.

Figure 3 (a loaded 1/4 wave vertical with wire radials) and Figure 4 (a modified 1/2 wave dipole) illustrate two examples of the antenna, ready for use. The system SWR is as sensitive to the counterpoise as it is to the loading coil tap. Quarter wavelength radials using color-coded 22 gauge jacketed wire are a simple and cost effective solution to the counterpoise problem. In any case, the exact position of the coil tap must be determined experimentally, as it depends on the counterpoise placement (the position of the radials) or the use of a second element to form a dipole. Radials that are suspended on a deck or table will not have the same coil tap position as those that are placed directly on the ground.

Tune-up is simple. Place the FT-817 (or any other transceiver) in a low power mode and set the meter to the SWR position (or use an in-line SWR indicator). In FM or CW (you will need a steady carrier) alternately adjust the tap and key the transmitter until you obtain an acceptable SWR. Do not

Table 1
Parts List for the FARApole Antenna
CI=Craftech Industries, **www.craftechind.com;** RS=RadioShack, **www.radioshack.com;** M=Mouser Electronics, **www.mouser.com.**

Description	*Quantity*	*Part Number*	*Source*	*Details*
7/16-14 PVC rod	1	D200-21-27	CI	10 inch (PVC threaded rod)
4-40 pan head × 5/8	1			Standard hardware item
4-40 flat washer	1			Standard hardware item
4-40 nylock nut	1			Standard hardware item
18 gauge bus wire	1	602-296-100	M	15 feet (tinned solid copper)
PL-259 connector	3	523-83-1SP	M	UHF coaxial connector
PL-258 coupler	1	523-83-1J	M	UHF coaxial female-female coupler
UG-175/U adapter	1	523-83-175	M	RG-58 reducer for PL-259
Telescoping antenna	1	270-1408	RS	72 inch
Shrink sleeve	1			1/2 × 2 inch
Mueller-type clips	2	13AC511	M	Large jaw type
Miscellaneous wire				10 inch wandering tap lead and radials

Figure 7—The FARApole used as a stationary mobile antenna with a magnet mount.

touch the antenna or wire lead assembly when transmitting. Figure 5 shows suggested tap positions. Short the coil with the clip assembly and adjust the overall length to approximately 55 inches for 6 meter operation.

There are two types of coil terminations, see Figure 1—A and B. In the dipole configuration (Figure 4) the second element is constructed with two "B" terminations (one at each end of the loading coil). Both elements of the dipole are joined with a UHF-type T connector.

You can mix and match— use the type B-B coil in series with the A-B coil for expanded (lower) frequency coverage. Use care, as the "B" terminations are RF "hot" on the connector shell and a longer coil assembly will place more strain on the coaxial connector. The "A" termination *always* connects to the radio or coaxial cable—as, here, the coil goes to the PL-259 center pin.

A Portable Mast

When the antenna is used in a dipole configuration a portable mast is handy; this can be seen in Figure 4. The mast is constructed from 1 inch PVC pipe and fittings; Figure 6 details the mast assembly. The T fitting is cut in half. A pipe coupling is attached to one of the mast pieces and the modified T to the other. Use a UHF-type coaxial T for the center of the dipole element, as outlined earlier, and electrical tape to hold it in position on the mast. You may wish to file a notch in the pipe T fitting for the cable to exit.

A Few Comments

In fussing with various types of portable antennas, I've noticed, on occasion, some RF feedback (RF output indicated on the meter when I wasn't talking). Winding the coax (I use 30 feet of RG-174/U cable) into a 5 or 6 turn, 4 inch diameter loop as a common mode choke at the radio's antenna connector was helpful. Small clip-on RF chokes also worked. I found Yaesu's tone-encoded microphone (MH-36) to be much more susceptible to RF feedback than the standard microphone (MH-31) supplied with the transceiver.

The orientation of the Yaesu FT-817 rear panel coax connector would not permit the antenna to be in a true vertical position when mounting the antenna to a right angle (UG-646/U) connector on the back of the radio. This was due to the index locking tabs on the mating connectors. I solved this problem by grinding the tabs off on the 90° connector with a Dremel tool.[4] The antenna looks better when it is truly vertical.

Fixed mobile operation (not in motion) with a magnet mount (Figure 7) was satisfactory on the higher frequency bands, but it required a "ground clip" to the car body on the 20 meter band for an effective counterpoise.

The antenna has only seen limited use outdoors, as it was winter on Cape Cod when it was completed. A number of successful coast-to-coast contacts were made and several contacts were made on 20 meters throughout the US and South America. It's lots of fun to use—even if it is cold outside!

The Falmouth Amateur Radio Association's Web site (**www.falara.org**) maintains some FAQ files on our TekTalk forum pages. Please check the Web site for any recent developments or modifications to the FARApole antenna. Thanks to our Webmaster, K1BI, for his support.

Notes

[1]FARA projects can be seen at the FARA Web site: **www.falara.org**.

[2]Available from the ARRL Bookstore. Order no. 6621. Telephone toll-free in the US 888-277-5289, or 860-594-0355, fax 860-594-0303; **www.arrl.org/shop/**; **pubsales@arrl.org**.

[3]Be careful when clamping the reducer in a vise. Do not clamp the threads. Too much pressure will distort the cylinder and it will then be difficult to thread into the PL-259 connector.—*Ed.*

[4]A 90° UHF male-female connector with a smooth interior shoulder with no locking tabs at the male end is available commercially, although it may be difficult to find.—*Ed.*

Photos by the author

Licensed since 1962, Jim Valdes, WA1GPO, holds an Amateur Extra class license. He is active on all of the amateur bands from 160 meters through 70 cm. Jim has a BS degree in electrical engineering and has worked for the Woods Hole Oceanographic Institution for almost 30 years. You can contact him at 63 Alderberry Ln, East Falmouth, MA 02536 or at **wa1gpo@arrl.net**. QST

By Joe Everhart, N2CX

From *QST*, April 2001

The NJQRP Squirt

This reduced-size 80-meter antenna is designed for small building lots and portable use. It's a fine companion for the Warbler PSK31 transceiver.

At one time, 80 meters was one of the more highly populated amateur bands. Lately, it has become significantly less popular because much DXing has moved to the higher frequencies and many suburban lot sizes are too small to accommodate a full 130-foot, λ/2 antenna for the band. That's unfortunate, because 80 meters has lots of potential as a local-communication band—even at QRP levels. The recently published Warbler PSK31 transceiver can serve as a great facilitator for close-in QRP communication without much effort.[1] What's really needed to complement the Warbler for this purpose is an effective antenna that fits on a small suburban plot. Because PSK31 (which the Warbler uses) is reasonably effective even with weak signals, we can trade off some antenna efficiency for practicality.

What's a Ham to Do?

I investigated a number of antenna possibilities to come up with a practical solution. One intriguing candidate is the magnetic loop. Plenty of design information for this antenna is presented in *The ARRL Antenna Book* and at a number of Web sites.[2, 3] To obtain high efficiency, however, the loop must be 10 feet or more in diameter and built from 1/2-inch or larger-diameter copper pipe. The loop needs a very low-loss tuning capacitor and a means of carefully tuning it because of its inherently narrow bandwidth. Another configuration, the DCTL, may be a solution, but it's likely not very efficient.[4]

An old standby antenna I considered is the random-length wire worked against ground. If it is at least λ/4 long (a Marconi antenna) or longer, it can be reasonably efficient. Shorter lengths are likely to be several S units down in performance and almost any length end-fed wire needs a significant ground system to be effective. Of course, you may not need much of a ground with a λ/2 end-fed wire, but it's as long as a center-fed dipole.

Vertical antennas don't occupy much ground space, but suffer the same low efficiency as the end-fed wire if they are practical in size.

Probably the easiest antenna to use with good, predictable performance is the horizontal center-fed dipole. Unfortunately, as mentioned earlier, the usual 80-meter λ/2 dipole is too large for many lots. But all is not lost! The dipole can be reduced to about a quarter wavelength without much sacrifice in operation (see the sidebar, "Trade-Offs"). Furthermore, if the dipole's center is elevated and the ends lowered—resulting in an inverted **V**—it takes up even less room. This article describes just such a dipole: the NJQRP Squirt.

V for Victory

You can think of the Squirt as a 40-meter, λ/2 inverted-**V** dipole being used on 80 meters. Figure 1 is an overall sketch of the antenna; Figure 2 is a photograph of a completed Squirt prior to erection. The Squirt has two legs about 34 feet long separated by 90° with a feed line running from the center. When installed, the center of the Squirt should be at least 20 feet high, with the dipole ends tied off no lower than seven feet above ground. This low antenna height emphasizes high-angle NVIS (Near Vertical-Incidence Skyware) propagation that's ideal for 80-meter contacts ranging from next door out to 150 or 200 miles. And

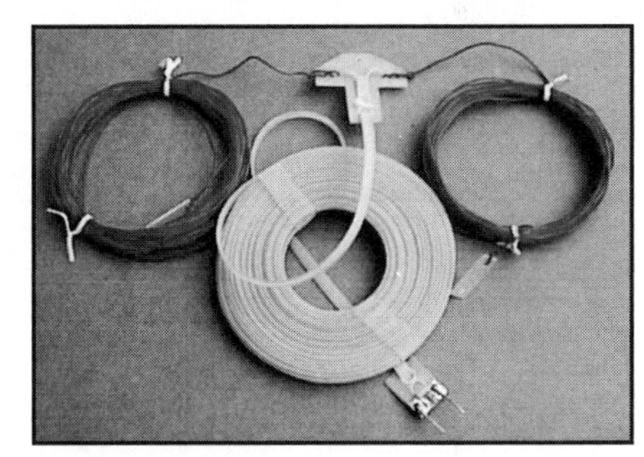

Figure 2—An assembled Squirt ready for installation.

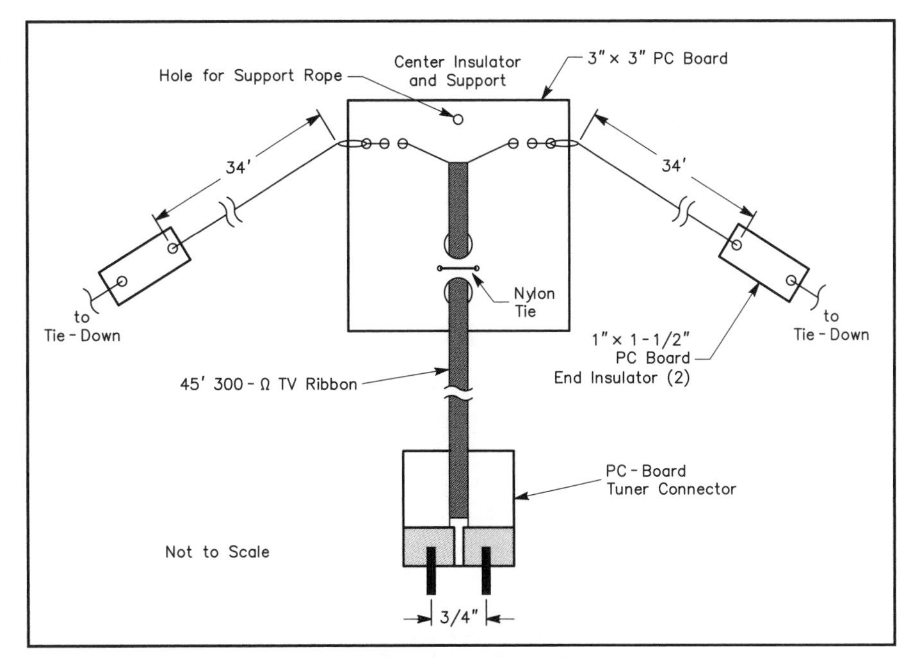

Figure 1—General construction of the 80-meter Squirt antenna.

[1]Notes appear on page 6-10.

that's where 80 meters shines! With the Squirt's center at 30 feet and its ends at seven feet, the antenna's ground footprint is only about 50 feet wide.

One nice feature of a λ/2 center-fed dipole is that its center impedance is a good match for 50- or 75-Ω coax cable (and purists usually use a balun). Ah! But the Squirt is only λ/4 long on 80 meters, so it *isn't* resonant! Its feedpoint impedance is resistively low and reactively high. This means that feeding the antenna with coax cable would create a high SWR causing significant feed-line loss. To circumvent this, we can feed the antenna with a low-loss feed line and use an antenna tuner in the shack to match the antenna system to common 50-Ω coax cable. I'll have more to say about the tuner later.

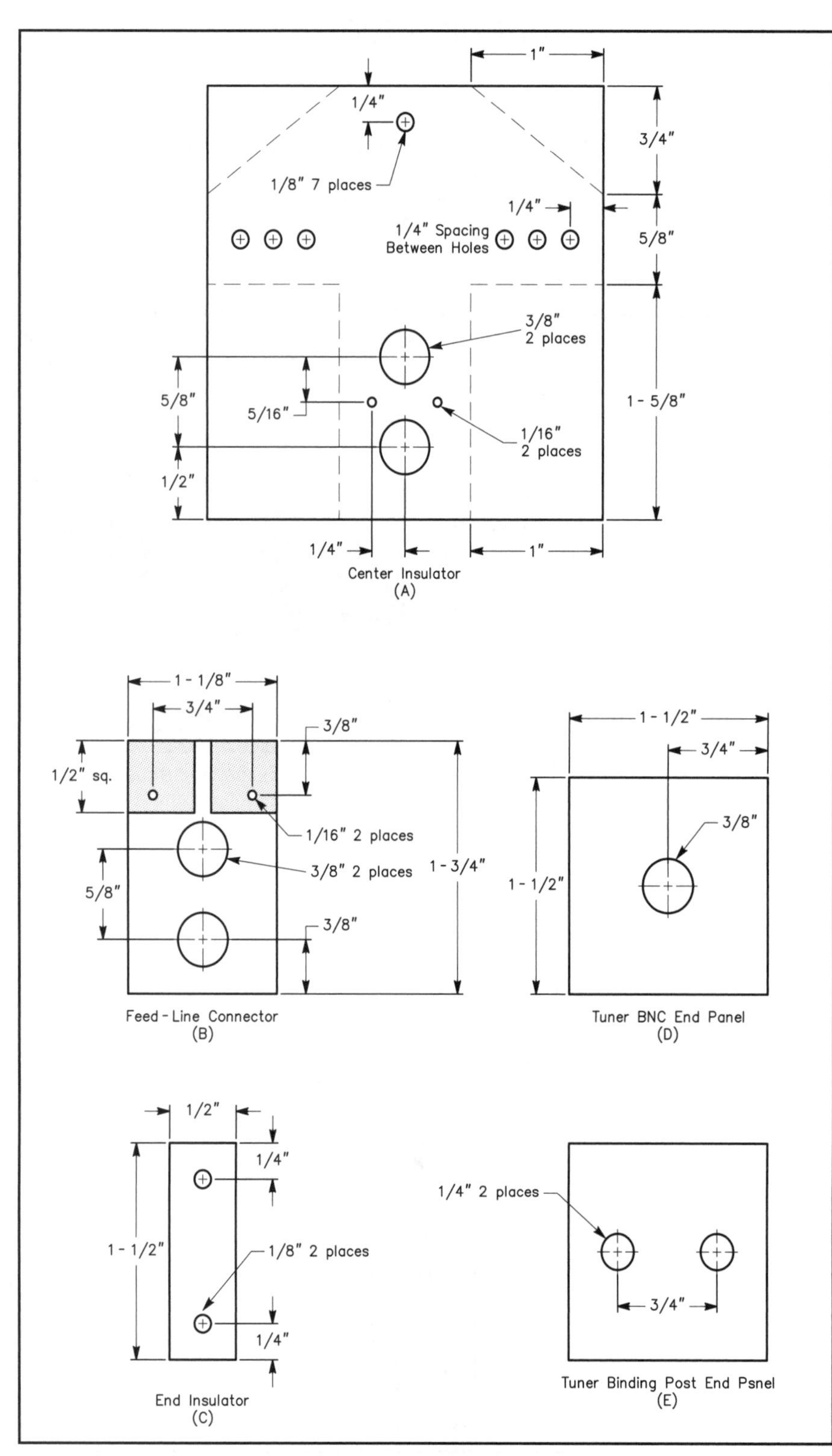

Figure 3—Hole sizes and locations for the various PC-board pieces. See Note 5.

I use 300-Ω TV flat ribbon line for the feed line. Although a better low-loss solution is to use open-wire line, that stuff is not as easy to bring into the house as is TV ribbon. Using TV ribbon sacrifices a little transmitted signal for increased convenience and availability. If you feel better using open-wire line, go for it!

Using Available Materials

It's always fun to see what you can do with junkbox stuff, and this antenna is one place to do it. See the "Parts List" for information on materials and sources.[5] For instance, the end and center insulators (see Figure 1) are made of 1/16-inch-thick scraps of glass-epoxy PC board. For the antenna elements, I use #20 or #22 insulated hookup wire. Although this wire size isn't recommended for use with fixed antennas, I find it entirely adequate for my Squirt. Because it's installed as an inverted **V** antenna, the center insulator supports most of the antenna's weight making the light-gauge wire all that's needed. The small-diameter wire has survived quite well for several years at N2CX. This is not to say, of course, that something stronger like #14 or #12 electrical house wire couldn't serve as well.

The 300-Ω TV ribbon can be purchased at many outlets including RadioShack and local hardware stores. Once again, if you want to use heavier-duty feed line, do so. The only proviso is that you may then have to trim the feeder length to be within tuning range of the Squirt's antenna tuner.

The End Insulators

I used 1/2×1 1/2-inch pieces of 1/16-inch PC board for the Squirt's two end insulators. As with everything else with the Squirt, these dimensions are not sacred; tailor them as you wish. If you use PC board for the end insulators, you have to remove the copper foil. This is easy to do once you've gotten the knack. Practice on some scraps before tackling the final product. The easiest way to remove the foil without etching it is to peel it off using a sharp hobby knife and needle-nose pliers. Carefully lift an edge of the foil at a corner of the board, grasp the foil with the pliers and slowly peel it off. You should become an expert at this in 10 or 15 minutes. Drill 1/8-inch-diameter holes at each end of each insulator for the element wires and tie-downs.

Tuner Feed-Line Connector

The tuner end of the feed line is terminated in a special connector. Because the TV-ribbon conductors aren't strong, they'll eventually suffer wear and tear.

Trade-Offs

One of the unfortunate consequences of shrinking an antenna's size is that its electrical efficiency is reduced as well. A full-size dipole is resonant with a feedpoint impedance that matches common low-impedance coax quite well. This means that most transmitter power reaches the antenna minus only 1 dB or so feed-line loss. However, when the antenna is shortened, it is no longer resonant. A *NEC-4* model for the Squirt shows that its center impedance on 80 meters is only about 10 Ω resistive, but also about 1 kΩ capacitive. This is a horrendous mismatch to 50-Ω cable, and feed-line loss increases dramatically with high SWR. The Squirt uses 300-Ω TV ribbon for the feed line with an inherently lower loss than coax. This loss is much less than if coax were used, but it's still appreciable. Calculated loss with 300-Ω transmitting feed line is about 7.7 dB (loss figures are hard to come up with for receiving TV ribbon) so the feed line used doubtless has more than that.

Although this sounds discouraging, it's *not fatal*. You have to balance losing an S unit or so of signal against not operating at all! Consider that the Squirt, even with its reduced efficiency, is still better than most mobile antennas on 40 and 80 meters. So for local communication (a low-dipole's forte), using PSK31 and the Squirt is quite practical.

If you don't already have an antenna, the Squirt's a good choice to get your feet wet when using PSK 31. Once you get hooked, you'll probably want a better antenna. If you have the room, put up a full-size dipole; you'll see the improvement right away. If you can't do that, use a lower-loss feeder on the Squirt, such as good-quality open wire.—*Joe Everhart, N2CX*

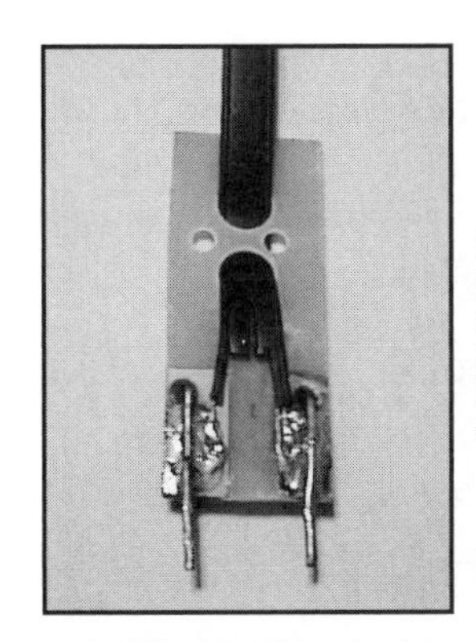

Figure 4—The pad side of the home-made feed-line-to-tuner connector.

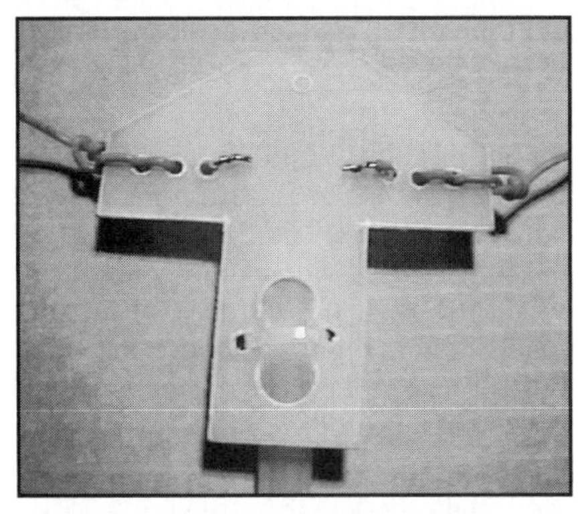

Figure 5—Here the feed-line-to-tuner connector is shown attached to the binding posts of the Squirt antenna tuner.

This connector provides needed mechanical strength and a means of easily attaching the feed line to the tuner. In addition to some PC-board material, you'll need four or five inches of #18 to #12 solid, bare wire. Refer to Figure 3 and the accompanying photographs in Figures 4 and 5 for the following steps.

Take a 1 1/8×1 3/4-inch piece of single-sided PC board and score the foil about 1/2 inch from one end; remove the 1 1/4-inch piece of foil. Now score the remaining foil so you can remove a 1/8-inch-wide strip at the center of the board, leaving two rectangular pads as shown in Figures 3B and 4. Drill two 1/16-inch-diameter holes in the copper pads spacing the holes about 3/4-inch apart. Drill two 3/8-inch holes at the connector midline about 5/8-inch apart, center to center, to pass the feed line and secure it.

Cut two pieces of #18 to #12 wire each about three inches long. Pass one wire through one of the 1/16-inch holes in the connector board and bend over about 1/4-inch of wire on the nonfoil side. Solder the wire to the pad on the opposite side and cut the wire so that about one inch of it extends beyond the connector. Repeat this procedure with the second wire. Next, strip about two inches of webbing from between the feed-line conductors and loop the feed line through the two 3/8-inch holes so that the free ends of the two conductors are on the copper-pad side. Strip each lead and solder each one to a pad. You now have a solid TV-ribbon connector that mates with the binding-post connections found on many antenna tuners. Figure 6 shows the connector mated with a Squirt tuner.

Center Insulator

Strip all the foil from this 3-inch-square piece of board. Use Figure 3A as a guide for the hole locations. The top support hole and the six wire-element holes are 1/8-inch in diameter; space the wire-element holes 1/4-inch apart. The feed-line-attachment holes are 3/8-inch diameter spaced 1/2-inch apart, center to center; the two holes alongside the feed-line-attachment holes are 1/16-inch diameter. These 1/16-inch holes accept a plastic tie to secure the feed line. I trimmed the insulator shown in Figure 2 from its original 3-inch-square shape to be more esthetic. Your artistic sense may dictate a different pattern.

Bevel all hole edges to minimize wire and feeder-insulation abrasion by the glass-epoxy material. You can do this by running a knife around each hole to remove any sharp edges.

Putting It All Together

The Squirt is simple to assemble. Once all the pieces have been fabricated, it should take no more than an hour or two to complete assembly. Begin with the center insulator. Cut each of the two element wires to a length of about 34 feet. Feed the end of one wire through the center insulator's outer hole on one side, then loop it back and twist around itself outside the insulator to secure it. Now loop it through the other two holes so that the inner end won't move from normal movement of the wire outside the insulator. Repeat the process for the other insulator/wire attachment. Separate several inches of the TV-ribbon feed-line conductors from the webbing; leave the insulation intact except for stripping about 1/2 inch from the end of each wire. Pass the TV ribbon through both 3/8-inch holes. Strip a 1/2-inch length of insulation from each dipole element, then twist each feeder wire and element lead together and solder the joints. It might be prudent also to protect the joint with some non-contaminating RTV or other sealant. Finally, loop a nylon tie through the holes alongside the feeder and tighten the tie to hold the feeder securely. A close-up of the assembled center insulator is shown in Figure 6.

Attach the end insulators to the free ends of the dipole wires by passing the wires through the insulator holes and twisting the wire ends several times to secure them.

So that the antenna/feed-line system can be tuned with the Squirt tuner, the 300-Ω feed line needs to be about 45 feet long. If you use a different tuner, you may have to make the feed line longer or shorter to be within that tuner's impedance-adjustment range.

Tuner Assembly

This tuner (see Figures 7 and 8) is about as simple as you can get. It's a basic series-tuned resonant circuit link-coupled to a coaxial feed line. At C1, I use a 20 to 200-pF mica compression trimmer acquired at a hamfest (you *do* buy parts at hamfests, don't you?), although almost any small variable capaci-

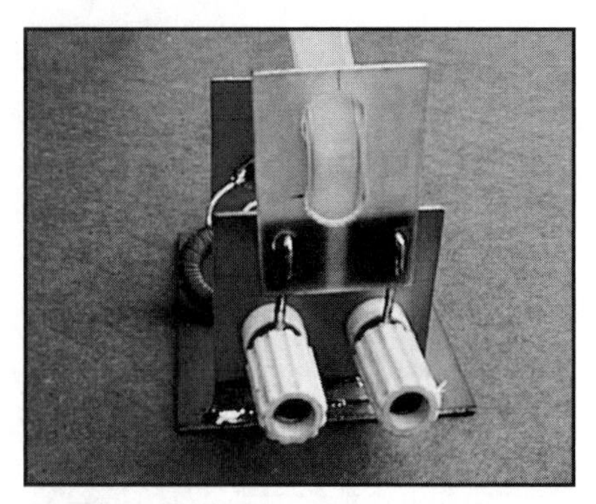

Figure 6—View of an assembled center insulator fashioned from a 3×3-inch piece of PC board from which all the foil has been removed.

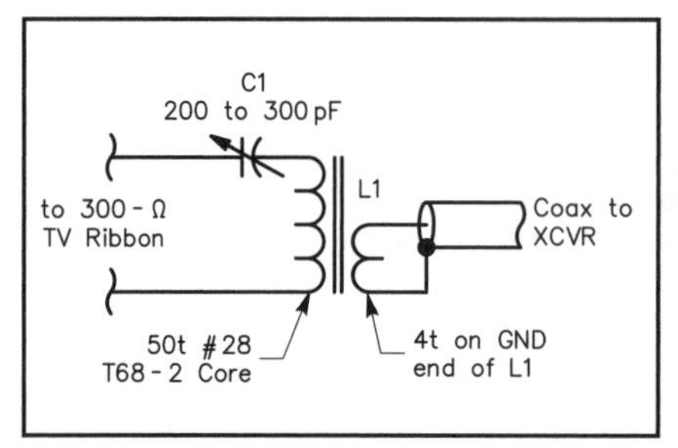

Figure 7—Schematic of the Squirt antenna tuner. See the accompanying Parts List.

Figure 8—This Squirt tuner prototype uses a 2×3-inch piece of PC board for the base plate, two 1½×1½-inch pieces for end plates and a ½-inch square piece as a tie point for the toroid and tuning capacitor.

Parts List

Squirt Antenna

Numbers in parentheses refer to vendors presented at the end of the list.

1—3×3-inch piece of 1/16-inch-thick glass-epoxy PC board for the center insulator (1)
2—½×1½-inch pieces of PC board for the end insulators (1)
1—1⅛×1¾-inch piece of PC board for the feed-line connector (1)
2—34-foot lengths of #20 (or larger) insulated hookup wire (2)
1—6-inch length of #16 (or larger bare) copper wire; scrounge scraps from your local electrician.
1—45-foot length of 300-Ω TV ribbon line (2)

Squirt Tuner

1—2×3-inch piece of PC board for base plate (1)
2—1½-inch-square pieces of PC board for end plates (1)
1—½-inch-square piece of PC board for the tie point (1)
1—200- to 300-pF (maximum) mica compression trimmer (3)
1—T68-2 toroid core (3)
2—Five-way binding posts (2)
1—55-inch length of #26 or 28 enameled wire (2 and 3)

Note: You can use 3/16-inch-thick clear Plexiglas for the Squirt's end and center insulators. Commonly used as a replacement for window glass, Plexiglas scraps can be obtained at low cost from hardware stores that repair windows.

Vendors

1. HSC Electronic Supply, 3500 Ryder St, Santa Clara, CA 95051; tel 408-732-1573, **www.halted.com**
2. Local RadioShack outlets or **www.RadioShack.com**
3. Dan's Small Parts and Kits, Box 3634, Missoula, MT 59806-3634; tel 406-258-2782; **www.fix.net/~jparker/dans.html**

tor of this value should serve. The inductor, L1, consists of 50 turns of enameled wire wound on a T68-2 iron-core toroidal form. An air-wound coil would do as well, although it would be physically much larger. Figure 8 shows the tuner built on an open chassis made of PC board. My prototype uses several PC-board scraps: a 2×3-inch piece for the base plate, two 1½×1½-inch pieces for each end plate (refer to Figure 3). A ½-inch square piece of PC board (visible just beneath the capacitor in Figure 8) is glued to the base plate to serve as an insulated tie point for the connection between the toroid (L1) and tuning capacitor (C1). The tuner end plates are soldered to the base plate to hold a pair of five-way binding posts and a BNC connector at opposite ends. L1 and C1 float above electrical ground, connected to the TV ribbon. One end of L1's secondary (or link) is grounded at the base plate and the coax-cable shield. The hot end of L1's secondary winding is soldered to the coax-connector's center conductor.

Tuner Testing

C1 tunes sharply, so it's a good idea to check just how it tunes before you attach the tuner to an antenna. You can simulate the antenna by connecting a 10-Ω resistor across the binding posts. If you use an antenna analyzer as the signal source, a ¼-W resistor such as the RadioShack 271-1301 is suitable. But if you use your QRP transmitter, you need a total resistance of 8 to 10 Ω that will dissipate your QRP rig's output, assuming here it's 5 W or less. Four RadioShack 271-151 resistors (two series-connected pairs of two parallel-connected resistors) provide a satisfactory load if you don't transmit for extended periods. Or, you can make up your own resistor arrangement to deliver the proper load. Adjust C1 with an insulated tuning tool to achieve an SWR below 1.5:1.

Once the tuner operation is verified using the dummy antenna, it's ready to connect to the Squirt. Tuning there will be similarly sharp, and a 2:1 SWR bandwidth of about 40 kHz or so can be expected as normal.

A Multiband Bonus

Although the Squirt was conceived with 80-meter operation in mind, it can double as a multiband antenna as well. The simple Squirt tuner is designed to match the antenna only on 80 meters. However, a good general-purpose balanced tuner such as an old Johnson Matchbox or one of the currently popular Z-match tuners (such as an Emtech ZM-2) will give good results with the Squirt on any HF band. The Squirt prototype was recently pressed into service at N2CX on 80, 40, 30, 20 and 15 meters for several months. It worked equally as well as a similar antenna fed with ladder line. Although no extensive comparative tests were done, the Squirt has delivered QRP CW contacts from coast to coast on 40, 20 and 15 meters and covers the East Coast during evening hours on 80 meters.

Build one! I'm sure you'll have fun building and using the Squirt!

Notes

[1]Dave Benson, NN1G, and George Heron, N2APB, "The Warbler—A Simple PSK31 Transceiver for 80 Meters," *QST*, Mar 2001, pp 37-41.

[2]R. Dean Straw, N6BV, *The ARRL Antenna Book* (Newington: ARRL, 1997, 18th ed), pp 5-9 to 5-11.

[3]**www.alphalink.com.au/~parkerp/nodec97.htm**; **www.home.global.co.za/~tdamatta/loops.html**

[4]**home.earthlink.net/~mwattcpa/antennas.html**

[5]Full-size templates are contained in SQUIRT.ZIP available from **www.arrl.org/files/qst-binaries/**.

You can contact Joe Everhart, N2CX, at 214 New Jersey Rd, Brooklawn, NJ 08030; **n2cx@arrl.net**.

Photos by the author

QST

By John Ceccherelli, N2XE

From *QST*, May 2004

The Dipole Dilemma

A successful QSO from a wilderness location is, more often than not, dependent on the antenna.

Erecting a dipole in the wild presents some challenges. Assuming you arrived at your destination under human power, you've brought everything you need—except the tower. Most people rely on natural structures for obvious reasons. This presents a huge problem when the destination is a high summit with a great view. That great view usually means no trees—the dipole dilemma.

The dipole is about as close to the perfect antenna as one can imagine. It has a little gain on the broadside and some useful nulls on axis. Get it up to a reasonable height and it presents a good match for coax and rig. It's light, simple, compact and foolproof. The dipole works well in many compromise situations and can be used in multiple configurations including as a flattop, inverted V, sloper and L. You can let the ends droop with reckless abandon and not suffer much performance loss.

Figure 1—The quintessential "dipole dilemma" location. The peak in the foreground is a favorite operating location. Unfortunately, none of the trees on top are over 6 feet tall.

Great Location, No Trees

The example shown in Figure 1 is typical of the dipole dilemma. Its features are a gorgeous knob longing for a radio setup and not a usable tree over 6 feet high. I suppose you could erect the dipole anyway and rationalize that it's really 1000 feet above average terrain. This has some merit as the radiation pattern is enhanced. The feed-point, however, does notice the ground right underneath. The radiation resistance of a low dipole may deviate markedly from 50 Ω. Simulations show that a low dipole should work fairly well, but I've never had very good luck with one.

A 33 foot telescoping fiberglass pole as a center support actually works well, but the pole weighs about 3 pounds and it's difficult to erect solo. If you've resigned to hauling another 3 pounds, you might as well engineer something more manageable than a 33 foot noodle.

The $^1/_4$ wave vertical antenna can perform well if it has an adequate ground system. Although verticals have a reputation for radiating equally poorly in all directions, they work very well in certain applications—near salt water is a well-understood example. Ground loss usually prevents a vertical from outperforming a dipole—even a low one. But a vertical will provide respectable performance and it's a whole lot easier to erect on a rocky summit.

QRP—It's Not Just Wimpy, It's An Education

I get a lot of flack for operating QRP mobile. My rig of choice is the Small Wonder Labs DSW. With 2 W out and no automatic gain control, it's a fun rig. After several thousand miles driving up and down the hills in the northeast US, it became clear that signals improved at the top of a hill. A vertical antenna on a high summit seemed to be a reasonable idea.

Some backyard testing confirmed that a Hamstick with counterpoise worked better than it does on my truck. A few trips to some local summits also confirmed that a Hamstick could pump out a respectable signal. With the stinger removed, the pieces are 4 feet long, a tad awkward for backpacking. It could be improved.

What Does the Simulation Say?

Experience suggested that a vertical outperformed a low dipole on hilltops. Rather than rely on anecdotal evidence, I turned to *EZNEC* for some answers. Figure 2 compares a flattop 20 meter half wave horizontal dipole up 33 feet with an inverted V and a half wave dipole 8 feet above flat ground. I consider the half wave dipole up 33 feet to be the gold

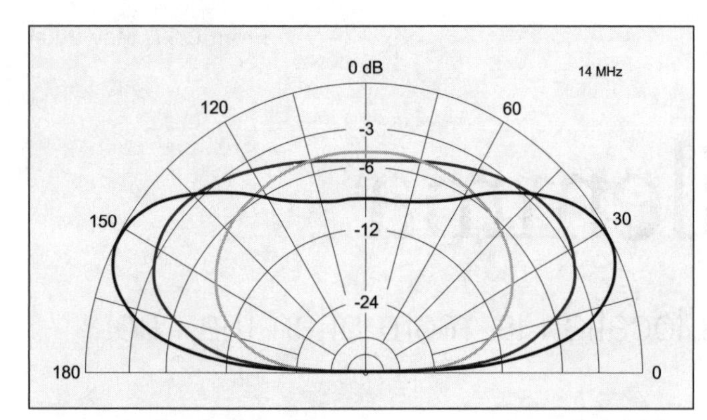

Figure 2—Comparison of 20 meter dipole antennas. A flattop horizontal half wave up 33 feet is shown in black. An inverted V up 33 feet is shown in blue. The flattop horizontal half wave up 8 feet is shown in green. All simulations are with poor conductivity soil.

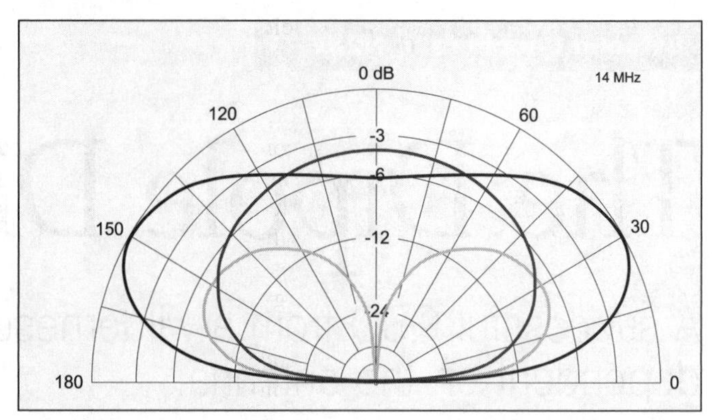

Figure 3—Comparison of the inverted V up 33 feet (black) with the low flattop horizontal dipole up 8 feet (blue) and the ¼ wave vertical (green). All use the poor conductivity, flat ground model.

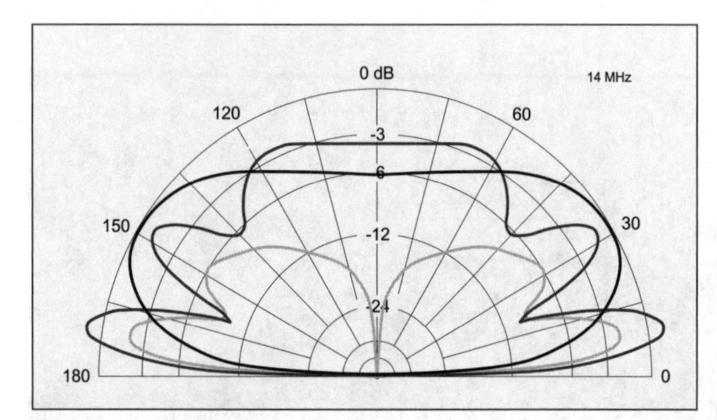

Figure 4—Comparison of the inverted V up 33 feet (black) with the low flattop horizontal dipole up 8 feet (blue) and the ¼ wave vertical (green). The inverted V is over flat ground for reference, the low dipole and vertical are over poor conductivity using the hill summit ground model.

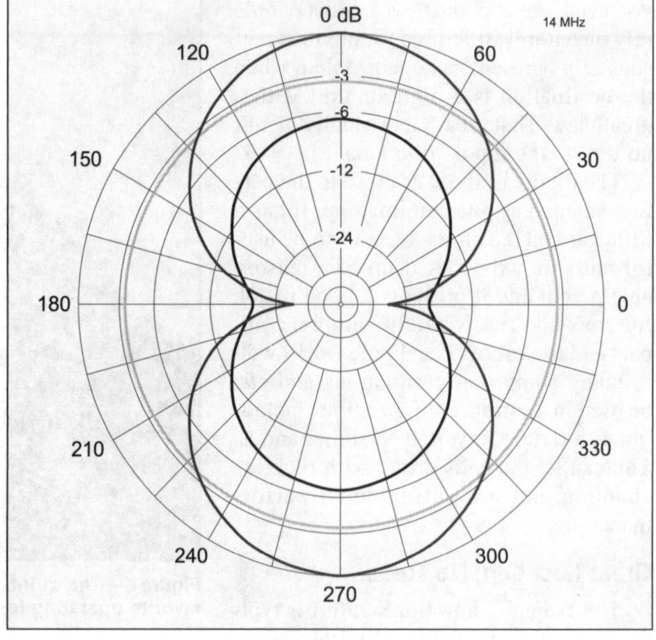

Figure 5—Azimuth pattern (10° elevation) of the inverted V over flat ground (black) and the low dipole (blue) and vertical (green)—both using the poor conductivity hill-summit ground model.

standard. It's hard to get any better than that on a backpacking excursion. The inverted vee, while not quite as good, is probably a more achievable configuration than the flattop, so that will be the 0 dB reference in the following figures. The dipole, even 8 feet up, is often difficult to install on a rocky barren summit. As Figure 2 shows, it doesn't radiate very well either.

Figure 3 shows a 20 meter ¼ wave vertical compared to the inverted V and low dipole over poor mountain soil. The vertical is about 2 dB better than the low dipole for radiation angles below 30°. The 2 dB difference is hardly noticeable and experience shows the vertical can outperform the inverted V by an S-unit or more at times, so something is amiss.

Ground Model

Correlating real life to a simulation can be frustrating. The most obvious problem with the previous simulations is the ground definition. Lobe elevation is determined by ground reflections away from the antenna. Vertical antennas are more sensitive to ground interaction up to 100 wavelengths away for low angles.

Unfortunately, it's not easy to simulate a hill with *EZNEC,* but a wire grid ground model might give a clue as to why the vertical seems to perform better than the simulation indicates.

Figure 4 shows the same antennas as Figure 3 but with a "hill" (30° slope and only 83 feet high due to *EZNEC* segment limitations) underneath them. The vertical exhibits a tremendous increase in low angle radiation. The low dipole exhibits even greater low angle radiation and it shows a nice lobe at 32°. Is the simulation accurate? Almost certainly not, but it does suggest that low angle radiation improves markedly for antennas on top of hills. This correlates well with my mobile experience but fails to explain why the vertical seems to do so well.

Azimuth Pattern

While the vertical does not show better simulation performance broadside to the low dipole, it does radiate equally in all directions. If the received signal is on (or near) axis to the dipole, the vertical will be a much better performer. Figure 5 shows the difference in azimuth pattern of the three antennas. This helps explain some of the positive field experience.

Polarization

The only other plausible explanation for the good field performance of the vertical is polarization. It's generally accepted that sky wave signals are randomly polarized.[1] In theory, a perfectly cross polarized signal would induce no energy into a linearly polarized antenna.

In side by side tests with a dipole and the vertical, the dipole usually wins on short skip (high angle sky wave) and it's a toss-up on DX (low angle sky wave). But when the vertical is better than the dipole, it's a lot better. Maybe sky wave signals are randomly polarized, but my guess is they're slanted toward vertical polarization more often than not.

I trust *EZNEC* more than my limited amount of testing. Maybe my judgment is a bit colored by the vertical's ease to erect. I also suspect that since my vertical was constructed by myself, psychologically, it just seems to work better than a low dipole. At least the simulation indicates that in some cases, it actually does.

The Fly Rod Vertical

Many hams I know are also fly fishermen, and some are not very good ones. As such, there's an ample supply of fly rod tubes available—the fly rods long since destroyed by car doors or other rod enemies. Some of these same folks have a similar track record with Hamstick antennas. This makes for a cornucopia of parts suitable for homebrew vertical antennas. Throw in some aluminum tubing, some leftover Hustler parts, and a really nifty, compact and portable vertical antenna is possible. The fly rod tube serves as both container and part of the radiating element.

I'm on version 3 of the fly rod vertical (FRV) as shown in Figures 6 and 7. Each version has gotten a progressively longer mast and a fatter, but shorter, carrying tube. Each version has tried to improve portability, ruggedness and performance and version 3 is close to optimum for my use. The mast is four sections, each 2.5 feet long of 1/2 inch OD, 0.083 inch thick

Figure 6—The Fly Rod Vertical in action on the summit of Little Whiteface Mountain near Lake Placid, New York. N2XE is operating an Elecraft K1 while son Greg keeps onlookers from tripping over guy lines.

Figure 7—The FRV is setup on that gorgeous knob shown in Figure 1. The FRV is set up for 40 meter operation. The antenna stands 17 feet tall in this configuration.

[1]B. Sykes, G2HCG, "The Enhancement of HF Signals by Polarization Control," *Communications Quarterly*, Nov 1990 (Reprinted from *Practical Wireless*).

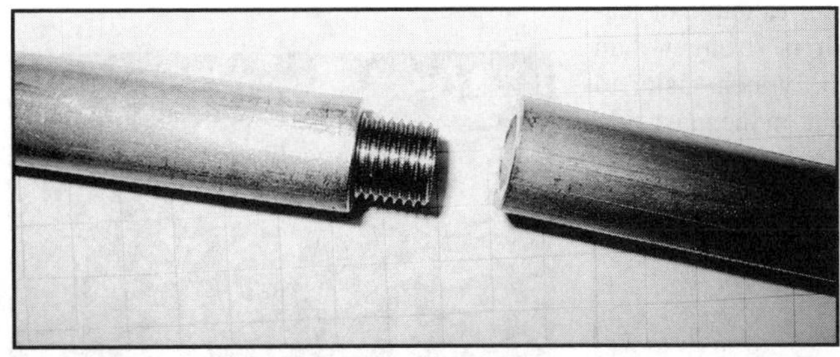

Figure 8—The FRV mast section connection. A 1 inch long, 3/8 × 24 TPI stud is on one end secured with Locktite (or strike with punch). The receiving end is tapped and the sections simply screw together.

Figure 9—Bottom detail of the fly rod vertical. Note the nylon washer isolating the ground radial and tripod system from the active element. The triangular plate with the tripod legs is a Hustler VP1 triband adapter.

Figure 10—The FRV RF cap assembly and tube with collar. Note the screws on the collar that provide good electrical contact from the collar to the tube.

walled aluminum tubing. The same tubing is used for the 10 inch tripod legs. All tubing is attached with 3/8 inch, 24 TPI studs (Figure 8). This also makes it easy to use Hustler resonators for any band. I suppose there are many different ways to make a mast but I've found that the thick wall, 1/2 inch tubing is about the minimum that can survive the high winds often found on hilltops.

Aside from the mast, the only other tricky part of the FRV is the tripod stand and RF feed arrangement. The tripod base is a Hustler VP-1 multiband adapter, originally intended to attach many Hustler resonators to a single mast. The center hole needs to be drilled out to 1/2 inch diameter to accommodate the insulating nylon washer (see Figures 9 and 10). An SO-239 antenna stud attaches to the VP-1 so that the VP-1 is on the shield side of the coax (ground) and the threaded tube cap is the hot side.

The tripod legs are made in the same fashion as the mast with 3/8 inch studs tapped into the tubing. Use a nut, finger tight, to attach the legs to the VP-1.

The Tube

If you're not a bad fisherman and don't know any, you can procure a fly rod tube brand spanking new. The best fly rod tubes in existence, in my opinion, are those made by REC Components (**www.reccomponents.com**). My latest version of the FRV uses a custom tube from REC. I had them make a tube with a screw cap and collar on both ends for a couple of reasons: the cap and collar is stronger than the plain end cap and it's much easier to attach the 3/8 inch long nut on the mast side of the tube. The cost was about $50, easily making the tube the most expensive component.

Since the collar is glued to the tube, you need to ensure good electrical contact. Three 4-40 screws, drilled and tapped into the collar will suffice (see Figure 10).

Since the mast (including the carrying tube) is 12.5 feet tall, a Hustler 17 meter resonator is needed to tune the vertical on 20 meters. In this configuration, and using low loss coax, my Elecraft K1 has no problem using its internal tuner to cover the 30, 20 and 15 meter bands. This may not be the most efficient way to operate but I've always been able to work what I can hear on those bands. In addition, you don't have to waste time switching resonators for band changes.

For 40 meters, the FRV needs either more metal in the air or some inductance to tune it to resonance. Figure 7 shows the FRV with a resonator made on a PVC tubing form.

Ground Radials

Probably the most confounding issue with vertical antennas is ground. The simulations show the ground scheme to be highly significant. In practice, I find it's not that big of a deal. As long as you're willing to accept that your antenna is only 60% to 75% efficient, you have a lot of latitude with the ground configuration. I've settled on six 20 foot wires. I've tried many different lengths and it just doesn't seem to make much difference. The ground wires are attached to the VP-1 with alligator clips and evenly fanned out as shown in Figures 9 and 11.

Figure 13—The disassembled Fly Rod Vertical. If you use a fat 2½ inch diameter fly rod tube, you can stuff it all in, including the feed line.

Figure 11—Close view of the FRV guy and ground wires. The ground consists of six 24 gauge, 20 foot wires attached to the VP-1 with alligator clips (two ground wires per clip).

Figure 12—N2XE working the Empire Slow Speed Net on 80 meters with the FRV in the rain-soaked Shenandoah Valley of Virginia. The signal report was 579 with 20 W from a Ten-Tec Argonaut 5. Not too shabby. The FRV blends in with the trees well and is difficult to avoid at night, even with a flashlight. The spiral reflective tape was added shortly after this trip.

Figure 14—All packed up and ready to go home.

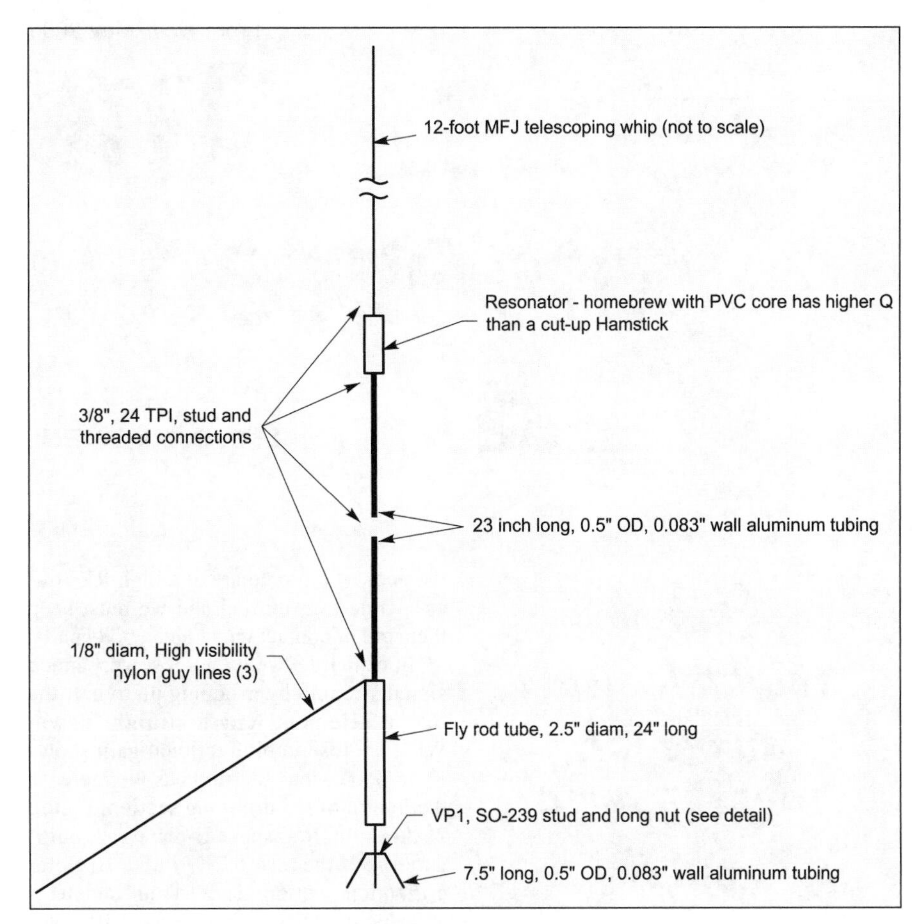

Figure 15—Detailed layout of fly rod vertical.

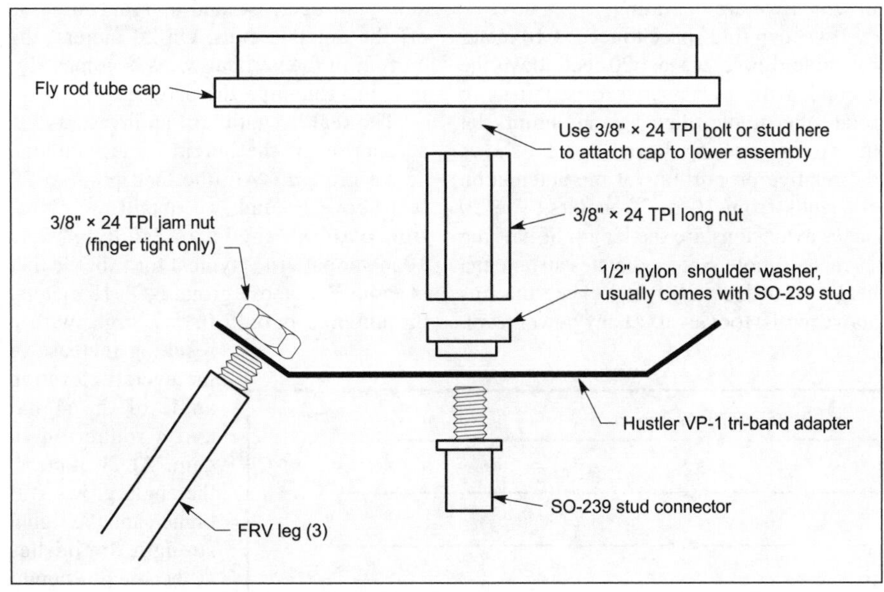

Figure 16—FRV base and tripod details.

A ground-mounted vertical should exhibit a feed-point impedance of about 36 Ω with a perfect ground. Mine always measure around 50 Ω so the difference must be 14 Ω—11 Ω of ground resistance and the rest, the loading coil loss. That's great for SWR but it means the efficiency is only about 70% and at the full QRP gallon, 1.4 W is keeping the coil and earthworms warm.

I'd much rather have a 36 Ω feed point impedance and a 1.5:1 SWR, but the excellent coax match allows me to use a 20 foot RG-178 feed line without much guilt. In addition, RG-178 stuffs into the tube nicely.

Finishing Touches

A good FRV would not be complete without a candy cane spiral of reflective tape. The spiral started out as a whimsical decoration. The last version I constructed had a beautiful green powder coated tube, and my wife convinced me not to spoil its crisp, clean appearance. On a recent camping trip (see Figure 12), however, everyone kept tripping over it in the dark. The reflective tape spiral has become an absolute necessity.

The Acid Test

Field experience with the FRV has been very good. The simulations indicate it should be a decent performing antenna, particularly on hilltops. Then again, stories of working rare DX with a light bulb are not uncommon. With that in mind, I had to put the FRV through a challenging test.

On July 12, 2003, I ventured back to the hill in Figure 1 with only a 40 meter Rock Mite and the FRV. Only 400 mW, no VFO, 72% antenna efficiency (worse on 40 meters) and RG-178 feed line seemed to be a pretty stiff test. Just to make it interesting, Mother Nature had the sun burp up a few flares, and the planetary K index was 7.

Undaunted, I trudged up the 1000 vertical feet and set up. Working less than half a watt, the best strategy is to listen so I listened. The band was not active. I did hear a few stations way off frequency. Two stations did call CQ on the Rock Mite frequency and I worked both Dennis, K1LGQ, with his Elecraft K1 and Brian, N1BQ, both in Maine and about 250 miles distant from me. I don't think a light bulb would have fared as well.

I've built several of these fly rod verticals and they perform quite well on all bands between 80 and 10 meters. Figures 13-16 provide the construction details. At 3 pounds, the FRV is heavier than any rig and battery I'm likely to put in a backpack but it has great potential to be optimized for weight.

The FRV sets up easily and is convenient to strap to a backpack. It doesn't exactly solve the dipole dilemma, but it does make it easy to get on the air.

All photos by the author.

John Ceccherelli, N2XE, loves QRP, CW and backpacking with both. When not on the air, John spends his time attempting to trisect angles, researching communication minutiae and collecting slide rules. He is a Senior Engineer for IBM in Hopewell Junction, NY and lives in nearby Wappingers Falls, NY. You may contact John at **n2xe@arrl.net**. QST

From *QST*, May 2005

The Inverted-U

L. B. Cebik, W4RNL

A simple rotatable field dipole for 20 through 10 meters.

The ARRL Field Day and other operations in the field tend to lean on three antenna principles—simplicity, small size and light weight. Complex assemblies increase the number of things that can go wrong. Large antennas are hard to transport and sometimes do not fit the space available. Heavy antennas require heavy support structures, so the overall weight seems to increase exponentially with every added pound of antenna.

In recent years, a number of light collapsible masts have hit the scene. When properly guyed with ropes, some will support antennas in the 5 to 10 pound range. Most are suitable for 10 meter tubular dipoles and as supports for wire antennas. The masts also allow the user to hand-rotate some antennas so that they are broadside to the desired target station. If we could only extend the range of the antenna to cover 20 through 10 meters, we might put these 20 to 30 foot masts to even better use. The inverted-U is an old idea that may provide one answer.

The Basic Idea

A dipole's highest current occurs within the first half of the distance from the feed point to the outer tips. We lose very little performance if we fold, bend or mutilate the outer end sections to fit an available space. If we start with a tubular 10 meter dipole—a little over 200 inches in overall length—we might add extensions for 12, 15, 17 or 20 meters, shaping them to suit the area in which we are operating.

If we can find space to erect a 10 meter rotatable dipole at least 20 feet above the ground, with a clear area to permit us to rotate the dipole, then we can simply let the extensions hang down. Figure 1 shows the relative proportions of the antenna on all bands from 10 to 20 meters. The 20 meter extensions are the length of half the 10 meter dipole. Safety dictates an antenna height of at least 20 feet to keep the tips above the 10 foot level. At any power level, the ends of a dipole are at a high RF voltage while transmitting and we must keep them out of contact with human body parts.

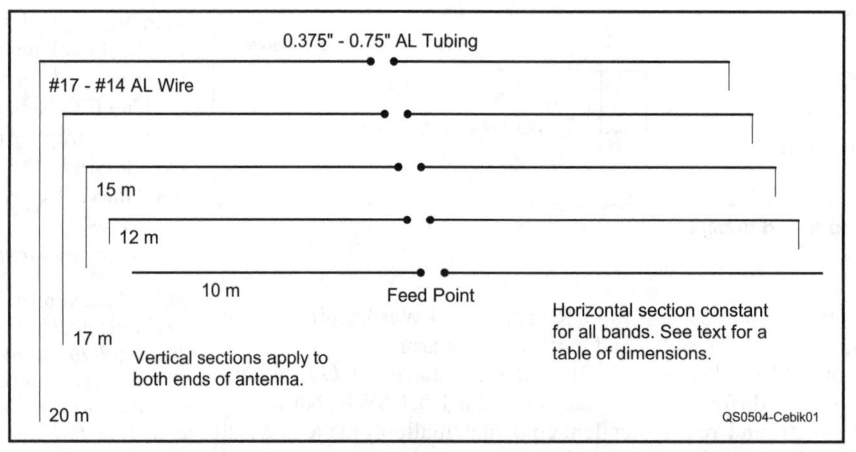

Figure 1—The general outline of the inverted-U field dipole for 20 through 10 meters. Note that the vertical end extension wires apply to both ends of the main 10 meter dipole.

In principle, we do not lose very much signal strength by drooping up to half the overall element length straight down. What we lose in bidirectional gain shows up in decreased side-nulls as we increase the length of the drooping section. Figure 2 shows the free-space E-plane (azimuth) patterns of the inverted-U with a 10 meter horizontal section. There is an undetectable decrease in gain between the 10 meter and 15 meter versions. The 20 meter version shows a little over a half dB broadside gain decrease and a signal increase off the antenna ends. On 20 meters, the current in the vertical wires becomes significant, rounding the pattern.

The real limitation of an inverted-U is a function of the height of the antenna above ground. With the feed point at 20 feet above ground, we obtain the elevation patterns shown in Figure 3. The 10 meter pattern is typical for a dipole that is about $^5/_8$ λ above ground. On 15 meters, the antenna is only 0.45 λ high, with a resulting increase in the overall elevation angle of the signal and a reduction in gain. At 20 meters, the angle grows still higher, and the signal strength diminishes as the antenna height drops to under 0.3 λ. Nevertheless, the signal is certainly usable. A full-size dipole at 20 meters would show only a little more gain at the same height, and the elevation angle would be similar to that of the inverted-U,

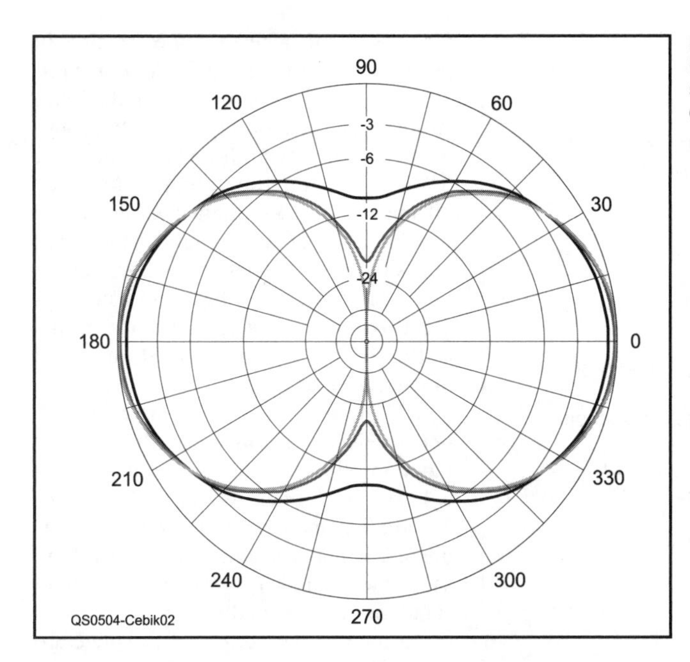

Figure 2—Free-space E-plane (azimuth) patterns of the inverted-U for 10 (green), 15 (red) and 20 (black) meters, showing the pattern changes with increasingly longer vertical end sections.

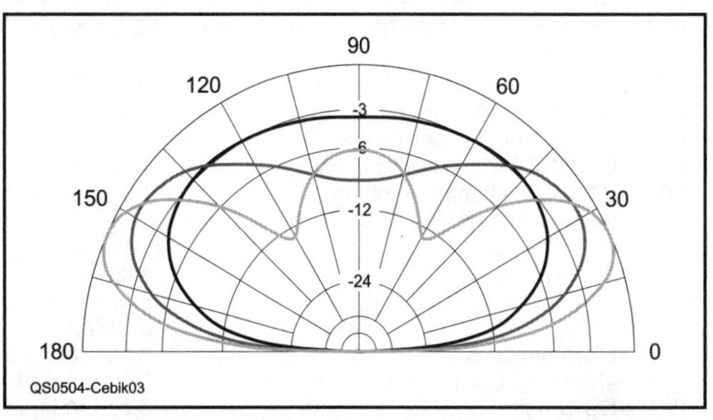

Figure 3—Elevation patterns of the inverted-U for 10 (green), 15 (red) and 20 (black) meters, with the antenna feed point 20 feet above average ground. Much of the decreased gain and higher elevation angle of the pattern at the lowest frequencies is due to its lower height as a fraction of a wavelength.

despite the difference in antenna shape.

As we decrease our frequency, there is no substitute for antenna height. Any horizontal antenna below about 3/8 λ in height will show a rapid decrease in low angle performance relative to heights above that level. If we raise the inverted-U to 40 feet, the 20 meter performance would be very similar to that shown by the 10 meter elevation plot in Figure 3. Table 1 summarizes the free-space and 20 foot performance characteristics of the inverted-U. Of special note is the fact that the feed-point impedance of the inverted-U remains well within acceptable limits for virtually all equipment, even at a height as low as 20 feet above ground. The SWR curves are also acceptably broad. Finding exact dimensions, even in special field conditions, becomes a non-critical task.

If we can accept the performance potential of a dipole for any band in the 20 meter to 10 meter range at the anticipated height of our mast possibilities, then the inverted-U provides a compact way of achieving that performance. The next step is fabricating one that we can use in the field.

Building a Field Version of the Inverted-U

Let's approach the construction of a field inverted-U in three steps: First, the tubing arrangement, second the center hub and feed-point assembly and third the drooping extensions. A parts list appears in Table 2.

1. The aluminum tubing dipole for 10 meters. Each half of the aluminum tubing dipole consists of three longer detachable sections of tubing and a short section mounted permanently to the feed-point plate, as shown in Figure 4. Let's consider each half of the element separately. Counting from the center of the plate—the feed point—the element extends 5 inches using 3/4 inch aluminum tubing. Then we have two 33 inch exposed tubing sections, with an additional 3 inches of tubing overlap per section. These sections are 5/8 and 1/2 inch in diameter, respectively. The exposed outer section is 30 inches long (with at least a 3 inch overlap) and consists of 3/8 inch diameter tubing.

Since the 5/8 and 1/2 inch sections are 36 inches long, you can make the outer 3/8 inch section the same overall length and use more overlap, or you can cut the tubing to 33 inches and use the 3 inch overlap. That much overlap is sufficient to ensure a strong junction while minimizing excess weight. When not in use, the three outer tubing sections will nest inside each other for storage. A 36 inch length for the outer section is a bit more convenient to un-nest for assembly. (I keep the end hitch pin on the 3/8 inch tubing as an easy way of pulling it into final position.) I prefer to use the readily available 6063-T832 aluminum tubing that nests well and has a long history of antenna service.

The only construction operation that you need to perform on the tubing is to drill a hole at about the center of each junction to pass a hitch pin clip. Obtain hitch pin clips (also called hairpin cotter pin clips in some literature) that fit snugly over the tubing.

Table 1
Anticipated Performance of the Inverted-U for 20 through 10 Meters

Using the tubular 10 meter dipole described in the text and #17 to #14 vertical wire element extensions.

	Free-Space			*20' Above Average Ground*		
Band	*Gain (dBi)*	*Resonant Impedance (Ohms)*	*Wire Length (inches) #17*	*Gain (dBi)*	*Elevation Angle (deg)*	*Impedance R ± jX (Ohms)*
10	2.1	73	—	7.6	24	65 – *j*2
12	2.0	71	16	7.2	27	67 – *j*8
15	1.9	64	38	6.4	32	69 – *j*8
17	1.7	55	62	5.7	38	65 – *j*4
20	1.4	41	108	4.8	50	52 + *j*4

Note: The wire length for the drooping ends is measured from the end of the 202 inch tubular dipole to the tip for #17 wire. Little change in length occurs as a function of a change in wire size. However, attachment of the extension to the element and special field conditions may require a few extra inches of wire.

Table 2
Parts List

Substitutions and alternative construction methods are possible and encouraged—as long as the overall antenna weight is not increased.

Qty	*Part*	*Comments*
6'	0.375" OD aluminum tubing	Two - 3' pieces
6'	0.5" OD aluminum tubing	Two - 3' pieces
6'	0.625" OD aluminum tubing	Two - 3' pieces
10"	0.75" OD aluminum tubing	Two - 5" pieces
4"	0.5" nominal (5/8" OD) CPVC	
50'	#17 wire	See Table 1 for the length of each piece. Aluminum preferred, but copper usable.
8	Hitch pin clips	Sized to fit tubing junctions.
1	4"×4"×1/4" Lexan plate	Other materials suitable.
2	SS U-bolts	Sized to fit support mast.
2	Sets SS #8/10 1.5" bolt, nut, washers	SS = stainless steel.
2	Sets SS #8 1" bolt, nut, washers	
2	Sets SS #8 0.5" bolt, nut, washers	
1	Coax connector bracket, 1/16" aluminum	See text for dimensions and shape.
1	Female coax connector	
2	Solder lugs, #8 holes	
2	Short pieces copper wire	From coax connector to solder lugs.

Note: 6063-T832 aluminum tubing is preferred and can be obtained from Texas Towers, **www.texastowers.com**, and other outlets. Lexan (polycarbonate) is available from McMasters-Carr, **www.mcmasters.com**, as are the hitch pin clips (if not locally available). Other items should be available from local home centers and radio parts stores.

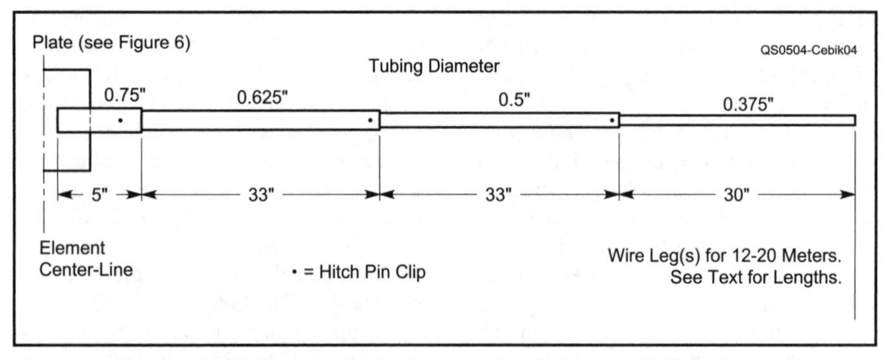

Figure 4—The general tubing layout for the inverted-U for each half element. The opposite side of the dipole is a mirror image of the one shown.

One size will generally handle about two or three tubing sizes. In this antenna, I used 3/32 inch (pin diameter) by 2 5/8 inch long clips for the 3/4 to 5/8 inch and the 5/8 to 1/2 inch junctions, with 3/32 by 1 5/8 inch pins for the 1/2 to 3/8 inch junction and for the final hitch pin clip at the outer end of the horizontal part of the antenna. Drill the holes (1/8 inch diameter) for the clips with the adjacent tubes in position relative to each other. I generally tape the junction temporarily for the drilling. Carefully deburr the holes so that the tubing slides easily when nested.

For this field antenna, the hitch pin clip junctions, shown in Figure 5, hold the element sections in position. The overlapping portions of the tube make the actual electrical contact between sections. Due to the effects of weather, junctions of this type are not suitable for a permanent installation, but are completely satisfactory for short-term field use. Good electrical contact requires clean, dry aluminum surfaces, so do not use any type of lubricant to assist the nesting and un-nesting of the tubes. Instead, clean both the inner and outer surfaces of the tubes before and after each field use.

Hitch pin clips are fairly large and harder to lose in the grass of a field site than most nuts and bolts. However, you may wish to attach a short colorful ribbon to the loop end of each clip. Spotting the ribbon on the ground is simpler than probing for the clip alone.

Each element half is 101 inches long, for a total 10 meter dipole element length of 202 inches. Length is not critical within about ±1 inch, so you may pre-assemble the dipole using the listed dimensions. However, if you wish a more precisely tuned element, tape the outer section in position and test the dipole on your mast at the height that you will use in the field. Adjust the length of the outer tubing segments equally at both ends for the best SWR curve on the lower 1 MHz of 10 meters. Even though the impedance will be above 50 Ω throughout the band, you should easily obtain an SWR curve under 2:1 that covers the entire band segment. However, you cannot perform this test until you construct a feed point and mounting plate.

2. *The center hub: mounting and feed-point assembly.* I constructed the plate for mounting the element and the mast from a 4×4×1/4 inch thick scrap of polycarbonate (trade name Lexan), as shown in Figure 5. You may use other materials as long as they will handle the element weight and stand up to field conditions.

At the top and bottom of the plate are holes for the U-bolts that fit around my mast. Since field masts may vary in diameter at the top, size your U-bolts and their holes to suit the mast.

The element center, consisting of two 5 inch lengths of 3/4 inch aluminum tubing, is just above the centerline of the plate (to allow room for the coax fitting below). CPVC tubing of 1/2 inch nominal size has an outside diameter of about 5/8 inches and makes a snug fit inside the 3/4 inch tubing. The CPVC aligns the two aluminum tubes in a straight line and allows for a small (about 1/2 inch) gap between them. When centered between the two tubes, the CPVC is the same width as the plate. A pair of 1 1/2 inch #8 or #10 stainless steel bolts—each bolt with washers and a nut—secures the element to the plate.

Note in the sketch that you may insert the 5/8 inch tube as far into the 3/4 inch tube as it will go and be assured of a 3 inch overlap. I drilled all hitch pin clip holes perpendicular to the plate. Although this alignment is not critical to the junctions of the tubes, it is important to the outer ends of the tubes when we use the antenna below 10 meters.

I mounted a single-hole female UHF connector on a bracket made from a scrap of 1/16 inch thick L-stock that is 1 inch on a side. I drilled the UHF mounting hole first, before cutting the L-stock to length and trimming part of the mounting side. Then I drilled 2 holes for 1/2 inch long #8 stainless steel bolts about 1 inch apart, for a total length of L-stock of about 1 1/2 inches. The reason for the wide strip is to place the bolt heads for the bracket outside the area where the mast will meet the plate on the back side. Note in Figure 5 that the bracket nuts are on the bracket-side of the main plate, with the heads facing the mast. The bracket-to-plate mounting edge of the bracket needs to be only about 3/4 inch wide, so you may trim that side of the L-stock accordingly.

With the element center sections and the bracket in place, I drilled two holes for 1 inch long #8 stainless steel bolts at right angles to the mounting bolts and as close as feasible to the edges of the tubing at the gap. These bolts have solder lugs attached for short leads to the coax fitting. Solder lugs do not come in stainless steel, so you should check these junctions before and after each use for any corrosion that may call for periodic replacement.

With all hardware in place, the hub unit is about 4×10×1 inches (plus U-bolts). It will remain a single unit from this point onward, so that your only field assembly requirements will be to extend tubing sections and install hitch pin clips. You are now ready to perform the initial 10 meter resonance tests on your field mast.

3. The drooping extensions for 12 through 20 meters. The drooping end sections consist of aluminum wire. Copper is usable, but aluminum is lighter and quite satisfactory for this application. Table 1 lists the approximate lengths of each extension below the element. Add 3 to 5 inches of wire—less for 12 meters, more for 20 meters—to each length listed.

Initially, I had hoped to use #14 aluminum wire. However, this material is becoming harder to find locally. A good substitute is common #17 aluminum electric fencing wire. Fence wire is stiffer than most wires of similar diameter, and it is cheap. Stiffness is the more important property, since we do not want the lower ends of the wire to wave excessively in the breeze, potentially changing the feed-point properties of the antenna while in use.

When stored, the lengths of wire extensions for 12 and 15 meters can be laid out without any bends. However, the longer extensions for 17 and 20 meters will require some coiling or folding to fit the same space as the tubing when nested. Fold or coil the wire around any kind of small spindle that has at least a 2 inch diameter. This measure prevents the wire from crimping and eventually breaking. Murphy dictates that a wire will break in the middle of an operating session. So carry some spare wire for replacement ends. The low cost of fencing wire (about $5 for 250 feet in my area) allows me to tote the entire spool with me to operating sites. All together, the ends require about 50 feet of wire.

Figure 6 shows the simple mounting scheme for the end wires. I push the straight wires through a pair of holes aligned vertically to the earth. I then bend the top portion slightly. To clamp the wire, I insert a hitch pin clip though holes parallel to the ground, pushing the wire slightly to one side to reach the far hole in the tube. The double bend holds the wire securely (for a short-term field operation), but allows the wire to be pulled out when the session is over or when I change bands.

Add a few inches to the lengths in Table 1 as an initial length for each band. Test the lengths and prune the wires until you obtain a smooth SWR curve below 2:1 at the ends of each band. Since an inverted-U antenna is full length, the SWR curves will be rather broad and suffer none of the narrow bandwidths associated with inductively loaded elements. Figure 7 shows modeled SWR curves for each band to guide your expectations. Unless your feed line is an exact multiple of a half-wavelength, the impedance that you might record on any of the antenna analyzer instruments may differ from the impedance at the feed point.

The impedance figures for a 20 foot height shown in Table 1 suggest that you should not require much, if any, adjustment once you have found satisfactory lengths for each band. However, leave enough excess so that you can adjust the lengths in the field, especially if you operate in an area where objects like trees and buildings are at different distances from the antenna than they are in your test setup. My initial tests required extension wires considerably longer than modeled, with the peak excess on 15 meters. Only during a second round of tests did I realize that my test mast is metal, ungrounded, and just about 22 feet long. The need for longer extensions likely resulted from coupling between the extensions and the mast, although nothing seemed amiss in reception tests. The pattern was clearly bi-directional.

Do not be too finicky with your SWR curves in the field. An initial test and possibly one adjustment should be all that you need to arrive at an SWR value that is satisfactory for your equipment. Spending half of your operating time adjusting the elements for as near to a 1:1 SWR curve as may be possible will rob you of valuable contacts without changing your signal strength is any manner that is detectable at the far end of the line.

Changing bands is now a simple matter. Remove the ends for the band you are using and install the ends for the new band. An SWR check and possibly one more adjustment of the end lengths will put you back on the air.

Some Final Notes

The inverted-U dipole with interchangeable end pieces provides a compact field antenna. Figure 8 shows the parts in their travel form, illustrating how compactly the antenna travels. When assembled and mounted at least 20 feet up (and higher is still better), the antenna will compete with just about any other dipole mounted at the same height. But the inverted-U is lighter than most dipoles at frequencies lower than 10 meters. It also rotates easily by hand—assuming that you can rotate the mast by hand. Being able to broadside the dipole to your target station gives the inverted-U a strong advantage over a fixed wire dipole.

You may experiment with other forms of construction according to your own skills and available materials. A wire version can be taped to bamboo poles, if you can obtain the material and find a means of connecting short transportable sections of bamboo. Try to avoid aluminum conduit and other heavy materials. Be certain

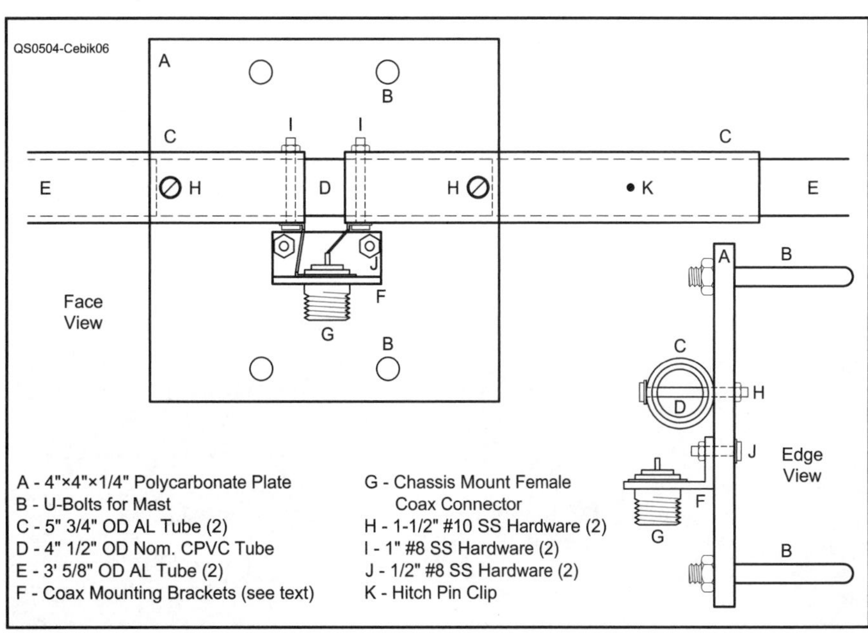

Figure 5—The element and feed-point mounting plate, with details of the construction used in the prototype.

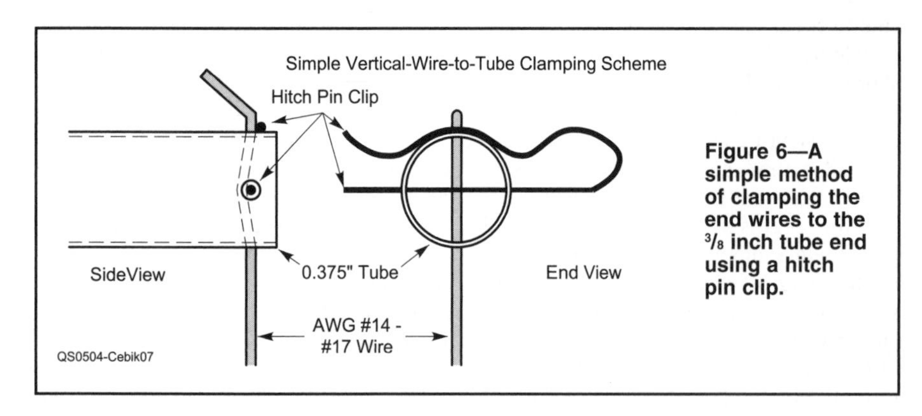

Figure 6—A simple method of clamping the end wires to the 3/8 inch tube end using a hitch pin clip.

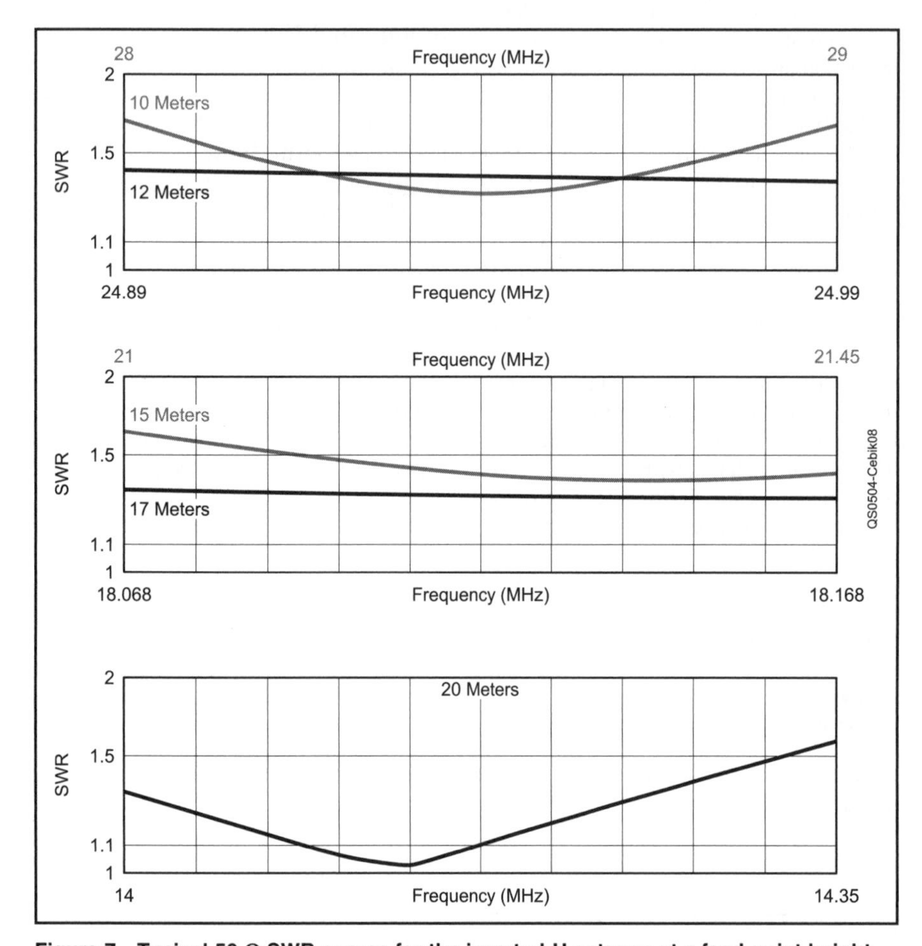

Figure 7—Typical 50 Ω SWR curves for the inverted-U antenna at a feed-point height of 20 feet.

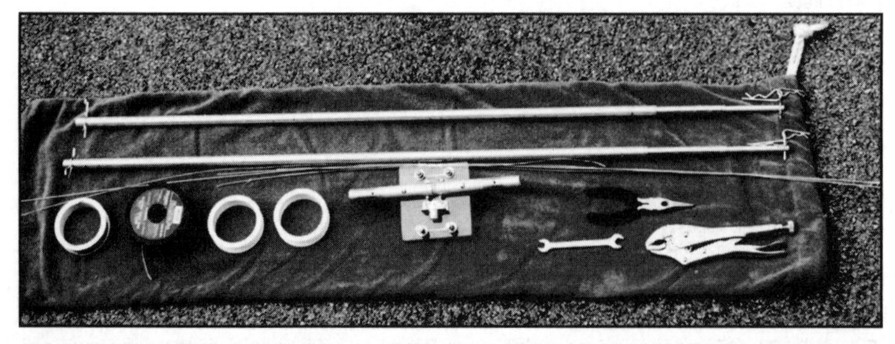

Figure 8—The entire inverted-U antenna parts collection in semi-nested form, with its carrying bag and the three dedicated tools for field assembly and disassembly.

that the drooping end wires are well secured to the 10 meter dipole ends both electrically and mechanically.

With a dipole having drooping ends, safety is very important. Do not use the antenna unless the wire ends for 20 meters are higher than any person can touch when the antenna is in use. Even with QRP power levels, the RF voltage on the wire ends can be dangerous. With the antenna at 20 feet at its center, the ends should be at least 10 feet above ground. For this reason, I have not tried to extend the inverted-U for use on 30 and 40 meters.

Equally important is the maintenance that you give the antenna before and after each use. Be sure that the aluminum tubing is clean—both inside and out—when you nest and unnest the sections. Grit can freeze the sections together, and dirty tubing can prevent good electrical continuity when the antenna is extended. Add a few extra hitch pin clips to the package to be sure you have spares in case you lose one.

You can carry all of the parts in any 3 foot long bag, either horizontal or vertical. A draw-string bag works very well. One of the photos shows the antenna pieces and their bag, along with the ribbon spools on which I store the wire extensions, the spool of extra wire, and the *field tools*. The tools that I store with the antenna include a wrench to tighten the U-bolts for the mast-to-plate mount and a pair of pliers to help me remove end wires from the tubing without cutting my fingers. The pliers have a wire-cutting feature in case I need to replace a broken end wire. A pair of Vise-Grip pliers makes a good removable handle for turning the mast. The combination of the Vise-Grip and regular pliers lets me uncoil the wire extensions for any band and give them a couple of sharp tugs to relax and straighten the wire.

Although the inverted-U is not the answer to every field antenna need, it will serve very well if it matches the kind of operating you do. Hitch pin clips simplify assembly and band changing in the field. At 20 feet and higher, the antenna will acquit itself very well. The entire antenna is inexpensive to build and easy to maintain. Those are reasonably good credentials for any antenna.

Licensed since 1954, L. B. Cebik, W4RNL, is a prolific writer on the subject of antennas. Since retiring from teaching at the University of Tennessee, LB has hosted a Web site (**www.cebik.com**) *discussing antennas—both theoretical and practical. He has written more than 15 books, including the ARRL course on antenna modeling. Serving both as a Technical and an Educational ARRL Advisor, he's also been inducted into both the QRP and QCWA Halls of Fame. LB can be reached at 1434 High Mesa Dr, Knoxville, TN 38938 or at* **cebik@cebik.com**. QST

By Markus Hansen, VE7CA

From *QST*, November 2001

A Portable 2-Element Triband Yagi

Have you ever dreamed about a portable beam you could use at your summer cottage, while camping or on Field Day? Dream no longer. This portable beam can be rolled up and stashed in your car's ski boot!

Several years ago I entered the ARRL November Sweepstakes CW contest in the QRP category, operating from a portable location. It turned out to be a very frustrating experience with only 3 W of output power and dipole antennas. After the contest I decided that the next time I entered a QRP contest it had to be with gain antennas.

My philosophy has always been to try to keep life as simple as possible. In other words, I look for the easiest way to accomplish a goal that guarantees success. Don't get me wrong: Dipoles work particularly well considering the time and effort put into making them. But adding a reflector to a dipole antenna increases the overall gain about 5 dB, depending on the spacing between the elements. This extra gain makes a significant difference, especially when you are dealing with QRP power levels. My 3-W transmitted signal would sound like a 9.5-W powerhouse just by adding another piece of wire! And it would be inexpensive too.

With Solar Cycle 23 in full swing, having an antenna with gain on 15 and 10 meters also became a consideration. Another parameter was the sale of the family van, which meant the new antenna had to fit into the ski boot of our car. Keeping these constraints in mind, I used a computer antenna-modeling program,

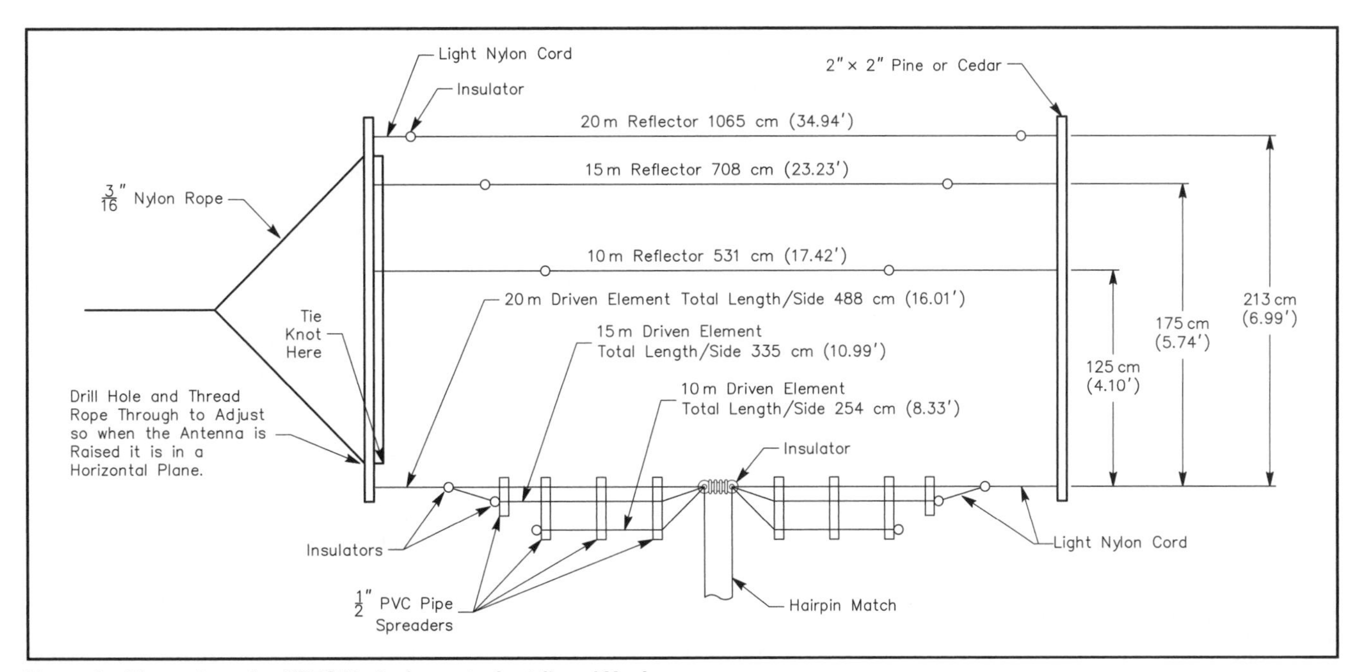

Figure 1—Dimensions for VE7CA's 2-element wire triband Yagi.

trying different design parameters to develop a triband 2-element portable Yagi using wire elements.

The basic concept comprises three individual dipole driven elements, one each for 10, 15 and for 20 meters tied to a common feed point, plus three separate reflector elements. The elements are strung

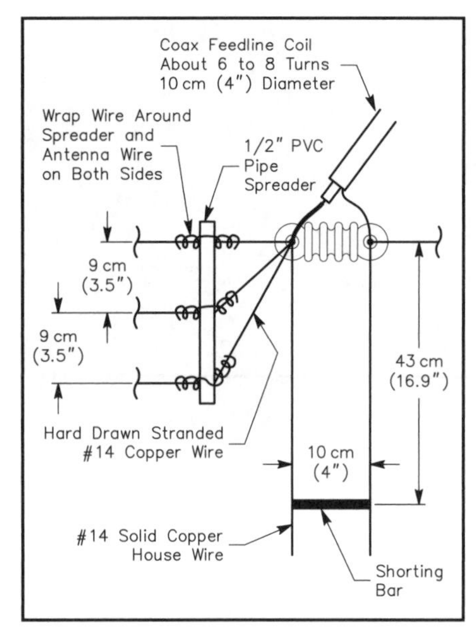

Figure 2—Close-up view of the feed point.

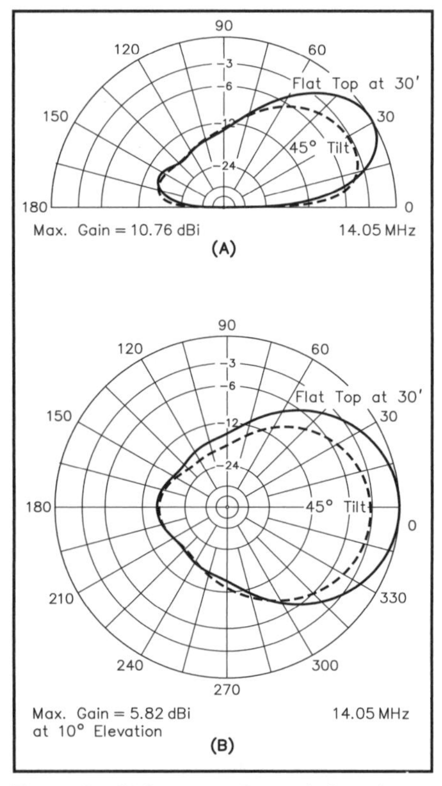

Figure 3—At A, comparison of elevation patterns for VE7CA Yagi as a horizontal flat top (solid line) and tilted 45° from vertical (dashed line). At B, comparisons of azimuth patterns for a 10° elevation angle.

between two 2.13-meter (7-foot) long, 2x2-inch wood spreaders, each just long enough to fit into the ski boot of the car. Use the lightest wood possible, such as cedar, pine or spruce to keep the total weight of the antenna as light as possible. Fiberglass poles would also work, or PVC pipe reinforced with maple doweling to ensure they don't bend. (Wood has the benefit of being easy-to-find and very affordable).

Adding a reflector element relatively close to the driven elements lowers the feed-point impedance of the driven element, so a simple hairpin match was employed to match the driven elements to a 50-Ω feed line. Figure 1 shows the layout and dimensions of the antenna.

The Hairpin Match

The matching system is very simple and foolproof. You should be able to copy the dimensions shown in Figure 2 and not need to retune the hairpin match, unless you plan to use the antenna in the top portions of the phone bands. The dimensions in Figure 2 produced a very low SWR—under 1.3:1 over the CW portions of all three bands. However, even in the lower portions of the SSB bands, the SWR doesn't rise above 2:1. SWR measurements were made at the end of a 25-meter (82-foot) length of RG-58 coax feed line.

Some may wonder why I used such a long feed line. First, when operating from a portable location it is better to be long than short. Nothing is more frustrating than finding that the coax you took along with you is too short. Further, when I change beam direction I walk the antenna around the antenna support, thus requiring a longer length than if I went directly from the antenna to the operating position.

If you are concerned about line loss you can run RG-58 down to the ground and larger-diameter RG-8 or RG-213 to the operating position. You may also find that in your particular situation a shorter length of coax will do. An 18-meter (59-foot) long piece of RG-58 has a loss of about 1 dB at 14 MHz, which is entirely acceptable considering the convenience of using coax cable.

Adjusting the Hairpin Match

If after raising the antenna the SWR is not as low as you want in the portion of the bands you plan to operate, first double-check to make sure that all the elements are cut to the correct length and that the spacings between the driven elements and reflectors are correct. Next you can adjust the hairpin match. Connect either an antenna SWR analyzer or a transmitter and SWR meter to the end of the feed line and pull the antenna up to operating height. Determine where the lowest SWR is on 15 meters. By moving the shorting bar on the hairpin match up or down you can adjust the lowest SWR point to the middle of the portion of the 15 meter band you prefer. If your preference is near the top end of 15 meters you may have to shorten the 15-meter driven element slightly. After adjusting the 15-meter element and hairpin match, adjust the 10 and 20 driven-elements lengths separately, without changing the position of the shorting bar on the hairpin match.

The hairpin match is very rugged. You can attach the feed line to it with tape, roll it up, pack the antenna away and even with the matching wires bent out of shape it just seems to want to work.

Antenna Support

Adhering to my constraint to keep things as simple as possible, I only use one support for the antenna, typically a tree. When the antenna is raised to its operating position it is a sloping triband Yagi. To achieve this, attach a rope to each end of the 2x2's to form a V-shaped sling, as shown in the Figure 1. Attach a length of rope to one sling and pull the antenna up a tree branch, tower or whatever vertical support is available. Tie a second length of rope to the bottom sling and anchor the antenna to a stake in the ground. By putting in two or three stakes in the ground around the antenna support, you can walk the antenna around to favor a particular direction. To change direction 180°, give the feed line a pull and the array will flip over. So simple but very effective!

Local or DX

One of the features of a sloping antenna is that you can adjust the take-off (elevation) angle. For example, if you are interested in North American contacts (whether for casual QSOs or the ARRL SS contest), then sloping the antenna away sideways from the support structure at 45° with the feed point approximately 8 meters (26 feet) above the ground, will yield a 20-meter pattern similar to Figure 3A. Here, the maximum lobe is between 10° and 60° in elevation. The pattern of the antenna in a flat-top horizontal configuration at 9.1 meters (30 feet) is overlaid for comparison. You can see that the tilted beam has better low-angle performance, but at higher angles has less gain than its horizontal counterpart. Figure 3B shows an overlay of the azimuth patterns for these two configurations at a 10° takeoff angle.

If DX is your main interest, then you want to position the antenna even closer to vertical to emphasize the lower elevation angles. Figure 4 shows the pattern on 20 meters when the antenna is tilted sideward 10° away from vertical, again compared with the other orientations in Figure 3A. The feed point is 6 meters above ground and the model assumes fresh water in the far field, which is the case at my portable location.

Remember that the radiation pattern is quite dependent on ground conductivity and dielectric constant for a vertically polarized antenna. A location close to saltwater will yield the highest gain and the lowest radiation angle. With very poor soil in the near and far field, the peak radiation angle will be higher and the gain less.

I have had the opportunity to test this out at my portable location. Using two trees as supports, I am able to pull the antenna close to horizontal with the feed point about 7 meters above the ground. In this position, with 20 meters open to Europe, I have found it difficult to work DX on CW with 3 W of output power. However, when I change the slope of the antenna so that it is nearly vertical I not only hear more DX stations, but I find it relatively easy to work DX.

I have tried this many times, since it is simple to lower one end of the antenna to change the slope and hence the radiation take-off angle. The sloping antenna always performs much better for working DX than a low horizontal antenna. Recently, I worked nine European countries during two evenings of casual operating, even though the highest end of the antenna was only about 10 meters high, limiting the slope to about 45°.

Figure 5 shows the elevation pattern on 28.05 MHz for the beam sloped 10° from vertical at 45° from vertical, with the feed point at 8 meters height, again compared with the beam as a flat top at 9.1 meters (30 feet). With a steeper vertical slope, the 10-meter elevation pattern has broken into two lobes, with the higher-angle lobe stronger than the desired low-angle lobe.

This demonstrates that it is possible to be too high above ground for a vertically polarized antenna. Lowering the antenna so that the bottom wires are about 2.5 meters (8 feet) above ground (for safety reasons) restores the 10-meter elevation pattern without unduly compromising the 20-meter pattern.

Portable It Is

A winning feature of this antenna is that it is so simple to put up, take down, transport and store away until it is needed again. When I am finished using the antenna and it's time to move on, I just lower the array and roll the wire elements onto the 2x2's. I put a plastic bag over each end of the rolled-up array and tie the bag with string so that the wires don't come off the ends of the 2x2's. I then put it in the ski boot of a car, or in the back of a family van and away we go. At home, it takes very little space to store and it is always ready to go—No bother, no fuss.

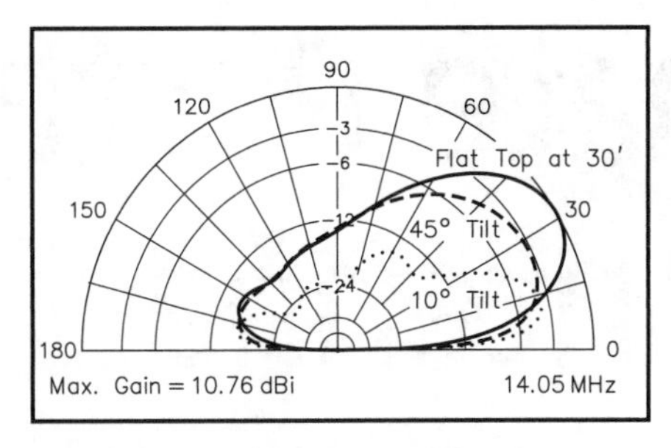

Figure 4—Comparison of elevation patterns for VE7CA Yagi as a horizontal flat top (solid line), tilted 45° from vertical (dashed line) and tilted 10° from vertical (dotted line).

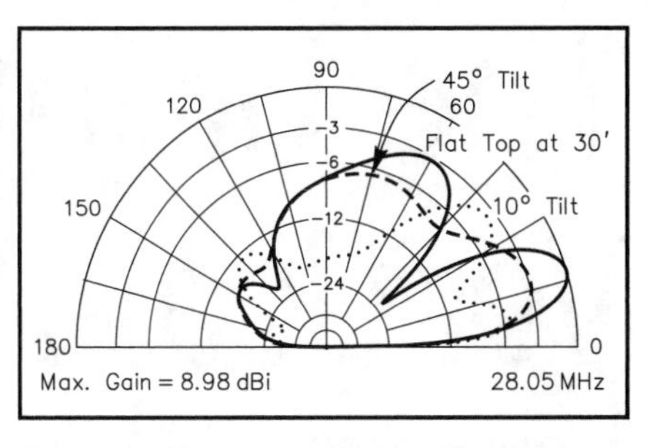

Figure 5—Same antenna configurations as shown in Figure 4, but at 28.05 MHz. On 10 meters, the flattop configuration is arguably best, but the 45° tilted configuration is not far behind.

Testimonial

How well does it work? It works very well. On location I use a bow and arrow to shoot a line over a tall tree and then pull one end of the array up as far as possible. For DX I aim for a height of 20 to 30 meters if possible. For the Canada Day, Field Day and Sweepstakes contests I aim for a height of about 15 meters. This antenna helped me to achieve First Place for Canada, in the 1997 ARRL CW Sweepstakes Contest, QRP category.

The ability to quickly change direction 180° is a real bonus. Late in the 1997 ARRL SS CW contest with the antenna pointed east I tuned across KH6ND. He was the first Pacific station I had heard during the contest and obviously I needed to work him. After trying many times to break through the pileup and not succeeding, I flipped the antenna over to change the direction 180° and then worked him on my next call. Figure 6 shows the azimuth pattern at 21.05 MHz for the beam mounted with a 10° slope from vertical. There is a very slight skewing of the azimuthal pattern because the slope away from purely vertical makes the antenna geometry asymmetrical.

VE7NSR, the North Shore Amateur Radio Club, has used this antenna sloped at about 45° for the last two years on 20 and 10 meters on Field Day with good success. The title photo shows theantenna attached to a tower during Field Day.

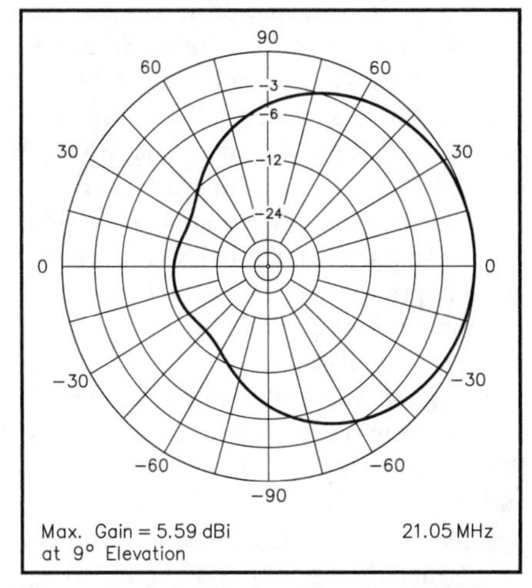

Figure 6—Azimuthal pattern for VE7CA Yagi tilted 10° from vertical on 15 meters.

As they say, the proof is in the pudding. If you need a 20 to 10 meter antenna with gain, this has to be one of the simplest antennas to build, and it will work every time!

Markus Hansen, VE7CA, was first licensed as VE7BGE in 1959. He has been a member of ARRL since he received his license. His main interests include DX, collecting grids on 6 meters, contesting and building his own antennas and various types of ham-radio equipment. He is also an ardent CW operator. Markus has had two previous articles published: "The Improved Telerana, with Bonus 30/40 Meter Coverage," in The ARRL Antenna Compendium Vol 4 *and "Two Portable 6-Meter Antennas" in* The ARRL Antenna Compendium Vol 5. *You can contact Markus at 674 St Ives Cres, North Vancouver, BC V7N 2X3, Canada, or by e-mail at* **ve7ca@rac.ca**.

You can download the EZNEC input-data files as ***VE7CA-1.ZIP*** *from ARRLWeb* (**www.arrl.org/files/qst-binaries/**).

QST

The Ultimate Portable HF Vertical Antenna

Phil Salas, AD5X

From *QST*, July 2005

I've had a tremendous response to the portable antenna project published in the July 2002 *QST*. Many folks had problems locating some of the parts, however. It seems that sprinkler-system parts are not as common in other sections of the country as they are in Texas! Also, the original article required that you drill and tap screw holes in brass couplings and the sprinkler-system risers. I wanted to eliminate the screw-hole tapping to simplify the construction.

Over the years, the antenna evolved from the original design through changes that used fiberglass, aluminum tubing, and then brass tubing. The most recent design is longer, lighter and more compact (when disassembled) than the original antenna. It is also easier to build, and easier to the find parts. Band coverage is also increased to include 60 meters, as well as 40 through 10 meters!

The Ultimate Portable Antenna is designed for easy transport. It breaks down into multiple mast sections, a whip section, an air-wound center loading coil section, and a small base support. No piece is longer than about 20 inches, so it will easily fit into most suitcases. The lead photo shows the disassembled antenna, including guys and radials. The fully assembled antenna has a length of almost 16 feet! See Figure 1. That's me, standing next to the antenna set up on my front lawn, in Figure 2.

The key to efficiency for short antennas (such as a loaded vertical less than a quarter wavelength) is the length. The longer the antenna, the greater the radiation resistance and therefore the less impact you have on efficiency due to ground and loading coil losses. This antenna is almost $^1/_4$ λ on 20 meters. For operation on 17 through 10 meters, you will shorten the antenna to $^1/_4$ λ. The length of the complete antenna minimizes the required loading coil for 60, 40 and 30 meters. So let's build it!

AD5X

The complete unassembled antenna.

AD5X

Figure 1—The complete antenna setup in the author's front yard.

AD5X

Figure 2—Standing next to the antenna to illustrate the height of the tubing sections and coil.

Gathering the Parts

You can find most of the parts for this antenna at your local hardware store. The loading coil, coil taps, 10 foot telescoping whip, and SO-239 are available from MFJ (**www.mfjenterprises.com**). Be sure to mention this *QST* article when ordering the kit of parts (MFJ-1964-K) for a special price discount. Table 1 is the complete parts list.

A few notes about pipe sizes may be in order for anyone not familiar with plumbing fixtures. When you go to the hardware store to purchase the $^1/_8$ inch NPT nipples and couplers, don't expect to find anything measuring $^1/_8$ inch! The outside diameter of a $^1/_8$ inch pipe is 0.405 inches, or about $^{13}/_{32}$ inch. The standard threads are 27 turns per inch. (The $^1/_8$ inch designation comes from the approximate inside dimension of the pipe, although today you may find pipes with different wall thicknesses and the same outside dimensions. For this project you don't have to worry about the wall thickness of your fittings.) The NPT specification

Table 1
Parts List

Qty	Item
1	5 inch long × 2.5 inch diameter × 10 TPI air wound coil (MFJ-404-008)*
1	10 foot telescoping whip (MFJ-1954)*
1	SO-239 chassis mount connector (MFJ-610-2005)*
5	Coil clips (MFJ-605-4001)*
2	3 foot pieces of 3/8 inch diameter brass tubing (ACE Hardware) (McMaster-Carr 8950K581 — 6 foot length)**
1	3/8 inch diameter wood dowel. (36 inch length at Home Depot—only 3 1/2 inch needed.)***
1	3/4 inch PVC T — All of the PVC fittings are white schedule 40 pipe (Home Depot)
1	3/4 inch slip × 1/2 inch female pipe thread PVC adapter (Home Depot)
1	3/4 inch slip × 1/2 inch slip PVC adapter (Home Depot)
1	3/4 inch slip PVC plug (ACE Hardware)
1	1/2 inch NPT (male thread) to 1/8 inch NPT (female thread) brass adapter bushing (ACE Hardware)(McMaster-Carr 50785K64)
8	1/8 inch NPT brass couplings (Home Depot)(McMaster-Carr 50785K91)
4	0.7 inch long 1/8 inch NPT all-thread nipples (these are also called "close nipples") (Home Depot) (McMaster-Carr 50785K151)
2	#8 brass wing-nuts (ACE Hardware)
2	#8-32 × 3/4 inch brass machine screws (ACE Hardware)
2	#8-32 brass nuts (pack of 6) (Home Depot)
2	#8 copper-plated steel (or brass) split lock washers (ACE Hardware)
1	36 inch length of 1/8 inch diameter brass rod (Home Depot)
1	3/8-16 × 1 1/4 inch hex head bolt, zinc plated. Choose the longest you can that is threaded all the way to the head. (Home Depot)
1	3/8-16 × 12 inch hex head or carriage bolt, zinc plated. Choose the longest bolt you can find in this size if you can't find a 12 inch bolt (Home Depot)
1	3/8-16 coupler, zinc plated (ACE Hardware)
2	3/8-16 nuts, zinc plated (pack of 6) (Home Depot)
1	3/8 inch lock washer, zinc plated (pack of 10) (Home Depot)
4	#6 stainless steel 3/8 inch sheet metal screws (ACE Hardware)
3	#8 solder lugs (Home Depot)
6	#8 × 1 1/2 inch brass wood screws (pack of 6) (Home Depot)
90 feet	Wire (any gauge, insulated or not, for six 15-foot ground radials.)
1	Alligator clip and miscellaneous short pieces of connecting wire.

*The total retail price for all the MFJ parts is $53.80, plus shipping/handling. Mention this article when you order and MFJ will sell a kit containing the telescopic whip, coil, coil clips, and SO-239 connector for $39.95 plus $6 s/h. The kit is MFJ-1964K. You can order additional coil clips for $2.95 each.

You can use 3/8 inch aluminum tubing if you prefer. Aluminum tubing is about half the cost of brass, but you will either need to drill the brass couplings and aluminum tubing so they can be connected with stainless steel sheet-metal screws or solder them together using aluminum solder and a torch (see **www.solder-it.com).

***You can use 3/8 inch fiberglass rod instead of the wood dowel if you prefer, but fiberglass rod is more difficult to find. Check out bicycle flags and driveway marker stakes as potential sources of fiberglass rod.

The brass plumbing items are also available from McMaster-Carr, (**www.mcmaster.com**) a mail-order supplier with no minimum order requirement. They stock 6 foot lengths of 3/8 inch brass tubing as well as the close nipples, couplings and brass bushings. The Table lists part numbers to help you look up those items on the Web site.

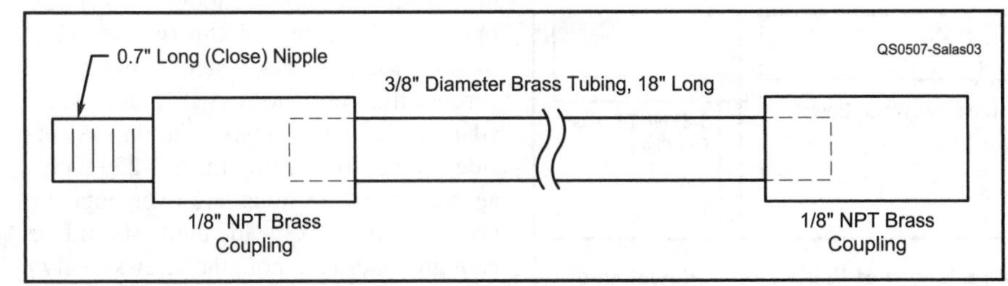

Figure 3—Assembly of the brass tubing sections.

means National Pipe Thread. There are several other thread specifications, but NPT is the most common.

I used 1/8 inch NPT close nipples for the antenna connectors. These are about 0.7 inches long, and are usually fully threaded over over the entire length. The PVC pipe fittings must be the white, schedule 40 PVC pipe. Do not use the thinner-walled pipe sometimes known as CPVC, which has a cream color or yellow tint. When fittings are designated as "slip," it means pieces are intended to be glued—they just slip together—rather than being threaded.

Brass Rod Preparation and Assembly

See Figure 3 for the assembly details. First cut three 18 inch sections of the 3/8 inch brass tubes with a hacksaw or tubing cutter and de-burr the tubing. The ends of the couplings that fit over the brass tubes must be reamed out with a 3/8 inch drill bit. Otherwise the couplings won't fit over the tubing. To do this, first screw a 1/8 inch NPT coupling on each end of a 1/8 inch NPT close nipple. Use wrenches to screw these together as tightly as possible. Next, clamp one of the couplings securely and ream out the opposite coupling with a 3/8 inch drill bit. (Use a drill press for this operation if at all possible.) Reverse, and ream out the other coupling. [See Figure 4 for one way to use a woodworker's clamp and a drill press.—*Ed.*] Now unscrew the couplings. One end of the nipples will break loose from one coupling, and the other end will stay tight in the remaining coupling. You'll now have a female and male end that will fit over each end of a section of brass tube, as shown in Figure 5. You will need four pair of these male/female brass connectors: three pair for the brass tubes and one pair

WR1B

Figure 4—A pipe coupling and nipple secured in a woodworker's clamp ready to drill out the end on a drill press.

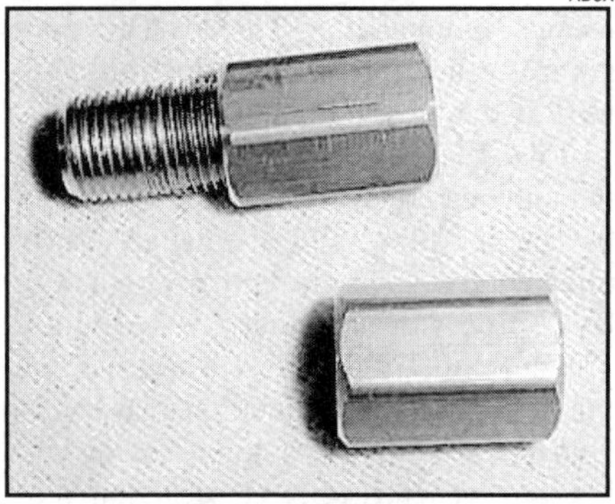

AD5X

Figure 5—After the pipe couplings have been drilled out and one coupling removed from the nipple, the pair is ready to be installed on the ends of an 18 inch length of brass tubing.

for the loading coil assembly. If you'd like, you can solder the nipple/coupling assemblies together. The assembly tends to be very tightly secured even without soldering, however.

Now insert the male/female brass pairs just constructed over all three of the 18 inch brass tubes and solder the couplings directly to the tubes. This is easily done with a large soldering iron, or even better, with a torch and silver solder. Solder-It has a nice small butane torch that works well. See **www.solder-it.com**.

Loading Coil Assembly

Slide $^1/_8$ inch NPT male/female coupling pairs over both ends of a $^3/_8$ inch diameter, $3^1/_2$ inch long wood dowel. You will need to drill a $^1/_8$ inch diameter hole completely through each of the $^1/_8$ inch NPT brass couplings and dowel as shown in Figure 6. Next cut two 3 inch lengths of the $^1/_8$ inch diameter brass rod. Insert one of these 3 inch sections through the holes on one brass coupling. Center the rod so that equal lengths are available on both sides of the coupling, and solder the rod to the coupling with a large soldering iron or torch. (Be careful not to burn the wood dowel with the torch!)

Now position a 3 inch length of the MFJ-404-008 coil such that the 3 inch brass rod just installed pokes through the last two turns on the coil. See Figure 7. Solder the coil turns to the rod. On the opposite end of the coil assembly, insert the remaining 3 inch brass rod through two adjacent turns on this end of the coil, through the brass coupling, and through the coil turns. Solder the coil turns to the brass rod, and then solder the brass rod to the coupling.

Next, indent every other turn on the coil with a small flat-head screwdriver. You should do this on opposite sides of the coil to give you plenty of adjustment capability.

Finally, on the end of the coil with the brass nipple (male end), solder a 6 inch piece of insulated wire terminated with an alligator clip. For extended outdoor use, you may wish to treat the wood dowel with varnish.

Top Whip

The MFJ-1954 10 foot telescopic whip comes with a standard $^3/_8$-24 mounting thread. While the whip mounting stud is not a correct match for the $^1/_8$ inch NPT coupling (27 turns per inch), I just thread the screw directly into the $^1/_8$ inch NPT coupling. Because the $^3/_8$ bolt is slightly smaller than the $^1/_8$ inch pipe outside diameter, and the pipe coupling is tapered inside, they will go together. [See the "Alternative Construction Ideas" sidebar for a way to match the threads, if you are uncomfortable with the mismatch.—*Ed.*]

Base Assembly

For the base spike, I've used a $^3/_8$-16 × 12 inch zinc-plated hex-head bolt. Only $1^1/_2$ inches of the bolt is threaded, and so I used the long smooth end of the bolt to go into the ground after cutting off the hex head. A damp cloth easily cleans the bolt after use. I also used a $^3/_8$-16 × $1^1/_4$ inch

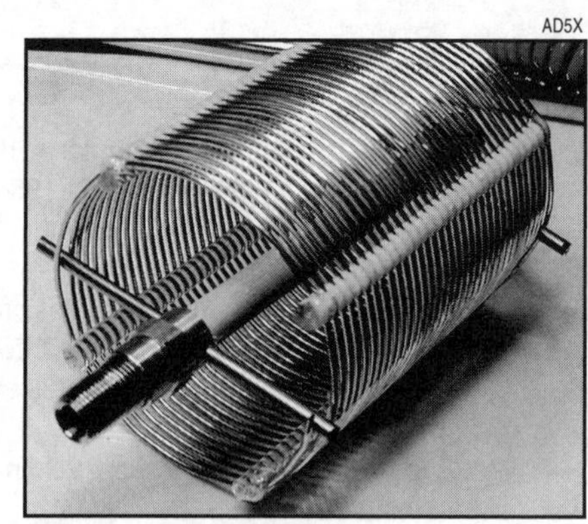

Figure 7—The loading coil assembly ready to be soldered.

zinc-plated hex-head bolt at the base of the PVC assembly, and a $^3/_8$-16 zinc-plated coupler to attach the $1^1/_4$ inch bolt to the 12 inch bolt, as shown in Figure 8. This way you can leave the long bolt off if you want to bolt the base assembly directly to a metal plate or trailer mount, or screw on the long bolt for ground mounting.

Referring to Figure 8, drill a $^3/_8$ inch diameter hole into the $^3/_4$ inch PVC plug used for the base support $1^1/_4$ inch bolt. Cut off about half of the length of the $^3/_4$ inch PVC plug to leave plenty of room inside the T for wiring. Solder a ground wire to the head of the $^3/_8$-16 × $1^1/_4$ inch bolt as shown, or use a $^3/_8$ inch solder lug. Insert the threaded end of bolt into the plug, and secure with a $^3/_8$-16 nut and a lock washer. If you wish, you can glue the plug in place with PVC pipe cement instead of using the #6 stainless steel sheet metal screws shown. The screws make changing the support assembly easy in case you should ever want to, though. You might even consider using screws to attach the other two adapters into the T.

To prepare the 12 inch bolt, cut off the hex head and round this end with a file. Screw a $^3/_8$-16 nut onto the threaded end, then add a lock washer and screw the $^3/_8$-16 coupler onto the threads so the bolt is about halfway through the coupler. Tighten the nut against the coupler (with a lock washer between the nut and coupler). This 12 inch bolt assembly can now be easily screwed onto the $1^1/_4$ inch bolt on the base assembly for ground mounting.

Now place the SO-239 over the $^1/_2$ inch hole in the $^3/_4$ to $^1/_2$ inch slip adapter and mark the location for the two #6-32 × $^3/_8$ inch long stainless steel sheet metal screws that will hold it in place. (The adapter I used had a hex head on the outside lip, and by turning the SO-239 so diagonal mounting holes are over opposite points in the hex head, there should be enough material to hold the screws easily.) You'll see that these holes will be right in

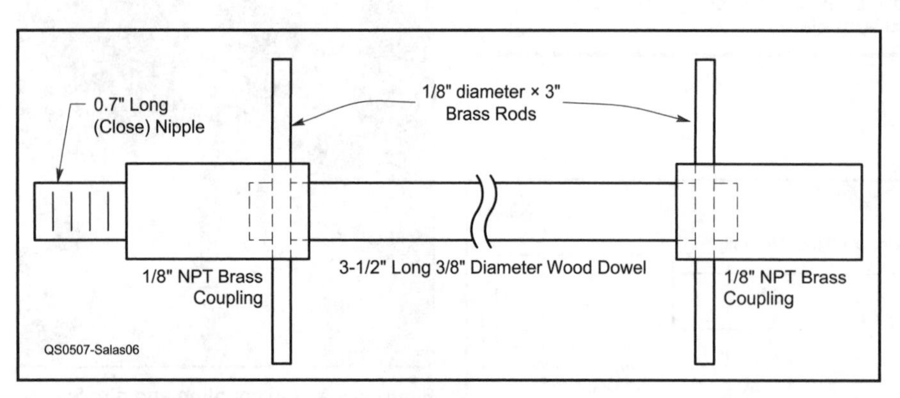

Figure 6—How brass couplings are installed on a length of $^3/_8$ inch wood dowel and $^1/_8$ inch brass rod in order to mount the coil.

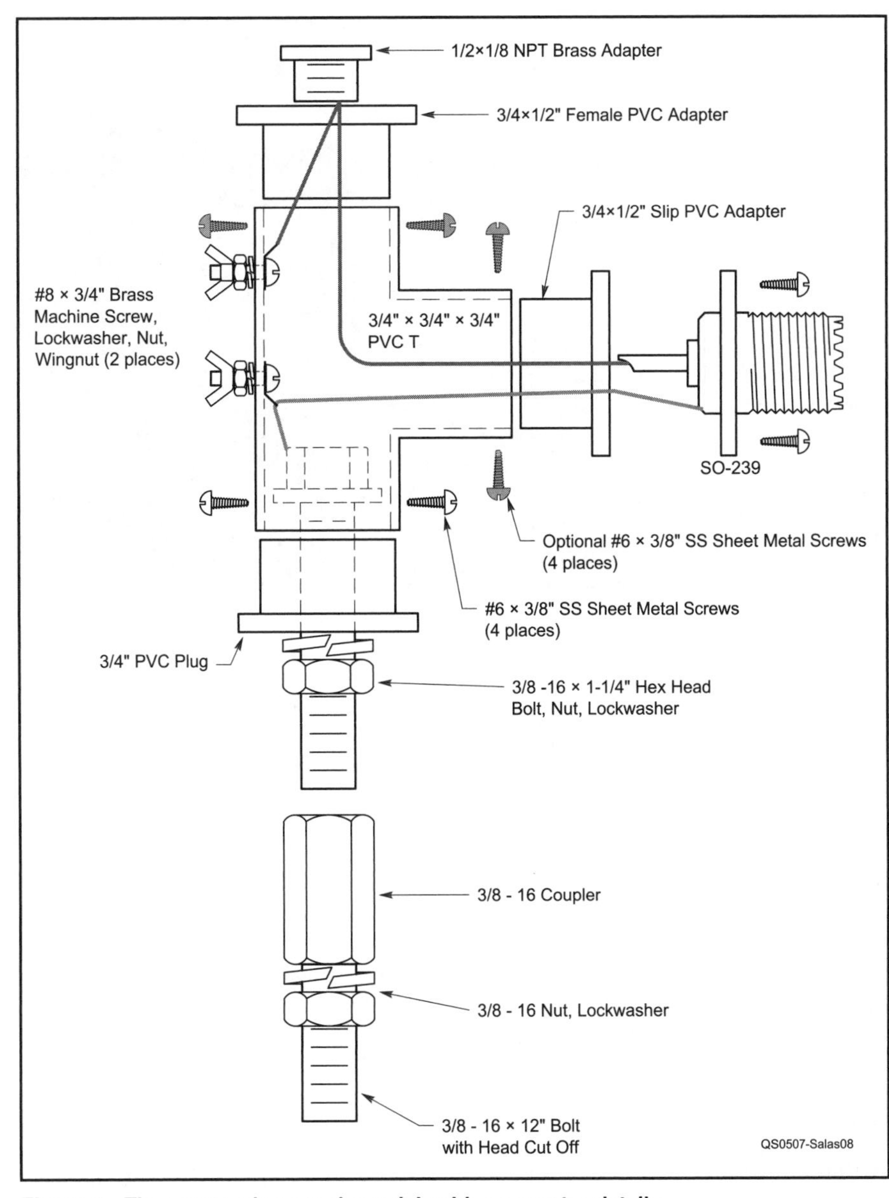

Figure 8—The antenna base and coaxial cable connector details.

the center of the PVC lip. Carefully drill two $^{5}/_{64}$ inch holes at these points. We will mount the SO-239 on the adapter later.

Place the $^{3}/_{4}$ inch PVC plug/spike assembly in the T and drill two $^{5}/_{64}$ inch diameter holes through the T and plug. [See sidebar Photo A for how I clamped the T in a woodworker's clamp to drill the holes. Try to drill straight through both sides in one pass to make alignment easier.—*Ed.*] Remove this assembly from the T and drill out these $^{5}/_{64}$ inch holes in the T to $^{9}/_{64}$ inch. Also drill out two holes in the SO-239 connector to $^{9}/_{64}$ inch, since the holes are not large enough to pass the #6 × $^{3}/_{8}$ inch sheet metal screws.

Use an $^{11}/_{64}$ inch drill bit to drill mounting holes for the #8-32 brass machine screws. I drilled the mounting holes close to the "top" and "bottom" of the T. Alternatively, if you align the bit with the inside edges of the side of the T and drill through the back of the T, you will be able to fit a screwdriver into the T to tighten this hardware.

Next we'll prepare the antenna interface at the top of the base. First, cut off part of the $^{3}/_{4}$ inch slip × $^{1}/_{2}$ inch female pipe thread PVC adapter so as to leave additional room in the T for wiring. Solder a piece of 14 gauge copper house wire directly to the inside lip of the $^{1}/_{2}$ × $^{1}/_{8}$ inch NPT brass adapter. You'll need a large soldering iron or a torch, since the brass adapter mass is pretty large. Screw this adapter tightly into the $^{3}/_{4}$ × $^{1}/_{2}$ inch PVC adapter.

Now solder a wire to the center conductor of the SO-239 connector as shown. This wire should be soldered to the wire stub on the $^{1}/_{2}$ × $^{1}/_{8}$ inch NPT brass adapter at the antenna interface, and then to the upper wing-nut assembly as shown. (Alternatively, you could use #8 solder lugs for these connections, and put the top #8-32 brass machine screw through the hole in the solder lug.) The $^{3}/_{4}$ × ½ inch PVC adapter can now be glued into place using PVC pipe glue. Solder a short piece of copper braid (from a piece of RG-58 cable) from the SO-239 ground (solder directly to the SO-239 body) to the brass ground screw, and finally to the wire soldered to the head of the 1$^{1}/_{4}$ inch bolt. (#8 solder lugs are a good alternative.) You can now complete the assembly of the base by inserting the PVC plug and 1$^{1}/_{4}$ inch bolt assembly into the T and installing the #6 stainless steel sheet metal screws as shown in Figure 8. Incidentally, the upper wing-nut assembly is used in case you need to add capacitive or inductive base matching should you want to improve the SWR on the lower bands. See Figure 9.

Ground Radial Network

The radial network is made up of six 15 foot radials, using 22 gauge insulated wire, though any gauge wire, insulated or not, can be used. I've found it best to make up three pairs of two wires each attached to a #8 spade lug on one end of each pair. This minimizes the hassle of deploying, and later rolling up, the radials. The three #8 lugs will attach to the ground screw on the base assembly. When the wires are rolled up, you should hold them together with twist-wraps. Solder a 1$^{1}/_{2}$ inch brass wood screw on the outer end of each radial. You can simply push these screws into the ground to hold the radials in place. Put a blob of hot glue on each wire/screw soldered interface to give it a little strain relief.

Guying

This antenna is self-supporting in a low breeze. In many cases, however, it will be necessary to guy the antenna because of its 16 foot length. For effective guying, I attached 9 foot lengths of nylon cord (3 pieces) just above the base of the 10 foot MFJ telescoping whip. Use a tie-wrap and close it just enough so that it won't slide over the base of the MFJ whip. Cut the 9 foot sections of nylon cord and heat the ends with a match to fuse the nylon so it won't unravel. Tie one end of a 9 foot section of nylon cord around each tie-wrap and secure with hot glue or epoxy. For the ground stakes, you can use the extra piece of brass tubing. (You only used 4$^{1}/_{2}$ feet of the 6 foot length.) Cut the remaining 18 inch piece of tubing into three 6 inch sections.

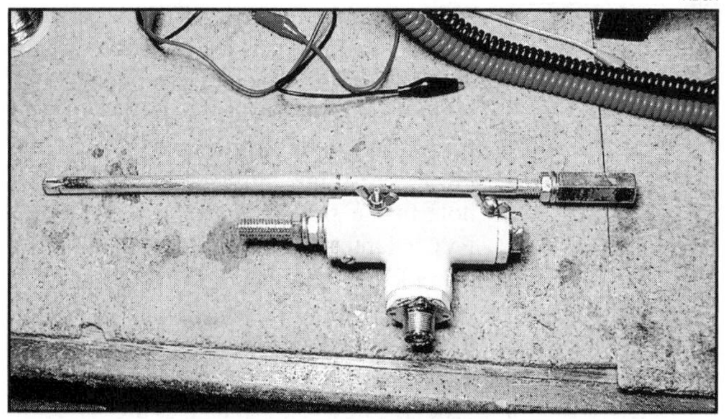

Figure 9—The completed antenna base and mounting spike.

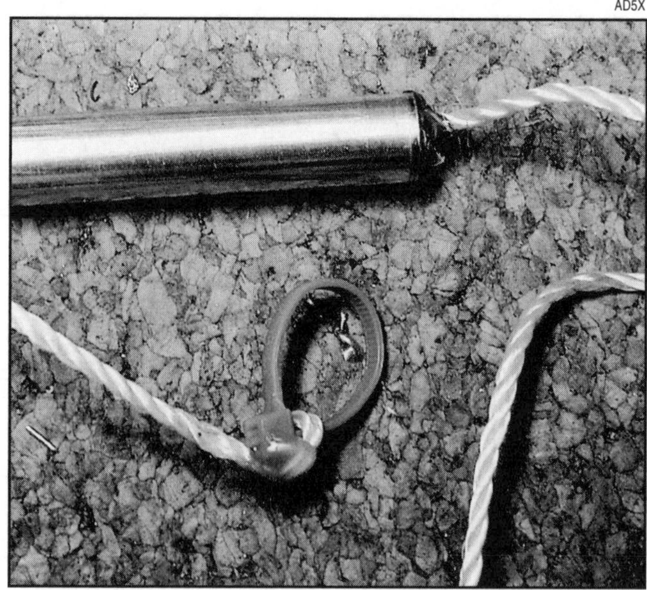

Figure 10—A set of 3 guys was made by using a wire tie (closed to slip over the top of the telescoping whip but not over the base of the whip). Hot glue was used to attach a nylon line to each wire tie. Hot glue was also used to attach the guy line to a left-over length of brass tubing.

Attach the end of the nylon cord without a tie-wrap to one end of each tube with hot glue. Also plug the open ends of the three tubes with hot glue. (You could also use some long spikes or bolts for stakes, if you are concerned that you won't be able to drive the brass tubing into hard ground.) For storage, wrap the nylon cord around each brass stake and hold it in place with masking tape. Figure 10 shows the details. See Figure 11 for a photo of the guys attached to the telescoping MFJ whip. When bolted to a trailer mount or plate, the antenna should really not need guying unless the wind is strong.

Antenna Assembly

To assemble the antenna, first screw the three brass-rod sections together, and then screw these into the top of the base assembly. Push the entire assembly firmly into the ground, keeping it as vertical as possible. Next, screw the loading coil and telescoping whip assemblies together. Slip the three guy tie-wrap/nylon cord assemblies over the whip and extend the telescoping whip. Screw this entire top assembly into the female end of the top brass tube. You only need to turn the brass fittings finger tight. Finally, push the guy rods in the ground and extend the six radials. Attach the common ends to the ground screw on the base assembly. Add coax, and you are ready to tune your antenna.

Antenna Tuning

Begin with 60 meters to determine the tap points on the coil. If your rig cannot tolerate a 2:1 SWR, you may need to add a 330 pF, 300 V, silver mica capacitor across the two wing-nut assemblies. See Figure 12 for a close-up view. This capacitor is fine for both 60 and 40 meters. The SWR on 30 meters will be closer to 1.7:1. If you need to improve this, use a 220 pF capacitor. You should not need any capacitors for 20 through 10 meters. Both my IC-706MKIIG and SG-2020 work fine into a 2:1 SWR.

To begin, use an antenna analyzer set to 5340 kHz to locate the coil tap point that gives the best SWR. Mark this tap point. Repeat the procedure at 5380 kHz. Move to 40 meters and repeat, again selecting two taps on 40 meters to give you the band coverage you desire. Repeat again for 30 meters—only a single tap is required for

Figure 11—A close-up of the guy lines attached to the telescoping whip. The coil and coil clips detail can also be seen.

Figure 12—The antenna base, with the ground radials and a matching capacitor attached to the brass screws on the PVC T.

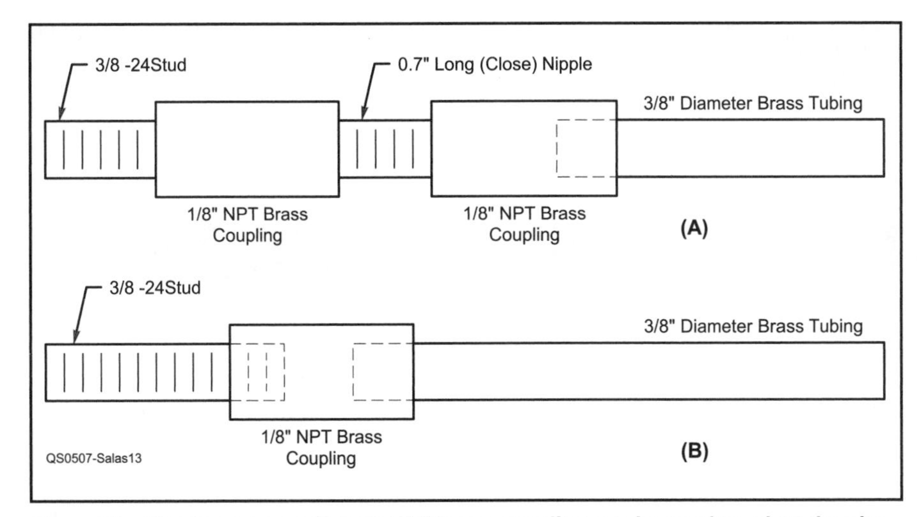

Figure 13—(A) shows how a 1/8 inch NPT brass coupling can be used to adapt the pipe threads to a 3/8-24 standard antenna mount. (B) shows an alternative attachment, if one only wants to attach the antenna to a standard mount, and not use the PVC base to mount the antenna on the ground.

this band. For 20 meters, you will find that only the top turn of the coil is necessary for resonance. The antenna is almost a quarter wavelength long on 20 meters.

For 17, 15, 12 and 10 meters, you will need to short out the loading coil and remove sections of the brass tube—then adjust the whip for resonance. On these bands, the antenna will be 1/4 λ long. Remove two brass sections for 17 and 15 meters, and all three sections for 12 and 10 meters. Use a permanent black marker pen to indicate the correct band positions on the telescoping whip.

Finally, attach the MFJ-605-4001 coil clips to the coil tap points determined above. You may wish to solder the clips in place to make the entire assembly a little more robust. From this point forward, you can just go back to these tap points, or re-adjust the top whip as necessary, and not have to worry about making SWR measurements.

Mounting Options

You can easily make a 3/8-24 threaded interface so that the antenna can be mounted on a standard 3/8-24 antenna mount. This would be useful for those with a standard ball mount on their car who want to use this extended-length antenna when stopped. As mentioned earlier, the 1/8 inch NPT thread is 27 turns per inch, and it is slightly larger than 3/8 inch diameter, with a slight taper. While the 3/8-24 standard stud will screw into an 1/8 inch NPT thread coupling, the 1/8 inch NPT nipple will not screw into a 3/8-24 threaded coupling. Therefore, an adapter is necessary if you want to mount this antenna to a standard 3/8-24 antenna mount.

One way to make an adapter is to purchase a 3/8-24 bolt and screw it tightly into a 1/8 inch NPT coupling. Cut off the head of the 3/8-24 bolt with a hacksaw and file carefully so that you don't damage the threads. Running a 3/8-24 die over the threads will clean up any damage you may have done. You can now either screw this assembly onto the 1/8 inch NPT nipple on the bottom brass tube section, as shown in Figure 13A, or screw the 3/8-24 bolt directly into the bottom 1/8 inch NPT coupling as shown in Figure 13B (if you'll never need the 1/8 inch NPT interface on the bottom antenna section).

Conclusion

Because of the interest in my original portable antenna, I've evolved that design into an antenna that is longer, lighter, more compact and easier to fabricate, and gives you more mounting options. You can also experiment with the antenna length. For example, you can remove a section or two, use more or fewer sections, decrease or increase section lengths, or place the loading coil in different positions. With the loading coil described in this article, you have quite a bit of latitude in the antenna length for a given band. For best efficiency though, try to keep the antenna as long as possible and the coil as high as possible.

Don't hesitate to make changes based on hardware availability. Try aluminum or copper tubing, or even wire wrapped 3/8 inch fiberglass or wood dowel. It's fun to design antennas "on the fly" while standing in the plumbing section of your hardware store. This makes for interesting discussions with the clerks, however!

Finally, I want to express my appreciation to Martin F. Jue, K5FLU, and Richard Stubbs, KC5NSZ, of MFJ, for working with me to provide a reasonably priced kit of parts to make this both an affordable and fun project to build.

Phil Salas, AD5X, is an ARRL Life Member. He's been licensed for 41 years, and enjoys HF operating (mostly CW). Phil's wife Debbie (N5UPT) and daughter Stephanie (AC5NF) are obviously very understanding of this hobby! Phil holds a BSEE from Virginia Tech, and an MSEE from Southern Methodist University, and is now fully retired after 33 years in the telecommunications industry. Phil can be reached at 1517 Creekside Dr, Richardson, TX 75081-2913 or at **ad5x@arrl.net** *if you have any questions or comments.*

By Phil Salas, AD5X

From *QST*, December 2000

A Simple HF-Portable Antenna

Tired of dragging that bulky old antenna tuner along on your vacation jaunts? Spare your suitcase and your pocketbook because this simple multiband wire antenna will get you on the air in a jiffy—with no extra gear required.

Every summer my wife (N5UPT), my daughter (AC5NF) and I spend about a week on Mustang Island off the coast of Corpus Christi, Texas. I always enjoy operating HF-portable when on vacation, and because Mustang Island is also known as IOTA NA092 (Islands On The Air, North American island number 92), getting on the air is even more fun! In case you're imagining typical DXpedition fare, you should know right from the start that we don't exactly rough it on Mustang Island. In fact, we always stay in a condo, which I request to be "the highest one available."

My first portable rig was a Kenwood TS-50, followed by an MFJ-9420 (see May 1999 *QST*). Last year I went deluxe and upgraded to an ICOM IC-706MKII. That little rig works dc to light—all bands and all modes, with goodies to boot. It it is an excellent choice for almost any type of portable operation.

I've experimented with several types of antennas on these outings—including Hamstick mobile whips, resonant dipoles and random-length wire dipoles fed through a tuner. I prefer resonant antennas so I don't have to worry about transporting and storing an antenna tuner. Of course, multiple dipoles or a handful of Hamsticks can take up a lot of room.

Last summer I used the multiband dipole described here with excellent results. If you're interested in a simple multiband wire that's easy to build and pack away, give this antenna a try.

The basic antenna covers all bands from 20-10 meters. You could increase its coverage, but the dimensions of a typical condo balcony seem to limit the lower frequency to 20 meters or so. If your operating site is larger, feel free to scale the antenna appropriately.

Basically, the antenna started as a full-size 20-meter dipole. I then inserted small in-line insulators to allow for multiband operation as shown in Figure 1.

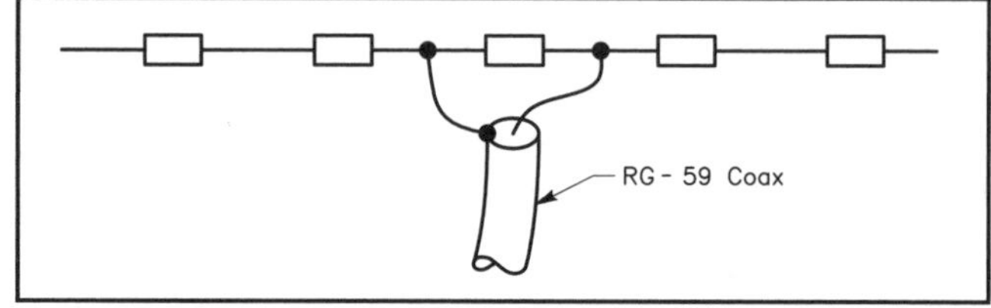

Figure 1—The concept began with a full-size dipole antenna that I "broke up" with small insulators.

The insulators are ³/₈-inch (diameter) by one-inch nylon spacers that can be found at most hardware stores. Each spacer is used as a "band switch" by drilling a small hole in each end and threading a short length of #14 bare wire (house wire) through each end, and attaching a short piece of wire terminated in an alligator clip. The clip, shown in Figure 2, is available at RadioShack stores (ask for part number 270-380).

I used #24 insulated wire for the dipole elements because it's lightweight and flexible. Obviously, any type of wire is fine. Use whatever you have on hand. The best way to determine the various segment lengths is to calculate the individual dipole lengths using:

L (feet) = 468/freq (MHz)

Tack solder the wire sections to the insulators, attach a feed line (RG-59 coax will do) and hang the dipole in a convenient place where it's easy to work on and adjust. Although the SWR meter method will work, to adjust the multiband dipole prop-

The entire antenna can be collapsed to a size that fits in the palm of your hand!

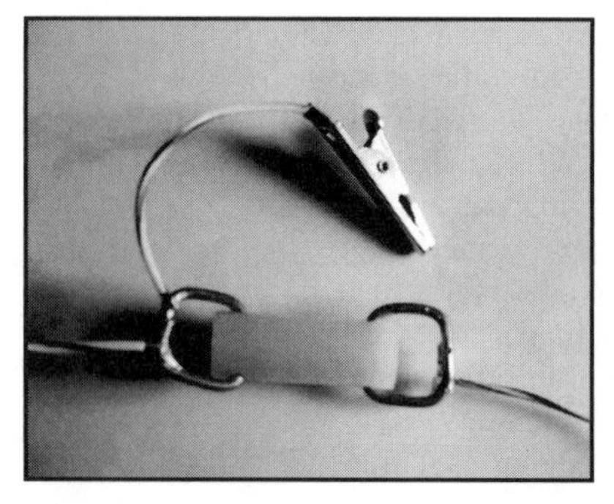

Figure 2— The band switches are constructed from nylon spacers, some wire and an alligator clip.

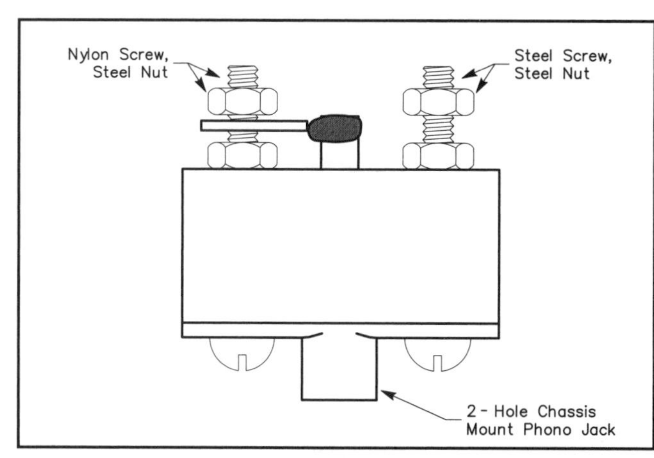

Figure 3 — I used an extra nylon spacer for the center insulator. I drilled the ends and attached a chassis-mount phono jack as shown. The nylon screw is used on one side to make sure that the phono jack's center conductor doesn't short to ground. I soldered #4 spade lugs to the inside ends of the 10-meter dipole elements so the dipole can be easily attached (and detached) to the center insulator. Feel free to use other center insulator designs as desired.

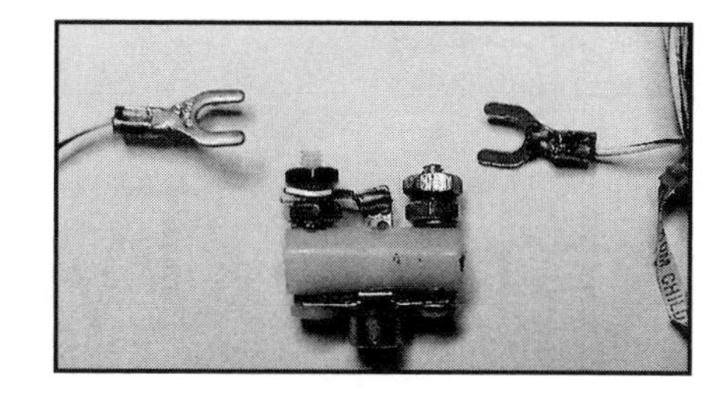

A photograph of my version of the center insulator.

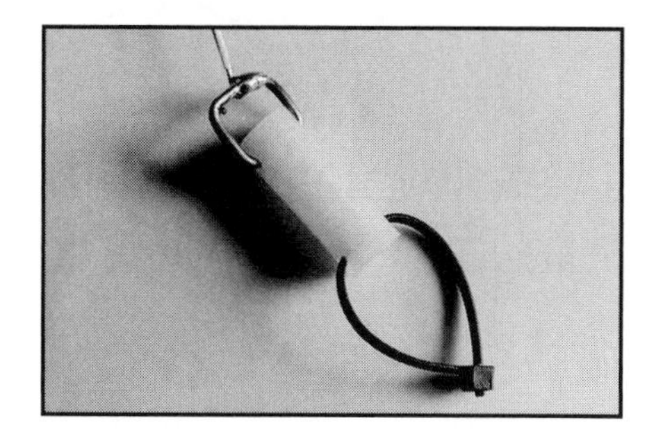

My design for the end insulator.

erly, beg, borrow or buy an antenna analyzer.

First, "unclip" all of the alligator clips and adjust the inner wire segments for the lowest SWR on your favorite part of the 10-meter band. The wires should be a bit long, so unsolder them on one end and trim them as follows:

New length = Original length × Measured low-SWR Frequency/Desired low-SWR frequency

Next, clip (attach) the inner pair of alligator clips and adjust the next segment for resonance on 12 meters using the formula and steps described previously. Continue this procedure for 15, 17 and 20 meters.

I know—you're adjusting your antenna low to the ground and your particular portable mounting location will undoubtedly vary. For our purposes it really doesn't matter. Most modern rigs can put out full power into a 2:1 SWR, so reasonable location-based SWR variations probably won't affect your rig's operation. If the SWR is really high, something's drastically wrong or you have the alligator clips set up for operation on the wrong band, etc. Incidentally, you can use a balun if you want to. I normally don't worry about feed line transformers when operating portable.

The antenna leg lengths I wound up with are shown below:

10 meters: 8 feet 3 inches on each side
12-10 meters: 10 inches on each side
15-12 meters: 1 foot 4 inches on each side
17-15 meters: 1 foot 8 inches on each side
20-17 meters: 3 feet 9 inches on each side

Each side is a total of 15 feet, 10 inches, for a total of 31 feet, 8 inches for the entire antenna.

Finally, if you want to electrically "shorten" your antenna, make the clip lead wires a little longer and wrap the excess wire around the insulators to make loading coils.

I used an extra nylon spacer for the center insulator. I drilled the ends and attached a chassis-mount phono jack as shown in Figure 3. The nylon screw is used on one side to make sure that the phono jack's center conductor doesn't short to ground. I soldered #4 spade lugs to the inside ends of the 10-meter dipole elements so the dipole can be easily attached (and detached) to the center insulator. Feel free to use other center insulator designs as desired.

Conclusion

If you need a simple portable antenna, spend an hour or two assembling this one. It's simple, cheap and a good performer. Simply adjust the clip leads for the desired frequency band and you're on the air—no tuner required! Sure, you have to make a quick trip to the balcony (or whatever) to change bands...but this is a vacation-oriented design, after all!

1517 Creekside Dr
Richardson, TX 75081
ad5x@arrl.net

By Rich Wadsworth, KF6QKI

From *QST*, February 2002

A Portable Twin-Lead 20-Meter Dipole

With its relatively low loss and no need for a tuner, this resonant portable dipole for 14.060 MHz is perfect for portable QRP.

My first attempt at a portable dipole was using 20 AWG speaker wire, with the leads simply pulled apart for the length required for a 1/2 wavelength top and the rest used for the feed line. The simplicity of no connections, no tuner and minimal bulk was compelling. And it worked (I made contacts)!

Jim Duffey's antenna presentation at the 1999 PacifiCon QRP Symposium made me rethink that. The loss in the feed line can be substantial, especially at the higher frequencies, if the choice in feed line is not made rationally. Since a dipole's standard height is a half wavelength, I calculated those losses for 33 feet of coaxial feed line at 14 MHz. RG-174 will lose about 1.5 dB in 33 feet, RG-58 about 0.5 dB, RG-8X about 0.4 dB. RG-8 is too bulky for portable use, but has about 0.25 dB loss. For comparison, *The ARRL Antenna Book* shows No. 18 AWG zip cord (similar to my speaker wire) to have about 3.8 dB loss per 100 feet at 14 MHz, or around 1.3 dB for that 33 feet length. Note that mini-coax or zip cord has about 1 dB more loss than RG-58. Are you willing to give up that much of your QRP power and your hearing ability? I decided to limit antenna losses in my system to a half dB, which means I draw the line at RG-58 or equivalent loss.

TV Twin Lead

It is generally accepted that 300 ohm ribbon line has much less loss than RG-58. Some authors have stated that TV twin lead has similar loss as RG-58, which is acceptable to me. A coil of twin-lead is less bulky and lighter than the same length of RG-58. These qualities led me to experiment with it. One problem is that its 300 ohm impedance normally requires a tuner or 4:1 balun at the rig end.

But, since I want approximately a half wavelength of feed line anyway, I decided to experiment with the concept of making it an exact electrical half wavelength long. Any feed line will reflect the impedance of its load at points along the feed line that are multiples of a half wavelength. Since a dipole pitched as a flattop or inverted V has an impedance of 50 to 70 ohms, a feed line that is an electrical half wave long will also measure 50 to 70 ohms at the transceiver end, eliminating the need for a tuner or 4:1 balun.

To determine the electrical length of a wire, you must adjust for the velocity factor (VF), the ratio of the speed of the signal in the wire compared to the speed of light in free space. For twin lead, it is 0.82. This means the signal will travel at 0.82 times the speed of light, so it will only go 82% as far in one cycle as one

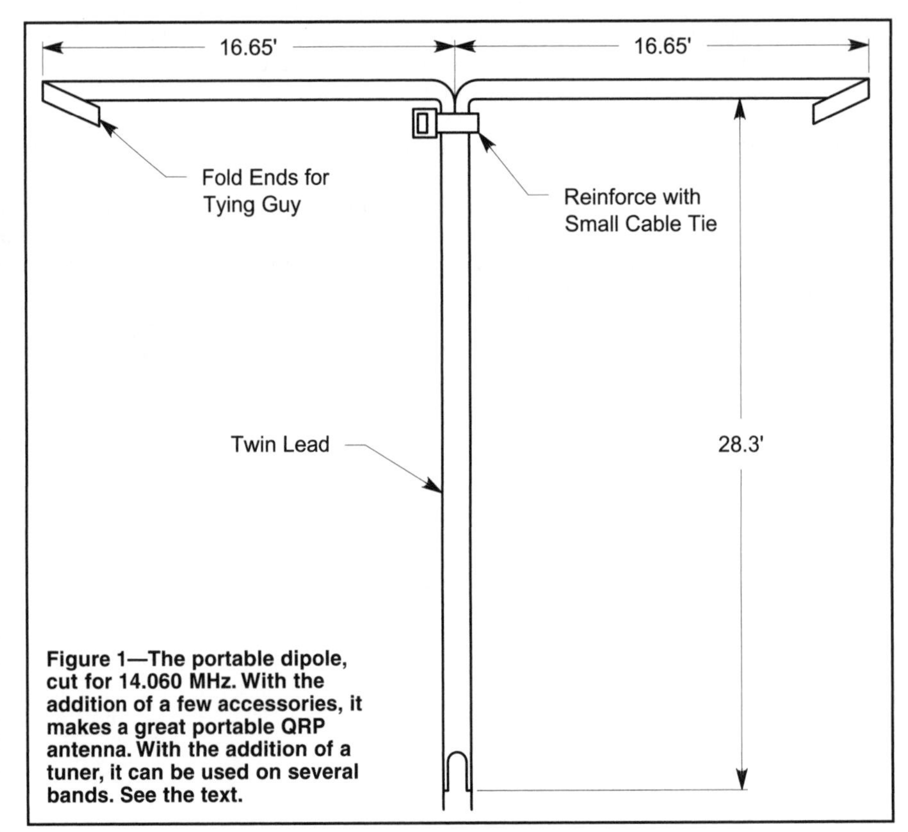

Figure 1—The portable dipole, cut for 14.060 MHz. With the addition of a few accessories, it makes a great portable QRP antenna. With the addition of a tuner, it can be used on several bands. See the text.

would normally compute using the formula 984/f(MHz). I put a 50 ohm dummy load on one end of a 49 ft length of twin lead and used an MFJ 259B antenna analyzer to measure the resonant frequency, which was 8.10 MHz. The 2:1 SWR bandwidth measured 7.76 to 8.47 MHz, or about 4.4% from 8.10 MHz.

The theoretical ½ wavelength would be 492/8.1 MHz, or 60.7 feet, so the VF is 49/60.7=0.81, close to the 0.82 that is published. A ½ wave for 14.06 MHz would therefore be 492×0.81/14.06 or 28.3 feet. I cut a piece that length, soldered a 51 ohm resistor between the leads at one end, and hoisted that end up in the air. I then measured the SWR with the 259B set for 14.060 MHz and found it to be 1:1. I used the above-measured 2:1 bandwidth variation of 4.4% to calculate that the feed line could vary in length between 27.1 and 29.5 feet for a 2:1 maximum SWR.

Now comes the fun part. With another length of twin lead, I cut the web between the wires, creating 17 ft legs, and left 28.3 feet of feed line. I hung it 30 feet high, tested, and trimmed the legs until the 259B measured 1:1 SWR. The leg length ended up at 16.75 feet. (Note: The VF determined above only applies to the feed line portion of the antenna.) There is no soldering and no special connections at the antenna feed point. I left the ends of the legs an inch longer to have something to tie to for hanging. I reinforced the antenna end of the uncut twin lead with a nylon pull tie, with another pull tie looped through it to tie a string to it for using as an inverted V. To connect the feed line to the transceiver, I used a binding post-BNC adaptor that is available from Ocean State.[1] My original intention of leaving the feed line free of a permanent connector was to allow connection to an Emtech ZM-2 balanced antenna binding post connectors. Since then I have permanently attached a short stub of RG-58 with a BNC, because I plan to either use it with my single band 20 meter Wilderness Radio SST, or with an Elecraft K1 or K2 with built-in tuner. I did this by connecting the shield to one side of the twin lead and the center conductor to the other side—no balun was used between the coax and twin lead.

After a year or so of use and further field testing, including different heights and V angles, I further trimmed the legs to a length of 16.65 feet. I found that the lowest SWR was usually obtained with the V as close to 90 degrees as I could determine visually.

[1]Ocean State Electronics, 6 Industrial Dr, Westerly RI 02891, tel 800-866-6626 or 401-596-3080; fax 401-596-3590; e-mail: **ose@oselectronics.com**.

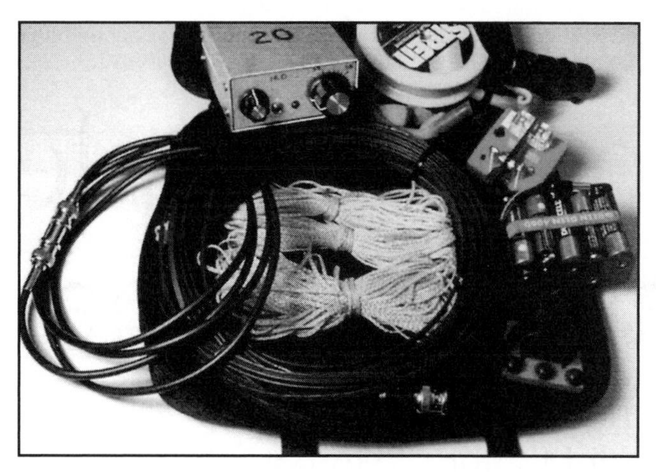

Figure 2—The author's portable station, including twin-lead dipole, 20-meter Wilderness Radio SST transceiver and support line. It all fits in the 8″×10½″×2″ Compaq notebook computer case.

Also, I found that the resonant frequency (or at least the frequency at which SWR was at a minimum) is lower if the antenna is closer to the ground, and vice-versa. For example, with the top of the V at 22 feet, the lowest SWR was measured at around 13.9 MHz, and with the top of the V at 31 feet, SWR was lowest at around 14.1 MHz. In both cases, SWR at 14.060 did not exceed 1.3:1.

I used Radio Shack 22 AWG twin lead that is available in 50 ft rolls. To have no solder connections, you need at least 45 feet. When I cut the twin lead to make the legs, I just cut the "web" down the middle and didn't try to cut it out from between the wires. It helps make the whole thing roll up into a coil, and the legs don't tangle when it's unrolled, since they're a little stiff. It turned out that the entire antenna is lighter than a 25 ft roll of RG-58. This antenna can be scaled up or down for other frequencies also. An even lower loss version can be made with 20 AWG 300 ohm "window" line, though the VF of that line is different and should be measured before construction.

How High?

Wait, you say—"After all that talk about having it a half wave up, you only have it up 28 feet." A 6 or 12 ft RG-58 jumper, available with BNC connectors from RadioShack, can be used to get it higher if the right branch is available. Since impedance at the feed point is 50 to 70 ohms, 50 ohm coax can be used to extend the feed line. I have used it in the field a few times as an inverted V, at various heights and leg angles, and used an SWR meter to double-check its consistency in different situations. SWR never exceeded 1.5:1, so I feel safe leaving the tuner home. For backpacking, I leave the SWR meter home, too!

And there's a bonus: Since it has a balanced feed line, it *can* be used with little loss as a multi-band antenna, with a tuner, from 10 to 40 meters. I quote John Heyes, G3BDQ, from *Practical Wire Antennas*, page 18: "Even when the top of the doublet antenna is a quarter-wave-length long, the antenna will still be an effective radiator." Heyes used an antenna with a 30 ft top length about 25 ft off the ground on 40 meters and received consistently good reports from all Europe and even the USA (from England). It will not perform as well at 40 meters as at 20 meters, however, though 10 through 20 meters should be excellent.

Testing, Testing

To test this theory, I recently worked some of Washington State's Salmon Run contesters and worked many Washington hams and an Ohio and a Texas station on 15 and 20 meters, with the antenna up 22 feet on a tripod-mounted SD20 fishing pole, using 10 W from an Elecraft K2 from central California. The K2 tuner was used to tune the antenna on 15 meters. Signal reports were from 549 to 599. Unfortunately, this was a daytime experiment and 40 meters was limited to local traffic.

At the 2001 Freeze Your Buns Off QRP contest, it was hung at 30 feet and compared to a 66 ft doublet up 50 feet on 10, 15 and 20 meters, using a K2 S-meter. There was little if any difference. At the 2001 Flight of the Bumblebees QRP contest I compared it, at 20 meters, to a resonant wire groundplane antenna with each antenna top at 20 feet and found it to consistently outperform the groundplane. I have concluded through these informal experiments that a resonant inverted V, when raised at a height close to or exceeding a half wavelength, produces the most "bang for your buck" and that extra length or height beyond that yields diminishing returns.

A ham since 1998, Rich Wadsworth, KF6QKI, is a civil engineer in private practice as a consultant. Since earning his license, he reports, that he has become obsessed with kits and homebrewing. You can reach Rich at 320 Eureka Canyon Rd, Watsonville, CA 95076; **richwads@compuserve.com**. QST

By Robert Johns, W3JIP

From *QST*, January 2001

A Ground-Coupled Portable Antenna

As the saying goes, "Imitation is the sincerest form of flattery." Here's a home-brewed antenna that proves the point.

This homebrew portable antenna for 40 through 6 meters is patterned after the ground-coupled design pioneered by Alpha Delta Communications, Inc.[1] Instead of using radials, this antenna employs a simple and very small grounding system that needs no tuning.

The antenna described here is a quarter-wave vertical sitting on a tripod base. The vertical mast and the tripod are each made of 2-foot-long telescoping sections of 3/4- and 5/8-inch-diameter aluminum tubing.[2] The mast itself resonates on 10 meters; lightweight aluminum tubing sections are added to the top of the mast to tune the antenna to 12, 15 and 17 meters.[3] These added tubing lengths can be installed vertically or horizontally. The antenna is fed at the top of the tripod, making the base a part of the radiating system. A bungee cord stretched from the top of the tripod to a stake in the ground keeps the structure stable.

Beneath the foot of each tripod leg is a grounding strip 2 1/2 inches wide and about 3 1/2 feet long, made of aluminum tape.[4] These strips are simply laid on the ground and form one plate of a capacitor coupling RF from the antenna to the ground. That's the whole grounding system! When I read about this in *QST* (see Note 1), I was skeptical, but intrigued. The arrangement is similar to that of a mobile antenna system in which the car body acts as one plate of a capacitor coupling RF to the road and ground. This grounding system works: The antenna radiates well and the SWR is reasonably low on all bands. (The tripod and grounding strips can also be used with any vertical element or mobile whip you have.) A loading coil added between the aluminum tubing mast and the flattop permits operation on 20, 30 and 40 meters. With the coil positioned this far up the antenna, the entire 10 feet of tripod and mast are unloaded radiators on all HF bands.

[1]Notes appear on page 6-38.

Building the Tripod

The top of the tripod, Figures 1 and 2, makes it easy to set up. The three 5/8-inch-diameter × 0.058-inch-wall aluminum tubes extending from the 1 1/2-inch PVC cap are permanently attached to it. To assemble the tripod, the legs slide over these tubes. A 3-inch-long, 3/8-inch carriage bolt passes through a hole in the top of the PVC cap to support the vertical element. This bolt also grips the 4-inch-long aluminum tubes inside the cap to form the three sloping legs of the tripod. See Figure 2 and its caption for details on how to make this top cap.

A 50-Ω coaxial feed line attaches to the antenna via an SO-239 chassis connector mounted on an aluminum angle bracket at the top of the PVC cap (see Figure 1). Make a 5/8-inch-diameter hole in the bracket to accept the coax connector body; you'll also need to drill four small holes for the connector's mounting hardware. The 3/8-inch bolt through the top of the cap keeps the aluminum bracket in place.

Figure 1—The top of the tripod with the bottom section of the mast connected to it. A bolt holds the three leg supports in the PVC cap slots. This bolt also passes through the 1 1/2 × 1 1/2-inch aluminum-angle piece that supports an SO-239 chassis connector for feed-line connection. A 1/4-inch hole in the cap top accepts the bungee-cord hook.

To assemble the tripod top, invert the cap so that you are looking down at the open end. Insert the carriage bolt through the 3/8-inch hole in the cap, through the mounting hole in the aluminum angle and add a lock washer and nut to the bolt. Initially, thread the nut about an inch onto the bolt so that the bolt is still loose and its head is out of the cap. Insert the three aluminum tubes into the slots in the wall of the cap and down against the carriage bolt where it passes through the hole in the cap. Tighten the nut so that the carriage-bolt head squeezes the tubes outward and into the slots. Once the nut is hand tight, wriggle each tube to seat it snugly with its tip into the countersunk hole with the bolt. Tighten the nut until the round wall of the cap is slightly deformed into a triangular shape.

Each tripod leg consists of a 0.058-inch-wall, 3/4-inch-diameter tube and a 0.058-inch-wall, 5/8-inch-diameter tube that fits inside the 3/4-inch tube. Each tube is two feet long; three can be made from 6-foot tubing lengths. Dimple each 3/4-inch tube about one inch from each end. The dimple acts as a stop and prevents the smaller tube from penetrating any farther. Form the dimples using a couple of firm hammer taps on a center punch placed against the tube. When joining the tubes, push a bit when inserting the smaller tube so that the side of the dimple holds the smaller tube in place.

Ground Strips

Although mating two strips of aluminum tape with their sticky sides together might seem like a routine job, it's probably the most difficult part of building this antenna! The adhesive is quite sticky and unforgiving, and handling the long strips can be messy. Get an assistant to help you with this task. You'll need three strips.

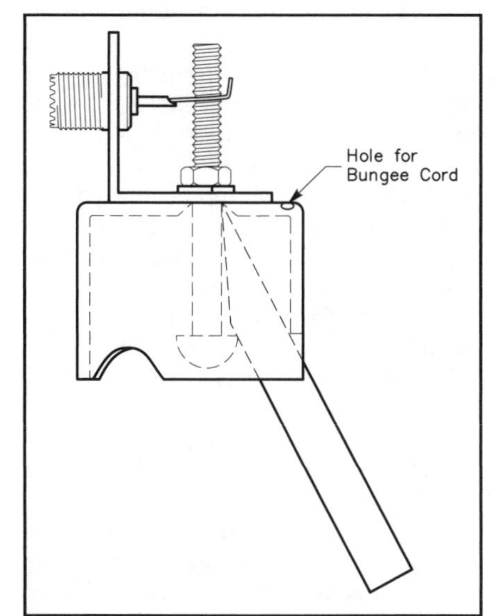

Figure 2—The tripod top cap. The three ⁵/₈-inch-diameter aluminum tubes are 4 inches long, cut at a 30° angle at the end within the cap. From inside the cap, countersink a ³/₈-inch hole in the cap top. This forms a trap that holds the ends of the aluminum tubes. Although only one of these leg supports is shown, all three are held between the bolt head, the three slots (either carved or filed in the wall of the cap) and the countersunk hole in the cap top. The slots in the cap are about ³/₈-inch deep and wide enough to receive the aluminum tubes. An easy way to lay out the slots is to use the fluted handle from an outdoor water faucet as a template. The handle fits nicely against the cap and has six flutes about the circumference allowing you to mark three equally spaced locations.

See Figure 3. Cut a 7-foot length of tape from the roll and lay it down, sticky side up, on the floor or a large table. Have your helper press a piece of heavy (#12) solid wire or a thin dowel across the width of the strip at the 3¹/₂-foot midpoint and hold it in place. Pick up one end of the strip and carry it over the midpoint, keeping it tight so that it doesn't sag and touch the lower half. Keep both ends of the strip aligned while your helper at the midpoint presses the top piece of tape against the lower, working their way toward you. Trim (or remove) the excess wire or rod and the ground strip is done. Don't worry if the strips aren't aligned perfectly.

The Mast

The 8-foot mast is made from two telescoping ³/₄- and two ⁵/₈-diameter × 0.058-inch-wall aluminum-tubing sections. Slot the ends of the ³/₄-inch tubes so that they can be tightened around the smaller tubes with hose clamps.[5] To insulate the bottom ³/₄-inch section from the ³/₈-inch bolt in the tripod that supports the mast, its lower end is equipped with a plastic insulator. As shown in Figure 4, the insulator is a 2-inch length of acrylic tubing. The lower end of the acrylic tube extends about a quarter inch below the aluminum tube and is slotted so that the mast can be tightened around the bolt. Drill a ⁵/₃₂-inch hole through the upper end of this insulator and the aluminum tube to pass a #6-32 bolt and nut to hold the insulator in place.

After mounting the SO-239 coax connector on the aluminum angle strip, solder a 2-inch length of #14 bare solid copper wire to the connector's center terminal and bend it close to the ³/₈-inch bolt in the tripod top. Then bend the wire up and parallel to the bolt and about a quarter inch from it. When the bottom section of the mast is placed over the bolt, place this wire between the aluminum tube and a hose clamp. As you tighten the clamp, it makes the electrical connection from the coax to the mast and squeezes the slotted aluminum tube and insulator tightly against the bolt.

With the mast on the tripod, an easy way to make frequency adjustments is to separate the mast from its bottom section and lower it to the ground. You can then reach the flattop and coil without tilting the mast. For this reason, I don't tighten this joint. I place a #6-32 bolt through the ⁵/₈-inch tube which is the second section of the mast, about one inch from its lower end so that it doesn't slide very far in. I still use a hose clamp over the ³/₄-inch tube, adjusting it to make a snug sliding fit for the upper mast.

For the top antenna sections, I use ⁵/₈-inch-diameter thin-walled aluminum tubing used for aluminum clothes poles. This material is lighter and cheaper than the 0.058-inch-wall tubing used for the tripod and mast, but is strong enough. Short tubing sections can be joined together using 2-inch-long sleeves made from the ³/₄-inch-diameter × 0.058-inch-wall aluminum tubing. You need two 2-foot, two 1¹/₂-foot and two 1-foot lengths of the ⁵/₈-inch thin-walled tubing, three couplings and a T joint to connect the

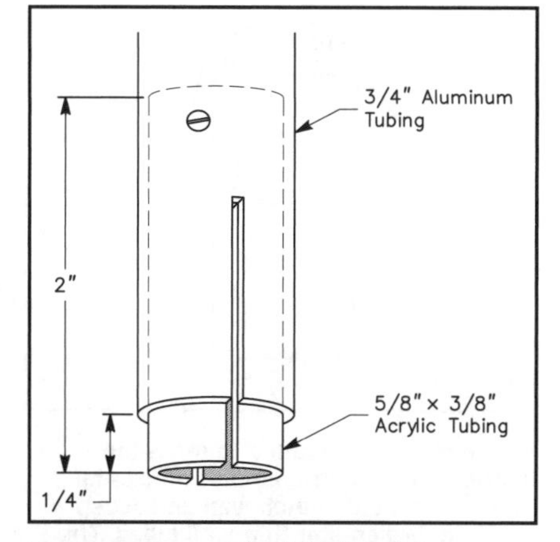

Figure 4—An acrylic (Plexiglas) tube insulates the antenna mast from the ³/₈×3-inch bolt that supports it. The tube has a ³/₈-inch ID and ⁵/₈-inch OD so that it slips over the supporting bolt and telescopes inside the lower ³/₄-inch mast section. Both the aluminum tube and the insulator tube are slotted using a hacksaw so they can be tightened around the bolt with a hose clamp. To mount a mobile antenna on the tripod, cut a 2-inch length of 1-inch-diameter acrylic rod and drill and tap one end to accept the ³/₈×16 coarse-thread bolt of the tripod and ³/₈×24 fine threads at the other end for the base of a mobile whip.

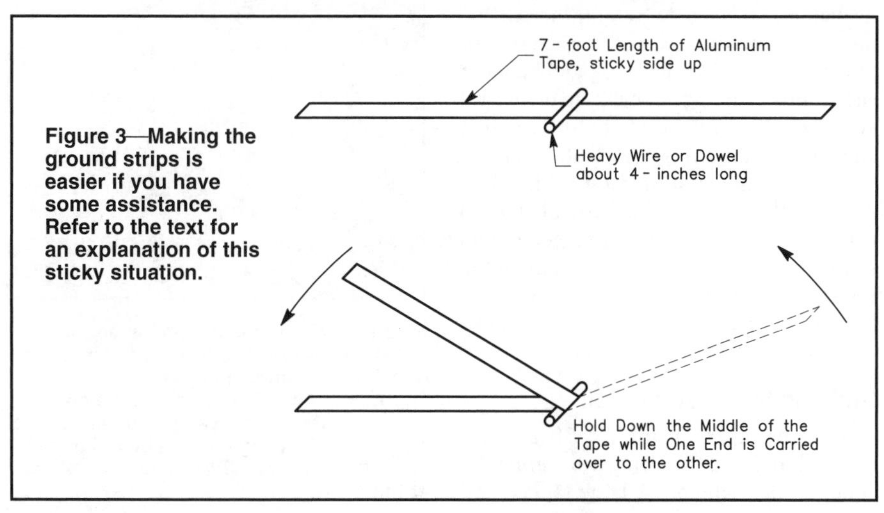

Figure 3—Making the ground strips is easier if you have some assistance. Refer to the text for an explanation of this sticky situation.

flattop to the mast or the top of the coil. See Figure 5.

Bungee Tie-Down

The antenna is quite light, and even with the wide base of the tripod it needs to be stabilized against wind gusts or someone tripping over the coax feed line. A bungee cord and a ground stake do an excellent job. The top of the tripod is about 3 feet high, so a 1/2- to 3/8-inch-diameter, 24-inch-long bungee cord works well. Any tent stake will do; drive it into the ground at an angle so it doesn't pull out easily. A special stake shaped like a large screw is ideal for this application.[6] It threads into the ground by hand and has a very low profile. (I leave the stake in the ground and my lawn mower doesn't even come close to striking it.) The stake won't go into hard, baked soil, however. For stability in such locales, or on pavement, hang some bricks, a rock or a jug of water beneath the tripod on the bungee cord.

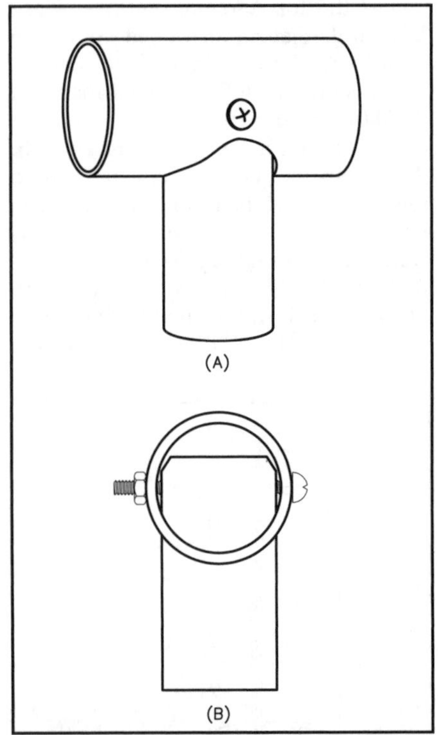

Figure 5—A T is needed to make the flattop. The 3/4-inch-diameter horizontal tubing has a 0.058-inch wall and accepts the 5/8-inch-diameter thin-wall tubes. The 5/8-inch-diameter vertical piece has a 0.058-inch wall and fits into the 3/4-inch tube at the top of the coil form. To make a 5/8-inch hole in the 3/4-inch tube, drill a hole then expand it with a 5/8-inch-diameter or larger countersink. (This process is heavy work for a countersink, so use a little lubricating oil.) Before drilling a 7/64-inch hole for a #6-32 bolt through the T, assemble the two pieces and squeeze them together tightly in a vise, making them perpendicular. To mount a flattop on the mast without the coil, first place a 3/4-inch-diameter coupling sleeve over the 5/8-inch-diameter top of the mast and fit the T into that coupling.

Table 1
Length of thin-wall tubes needed for operation on 10 through 17 meters.

Band (Meters)	*Length of Flattop (ft) and Number of Sections*	*Length of Vertical Top (ft)*
10	0	0
12	1 × 2	1.5
15	2.5 × 2	3.5
17	3.5 × 2	5.5

Antenna Operation on 10 through 17 Meters

For 10-meter operation, set up the tripod and place one end of a ground strip under each tripod foot. The ground strips may be laid in any direction. Adjust the mast to a length of about 7.4 feet. This length is quite a bit less than a quarter wavelength and I believe it's because of its closeness to the ground and the thickness of the tripod. No top hat is used on 10 meters. Adjust the mast length to resonate the antenna at your desired 10-meter frequency.

Table 1 provides lengths for the thin-wall tubes that you add to the antenna, either as a flattop or a vertical, for operation on 12, 15 or 17 meters. No change to the ground system is needed when changing bands. Table 1 assumes that you will leave the mast set for 10-meter operation. This simplifies band changing, such as moving from 10 meters to 15 meters and returning to 10 meters. These changes are quickly made by just adding the tubing lengths for 15 meters and removing them to return to 10 meters—no measurements, no tools.

6-Meter Antenna Operation

For 6-meter operation, the tripod must be insulated from ground and the mast reduced to a length of 52 inches from tripod to tip; see Figure 6. No ground-coupling strips are needed. Simple insulators can be made from 1/2-inch CPVC pipe and couplings. Cut three lengths of pipe about 4 to 6 inches long and hammer each into a coupling. Cementing them isn't necessary; they will be a tight fit. The other side of the coupling fits well over the 5/8-inch-diameter aluminum-tubing leg. Adding these insulators to the tripod resonates the antenna in the 6-meter band with good SWR. You can change the operating frequency by adjusting the length of the mast only—you don't need to adjust the size of the tripod.

Building the Loading Coil

For operation on the 20, 30 and 40-meter bands, a loading coil must be added to the antenna. A large tapped coil is shown in Figure 7; it tunes the antenna to 20, 30, or 40 meters and permits you to tune to the higher-frequency bands without changing the lengths of the top hat. The coil has 13 turns of #8 aluminum wire wound on a 4-inch styrene pipe coupling.[7, 8] This coil form is secured to a 7-inch-long, 1/2-inch-diameter CPVC pipe using 1 1/4-inch-long, #6-32 brass or stainless steel machine screws and nuts. I like to reinforce the 1/2-inch pipe by hammering a 2-inch length of 1/2-inch wood dowel into each end. This allows me to tighten the nuts and bolts without flattening the pipe. These bolts also secure the ends of the 13-turn coil. Using a marking pen, I made black marks on the coil to identify the fifth and tenth turns. The marks serve to locate the proper tap points without having to count coil turns each time.

Figure 6—Here, the antenna is set up for use on 6 meters. The tripod construction remains the same, using legs approximately 4 feet long, but the mast has been shortened. No ground strips are needed and the legs are insulated from ground by 1/2-inch CPVC pipe extensions at their feet.

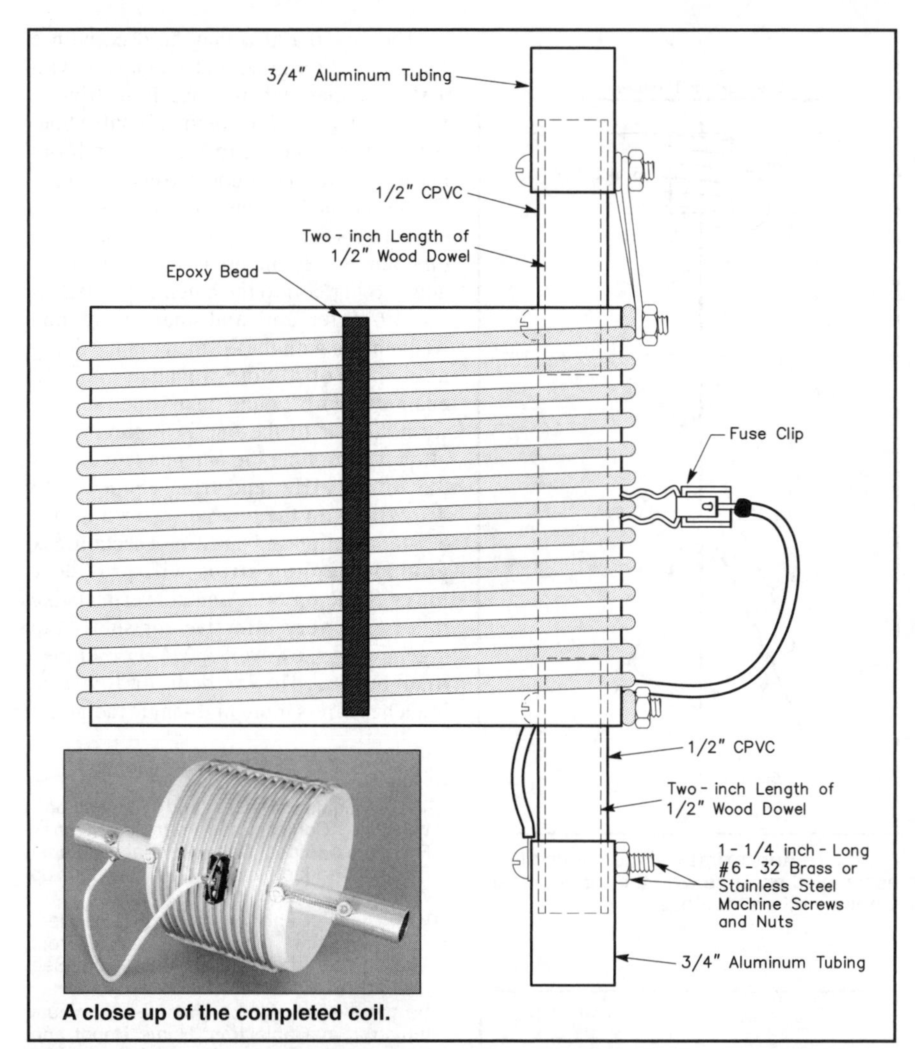

Figure 7—To make the loading coil, 13 turns of heavy aluminum wire are spaced to fill the form. Secure the coil ends using the same bolts that hold the plastic pipe inside the 4-inch styrene coil form. Mount the coil on the mast with a 3/4-inch-diameter aluminum sleeve at the bottom of the plastic pipe; the tap wire is also connected here. An identical sleeve at the top of this plastic pipe connects to the thin tubing for the top vertical section, or to an aluminum T to hold the flattop.

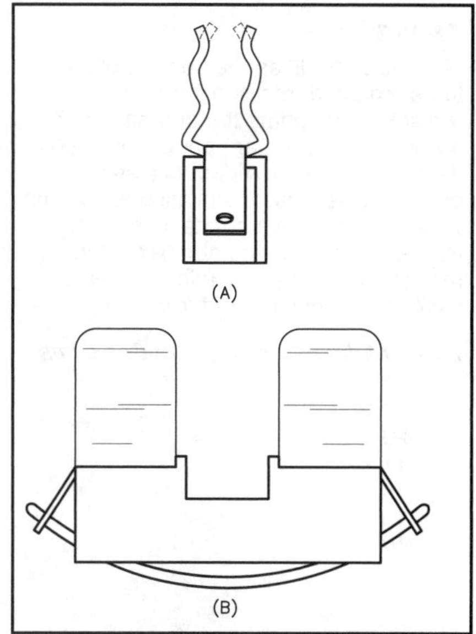

Figure 8—Making the tap connection to the coil. At A, the ends of the jaws of a 5-mm cartridge-fuse holder are bent inward (dotted lines) to grip the heavy wire of the coil. A side view of the fuse holder is shown at B. Bend the solder lugs at the ends of the fuse holder to accept a wire passing through them and beneath the fuse-holder base. When this wire is in place, bend the lugs farther up against the ends of the holder and solder them. Strip a 1/4 inch of insulation from the tap wire and solder it to the wire joining the lugs beneath the fuse holder. Round off any sharp points or rough edges with a file, because you'll be gripping this connector tightly for attachment to and removal from the coil.

Inside the styrene coil form is a ridge. Use a chisel or file to remove about a 1-inch-long section of this ridge to allow the CPVC pipe to lie flat against the inside of the form. Drill 7/64-inch holes at the ends of the styrene coil form and through the 1/2-inch pipe, then bolt them together as shown in Figure 7. Take a 16-foot length of aluminum wire, bend a loop at one end of it, attach the loop to one of the bolts and wrap the form as neatly as possible with 13 turns of wire, without bends, spacing the turns to fill the form. Wrap the end of the 13th turn around the bolt at the other end of the coil form and cut off the excess wire.

To tighten the wire on the form, clamp the form in a vise, grab the coil turns between both hands and progressively rotate the coil from one end to the other several times. This makes the turns tight enough to stay in place as you even out their spacing. To hold the turns in place permanently, run three ribs of epoxy the length of the coil. Use metal/concrete epoxy which has black resin and white hardener, making a dark gray mix that is easy to see against the white background of the coil form. One of these ribs is visible in Figure 7. To make nice straight ribs, first place strips of tape on each side of an intended rib location, apply the epoxy and remove the tape before the epoxy hardens.

Several types of alligator-clips will fit between the coil turns without touching neighboring turns, but I prefer to use a tap connection made from a fuse holder; see Figure 8.[9] After bending the fuse-holder-jaw tips, bend the jaws themselves to make them fit the wire tightly, but remain easy to attach and remove. Suit yourself as to how tight a grip they should have. Join the tap connector to the sleeve at the bottom of the coil form using a 9-inch length of stranded, insulated #14 copper wire, with a solder lug at the end. Use a similar piece of wire to join the top of the coil to the sleeve at the top of the coil form.

The sleeve at the coil bottom joins the coil to the mast. It is a 1 1/2-inch-long, 3/4-inch-diameter, 0.058-inch-wall aluminum-tubing piece. Insert the bottom of the 1/2-inch CPVC pipe halfway into this sleeve and drill a 7/64-inch hole through the sleeve and pipe. Fasten them together with a 1-inch-long, #6-32 brass or stainless steel machine screw and nut. The wire to the tap connector is attached with this same screw.

Antenna Operation with the Coil

To use the antenna on 40 through 10 meters, shorten the mast to 6 feet 2 inches and connect the coil to the mast. Atop the coil, add an element consisting of two horizontal 3 1/2-foot lengths of 5/8-inch-diameter tubing, or a single 7-foot vertical piece of tubing. With the full 13 turns of the coil, and part of an extra turn supplied by the tap wire, the antenna will likely resonate in the middle of the 40-meter band. To operate at the low end of the band, add a 1-foot length of tubing to one side of the flattop or the vertical tubing section. See Figure 9 for approximate dimensions of the assembled antenna.

It may seem as though Table 2 has some errors because it lists a greater

Table 2

This table identifies the number of coil turns (counted from the top of the coil) required to resonate the antenna on the 40- through 10-meter bands. These coil-tap settings are provided as a starting point only because installation conditions vary. To raise the antenna's operating frequency, reduce the number of turns used; to lower the operating frequency, increase the number of turns.

Band (Meters)	*Number of Coil Turns*
40	13
30	7.1
20	3.1
17	2
15	5
12	7
10	13

number of coil turns for operation on 15, 12 and 10 meters than for 17 meters! You're right—something strange is going on. It's because there are *two resonant frequencies* for each setting of the coil tap. Figure 10 shows the two paths that RF can take in the antenna. The upper part of the coil and the top hat provide the lower frequencies; the lower half of the coil provides the higher frequencies. A coil this large has considerable capacitance to free space, so it's not just an end-loading inductor at the higher frequencies. The antenna bandwidth is good, the SWR low and the antenna performs well on these bands. The charm of this coil system is that you can change bands by just moving the tap on the coil, without any adjustments to the mast length or the flattop. And a bonus: With 13 turns on the coil, the antenna works on 40 and 10 meters simultaneously.

The coil settings of Table 2 may need some minor adjustments if a vertical top section is used instead of the flattop. In general, the SWR is lower with the flattop and the antenna is easier to handle.

Power-Handling Capability and Safety

Because of the large coil and tubing used, you might be tempted to run high power with this antenna. I suggest you don't. The antenna may take it, but people can't. At high-power levels, dangerous RF voltages on the antenna are within range of physical contact. I have used the antenna at a 100-W level, but even that requires care and supervision.

Other Possibilities

With the tapped coil, this antenna can be tuned to any frequency from 7 to 40 MHz when operated on the ground-coupled tripod, and up to 110 MHz with the tripod insulated from ground.

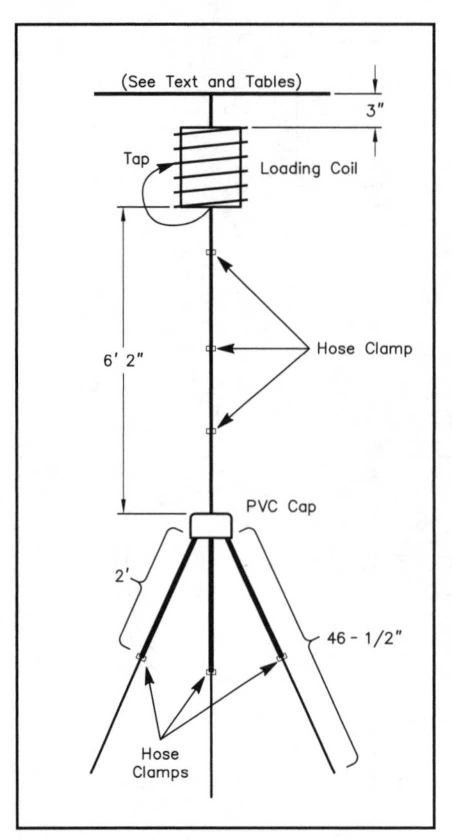

Figure 9—Approximate dimensions of the assembled antenna with the tripod, mast, loading coil and top hat.

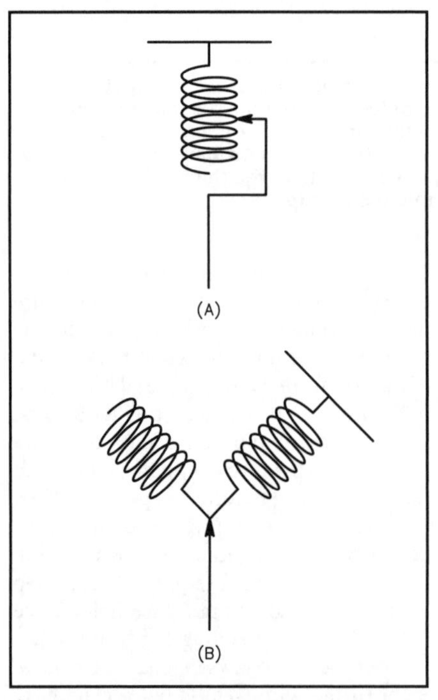

Figure 10—At A, the upper part of the antenna includes the coil, the adjustable tap and the top hat. The bottom of the coil is free and not connected to anything else. At B, this has been redrawn to show the two antenna circuits with the two resonant frequencies that are present. The upper half of the coil has a lower resonant frequency because of the length of the top hat above it.

The antenna also may be used with a longer mast for greater efficiency, or with a shorter mast when space is restricted. Even though the short version is only about 6 feet high, you can't use it indoors because it must be coupled to earth ground. The taller antenna gets out better, but band changing is more complicated. If operation on 75 and/or 80 meters is a must, you can add another coil to the antenna just below the 40-meter coil and change antenna frequencies with the 40-meter tap. Adding a coil made of 20 close-wound turns of #12 enameled wire wound on a 4-inch styrene form similar to the one in Figure 7 will allow you to tune the antenna from about 3.5 to 3.8 MHz, and from about 3.8 to 4.1 MHz with the top hat reduced to one 3.5-foot section and one 2-foot section. Six ground-coupling strips will provide a lower SWR on 80. A small vertical like this is not very effective for short-skip ragchewing, however. A $\lambda/4$-wire draped over bushes, flower beds or low tree branches offers more high-angle radiation.

Notes

[1]Rick Lindquist, N1RL, and Steve Ford, WB8IMY, "Compact and Portable Antennas Roundup," 'Alpha Delta Outreach/Outpost System,' Product Review, *QST*, Mar 1998, pp 72-73.

[2]Twelve feet of each tubing size is needed. The aluminum tubing is available from Texas Towers and Metal and Cable Corp. See their ads elsewhere in this issue.

[3]The thin-walled 5/8-inch-diameter aluminum tubing is available from Home Depot and hardware stores as aluminum clothes poles, each about seven feet long.

[4]Adhesive-backed aluminum tape 2 1/2 inches wide is available from Home Depot stores in the heating-vent section.

[5]You may want to consider using an antioxidant at the tubing joints. Antioxidant compounds available from electrical wholesale supply houses, Home Depot and hardware stores include Noalox (Ideal Industries Inc, Becker Pl, Sycamore, IL 60178; tel 800-435-0705, 815-895-5181, fax 800-533-4483) and OX-GARD (GB Electrical, 6101 N Baker Rd, Milwaukee, WI 53209; tel 800-558-4311). Use either sparingly; a thin coat is sufficient.—*Ed.*

[6]Aluminum angle 1 1/2 × 1 1/2 × 1/16-inch thick is available from hardware and Home Depot stores. The green plastic ground stake that threads into the ground has the name "Twizelpeg" stamped into it, and is available at camping supply stores.

[7]The #8 aluminum wire is RadioShack #15-035.

[8]The coupling is available from Home Depot in the drainage pipe section, and also from large plumbing or swimming pool distributors. The couplings are actually 4 1/2 inches in diameter and made from polystyrene, a very low-loss insulator.

[9]RadioShack #270-738.

Bob Johns, W3JIP, is an old gadgeteer who likes to play with antennas and coils. You can contact Bob at PO Box 662, Bryn Athyn, PA 19009; **ksjohns@email.msn.com**.

Photos by Joe Bottiglieri

By Robert H. Johns, W3JIP

From *QST*, January 1999

Roll Your Own Dipole

Tired of dealing with tricky, convoluted multiband dipoles and **V**s for portable operation? Why not simply wind the excess wire on spools? It makes sense logically—and electrically—as you'll see here!

Although superstitious hams may tell you that coiling wire at the ends of dipole or **V** antenna legs is some how taboo, winding the excess wire on small spools is a convenient and effective way to make antennas that are physically and electrically adjustable.

To make a dipole antenna that can be easily lengthened or shortened, simply wind the unused wire at the ends of the elements onto spools. If you use insulated antenna wire, the coils act as high-impedance chokes that have little effect on the antenna. Uninsulated "end coils" are "blobs of conductor"—small capacitance hats at the ends of the wire elements.

Figure 1 shows a portable dipole with a center insulator and two spools of wire, each containing about 65 feet of insulated, stranded copper wire. By unwinding the proper lengths, a dipole for any band from 6 through 80 meters can be produced. And by configuring the system as an inverted **V**—with the ends close to the ground—it's easy to change bands. "End spooling" also makes it easy to adjust feed points and leg lengths for off-center-fed dipoles.

Construction

Insulated wire is preferred for portable antennas. In addition to increased electrical safety, the insulation minimizes the effects of wet bushes or trees that antenna wires must often pass through. The spools in the photo are from Home Depot, which sells #12 and #14 stranded copper wire in 50 and 100-foot lengths. Smaller spools are available from RadioShack. I prefer the larger spools because they're easier to wind. Three-quarter-inch wooden dowels make good handles and axles, and a short nut-and-bolt makes a crank handle on the outer edge of a spool. A loop of bungee cord wrapped around the spool, as shown in Figure 1, will prevent the wire from unwrapping.

It's convenient to mark the spooled wires so it's easy to determine exactly how much wire has been unwound. I mark each foot with a permanent marker pen, place a black electrical tape "flag" every five feet and a bright yellow numbered flag every 10 feet.

Any reasonable center insulator will do. The one in the photo was made from a small PVC cap.

Inverted V center insulators use a rope or line to support the weight of the antenna elements and the feed line. I use 1/8-inch nylon or polypropylene rope for the main support line (and for the guy lines at the ends of the antenna). I simply throw a line over a high tree branch or other available support to raise the center insulator skyward. Of course, scout the area carefully beforehand and make sure there aren't power lines nearby.

Be sure to attach the guy ropes several feet in from the ends of the antenna elements to allow for easy adjustments and length changes. Figure 2 shows an easy knot to tie for just such an installation. Figure 3 shows variations on what to do with the extra wire. You can stretch it out along the guy rope, fold it back and hang it from the antenna, or run it off to some bush or tree in another direction. The idea is to have it readily accessible from the ground.

Harmonics

For portable operations it'd be nice to have a lower-frequency antenna that can work effectively at higher frequencies. Thanks to the harmonic nature of antennas and amateur bands, these double-duty combos can work on 40 and 15 meters, or 75 and 10 meters, for example. This is possible because half-wave dipoles are reso-

Figure 1—A portable dipole or inverted V antenna. The wire is unrolled from the spools as needed while the rest of the wire remains coiled at the ends. A short transmission-line matching section is connected to the center insulator.

Table 1

Leg Lengths and Resonant Frequencies of Inverted V antennas with Element Ends Approximately Six Feet Above Ground. Numbers in Parentheses are SWRs at the Indicated Frequencies.

Leg length	*Frequency MHz (SWR)*			
(feet)	*Fundamental*	*3rd Harmonic*	*5th Harmonic*	*7th Harmonic*
21	**10.1 (1.2)**	33.0 (1.3)		
23	9.3 (1.3)	**28.7 (1.6)**		
30	**7.2 (1.2)**	22.2 (1.7)		
31	7.0 (1.3)	**21.5 (1.4)**		
48	4.5 (1.1)	**14.0 (1.2)**	23.6 (1.0)	34.3 (1.2)
53	4.0 (1.0)	12.6 (1.2)	**21.4 (1.0)**	31.1 (1.2)
56	**3.8 (1.1)**	12.0 (1.4)	20.5 (1.2)	**28.4 (1.1)**
64	3.4 (1.1)	10.9 (1.3)	**18.3 (1.2)**	26.4 (1.2)

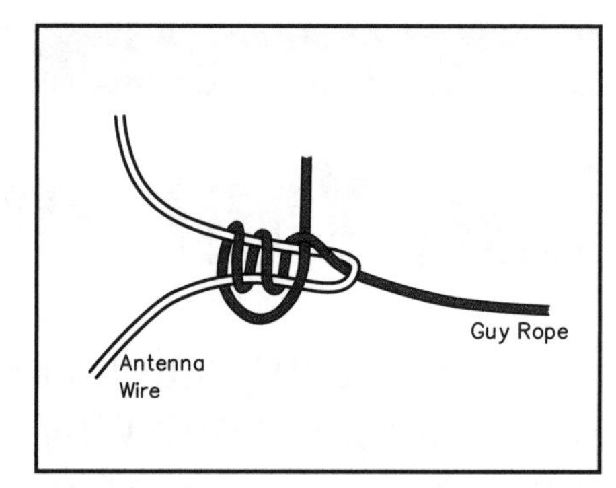

Figure 2—A knot to attach guy ropes to antenna wires.

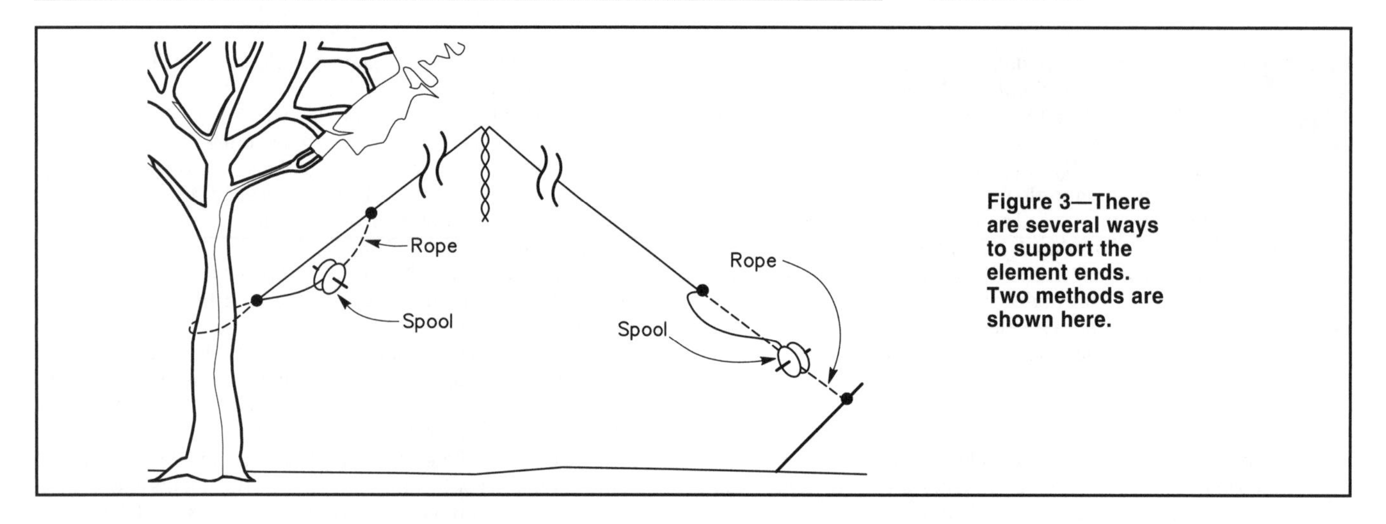

Figure 3—There are several ways to support the element ends. Two methods are shown here.

nant at odd multiples of their fundamental frequencies.

There are, however, two difficulties in using a 40-meter dipole on 15 meters. The 15-meter resonant frequency will be slightly above the band, and the SWR there will be about 2 to 1. An inverted **V** with adjustable end coils takes care of the first problem. Simply lengthen the antenna a bit when going from 40 to 15 meters.

Fixing the high SWR is also possible. A short segment (about 6 feet) of transmission line can be added between the antenna and the 50-Ω coax feed line. Its impedance is probably somewhere between that of the coax and the higher impedance of the 3/2-wavelength antenna on 15 meters. This transmission line is shown in Figure 1. It's a twisted pair of #14 or #12 stranded insulated copper wires with an SO-239 coax connector at the lower end. At the top, each wire is connected to one dipole leg at the center insulator.

I have used many different kinds of wire for these transformer sections, and the insulation type isn't critical. This simple addition reduces the SWR of HF dipoles and **Vs** while operating on odd harmonics. See Table 1.

Operation

Table 1 shows leg lengths for three inverted **Vs** with fundamental frequencies in or near the 30, 40, and 80-meter bands. The frequencies of the odd harmonics are also shown, as are the SWRs (in parentheses) measured by an MFJ Model 249 Antenna Analyzer. Table 1 is useful for determining leg length changes necessary when switching between fundamental and harmonic frequencies.

For example, if you've been operating in the 40-meter phone band (with your 40-meter **V** or dipole), you would add one foot to each leg to operate on 15 m. The lengths in the table are a starting point. Height and ground conditions at your location will influence your results.

Notice that an 80-meter inverted **V** provides access to five bands with only minor changes in leg length, plus the ability to move anywhere in the 75/80 meter band. The bandwidths of the harmonic bands are very broad.

So, instead of cutting and testing several dipoles for Field Day or your next radio outing, why not "roll your own" truly versatile antenna?

Box 662
Bryn Athyn, PA 19009
ksjohns@mindspring.com

By Robert Victor, VA2ERY

From *QST*, July 2001

The Miracle Whip: A Multiband QRP Antenna

Want to hold the world in the palm of your hand? Tired of packing a suitcase-size antenna for your hand-held, dc-to-daylight transceiver? The Miracle Whip, a self-contained wide-range antenna made from inexpensive parts, can give you the flexibility you need to be truly free—no ground required!

One of my favorite radio fantasies started with Napoleon Solo—the man from U.N.C.L.E. He'd be in a tight spot, say, under fire from a crack team of THRUSH nannies in miniskirts, and he'd reach into his pocket and pull out the world's niftiest radio. It was about the size of a pack of cigarettes and had a two-inch whip antenna. He'd call up Control—who could be anywhere in the world at that particular moment—and try to muster some help. Control, of course, would dish out a number of droll comments about Solo's regrettable tendency to get into any number of tight spots, whereupon Napoleon would dial up partner Illya Kuryakin, on the other side of the room, and ask him to shoot back. The nannies, twittering like squirrels in a dog pound at having their pillbox hats punctured, would retreat in disarray. End of episode.

The mini rig was a prop, of course, and I realized even then that such a radio could never work. Short of satellite support (which would come soon enough) or a new understanding of the universe (which may or may not come), a two-inch whip on a hand-held HF transceiver might get a signal across a room, but not around the world.

Since then, often during evenings spent at a campground picnic table, I continued to think about what it would be like to have such a handy radio. I often visualized a book-size, multi-band rig powered by internal batteries; something that would be practical for cycling, hiking or working skip from any nearby picnic table. It was easy to imagine the rig rendered in such a portable package, but I could never get the two-inch whip to work—even in my mind. I guess I couldn't set aside *all* the laws of physics. Agent Solo (or his heirs and assigns) would be forever doomed to throwing wires into trees.

Although nobody seemed to consult me, the radio of my daydreams appeared on its own. When I saw the first magazine ads for Yaesu's FT-817 low power (QRP) transceiver earlier this year, I was delirious—it was *exactly* the rig I'd been fantasizing about. I dug out my credit card, told my wife I was ordering an Ab-Rocker and called my buddy Angelo at Radioworld in Toronto ("...but Honey, we can't send it back, we'll lose money on the restocking charge...").

Rig in hand, the man from U.N.C.L.E. was *still* in my thoughts. He wanted *his* radio, or at least something like it. He wanted an antenna that plugged into the back of his (my) new '817 so he could easily brandish it when in desperate need, without having to find a tree, when there were *nannies*. I tried to explain about antennas, but he merely gave me that pained, condescending look usually reserved for conversations with Control.

What might actually work here? A telescoping whip perhaps, around 50 inches long, with some kind of loading system so the antenna could cover all the HF bands. I'd have to stay away from "interchangeable" coils (Solo wouldn't want the hassle), and I'd have to produce some kind of workable results. Efficiency might be measured in the single digits on some bands, yet DX *had* to be a possibility.

What I came up with is definitely fit for an U.N.C.L.E. operative. It's a 48-inch telescoping whip with a homebrew loading and mounting device. Physically, it's portable and practical, and looks *secret agent cool* on the Yaesu. I finished construction just as a contest weekend was starting, so I got to try it out under ideal conditions.

Although my QRP signal didn't burn out anyone's receiver, I'm pretty satisfied with the results. I spent about fours hours on HF during this particular contest and had scads of contacts on 10, many on 15 and 20, and a couple on 17—almost all overseas! I also worked four stations on 40 (within about 400 miles) and managed one contact with a local operator on 80 meters. The rig was sitting on my desk—indoors—and the whip was plugged into the back of the radio, which was ungrounded. That's definitely a worst-case scenario! Because I figured it would take a miracle for a rig-mounted antenna to work DX, I christened my creation the "Miracle Whip"!

In Theory

The heart of this design is in the loading system, which is made from readily available parts and costs about $30 for the whole works (less if you have the proverbial well-stocked junkbox). Here's the theory...

There are three ways (that I can think of) to load a length of wire on a particular frequency. The first is to make the wire a quarter of a wavelength long, which makes it resonant at the desired frequency. This works because the feed point impedance of a quarter-wave wire (assuming you have a counterpoise) is about 50 Ω, which matches the coaxial output found on most rigs. Unfortunately, the shortest wavelength I'd be using was 10 meters, and a quarter of that is about eight feet, so this method wasn't an option.

The second way is to place a loading coil somewhere along the length of a wire that's shorter than a quarter wavelength at the desired frequency. You can place the coil at the base (base-loaded), somewhere in the middle (center-loaded) or at the top (end-loaded). Very simply, the loading coil makes up for the "missing" wire and forms a resonant circuit at the desired frequency.

How does it work? If you graphed the impedance of a quarter-wave antenna along its length, you'd see a continuous curve, with a low impedance at the feed end and high impedance at the far end. If you can imagine removing a section anywhere along the length of the antenna, you'd create a gap in that curve. The loading coil performs the impedance transformation required to bridge the impedance "gap" created by the missing section, allowing the use of a physically smaller (shorter) antenna.

A third method of achieving an impedance transformation is by using a transformer instead of a coil. A transformer is, after all, a device for matching different impedances! The hitch with this technique is that a transformer, unlike a loading coil, isn't a series device; it needs to be fed in parallel and usually "against" the antenna ground. Because of this factor, transformers must be used at the feed point.

Because Napoleon wouldn't like to swap loading coils to change bands, method three would have to be used. I figured an adjustable loading device would have to be placed at the base of the antenna, anyway, so a transformer seemed like a good possibility.

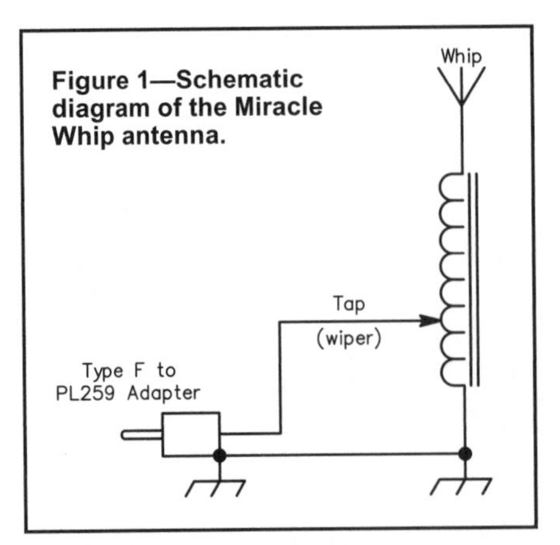

Close-up view of the inside of the Miracle Whip clearly showing the core and wiper.

The Autotransformer

Most of us are familiar with the broadband transmission-line transformers often used as baluns. They can be made somewhat adjustable with clever switching arrangements, but they're always limited to whole-numbers-squared ratios such as 1:1, 4:1, 9:1, 16:1 and so on. I worried that this limitation wouldn't allow for enough adjustment flexibility.

Thankfully, there's another kind of broadband RF transformer that *can* perform a match like this—the autotransformer—and it isn't limited to natural-square ratios. Although it's theoretically not as efficient as a transmission-line transformer, in practice it works quite well. The efficiency usually suffers as you apply more power (because of core losses), but at QRP power levels (5 W or less), those losses are minimal. With a little seat-of-the-pants engineering I came up with a way to make an autotransformer more-or-less continuously variable, which was exactly what I needed to use the same whip and matching unit over such a wide range of frequencies.

An autotransformer works like a conventional double-wound transformer as shown in Figure 1. The bottom part, where the input connects, represents the primary, and the entire coil, with the whip on the end, acts as the secondary. The impedance transformation is the square of the ratio between these two virtual sets of windings (turns). As the slider moves it taps the transformer, varying the ratio between the primary and secondary, providing (hopefully) the right match on each band.

This arrangement looks a bit like a series loading coil with a sliding tap, but if you look closer you'll see that we're applying the signal *across* the coil, which is connected to the signal source and to ground. The antenna winding—in effect the whole winding—is also across the output (the whip) and ground. Thus, we really do have a transformer as opposed to a loading coil, and the device does indeed transform our feed impedance into our whip impedance in a variable manner.

If you'd like to do a thought experiment, imagine exchanging the signal source and the ground, putting the ground on the tap and the source at the bottom. You'll see that the ratio is now different for any given tap position because the ground is now farther up the coil, which changes the number of windings on the antenna side. You'll also see that you've reversed the phase of the output. If you build and test this, you'll confirm this result.

Construction

I'm no machinist, so it was challenging for me to figure out how to homebrew the mechanics of the Miracle Whip. When I have no idea how to create what I need, a wander through the local surplus shop will occasionally provide inspiration.

I did just that, and happened to find a wire-wound rheostat that looked like it was designed for just this project. It had the perfect wiper-and-brush mechanism that I'd need to make the sliding tap, and the resistance winding and the coil form it was wound on looked a lot like a toroidal transformer, which gave me some confidence that the unit could be adapted for my needs. It worked well, so here's how to build your own transformer out of a similar rheostat.

I've located some common commercial rheostats made by Ohmite that you can order from any of several suppliers. Go to the Ohmite web site at **www.ohmite.com**, click on "distributors" and choose one near you (or order from the Allied site in the parts list). These rheostats are supplied in many resistance values, but because you won't be using the resistance winding you can take anything that's in stock that's the correct physical type. These are identified as Ohmite part number RES*xxx*, with the "*xxx*" being the resistance. Typical values are shown in the parts list.

I'm going to go into quite a bit of detail on the construction of this device, but don't be intimidated—the whole process is straightforward and shouldn't take more than a couple of hours.

Start building by stripping the rheostat. You'll use the central shaft, which has a spring-loaded wiper and brush, its associated hardware and the collar/tube in which the shaft rotates. You can toss the resistance winding into your junk box. To get these parts free you'll need to unscrew the collar-retaining nut and remove the C-clip that holds the shaft in the collar. Don't lose the C-clip and be careful not to stress the wiper spring and its contact. The brush is held in its seat on the wiper by pressure alone, so when you take it apart, expect the brush to dangle on its pigtail.

Winding the Transformer

The transformer (Figure 2) is created by winding about 60 turns of #26 enameled wire onto the ferrite core specified in the parts list. I say "about" 60 turns because the number of turns isn't critical. A loading coil would need exactly the right number of turns on exactly the right core for consistent performance, but because our device is a broadband transformer, we're only concerned with the appropriate ratios between the primary and the secondary. Because the windings ratio of the finished unit will be adjustable anyway (that's why we're building it, right?), the number of windings isn't overly critical.

That said, you should shoot for about 60 turns; one or two less or more shouldn't be a problem. What you *do* want are uniform windings that are tight on the core, regularly spaced, with a bit of room between the windings (so the brush will contact only one at a time) and a gap of 30 degrees or so where there are no windings at all.

Why the gap? The rheostat, as originally manufactured, has stops to prevent rotation beyond the ends of the windings, but we lose those stops when we discard the original mounting. The gap will give you a good "feel" for when you've reached the beginning or end of your windings as you tune, so you'll know where you are. (If you think of a better solution, let me know.)

Spread some non-corrosive glue (Elmer's wood glue works fine) on the bottom and rim of the core to hold the windings in place and let the assembly dry completely before proceeding. Use a piece of fine sandpaper or emery cloth to carefully remove the enamel from the wire in the area where the wiper will make contact. You can eyeball this area by temporarily placing the wiper on the core with the shaft centered through the hole in the core.

Mounting

Here's the only tricky part of the project—mounting the core, the wiper and the shaft so the wiper contacts the coil windings with a suitable pressure. If the wiper is too high above the windings, you won't get good contact; if it's too low, adjustment will be difficult and you might tear the brush and perhaps even the windings. That said, it's not *that* difficult to get this right. Look at Figure 3 to understand the mechanics.

Cut a square of perfboard about 1$^1/_2$ inches to a side and drill a hole dead center to accept the shaft collar. Center your newly wound core over the hole. Slide the wiper and shaft into the collar and install the C-clip. Insert the wiper/shaft/collar assembly through the core and the hole in the perfboard with the wiper positioned to contact the windings. Pull on the shaft and collar from the opposite side of the perfboard to see how things fit. If the shaft collar flange bottoms out on the perfboard *and* the wiper is contacting the windings with a reasonable-but-not-excessive force (there's still some spring travel in the wiper), you're home free.

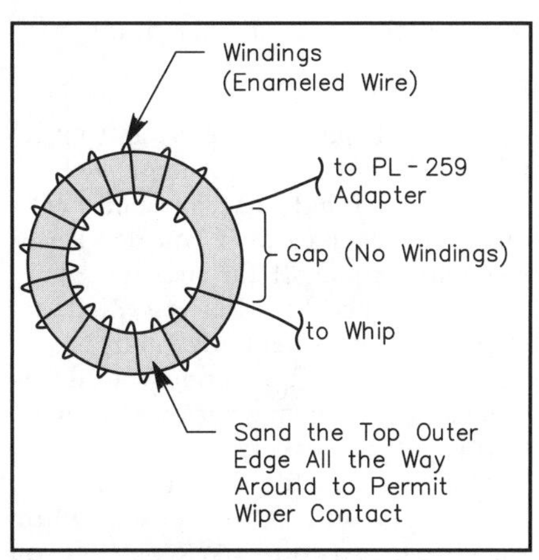

Figure 2—Winding the ferrite core with approximately 60 turns of #26 enamel wire. Note that you must sand the windings along the top outer edge to remove the enamel coating so that the brush can make contact.

Parts List

- Wire-wound rheostat—Ohmite # res100, res250, res500, res1000 or similar (available from Allied Electronics, **www.alliedelec.com**, about $20 each).
- Core—Palomar F82-61 or similar (available from Palomar Engineers at **www.palomar-engineers.com**; about $1.60 each).
- Whip, wire, PL-259, etc
- Enclosure—Hammond #1551HBK or similar.
- F-female to PL-259 adapter—RadioShack 278-258.

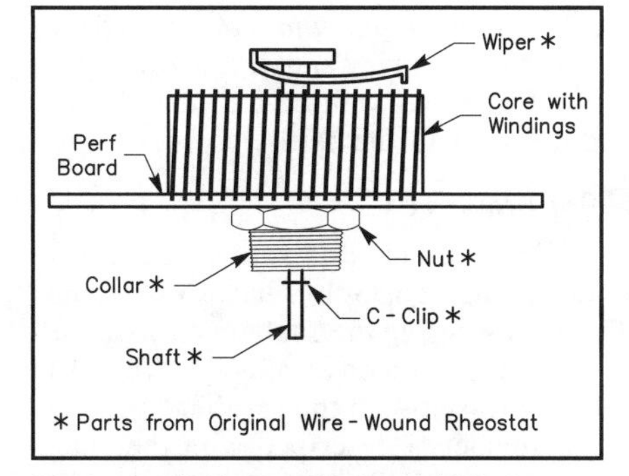

Figure 3—Side view of the modified rheostat assembly. The wiper and brush make contact with the core windings, jumping from one winding to another as you turn the wiper shaft.

If the wiper spring is bottoming out before the shaft collar flange is firmly seated on the perfboard, you'll need to insert one or more washers between the flange and the perfboard until the fit is right. This happened to me, and I wound up cutting a washer from a piece of transparent Mylar to get a good fit.

On the other hand, if the shaft collar flange bottoms out on the perfboard but the wiper contacts the windings only lightly (or not at all), you'll need to elevate the core above the perfboard by shimming underneath the core. You can do this by cutting a core-shaped ring of glueable, non-metallic material that's the right thickness, and gluing it under the core to raise it enough to get good contact between the wiper and the windings.

Fortunately, the wiper spring has a good deal of travel, so this adjustment isn't too difficult. Don't rush it, however, and spend enough time to set this up properly.

Once the adjustment's right, glue the core permanently to the perfboard, centering it over the hole and set it aside to dry. You can then insert and fasten the mounting collar with its nut. Finally, remove the C-clip from the shaft and extract the shaft and wiper for the next step.

The Brush

The original brush is quite wide for our purposes, so we need to file it down so it forms a flattened point that will contact only one winding at a time. You're going to file the sides and top to shape the contact area like a wedge with a flattened point. Check out Figure 4 to see what I mean.

Use a fine-tooth file and go slowly. The brush material is quite soft and you don't want to go too far. After shaping, use the file to cut a shallow groove across the middle of the point. This helps the point seat solidly when it settles over a winding. Make sure to round the edges as shown so the brush doesn't hang up when stepping over the windings.

After this you're ready to insert the shaft and wiper into the collar and replace the C-clip on the shaft to hold it in place.

Assembly

All that remains is to install your completed transformer assembly, a PL-259 coaxial connector and whip in a suitable enclosure. The transformer unit and the coaxial connector should be mounted so they don't interfere with each other, and the whip mounts on the top of the box. Eyeball the positions before you drill any holes. That done, drill all three required holes in the appropriate locations.

Panel-mount PL-259s are few and far between, but I managed to find something

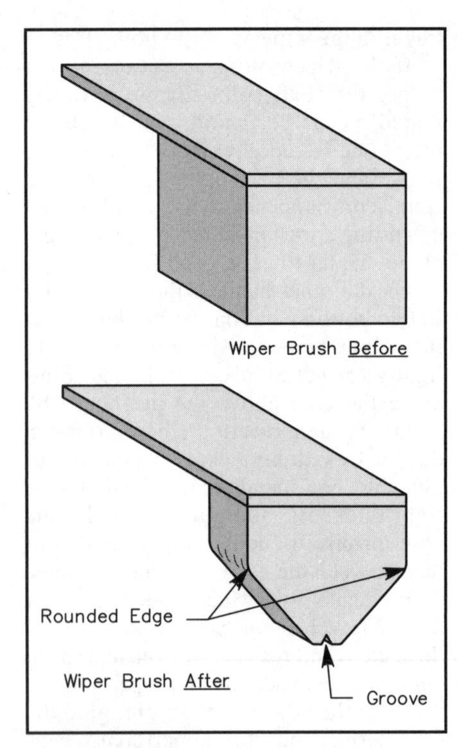

Figure 4—The normally flat-edged wiper brush must be gently filed to a rounded point (with a narrow groove) and rounded corners.

suitable. It's an "F-female to PL-259" adapter sold by RadioShack, part number 278-258. There's no way to solder to the inside (the F-type end of this adapter), but it's designed to make good contact with a piece of solid wire inserted straight into the end (like a cable-TV connector), so cut a short length of solid copper hookup wire, remove the insulation and stick it in the hole. You'll solder a lead from this to the transformer wiper lead. Your ground connection can be provided by using an appropriately sized lug washer (if you can find one) or by slipping another stripped lead under the connector nut as you tighten it down.

Mount the transformer in the box by inserting the shaft collar through the mounting hole and install the retaining nut.

My whip is 48 inches long and came from a surplus store. It looks like it might once have been part of a "rabbit ear" assembly. I chose it because it's beefy and because it had a swivel mount that would allow swinging the antenna to a horizontal or vertical orientation. Mount yours to the top edge of the box, making a connection in whatever fashion required; a stripped lead or lug under the mounting screw should do just fine.

Wire things up as per the diagram and remember to use a thin, flexible lead to make the connection between the wiper the Type-F end of your PL-259 adapter. Make sure there's plenty of slack. You'll want this to move freely, without strain. Screw on the cover and plug it in!

Operating

Select a band and tune the antenna by rotating the wiper while listening to band noise or a signal. The antenna peaks nicely on receive, so if you don't hear something right off the bat, something needs to be checked. You may find that the whip works better horizontally or vertically. Listen and experiment to determine how the antenna performs with the station you're working.

Once peaked for maximum receive signal, transmit at low power while watching the FT-817's SWR meter. If you have significant reflected power, rotate the slider a little to one side or the other and try again. You can feel each "detent" as you step from winding to winding. You might not get a perfect match on the lower bands because the impedance transformation ratios jump rather quickly at the bottom end of the transformer, but you should get something that's workable. I get 1:1 on 20, 15 and 10, and about 2:1 on 40 and 80 meters. Remember that your transmission line is about two inches long, so SWR-induced line losses aren't a consideration—you're mainly looking for reasonable loading.

A few words to the wise: *always tune at the lowest power setting and never attempt to transmit at higher power unless you see a decent match*. And, as mentioned before, the antenna peaks nicely on receive, so if you don't hear a peak, investigate and fix things before you transmit!

Once peaked, you're ready to switch to higher power and talk to someone. Remember that you're working QRP *with a compromise antenna*. A little patience will go a long way and, like a glider pilot or a fisherman, waiting for the right conditions is half the battle.

Performance

I'm not sure this setup would have saved Napoleon Solo's bacon every time, but considering the challenges of operating a QRP rig with an attached whip, I'm very pleased with the results. I've made the contacts I described with the whole kit and caboodle sitting on my desktop, without any sort of ground or counterpoise. In fact, adding a ground might make impedance matching considerably more difficult.

Obviously, the antenna performs better at higher frequencies. On 10 meters it's about an eighth of a wavelength—which isn't bad. As you go down in frequency the antenna is electrically shorter and less efficient. But it loads and radiates all the way down to 80 meters, which is the design goal, and it *will* make contacts there, given the right conditions.

Exterior view of the Miracle Whip housing.

Six, Two and More and More

Although I didn't design it to do so, the antenna works great on 6, 2 and even 440. The trick is to set the wiper to the very last turn—in effect providing a direct connection to the whip—with the transformer simply acting as a choke to ground. You can then slide the whip in or out to approximate a quarter wavelength for whatever band you're on. In this case the antenna is full size, so there's no compromise at all!

The autotransformer principle should also be applicable to a general-purpose, random-wire tuner. I think I'll play around with this. If you're an intrepid experimenter, I invite you to do the same and let me know what you find.

This antenna should work with just about any QRP rig, homebrew or store-bought. The only proviso is that, although the TX outputs of almost all rigs are designed to work into 50 Ω, the receiver inputs may prefer other impedances. Receiver input impedance is far less critical for most applications, however, so this may not be much of a handicap.

I'm completely satisfied with my first Miracle Whip—so much so that I plan to offer a commercial version to the amateur community (on the Web see **www.miracleantenna.com**).

With the Miracle Whip I've realized a radio dream I've had for many years: working the world with a self-contained, hand-held station. I haven't yet tested it on a picnic table, but my desk is a pretty fair substitute. I'm expecting Napoleon to knock off the condescension once we get out to the campground.

The Miracle Whip trades efficiency for size and portability, so don't expect, well, miracles. But if you want a system that can work DX from a picnic table, an ocean view or a mountaintop, this one does the trick. Now, Mr. Solo, about those nannies...

You can contact the author at 1220 Bernard St, No. 21, Outremont, QC, H2V 1V2, Canada; **Lebloke@attcanada.ca**. QST

FEEDBACK

Please use this form to give us your comments on this book and what you'd like to see in future editions, or e-mail us at **pubsfdbk@arrl.org** (publications feedback). If you use e-mail, please include your name, call, e-mail address and the book title, edition and printing in the body of your message. Also indicate whether or not you are an ARRL member.

Where did you purchase this book?

☐ From ARRL directly ☐ From an ARRL dealer

Is there a dealer who carries ARRL publications within:

☐ 5 miles ☐ 15 miles ☐ 30 miles of your location? ☐ Not sure.

License class:

☐ Novice ☐ Technician ☐ Technician with code ☐ General ☐ Advanced ☐ Amateur Extra

---

Name ______________________ ARRL member? ☐ Yes ☐ No

Call Sign ______________________

Daytime Phone ( ) ______________________ Age ______________________

Address ______________________

City, State/Province, ZIP/Postal Code ______________________

If licensed, how long? ______________________ e-mail address: ______________________

Other hobbies ______________________

Occupation ______________________

For ARRL use only	MQRP
Edition	1 2 3 4 5 6 7 8 9 10 11 12
Printing	1 2 3 4 5 6 7 8 9 10 11 12

From ______________________

____

____

Please affix postage. Post Office will not deliver without postage.

EDITOR, MORE QRP POWER
ARRL—THE NATIONAL ASSOCIATION FOR AMATEUR RADIO
225 MAIN STREET
NEWINGTON CT 06111-1494

--------------------- please fold and tape ---------------------